AF352887

Geo-Environmental Approaches for the Analysis and Assessment of Groundwater Resources at Catchment-Scale

Editors

Evangelos Tziritis
Andreas Panagopoulos

MDPI • Basel • Beijing • Wuhan • Barcelona • Belgrade • Manchester • Tokyo • Cluj • Tianjin

Editors
Evangelos Tziritis
Soil and Water Resources
Institute
ELGO-DIMITRA
Thessaloniki
Greece

Andreas Panagopoulos
Soil and Water Resources
Institute
ELGO-DIMITRA
Thessaloniki
Greece

Editorial Office
MDPI
St. Alban-Anlage 66
4052 Basel, Switzerland

This is a reprint of articles from the Special Issue published online in the open access journal *Water* (ISSN 2073-4441) (available at: www.mdpi.com/journal/water/special_issues/Geo-environmental_groundwater_catchment).

For citation purposes, cite each article independently as indicated on the article page online and as indicated below:

LastName, A.A.; LastName, B.B.; LastName, C.C. Article Title. *Journal Name* **Year**, *Volume Number*, Page Range.

ISBN 978-3-0365-4372-7 (Hbk)
ISBN 978-3-0365-4371-0 (PDF)

Cover image courtesy of Evangelos Tziritis

Contents

About the Editors

Evangelos Tziritis

Dr Evangelos Tziritis is a senior research scientist at the Soil and Water Resources Institute of the Hellenic Agricultural Organization (ELGO-DIMITRA). His main research domain is focused on environmental hydrogeochemistry and relevant water resources aspects. He has significant experience in various methodological approaches for geo-environmental analysis at the catchment-scale level (e.g., hydrogeochemical modelling, environmental indicators, hydrological isotopes, groundwater vulnerability). He also has expert knowledge on environmental monitoring of water resources and groundwater salinization. His overall record of achievements includes several years of experience in geo-environmental projects of basic and applied research in liaison with private firms, stakeholders, and academia. He has participated in more than 20 projects and has published more than 60 scientific papers (h-index 15) in peer-reviewed journals and international conferences. He is a regular reviewer of more than 30 scientific journals of the Science Citation Index (SCI). Currently, he is the coordinator of the MEDSAL Project (PRIMA Programme), focusing on the multi-sourced salinization of coastal aquifers.

Andreas Panagopoulos

Dr. Andreas Panagopoulos is a Hydrogeologist, Research Director at the Soil & Water Resources Institute (SWRI) of the Hellenic Agricultural Organisaton. His main scientific interests concern water resources management, monitoring, modelling and policy, hydrogeology and climate change impact assessment in water bodies. With more than 25 years of working experience, he has participated in more than 40 national and international research projects, in most of which he was scientifically responsible for SWRI. He has worked as a consultant for European and Greek partnerships in the implementation of development studies in Greece and Turkey (EUROPAID) concerning the protection and management of water resources in the framework of the implementation of European Directives. He is a member of the National Water Council and a nominated national representative in the CIS WG groundwater of the EU. He is also the co-founder of Pinios Hydrologic Observatory, which is included in International Long Term Ecological Research (ILTER) sites and the European Network of Hydrological Observatories (ENOHA) network. He has been the Director of Land.

Editorial

Geo-Environmental Approaches for the Analysis and Assessment of Groundwater Resources at the Catchment Scale

Evangelos Tziritis * and **Andreas Panagopoulos** *

Soil and Water Resources Institute, ELGO-DIMITRA, 57400 Thessaloniki, Greece
* Correspondence: e.tziritis@swri.gr (E.T.); a.panagopoulos@swri.gr (A.P.)

Citation: Tziritis, E.; Panagopoulos, A. Geo-Environmental Approaches for the Analysis and Assessment of Groundwater Resources at the Catchment Scale. *Water* **2022**, *14*, 1085. https://doi.org/10.3390/w14071085

Received: 2 March 2022
Accepted: 16 March 2022
Published: 29 March 2022

Publisher's Note: MDPI stays neutral with regard to jurisdictional claims in published maps and institutional affiliations.

Groundwater resources constitute nearly one-third of the globe's freshwater resources. They are widely used in national economies for various key purposes, including domestic water supply, industrial uses, irrigation, medicinal uses (balneotherapy), and energy applications (low- and high-enthalpy geothermal energy). Moreover, they often form the only supply to dependent ecosystems and give rise to spectacular underground karstic formations (lagoons, river canyons, etc.). In addition, they provide the necessary mechanical strength to lose formations that host phreatic aquifers, thus preventing matrix compaction and subsidence at the ground surface that causes devastating deformations and extensive damage to infrastructures. Hence, groundwater's quantitative sufficiency and high quality are paramount to secure environmental sustainability and socio-economic prosperity. Nevertheless, groundwater systems are subject to various factors that are adverse to or deteriorate their inherent characteristics. Such factors could be geogenic or anthropogenic; geogenic factors are mainly related to natural water–rock interaction processes, whereas anthropogenic factors are primarily associated with agricultural practices, industrial activities, urbanisation, landfills, domestic effluents, and aquifer overexploitation.

Another critical aspect of groundwater management is that the proper management of groundwater is difficult after contamination. On the other hand, environmental monitoring projects that provide critical ground-truth values for groundwater quality are not always feasible due to a lack of personnel, funding, and time. In this respect, scientists and decision makers seek alternative strategic tools to spatially identify threats and subsequently design and implement groundwater protection measures towards practicing sustainable groundwater management. These tools are often diverse but interlinked methodologies, such as hydrogeological and hydrogeochemical modelling, environmental isotopes, environmental indicators, geostatistics, and artificial intelligence.

The complexity of different hydrological and hydrogeological setups, hydrodynamic patterns, site specifications, and the wide variability of internal and external factors and/or processes at the level of catchment scale necessitates combined approaches that integrate robust methods, thus leading to more accurate and reliable outcomes towards sustainable groundwater management. Sound knowledge of a studied groundwater system may reduce uncertainties in the prediction of its future evolution, thus enabling better management and protection whilst limiting the need to hypothesise. In line with the above goal, this Special Issue aims to provide successful applications or new insights on the stand-alone or joint considerations of groundwater resources assessment and characterisation methods and explore new state-of-the-art methodological concepts in light of a rapidly changing environment.

This Special Issue of the journal *Water* comprises 12 papers with contributions of more than 50 authors originating from 10 countries, all of which deal with various geo-environmental methods and tools. The papers include six feature papers and four editor's choice papers.

Two of these papers apply hydrogeochemical tools and methods to robustly evaluate groundwater resources quality and hydrodynamic regime. Specifically, Vrouhakis et al. [1]

used combined hydrogeochemical and hydrodynamic characterisation to assess key aspects related to groundwater resources management in a highly productive agricultural basin of central Greece. A complementary array of tools and methods, including graphical processing, multivariate statistics, and environmental isotopes, was applied to a comprehensive dataset of physicochemical analyses and water level measurements. The outcomes proved valuable in the progression towards sustainable management of groundwater resources. The results provide spatial and temporal insights into significant parameters, sources, and processes that, as a methodological approach, could be adopted in similar cases of other catchments. Vasileiou et al. [2] investigated the hydrogeochemical processes and natural background levels (NBLs) of chromium in the ultramafic environment of Vermio Mountain, Western Macedonia, Greece. The holistic methodology proposed in this paper may be implemented in similar cases at the catchment scale to assess geogenic and anthropogenic Cr sources that degrade groundwater quality.

Falling within the broad category of groundwater quality, two papers deal with the significant problem of groundwater salinisation, which is crucial to address in arid and semi-arid coastal areas. More specifically, Bonamico et al. [3] applied an integrated hydrogeologic and geochemical methods approach to describe freshwater–saltwater interactions in a coastal aquifer in the Ostia Antica archaeological park Roma, Italy. Their assessments were based on a water monitoring program that included the installation of multiparametric probes in wells, with continuous measurement of temperature, electrical conductivity, and water table level. Field surveys, water sampling, major elements, and bromide analyses were carried out to understand the detailed stratigraphic settings of the area. The authors used oxygen and carbon isotopic signatures of calcite from well sediments and evaluated major elements and Br to determine the salinisation sources and the processes of gas–water-rock interaction. The second paper focuses on assessing the drought index as an indicator of groundwater salinisation. The research by Alfio et al. [4] was performed on the Salento aquifer (southern Italy), where in recent decades, groundwater depletion and salinisation worsened because of the increased frequency of droughts, as revealed by the documented data derived from the analysis of the Standardised Precipitation Index (SPI) calculations during 1949–2011 based on monthly precipitation levels. Groundwater level series and chloride concentrations collected over the extreme drought period 1989–1990 allowed a qualitative assessment of groundwater behaviour, highlighting the concurrent groundwater drought and salinisation.

Two other papers deal with the conceptualisation and application of index and overlay methods to assess groundwater vulnerability, considered a significant, proactive measure towards successful decision making and rationale management. Kim et al. [5] evaluated the vulnerability to seawater intrusion by classifying the existing GALDIT method into static parameters (groundwater occurrence (G), aquifer hydraulic conductivity (A), and distance from shore (D)) and dynamic parameters (height-to-groundwater level above sea level (L), impact of existing status of seawater intrusion (I), and aquifer thickness (T)). Data indicating averages of measurements over a 10-year period for each month were used, representing the seasonal characteristics of local water cycles. To reflect subtle monthly variations, the range of scores was divided into deciles to capture the temporal dynamics of seawater intrusion. The proposed modified method can determine where to apply countermeasures to vulnerable coastal areas and develop water resources management plans considering vulnerable seasons. Vrouhakis et al. [6] assessed the intrinsic perspective of groundwater vulnerability in central Greece's highly productive agricultural area. A novel index-based method (RIVA) was applied to the Tirnavos basin to assess susceptibility to surface-released contamination. Data from field surveys, previous studies, and the relevant literature were used to calculate factors constituting the RIVA method, which was demonstrated to be a data-intensive and efficient method, thus a sound investment to reach highly accurate results. Overall, RIVA proved to be a robust tool for reliable groundwater vulnerability assessments and could be further exploited for risk assessment and decision-making processes in the context of groundwater resource management.

Fuentes-Arreazola et al. [7] estimated aquifer parameters in the Mexican wine-producing region Guadelupe Valley from fluctuations in levels of groundwater. They proposed an alternative tool with significant advantages in studying the groundwater-level response due to variations in pore pressure caused by internal deformation of the aquifer structure induced by barometric pressure and solid Earth tide. This analysis reveals helpful insights that can help to establish a framework to design and assess management strategies for groundwater resources in similar cases.

Two other papers provided insights into the application of modelling in groundwater resources. Lyra et al. [8] presented an integrated modelling system to evaluate the availability of water resources in coastal agricultural watersheds. Their modelling system was made from an ensemble of surface and groundwater hydrology models, crop growth/nitrate leaching, contaminant transport, and seawater models. Its efficacy to simulate the quantity and quality of water resources was tested at the Almyros basin in Thessaly, Greece. The proposed modelling system could be used as a tool for the simulation of water resources management and climate change scenarios. Further research on the plausibility of modelled nitrate concentrations in the leachate on the scale of federal states in Germany was performed by Wolters et al. [9]. This research aimed to model nitrate concentrations in leachates as a robust tool for water resources management, in line with the requirements enforced by the EU Water Framework Directive. The validity of simulations was checked against values from 1119 preselected monitoring stations from shallow springs and aquifers filtered near the surface with oxidising properties. The case study revealed that the applied model system (RAUMIS–mGROWA–DENUZ) can reliably represent interrelationships and influencing factors that determine simulated nitrate concentrations in the leachate. Moreover, it was demonstrated that observed nitrate concentrations in groundwater may provide a solid source of data for checking the plausibility of modelled nitrate concentrations in leachate in cases in which certain preselection criteria are applied.

Pisinaras et al. [10] studied the effects of agricultural water management in a Mediterranean coastal aquifer under current and projected climate change conditions. Their research focused on the coastal delta plain of River Pinios, central Greece. Such areas are significant for the Mediterranean region because of their high soil fertility and agricultural productivity. Nevertheless, they also constitute fragile systems in terms of water resource management due to the interaction of underlying aquifers with the sea. Soil and Water Assessment Tool (SWAT) and SEAWAT models were combined to simulate the impact of current practices in water resource management on the main groundwater budget components and groundwater salinisation of the shallow aquifer developed in the area. Moreover, the potential impact of climate change was investigated using projected data gathered from the Regional Climate Model for two periods (2021–2050 and 2071–2100) and two sea-level rise scenarios (increasing by 0.5 and 1 m).

The combined impact of the hydrological and socio-economic perspectives on the sustainability of groundwater resources was examined by Oke and Alowo [11]. Their research presented a spatial interpolation of the anticipated impact of the above factors on groundwater systems and predicted the sustainability of the Modder River catchment in South Africa. The results were presented with sustainability maps indicating areas with differing groundwater dynamics in the catchment. The key finding in this paper may assist groundwater managers and regulators to effectively plan groundwater resources utilisation, especially with regard to the prevention of licencing and overpumping practices.

Finally, a relatively new and challenging field of joint tools for hydrological sciences was addressed by Namous et al. [12]. Their research focused on the spatial prediction of groundwater potentiality in a large semi-arid karstic mountainous region by using a combination of machine learning models, such as random forest (RF), logistic regression (LR), decision tree (DT), and artificial neural networks (ANNs). A total of 24 groundwater influencing factors (GIFs) were selected based on a multicollinearity test and the information gain calculation. The results of the groundwater potentiality mapping were validated using statistical measures and the receiver operating characteristic curve (ROC) method.

Compared with individual models, the combined models proved the most stable and suitable tools to map groundwater potentiality in mountainous aquifers, based on success and prediction rate.

The sustainability and environmental welfare of human civilization are based on utilising sufficient volumes of water with acceptable quality. As the impacts of climate change intensify, water resources safety becomes of primary concern. Groundwater resources impose extra challenges regarding their management, as these sources are invisible, thus needing proactive and carefully designed measures. World Water Day for the year 2022 is devoted to making the invisible, visible. This Special Issue provides a series of papers that propose state-of-the-art methodologies, technologies, and approaches that exactly contribute to this goal.

Author Contributions: E.T. and A.P. have equally contributed to the original draft preparation, review, editing and proof-reading. All authors have read and agreed to the published version of the manuscript.

Funding: This research received no external funding.

Acknowledgments: The authors appreciate the efforts of the *Water Journal* editors and publication team at the MDPI and the anonymous reviewers for their invaluable comments.

Conflicts of Interest: The authors declare no conflict of interest.

References

1. Vrouhakis, I.; Tziritis, E.; Stamatis, G.; Panagopoulos, A. Hydrogeochemical and Hydrodynamic Assessment of Tirnavos Basin, Central Greece. *Water* **2021**, *13*, 759. [CrossRef]
2. Vasileiou, E.; Papazotos, P.; Dimitrakopoulos, D.; Peraki, M. Hydrogeochemical Processes and Natural Background Levels of Chromium in an Ultramafic Environment. The Case Study of Vermio Mountain, Western Macedonia, Greece. *Water* **2021**, *13*, 2809. [CrossRef]
3. Bonamico, M.; Tuccimeni, P.; Mastrorillo, L.; Mazza, R. Freshwater–Saltwater Interactions in a Multilayer Coastal Aquifer (Ostia Antica Archaeological Park, Central ITALY). *Water* **2021**, *13*, 1866. [CrossRef]
4. Alfio, M.R.; Balacco, G.; Parisi, A.; Totaro, V.; Fidelibu, M.D. Drought Index as Indicator of Salinization of the Salento Aquifer (Southern Italy). *Water* **2020**, *12*, 1927. [CrossRef]
5. Kim, I.H.; Chung, I.M.; Chang, S.W. Development of Seawater Intrusion Vulnerability Assessment for Averaged Seasonality of Using Modified GALDIT Method. *Water* **2021**, *13*, 1820. [CrossRef]
6. Vrouhakis, I.; Tziritis, E.; Stamatis, G.; Panagopoulos, A. Groundwater Vulnerability Analysis of Tirnavos Basin, Central Greece: An Application of RIVA Method. *Water* **2022**, *14*, 534. [CrossRef]
7. Fuentes-Arreazola, M.; Ramirez-Hernandez, J.; Vazquez-Gonzalez, R.; Nunez, D.; Diaz-Fernandez, A.; Gonzalez-Ramirez, J. Aquifer Parameters Estimation from Natural Groundwater Level Fluctuations at the Mexican Wine-Producing Region Guadalupe Valley, BC. *Water* **2021**, *13*, 2437. [CrossRef]
8. Lyra, A.; Loukas, A.; Sidiropoulos, P.; Tziatzios, G.; Mylopoulos, N. An Integrated Modeling System for the Evaluation of Water Resources in Coastal Agricultural Watersheds: Application in Almyros Basin, Thessaly, Greece. *Water* **2021**, *13*, 268. [CrossRef]
9. Wolters, T.; Cremer, N.; Eisele, M.; Hermann, F.; Kreins, P.; Kunkel, R.; Wendland, F. Checking the Plausibility of Modelled Nitrate Concentrations in the Leachate on Federal State Scale in Germany. *Water* **2021**, *13*, 226. [CrossRef]
10. Pisinaras, V.; Paraskevas, C.; Panagopoulos, A. Investigating the Effects of Agricultural Water Management in a Mediterranean Coastal Aquifer under Current and Projected Climate Conditions. *Water* **2021**, *13*, 108. [CrossRef]
11. Oke, S.A.; Alowo, R. Groundwater of the Modder River Catchment of South Africa: A Sustainability Prediction. *Water* **2021**, *13*, 936. [CrossRef]
12. Namous, M.; Hssaisoume, M.; Pradhan, B.; Lee, C.W.; Alarmi, A.; Elaloui, A.; Edahbi, M.; Krimissa, S.; Eloudi, H.; Ouayah, M.; et al. Spatial Prediction of Groundwater Potentiality in Large Semi-Arid and Karstic Mountainous Region Using Machine Learning Models. *Water* **2021**, *13*, 2273. [CrossRef]

Article

Groundwater Vulnerability Analysis of Tirnavos Basin, Central Greece: An Application of RIVA Method

Ioannis Vrouhakis [1,2,*], Evangelos Tziritis [2], Georgios Stamatis [1] and Andreas Panagopoulos [2]

[1] Mineralogy and Geology Laboratory, Sector of Geological Sciences Department of Natural Resources & Agricultural Engineering, Agricultural University of Athens, Iera Odos 75, 11855 Athens, Greece; stamatis@aua.gr

[2] Soil & Water Resources Institute, Hellenic Agricultural Organisation "Demeter", Gorgopotamou Street, 57400 Thessaloniki, Greece; e.tziritis@swri.gr (E.T.); a.panagopoulos@swri.gr (A.P.)

* Correspondence: i.vrouhakis@swri.gr; Tel.: +30-2310798790

Abstract: A novel index-based method (RIVA) for assessing intrinsic groundwater vulnerability was applied to Tirnavos basin (central Greece) to assess the susceptibility to surface-released contamination. Data from field surveys, previous studies, and literature were used to calculate the factors that compile the RIVA method. The aggregated results delineated the spatial distribution of groundwater vulnerability from very low to very high. The modelled results were successfully validated with ground-truth values of nitrates obtained from 43 boreholes. Overall, the modelled and the monitored values match more than 80%, indicating the successful application of the RIVA method. Few deviations were observed in areas dominantly affected by lateral crossflows and contamination from adjacent areas. RIVA proved an efficient method in terms of accuracy, data intensity, and investment to reach highly accurate results. Overall, RIVA proved to be a robust tool for reliable groundwater vulnerability assessments and could be further exploited for risk assessment and decision-making processes in the context of groundwater resource management.

Keywords: aquifer; groundwater; intrinsic vulnerability; RIVA method; index-overlay method; Tirnavos basin

Citation: Vrouhakis, I.; Tziritis, E.; Stamatis, G.; Panagopoulos, A. Groundwater Vulnerability Analysis of Tirnavos Basin, Central Greece: An Application of RIVA Method. *Water* **2022**, *14*, 534. https://doi.org/10.3390/w14040534

Academic Editor: Pankaj Kumar

Received: 19 January 2022
Accepted: 9 February 2022
Published: 11 February 2022

Publisher's Note: MDPI stays neutral with regard to jurisdictional claims in published maps and institutional affiliations.

1. Introduction

Groundwater accounts for nearly 99% of the total volume of freshwater presently circulating on our planet [1]. This overwhelming percentage, the lower susceptibility to pollution, and the large storage capacity of groundwater compared to surface water highlight its paramount importance at a global socio-economic level. Its significance, however, requires the most outstanding possible effort to protect it. A significant task towards this goal is the assessment of groundwater intrinsic vulnerability, a practice that suggests areas where priority and special attention should be given to the protection and overall management of groundwater resources. However, assessing the vulnerability of groundwater to adverse effects of human impacts is one of the most critical problems in applied hydrogeology [2]. The anthropogenic agricultural activities are often responsible for overdraft, groundwater quality deterioration, and increasing vulnerability. Due to level decline and quality degradation, sustainable development plans are needed to protect these resources [3].

In general, in most parts of the world, groundwater vulnerability assessment is based on (i) process-based methods [4–6], (ii) statistical methods [7–9], and (iii) overlay and index methods [10–13]. The limitations of process-based methods are adequate data and quality to capture the physical, chemical, and biological reactions from the surface to the uppermost aquifer. The statistical methods focus on the uncertainty by minimizing the error and using parameters' coefficients instead of weights. A possible drawback of these methods is the required monitoring data, which is essential. These methods are only applicable to those regions where similar factors govern groundwater contamination. The overlay and index methods are the most suitable for groundwater vulnerability assessment, overcoming all the limitations mentioned above [14]. They focus on critical factors potentially controlling contaminant transport, and they are relatively cost-effective and adaptable to on-site specific conditions. Moreover, they demand minimum data to produce outcomes that can directly facilitate the decision-making processes [15]. The basic steps of these methods include raw data analysis, ranking of features on maps, integration of maps, and classification of the integrated map based on an index. These methods can be applied from a regional to global scale; nevertheless, they should be supplemented with field visits and on-site validation to produce reliable results [16].

Due to the advantages above and their ability to be easily used as strategic tools, the scientific community has increased interest. Some examples of this type of groundwater vulnerability assessment are DRASTIC (D: aquifer depth, R: recharge rate, A: aquifer lithology, S: soil type, T: topography, I: impact of vadose zone, C: aquifer hydraulic conductivity) [17], GOD (Groundwater occurrence, Overall lithology of aquifer, and Depth to groundwater level) [18], AVI (Aquifer Vulnerability Index) [19], SEEPAGE (System of Early Evaluation of Pollution Potential of Agricultural Groundwater Environments) [20], EPIK (Epikarst, Protective cover, Infiltration conditions, and Karst network development) [21], GALDIT (G: Groundwater occurrence, A: Aquifer hydraulic conductivity, L: Height of groundwater level, D: Distance from the shore, I: Impact of existing status of seawater intrusion, T: Thickness of the aquifer) [22], RISKE (Rock of aquifer media, Infiltration, Soil media, Karst, and Epikarst) [23], and a global risk approach [24].

Two types of aquifer vulnerability are considered. The first refers to intrinsic vulnerability, which considers the system's inherent properties, such as the properties of the vadose zone or the recharge conditions, etc. The second and more complex one is specific vulnerability. In addition to the intrinsic, it considers the properties of a specific contaminant or group of contaminants [25]. This research focuses on intrinsic vulnerability using a relatively new method (RIVA) [26]. The selected method is based on the successful concept of the European approach [27] and incorporates additional elements that provide more realistic and representative results. The main advantage of RIVA over similar methods is that it can be applied to all types of groundwater bodies independently of the specific conditions, lithologic phases, and aquifer typology of each area. This way allows uniform evaluation and comparison between hydrogeological systems with different characteristics. To this aim, the Tirnavos alluvial basin area is considered an appropriate test site for its application, as it includes an intensively cultivated area with diverse geological and hydrogeological characteristics. The proximity to the adjacent karstic system, combined with the occurrence of significant tectonic structures, further increases the aquifer system's vulnerability potential.

2. Case Study Area

Thessaly Plain in central Greece is the largest alluvial basin of the country and is divided, through the mid-Thessaly hills, into two sub-basins, the western Thessaly basin and the eastern Thessaly basin, developed in an NW-SE direction as part of the broader tectonic trough. The case study area where the RIVA method was applied is located in the northwest part of the wider eastern Thessaly basin (Figure 1). The hydrological perspective includes parts of the Titarisios River basin to the north and the Pinios River basin to the south. The total area is estimated to be 688 km^2, including the alluvial Tirvanos basin and part of its surrounding geological formations (Figure 2), which are hydrogeologically interdependent. The perimeter of the study area is 105 km, and the mean altitude is 160 m. The smallest and largest morphological gradients recorded are 0 and 62.57%, respectively, with a mean slope of 10.5%.

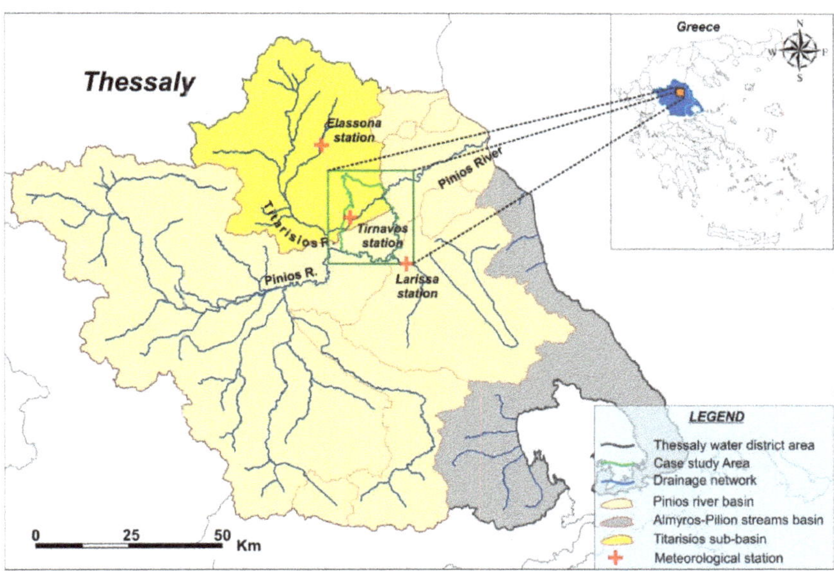

Figure 1. Location map of the case study area where the RIVA method has been applied.

The Tirnavos area is characterised by a typical Mediterranean climate, with annual rainfall from 400 mm to 600 mm, distributed almost entirely during the wet hydrological period, without significant summer precipitation. Larissa station is the closest one to the study area regarding the meteorological data, with continuous and reliable data over several years. Adjacent meteorological stations, used later in the methodology, are in Elassona and Tirnavos.

It is primarily a rural area covered by agricultural land, where intensified agricultural activities, both cultivation and livestock, are a significant source of groundwater contamination by nitrogen compounds. Manure waste and the often excessive and improper use of nitrogen fertilizers, aiming to improve agricultural production, lead to the occurrence of elevated concentrations of nitrates in groundwater [28]. Figure 3 shows the percentage distribution of land use according to CORINE categorization [29], suggesting that about 80% of land use consists of agricultural areas (arable land and permanent crops). The most common irrigation methods are drip irrigation and sprinkler, mainly using groundwater resources.

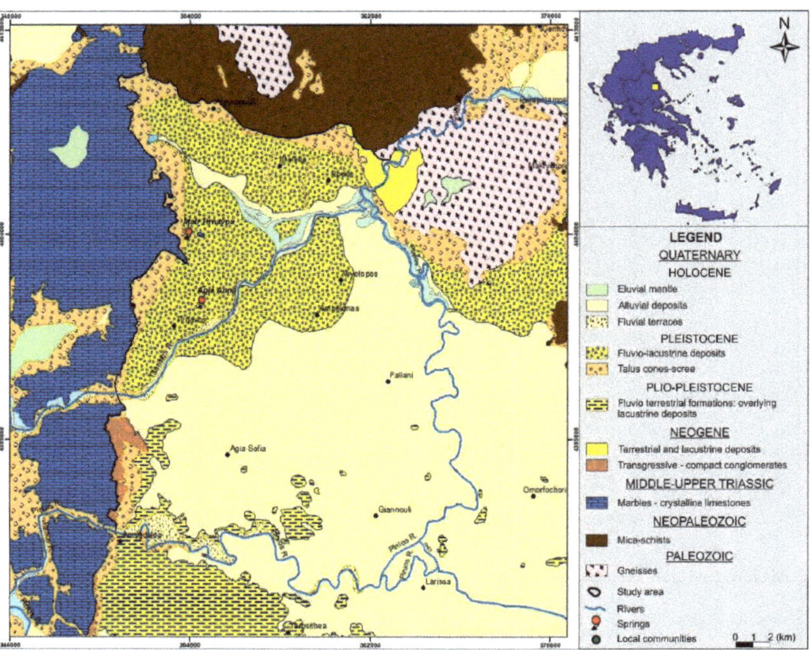

Figure 2. Geological map of the case study area, based on [30,31].

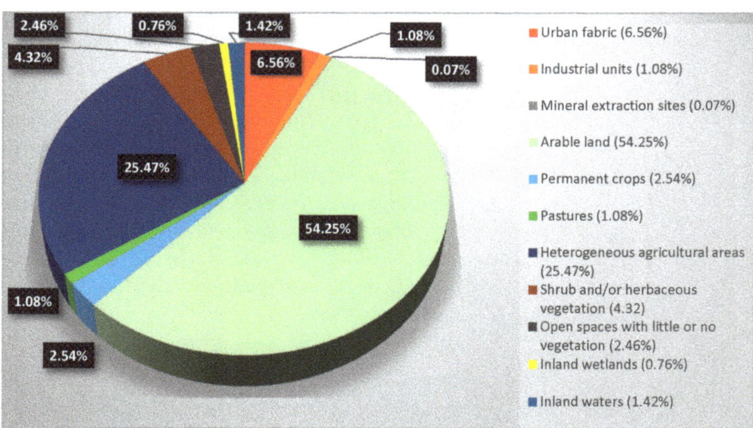

Figure 3. Land-uses distribution at the study area.

Three local irrigation organisations (Tirnavou, Ambelona, Agia Sofia) cater for irrigation water to producers through collective networks within their districts. Nowadays, all operating networks are pressurised, while until recently, an extensive gravitational network operated on the eastern part of the Ampelona district utilizing the karstic spring of Mati Tirnavou until 1998, when it was abandoned because of diminishing spring discharge. Despite the collective irrigation works, several privately owned wells exist and operate to cover irrigation demands of land that does not fall into the jurisdiction of the aforementioned collective networks, mainly in the northern and eastern ends of the study area [32].

Quaternary alluvial formations fill the boundaries along the southwest part of the basin with Neogene marls and sandy-clay deposits, whilst the western margins consist of karstified marbles of the middle-upper Cretaceous. The crystalline bedrock is composed of mica-schists and gneisses of the upper Palaeozoic and Paleozoic age, respectively, forming the northern boundary of the basin (Figure 2). Two major springs (Mati Tirnavou and Agia Anna, Figure 2) emerge at the contact of the karstified system with the alluvial deposits. Pinios and Titarisios rivers flow across the basin, which, as already mentioned, hydrologically is part of the wider Pinios River basin.

The Quaternary deposits host an unconfined aquifer near the talus cone of Titarisios at the northwest, which, towards the central parts of the basin, sinks under a sequence of clay layers that form an aquitard [32–34] and have a maximum thickness of over 550 m at their central parts [32,35]. Under this setup, confining conditions occur, while the phreatic aquifer has been seriously affected due to systematic over-abstraction and is almost depleted [28,33]. The marbles at the western margins of the basin host a karstic aquifer of great potential, which recharges the alluvial system by lateral crossflows [28,33,34,36].

In the western part of the basin, an extensive marginal cone exists, through which the alluvial system receives significant amounts of recharge as crossflow through the Titarisios River gorge sediments. To the south, a smaller volume recharges the aquifer system as crossflow from the Pinios River gorge sediments [32].

Crossflow from the crystalline bedrock at the northern margins of the basin also occurs but is of minor importance. The southern extent of the karst system and from the mid-Thessaly hills recharge the central plain parts by crossflows from the southwest and the southern part of the area; however, they have a lower potential than that of the northern parts due to the existence of marls.

Based on the above data, it is perceived that the specific area is characterised by increased complexity, as reflected by the geological and hydrogeological conditions, tectonics, hydrodynamics, and land and water use. In addition, the area lacks regional planning and management of natural resources, leading to considerable deterioration of existing groundwater reserves. For this reason, RIVA was selected as an appropriate method of intrinsic vulnerability assessment to be implemented in the Tirnavos basin.

3. Materials and Methods

RIVA is the acronym of the four key factors considered in assessing intrinsic vulnerability, each of which has a distinguished control over groundwater vulnerability. The first factor is the recharge (R), which refers to the overall assessment of the effect of recharge conditions. The second one is the infiltration (I factor), which considers the soil's infiltration, subsequently affecting deep percolation to the saturated zone. The third factor relates to the protective cover and specifically to the impact of the vadose zone (V factor). The last factor is focused on the aquifer characteristics (A factor) and relates to the potential contaminant migration within the saturated zone as assessed by the hydraulic conductivity of the aquifer. Each of the four factors affects groundwater vulnerability individually and independently of the potential interactions. The final assessment of each factor is related to five classes of vulnerability (Figure 4), from very low (VL) to very high (VH), respectively [26]. The cumulative effect of these four factors constitutes the final assessment of intrinsic groundwater vulnerability, according to the relationship

$$i = R + I + V + A \tag{1}$$

The contribution of each factor in Equation (1) is not equivalent due to the different degrees of importance. Hence, each one is multiplied by a weighting factor, which reflects its significance to the result. The weighting factors (a, b, c, d), which were calculated with the use of Analytic Hierarchy Process (AHP) and proposed by [26], are a = 0.40 for V factor, recognizing that is the most important, followed by the I factor with b = 0.30, c = 0.15 for R factor, and d = 0.15 for A factor (a + b + c + d = 1). V factor is the most important one in shaping the overall vulnerability, as it assesses the buffering capacity of the geological medium in constraining deep percolation of the potential pollutants to the saturated zone, the flux of which is directly related to their leaching rate potential as defined by factor I, which holds the second highest weight. Weighting factors' values are thus deduced as a fusion of the scientific approaches and documented viewpoints proposed by a number of researchers in the international literature [26].

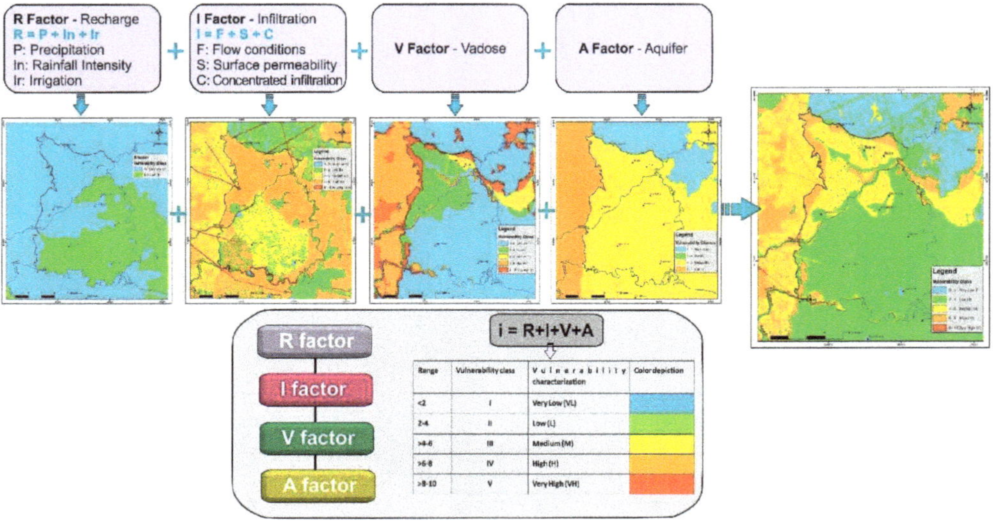

Figure 4. Schematic presentation of RIVA method (according to [26]).

In Table 1, the data used for the calculation of each factor and sub-factor of the RIVA method as well as their source of origin are presented. The tables of each method's classes and ratings can be downloaded from the link at the end of this paper as a supplementary document. The analytical description of RIVA's rationale and conceptualization per factor is beyond the scope of this research and may be reached through the original application of the method [26].

Table 1. Aggregate data used for RIVA method implementation in Tirnavos basin.

Factor	Sub-Factor	Material	Source
R	P	Rainfall data series	[37,38]
	In	Rainfall data series	[37,38]
	Ir	Spatial data of irrigated fields	[39]
		Dominant irrigation methods	Local Land Reclamations Organizations (Tirnavou, Agia Sofia, Ambelona) personal communication
I	F	Digital Elevation Model	[40]
		Land use	[39]
	S	Soil map	[41]
		Degree of karstification	[28,32]
		Geological map	[30,31]
	C	Main fault zones	[42,43]
		Hydraulic interactions	[28,32]

Table 1. *Cont.*

Factor	Sub-Factor	Material	Source
V	-	Piezometric data (alluvial)	[28,34,36]
		Bottom of confining aquifer	[44]
		Piezometric data (karst)	[45]
		Metamorphic formations weathered mantle	[28]
A	-	Hydraulic conductivity (Alluvial aquifer)	[33,44]
		Hydraulic conductivity (Marbles, Gneisses, Schists)	[46–49]

4. Results

4.1. R Factor Description

R factor refers to the assessment of the internal vulnerability of the groundwater body as a function of the total surface recharge it receives and which can reach the aquifer through percolation under certain conditions. It includes three sub-factors [26], which are expressed through Equation (2):

$$R = P + In + Ir \tag{2}$$

where

R = recharge factor
P = precipitation sub-factor
In = rainfall intensity sub-factor
Ir = irrigation recharge sub-factor

As in the case of Equation (1), the contribution of each factor in Equation (2) is not equivalent. The adopted weights are 0.5, 0.3, and 0.2 for P, Ir, and In, respectively, which are embedded in the intermediate calculations of each factor and are not shown in the initial equation. The cumulative effect of these three sub-factors constitutes the final assessment of the R-factor (Equation (2)), which is linked to the intrinsic vulnerability of the groundwater system according to the classification shown in Figure 4.

4.1.1. P Sub-Factor Calculation

Sub-factor P refers to the total precipitation depth received at the examined area over a year (y), in millimetres (mm). The representative precipitation value is deduced as the mean annual value over a sufficient time, e.g., over 20 years. If more than one station is available, the final figure is calculated by spatially integrating precipitation distribution of individual stations over the examined area, employing an appropriate interpolation scheme. The corresponding P factor values concerning precipitation ranges, according to the authors [26], are as illustrated in Table S1 in the supplementary documents.

In the case study area, there are three mereological stations (Figure 2) that were taken into consideration in the P sub-factor calculations: Larissa and Tirnavos stations located within the examined area, at 74- and 92-m altitude, respectively, and Elassona station at 314-m altitude, being representative of the higher elevation parts included in the examined area. For each station, precipitation data for a period of 30 years (1989–2018) were used to obtain a mean annual precipitation value for each of them. The calculated values were 422.2, 485.7, and 520 mm for Larissa, Tirnavos, and Elassona stations. Based on class values tabulated in Table S1, a P value of 2.5 is assigned to all three stations, and therefore, no spatial distribution is required.

4.1.2. In Sub-Factor Calculation

Except for the total precipitation, aquifer recharge depends on the intensity of the precipitation, which represents the amount of water received in a specific period. The effect of this parameter is difficult to estimate, as various factors, such as soil texture, root system, moisture state of the soil, the discontinuities of the geological formations, etc., form different conditions [50]. Nevertheless, it is generally accepted that an increase in rainfall intensity eventually increases the recharge and consequently the infiltration [51,52], thus affecting the vulnerability of the groundwater system. Regarding its calculation, the approach of the RIVA method is a compilation of approaches followed in similar methods [53,54]. The In values are calculated through Equation (3) [26]:

$$In = \Sigma P / \Sigma d \tag{3}$$

where

In = intensity value of rainfall
ΣP = total precipitation at a given period (mm)
Σd = total number of rainfall days in the same period (days)

The rain intensity values for all stations were in the same range according to Table S2 in the supplementary documents (5.8 for Larissa, 8.9 for Tirnavos, and 9.5 for Elassona). Hence, the same In value (0.4) and vulnerability class (II) are assigned to the stations, and in this case, no spatial distribution method needs to be applied.

4.1.3. Ir Sub-Factor Calculation

The Ir sub-factor represents the effect of irrigation on groundwater vulnerability and constitutes a parameter that has not been considered in vulnerability by any other method. This is possibly due to the difficulty of estimating the volume of irrigation water used and its spatial distribution, as several variables are included that are not easy to quantify. However, the importance of the irrigation effect in groundwater vulnerability is recognized, especially in intensively cultivated regions, such as the examined area. Hence, RIVA, considering the irrigation impact through a qualitative approach based on the appraisal of the mean overall performance, as shown in Table S3 (Supplementary document) [26], is deemed appropriate for assessing vulnerability in the examined area.

The irrigated fields at the case study area are spatially distinguished based on data from the Greek Payment Authority of Common Agricultural Policy Aid Schemes [39]. For the categorization of the irrigated fields in accordance to the classification referenced in Table S3, data retrieved from the Local Land Reclamation Organizations of the region were used along with assessments carried out during in situ visits to the area, accounting also for the irrigation methods employed in each part of the basin, in cases where no specific data were available. Thus, drip-irrigated fields were assumed to apply nominal irrigation doses and result in minimal leaching rates due to the high efficiency of the method. This is in contrast to sprinkler irrigation, where due to low water use efficiency, significant water losses occur that lead to considerable leaching rates, and therefore, increased irrigation doses are applied to satisfy crop water demands.

Based on this spatial data distribution according to the classes discussed (Table S3), the Ir sub-factor spatial distribution map was extracted, as illustrated in Figure 5.

4.2. I Factor Description

The I factor refers to assessing the intrinsic vulnerability of a groundwater body as a function of the surface infiltration conditions upon which deep percolation to the geological layers and eventually groundwater recharge depends. Its calculation is based on Equation (4), and its classification is shown in Table S1.

$$I = F + S + C \tag{4}$$

where

I = infiltration factor

F = flow conditions sub-factor

S = permeability of the surface medium

C = concentrated infiltration sub-factor

In Equation (4), F and S sub-factors have equal weighting factors (0.5), while sub-factor C does not contribute equally. Sub-factor C refers to the concentrated flow that could cause increased infiltration due to specific surface structures, such as epikarst, river beds, and tectonic contact. It is considered a critical aspect, which may induce maximum vulnerability to the groundwater body if existing.

4.2.1. F Sub-Factor Calculation

The F sub-factor relates to the surface water flow conditions, which affect infiltration to the saturated zone and consequently vulnerability of the groundwater system. Its calculation is determined by two parameters, the topographic slope (s) and vegetation (v). Based on the method's developers [26], the classification of "s" is made according to the approaches of [55–57], as shown in Table S4 in the supplementary documents.

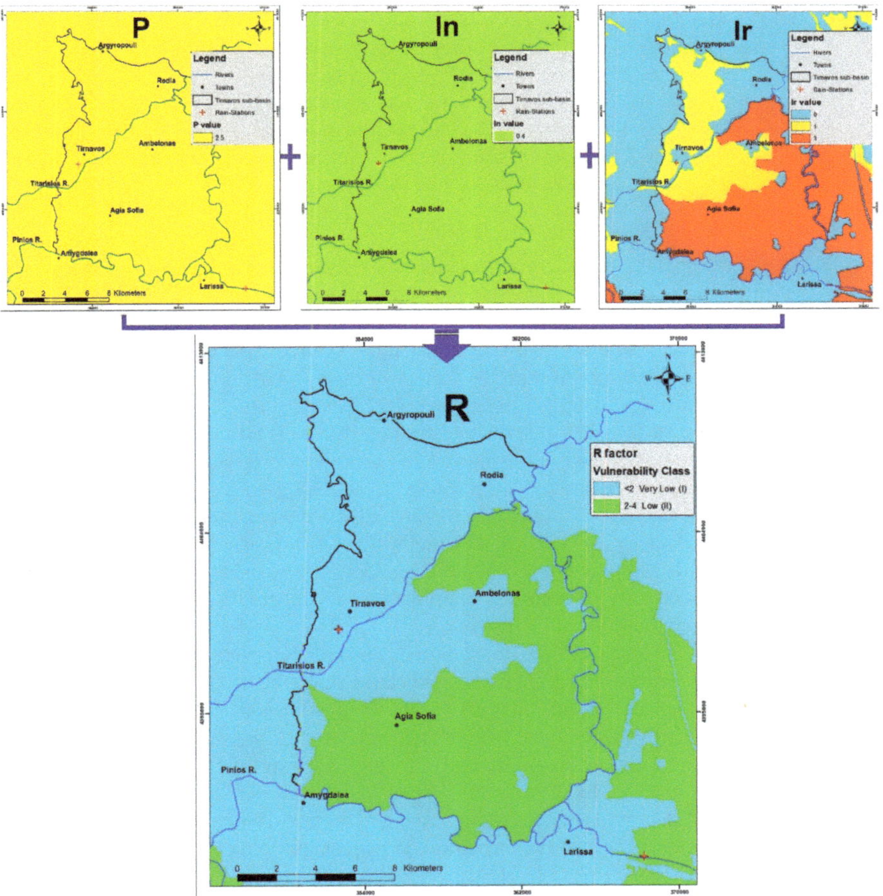

Figure 5. Spatial distribution maps of P, In, and Ir sub-factors in the Tirnavos basin and the R factor map compiled as a synthesis of the sub-factors above.

The vegetation (v) parameter is used as a slope correction parameter, as the denser it is, the more it includes plants with a well-developed root system, the more the surface runoff is prevented, and the more the surface infiltration is favoured, creating conditions of higher vulnerability. Vegetation is classified into three general groups: (a) forest areas (high vegetation), (b) cultivated and grassland areas (low vegetation), and (c) absence of vegetation or sparse vegetation.

Classification of vegetation was made according to the widely acceptable CORINE categorization [29] and shown in Table S5 (Supplementary document), where only Categories 2 (agricultural areas) and 3 (forests and semi-natural areas) related to vegetation are considered relevant. Based on the method, for the rest of the CORINE categories, i.e., 1 (artificial areas-artificial surfaces), 4 (wetlands), and 5 (water bodies), it is assumed that no surface flow occurs; thus, by definition, the sub-factor F takes the value zero (0).

Based on older studies [58,59], the effect of the vegetation parameter (v) was assessed comparatively for all considered land covers; regarding forest vegetation (forest), the latter assumed to result in the least total soil loss and runoff. High-vegetation areas (forest) may cause a decrease of surface runoff coefficient up to 88%, while low-vegetation areas (pastures) cause up to 44% decrease, respectively. The result constitutes the final vulnerability class of the sub-factor F, which takes values from 1 (very low vulnerability— V.L.) to 5 (very high vulnerability—V.H.) (Table S6, Supplementary document). Therefore, the final value of the sub-factor F to be used in Equation (4) will be deduced empirically from the combination of slope parameter (s) and vegetation parameter (v), according to Table S6 in the supplementary documents.

Based on the slope (s) parameter distribution (Table S4) as derived by the digital elevation model of the study area [40] and the vegetation (v) spatial distribution in accordance to the classes discussed (Table S5), the F sub-factor spatial distribution values were calculated regarding Table S6, as illustrated in Figure 6.

4.2.2. S Sub-Factor Calculation

The S sub-factor accounts for the permeability of surface geological formations and is directly proportional to the vulnerability of groundwater systems. The higher the permeability, the greater the vertical infiltration (percolation) and therefore the more significant recharge that will potentially reach the saturated zone of the groundwater system. RIVA regards surface formations that occur up to 1.5 m below surface and control surface/subsurface flow [26].

Surface formations are classified into soils and consolidated geological formations. Soils are considered the upper loose earth horizons, including topsoil, developed over nonconsolidated geological formations (e.g., Neogene formations, Quaternary deposits). As consolidated geological formations, RIVA considers those that are lithified and constitute the underlying bedrock. The method assumes that soils of considerable thickness are not developed over the consolidated geological formations. Therefore, the surface hydrological conditions are controlled mainly by the permeability of the consolidated formations and not by the soil compartment.

S values for soils are shown in Table S7 (Supplementary document) based on the U.S. Department of Agriculture classification that is based on their texture [60], whereas, for the geological formations, S values are deduced based on their permeability, as proposed by the British Geological Survey and presented in Table S8 (Supplementary document) [61]. Permeability is a property that is not easily determined accurately, as it is affected by several factors (e.g., degree of fracturing, tectonic stress, karstification, alternations with horizons of different permeability, homogeneity, isotropy, etc.), which may more or less change the original nature of the formation. However, guidelines for a more general characterization framework are not limited and provide a range of values, as shown in Table S8. The lower values correspond to solid (unaffected) formations, while the higher reflect the factors above effect that eventually increases their permeability and thus their vulnerability class. The final calculation of the S sub-factor and its spatial distribution map results from compiling

F values for the soils and F values for the consolidated geological formations. Note that for each cell of the generated map, a value S of soil or formation is assigned and not an aggregated value of two individual values of soil and formation.

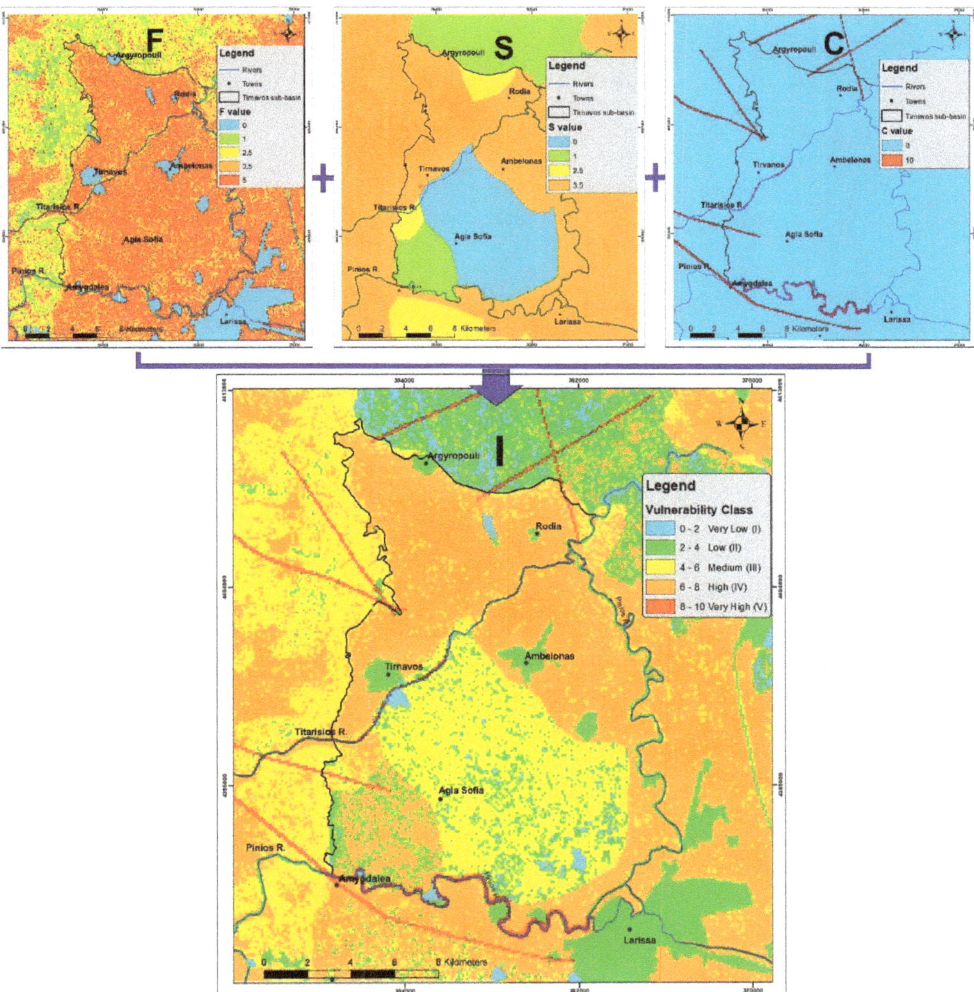

Figure 6. Spatial distribution maps of F, S, and C sub-factors in the Tirnavos basin. According to the I factor, the final map of vulnerability classes results from the composition of the three above mentioned thematic maps.

Regarding the Tirnavos basin, Quaternary deposits based on the geological map (Figure 1) were classified based on Table S7 categories, considering the soil study of the area [41] in which soil units were assigned with soil and geological and topographic criteria. In addition, Neogene formations at southwest and northeast margins and marbles at the west and the crystalline bedrock, composed of mica-schists and gneisses at the northern boundary of the basin, are classified based on Table S8. In addition, taking into account the degree of karstification of the marbles [28,32,34], vulnerability class III is assigned to them, and class II is assigned to the Neogene marls (resulting from their permeability based on the same studies). The above data processing composed the distribution map of the S sub-factor, as shown in Figure 6.

4.2.3. C Sub-Factor Calculation

Sub-factor C addresses the special cases of concentrated flow in fractured/discontinuities media that may impact the vulnerability characterization attributing the maximum vulnerability (very high) value to factor I. This sub-factor refers to the spatially concentrated because of specific surface features, which results in increased infiltration and thus maximum aquifer vulnerability. These features may include (a) epikarst; (b) drainage patterns, which are documented, that is, in hydraulic relation with the aquifer; (c) sinkholes; and (d) tectonic structures (e.g., faults, overthrusts). Infiltration is significantly favoured within the influence zone of the above features, causing eventually very high vulnerability.

As the exact orientation of the impact zones is not possible to be defined due to variable influencing factors that require complicated modelling approaches, RIVA proposes using an approximate impact zone of 100 m around critical surface features, to which a maximum score of 10 (V.H. vulnerability) is attributed [26]. If such structures do not occur (therefore no influence zones), the C value is by definition negligible. Hence, the C sub-factor constitutes an on-off (0–10) feature.

In the case of the Tirnavos basin, there are no indications of the existence of developed epikarst or sinkholes. However, some significant fault zones [42,43] can increase the infiltration and, consequently, the aquifer vulnerability (C values map, Figure 6). Furthermore, in earlier studies [28,32], it is suggested that in certain parts of the Pinios and Titarisios riverbeds, the hydraulic relationship between rivers and the groundwater system occurs. Regarding the Pinios River, confirmed hydraulic relation exists from its entrance to the Tirnavos basin up to Larissa city. In contrast, for the Titarisios River, significant hydraulic interaction occurs along the western margins of the basin, where extensive talus cones are formed (C values map, Figure 6).

4.3. V Factor Description

Factor V (vadose zone) accounts for the protection provided by the vadose zone as a function of the nature of its geological formations and its total thickness, which is directly related to the piezometric level. Factor V takes values from I (very low vulnerability) to V (very high vulnerability) according to the grading of Figure 4. As mentioned, it differentiates from the I factor because it regards the part below 1.5 m from the surface (upper soil horizons). In this context, the V factor may include (a) the soil's underlying non-lithified geological formations and/or strongly weathered zones of bedrock and (b) the bedrock (lithified geological formations).

Calculation of the V factor is performed through the modification and compilation of previous approaches [14,61–63] as follows [26]:

1. Initially, each geological formation of the vadose zone of the study area is classified according to its dominant lithological type, prior to any secondary effects (e.g., karstification), and is attributed a "reference layer (ly) value" based on its permeability range, as shown in Table S9 in the supplementary documents.
2. The "ly" values are multiplied by the fracturing or karstification factor (f), corresponding to an internal modification of the initial value "ly" due to secondary effects that impact permeability. The "f" factor derives from assessing the fracturing/karstification degree of the considered geological formation (only for the lithified) based on the values of Table S10 in the supplementary documents.
3. The derived product is multiplied by the total thickness of the formation in meters to provide the final value of the protective cover (pc), which corresponds to a class of V factor, shown in Table S11 (Supplementary document). If the vadose zone consists of more than one layer in the vertical dimension, each formation is calculated individually as described, and then, all are summed up to calculate the final "pc" value.

From the above methodology, the critical point is to estimate the thickness of the formation (or formations) that shape the vadose zone. For this purpose, the following approach was adopted. For the phreatic aquifer zone within the alluvial basin, the piezometric level distribution as deduced from field measurements was considered [28,34,36], along with the compiled digital elevation model (DEM) (subtracting piezometric map contours from DEM). When only a confined aquifer is active, the thickness of the vadose zone is based on the bottom of the confining layer as mapped by older foundation studies in the region [44], subtracted from the DEM. For the karstic domain considered, the piezometric levels provided by older studies were considered [45], under the assumption that no significant changes have occurred in the spatial distribution of the groundwater levels in this environment. If they have occurred, they would have related to groundwater level decline, leading to an increased thickness vadose zone and consequently lower vulnerability assessment. Therefore, this assumption provides a rather conservative calculation, favouring the area's environmental protection. No piezometric data exist for the crystalline bedrock at the northern boundary of the basin. However, a low-capacity aquifer is reported to occur in the weathering mantle of the metamorphic formations, the thickness of which does not exceed 30 meters [28]. Therefore, it is reasonable to assume that groundwater level will not occur at a depth greater than that. Hence, this value is used as a baseline for the vadose zone thickness in this environment.

The compiled map for vadose zone thickness, based on the above data, was multiplied by the map synthesized as the product of ly and f (ly × f), based on Tables S9 and S10, and the geological map of the area, in order to deduce a protection cover (pc) map.

4.4. A Factor Description

The A factor (aquifer) refers to the easiness with which a potential contaminant will travel within the saturated zone of an aquifer as a function of its hydraulic conductivity. Factor A takes values from 1 for an aquifer of very low hydraulic conductivity (very low vulnerability—V.L.) to 5 for an aquifer of very high hydraulic conductivity (very high—V.H.) according to the classification in Figure 4.

The value of factor A is calculated based on the quantitative or qualitative assessment of the hydraulic conductivity of the aquifer. In the case of measured values, the link between hydraulic conductivity and vulnerability class (A factor values) is shown in Table S12 in the supplementary documents, based on a modified, generic, yet widely accepted conductivity classification according to [64].

For the alluvial basin, the values of the hydraulic conductivity of the geological formations were derived based on the results from the pumping test analyses [33,44] and the geo-hydrogeological characteristics of the aquifer. Moreover, the hydraulic conductivity values regarding the marbles at the west and metamorphic formations (gneisses, mica-schists) at the N-NE were based on bibliographic references [46–49].

4.5. Compilation of Results

The calculations of factors, sub-factors' values, and spatial distribution were performed in ArcPRO® software to extract the digital maps, employing the corresponding equations and attributing class values as referenced in the mentioned Tables above. All data were transformed to raster grids with cell size 100 m × 100 m, which covered the case study area of the Tirnavos basin. The derived maps for each sub-factor, including the weighting factors and the cumulative map for main factors, are depicted in Figures 5–8.

Two vulnerability classes are assigned for the R factor: class I for the greater part of the study area and class II at the central and southeast part of the study area, as shown in Figure 5.

Regarding factor I, all vulnerability classes emerged in the spatial distribution map (Figure 6) following the raster calculations of its three sub-factors. The highest vulnerability class (V) was assigned along the traces of significant fault zones and surface-ground water system interaction zones. The class (IV) occupies the highest area percentage, and its main distribution is at the alluvial basin and the eastern part of the case study area, excluding its central part. The central part of the alluvial basin and most of the karstified domain are characterised by medium vulnerability class (III), while at the north and southeast, the low class (II) prevails. The lowest vulnerability class (I) has only isolated occurrences at the central and eastern parts of the area.

In the part west of Agia Sofia local community, low- and high-class alternations are presented, while the intermediate medium class is absent, as observed at the I factor distribution map. This fact is probably due to Neogene formations in the specific area and the southern part. However, what differentiates the two parts, is the topographic slope, which is lower in the south than in the west of the Agia Sofia village, where a slope break exists from the mountainous terrain to the alluvial basin. That justifies the appearance of the medium class in the southern part in contrast to the area west of the Agia Sofia village.

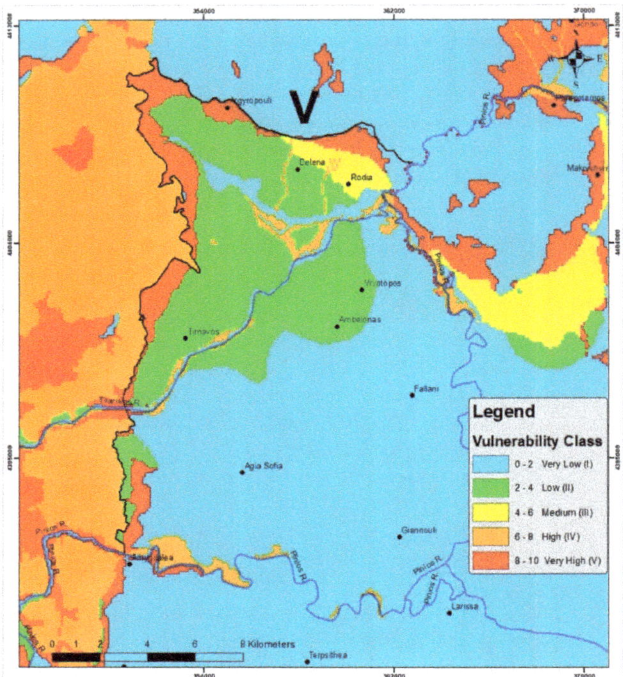

Figure 7. Spatial distribution of vulnerability classes according to the V factor in the study area.

In order to compile the V factor distribution map, based on Table S11, the pc values were attributed to a vulnerability class/characterization, corresponding to a V factor value. The map produced after this exercise is illustrated in Figure 7. In this map, the full range of vulnerability classes is also displayed.

Following the approach mentioned in the A factor description paragraph, a value of hydraulic conductivity correlated to vulnerability class was attributed to Table S12. Ultimately, a value of A factor was assigned to each formation and known permeability value point, and subsequently, data values were spatially interpolated with Inverse Distance Weighting (IDW) method to obtain the spatial distribution map for factor A (Figure 8).

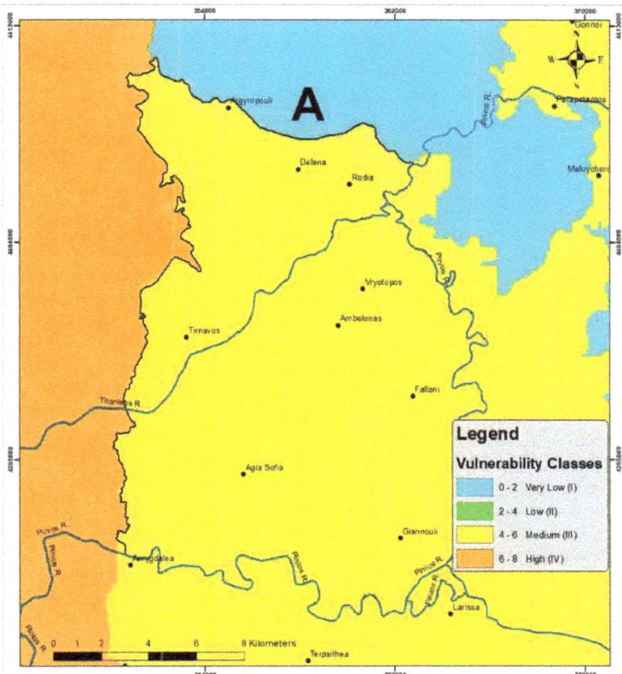

Figure 8. Spatial distribution of vulnerability classes according to the A factor in the study area.

Since vulnerability maps were created for all four factors (R,I,V,A), groundwater intrinsic vulnerability (i) was estimated based on Equation (1), considering the corresponding weighting factors. The final calculations between raster images were performed following a normal grid 100 × 100 m cell distribution with ArcPRO GIS software. The value of any cell of the final map was derived by summing up the individual values of corresponding cells of all factor maps, according to Equation (1). The final intrinsic vulnerability map, which was produced following this procedure, is illustrated in Figure 9. Based on this map, the central and southeast parts of the case study area, where alluvial deposits dominate, are characterised by low vulnerability (II), occupying 48.75% of the total area. Comparing this map with the V factor distribution map, it is concluded that protecting the vadose zone in this area is highly effective since factor F is assigned the highest weighting factor. The karstic area at the west and northwest parts of the alluvial basin shows medium (III) to high (IV) vulnerability, occupying 24.04% and 14.14% of the examined domain. This fact reflects the sensitivity of the nature of the formations in these areas, as they are karstified marbles at the west side and conglomerates and talus cone at the northwest margins of the alluvial basin. The dominance of the high (IV) class at the west and north boundaries of the plain areas is reasonable because of the lithology and topography of the existing transition zones. The north and northeast boundaries of the study area appear to have the lowest vulnerability class (I), occupying 13.01% of the total area and reflecting the crystalline bedrock, which is mica-schists and gneisses, respectively. It should be noted that the areas of very high (V) vulnerability occupy a minimum percentage (0.07%) of the total area and are placed on the tectonic structures between Pinios and Titarisios Rivers.

4.6. Validation

The validation procedure is critical for the initial vulnerability assessment [27]. The most common approach, particularly for verification of assessments done with overlay and index methods, is to compare the vulnerability map with the actual occurrence of some common contaminant in groundwater [65,66]. Considering the initial vulnerability concept and the specific characteristics of the study area, validation was performed by comparing the monitored values of nitrates with the modelled vulnerability, as defined by the spatial distribution of the final vulnerability map [26] (Figure 9).

Nitrate values from 43 boreholes were used to create the spatial distribution map of nitrates in the Tirnavos basin, based on previous surveys and sampling campaigns [28,34,36]. Each value is the average result of four sampling campaigns at wet and dry seasons from September 2016–April 2018. These average values of nitrate concentrations were classified into ranges as follows: 0–20 mg/L, class I (very low); 20–40 mg/L, class II (low); 40–50 mg/L, class III (medium); 50–70 mg/L, class IV (high); >70 mg/L, class V (very high). Based on the above classification, which constitutes a modified version of the groundwater nitrates concentration classification proposed in the framework of the Nitrates Directive [67], a spatial distribution map of nitrates' classes was compiled (Figure 10). Nitrates' classes distribution map was subsequently compared to the final vulnerability map, a result of summing RIVA factors (Figure 9). The subtraction of these two maps yields (modelled values—monitored values) the validation map (Figure 11).

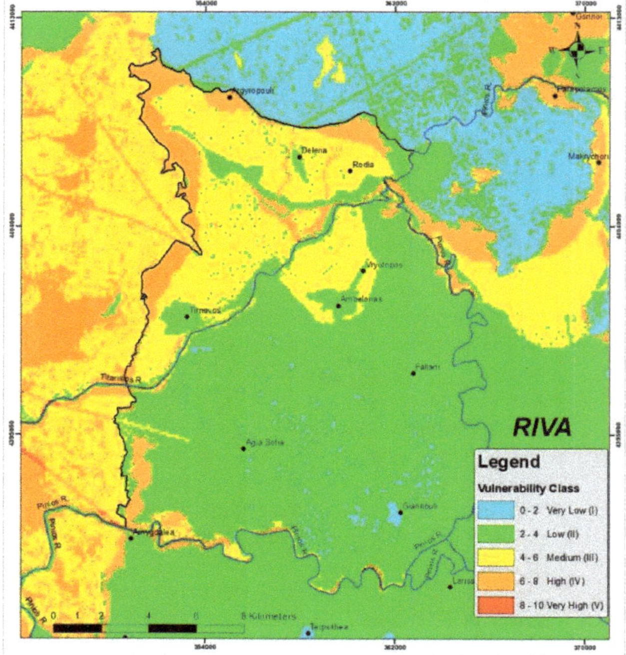

Figure 9. Spatial distribution of vulnerability classes according to RIVA method in the study area.

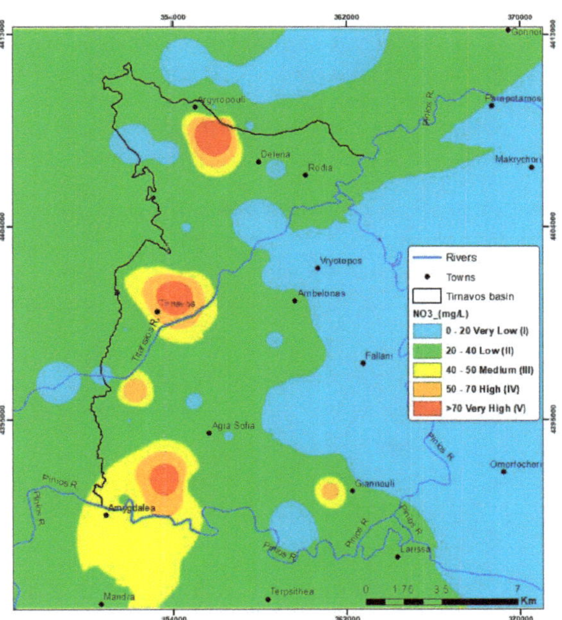

Figure 10. Spatial distribution map of nitrates concentrations (average values of four periods, September 2016–April 2018).

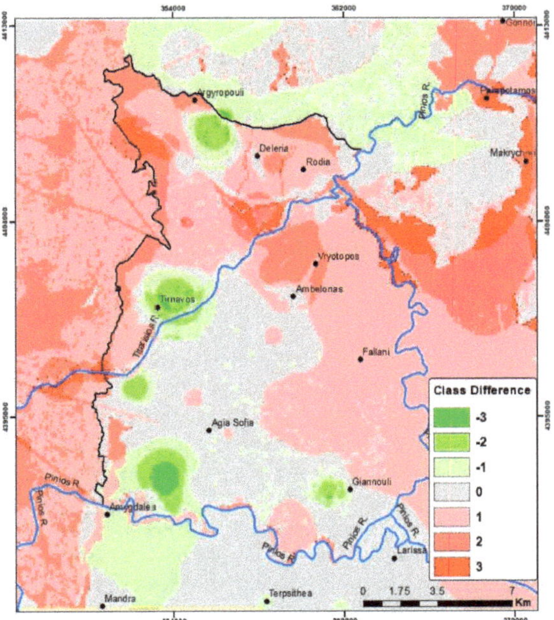

Figure 11. Validation map of the RIVA application method in the Tirnavos basin. The class difference between modelled and monitored values, depicted.

In Table 2, the analytical results of the validation procedure are shown. Based on this table, 29.14% of the total area presents a perfect match between modelled and monitored values (difference class 0). Overall negative difference (underestimating, lower modelled than monitored values) presents 13.20% and positive difference (overestimating, higher modelled than monitored values) 57.65%. According to the original validation of the method [26], a difference of one class (−1 to 1) is accepted as a very good match between the modelled and monitored values. If that percentage is equal to or greater than 80%, then the modelled results are successful compared to the ground truth values.

Table 2. Difference of vulnerability class between modelled and monitored values.

Modelled-Monitored Class	Percentage
−3	0.33
−2	1.61
−1	11.27
0	29.14
1	41.11
2	13.82
3	2.72
	100.00

According to the validation performed in the present research, the total area with a difference from −1 to +1 between the modelled and monitored values is 81.52%; thus, the validation may be regarded as successful. Continuing, 96.95% of the total area presents a difference of two classes (−2 to 2), and only 3.05% exhibits a difference of three classes (−3 or +3).

The largest class differences (−3) where RIVA underestimates the potential vulnerability appear in a few individual sites related either to point-source contamination or to the effect of migrating contaminants through lateral crossflows. The latter is probably the case for the deviations in the alluvial basin related to zones of lateral contaminant fluxes and especially so along the southwestern-most edge of the basin. By definition, aquifer vulnerability accounts for the vertical susceptibility of the system and does not account for any lateral crossflows of contaminant plumes from adjacent hydrogeological units. On the other hand, the infiltration of N-containing pollutants from surface water and the transport of nitrate contaminants through soil and groundwater occurs via a series of complex chemical and hydraulic phenomena [68]. As a result of these complex procedures, a horizontal migration of the contaminants is frequently dominant. Hence, practical validation with measured contaminant values at the saturated zone should only be performed at hydrologically "closed" systems without any hydraulic connections with other units (surface or underground); if not, then potential migrations of contaminant plume(s) must be taken into account prior to validation [69].

Indeed, based on the geometry of the groundwater system and its hydrodynamic evolution, as these were discussed in the earlier parts of this work, the area where the highest deviation between modelled and monitored values coincides with the parts of the system where contaminants' migration is influenced by a conurbation of factors where not only percolation is important, but lateral flows do play a significant role. Moreover, it needs to be taken into account that the results of water sampling on nitrate concentrations do not always reflect the regional background regarding contaminant's state, but instead, they may indicate local factors of N-bearing compounds mismanagement. Hence, deviations between the RIVA and nitrates distribution classes are sufficiently explained.

5. Discussion

Insight to the factors used

The effect of precipitation and irrigation on groundwater vulnerability was examined by calculating the R factor in the study area. The mean annual approach is more efficient for strategic planning at a regional scale, whilst the seasonal approach can flag vulnerability aspects that are significant for local or regional scale risk assessment uses [70,71]. Although data from three meteorological stations located at different altitudes were used, both the mean precipitation (P sub-factor) and its intensity (In sub-factor) did not form separate vulnerability classes based on the various ranges that have been defined.

Irrigation (Ir sub-factor) as a factor of vulnerability influence has not been considered in any other method yet. In rural areas where agriculture forms the key socio-economic activity, irrigation is indeed an overlooked parameter of paramount importance in assessing the vulnerability of a groundwater system, as clearly demonstrated through the performed analysis. Especially in the absence of multiple classes for the other 2 sub-factors shaping R factor (P + In), the irrigation sub-factor considerably influenced the spatial distribution of the R factor. Given that the average field size at the study area is rather small (ca 1 ha), it is thought that should irrigation system data occur at the field scale, a more detailed spatially diversified distribution of the R factor would have been yielded. In turn, such an approach would provide a higher spatial resolution assessment of the system's vulnerability concerning this particular sub-factor. On a regional scale, however, this differentiation would have not led to considerably different results apart from the potential for even better agreement between the vulnerability map and the ground-truth nitrate concentrations and, in particular, the ability to better explain hot spots of the nitrates concentration distribution map. However, in the absence of such detailed data, the dominant irrigation system per local irrigation organisation inherently incorporates the assumption of the least water-efficient irrigation system used. Hence, the higher irrigation water is applied on the field, which subsequently leads to a relative over-estimation of water use and, therefore, higher assessments of vulnerability due to this sub-factor. Overall, the approach followed results in a more conservative estimate of vulnerability, i.e., a more environmentally sound approach.

Assessing I factor incorporates evaluating the combined action of slope and vegetation, which constitutes a straightforward procedure based on the principle that the lower the slope and the denser the vegetation, the greater the class of attributed vulnerability. It also includes evaluating permeability of the top 1.5 m of soils and consolidated geological formations, as assigned with the relevant tables developed by the method. Last, it incorporates the effect of linear features, including tectonic lines [72,73] and hydrologically interacting river stretches [74,75]. As a whole, the subjectivity of this factor is rather limited due to the wide classes distinguished for the evaluation of each sub-factor and the overall sound principles it adopts in the definition of the considered sub-factors.

As already stressed, V factor constitutes the most influential factor in assessing the examined system's vulnerability, and this is acknowledged by assigning to the final vulnerability index calculation formula a very high weighting factor (a = 0.4). Hence, it is rather critical to evaluate this factor carefully and based on reliable information to avoid misconceptions on the final spatially distributed vulnerability product. Out of the three considered sub-factors, the dominant lithological type (ly) and the corresponding fracturing or karstification degree (f), which represent the bulk permeability values per formation and the alterations imposed due to secondary deformations, are safely assessed utilizing bibliographic references [28,61,63], geological maps [30,31], field observations [28,34,36], and the adopted classification scheme of RIVA, as earlier presented [26]. Calculation of the protective cap, which forms the third sub-factor (pc), is the most critical and presents the highest risk for erroneous calculations, especially when groundwater system vulnerability assessments are attempted for complex geometry environments, as the examined one is. However, the adopted approach and utilization of the detailed data available on the system's geometry for most of its parts ensures that performed assessments are reliable and, in fact, shifted towards the conservative margin of analysis. Thus, calculations tend to accept the minimum thickness of protecting cap in the few parts of the basin where relevant data for this sub-factor are not detailed. As with the I factor, this provides a safety margin in favour of environmental protection, assuming the worst-case scenario on the geometry of the considered system.

Last, factor A reflects the influence of hydraulic conductivity on intrinsic vulnerability, and its spatial distribution is deduced based on pumping test analyses, augmented by region-specific and more generic but well-established bibliographic references [46–49] transposed to wide range classifications, as earlier presented. Hence, the margin of error in assessing the spatial distribution of this factor is limited.

Validation of the method was performed on the basis of a considerable population of real monitoring data that cover dry and wet hydrological conditions for two years and are typical of an agriculture-related pollutant. Considering the particularities of the studied region as it is reflected by geological structure, hydrogeological setup, and regional hydrodynamic evolution mechanisms, validation results are satisfactory, demonstrating the validity and efficacy of RIVA in similar complex environments. Spatially distributed zones of different vulnerability classes are delineated accurately and reliably even though the tested hydrogeological environment is characterised by a high degree of complexity.

Observed deviations relate mainly to intrinsic vulnerability assessment methods' inherent limitations in accounting for horizontal flow driven migration of pollutants. Moreover, it is expected that the produced results would have been further improved if monitored data utilized for validation were originating from a single aquifer of the examined system rather than reflecting an average concentration of the entire system. That is the result of the construction characteristics of most production wells in the studied region, which tap the entire groundwater system that consists of multiple aquifers, the hydrochemical evolution of each one of which is controlled by different mechanisms. Last, the highest deviations noted reflect hot spots areas that are most likely related to one or more of the following reasons: (a) N-compounds mismanagement at the vicinity of the well heads, (b) systematic over-irrigation leading to excessive leaching of contaminants, and (c) poor construction characteristics of production wells that enable migration of contaminants to deeper hydrogeological strata through the wells' gravelpack.

In total, implementation of RIVA in the studied region provides a valuable tool for regional planning and management of natural resources that the area lacks, leading to considerable deterioration of existing groundwater reserves. Compared to other approaches, the selected method enables uniform consideration and evaluation of the entire region structured of highly contrasting character lithological typologies, rendering vulnerability assessment easier and comparable amongst the existing geological media. Of particular importance is the consideration of irrigation water management, which leads to a more comprehensive assessment of the essence in rural areas where intensified agriculture is practised. Even though irrigation is an externality, being an established condition, it directly influences vulnerability and should be given a thorough, reliable, and pragmatic assessment. In several parts of the world, data scarcity is a fact. Still, reliable vulnerability assessments are imperative and may not await appropriate state-of-the-art data collection. RIVA is not a data-intensive method that can easily retrieve information based on sound geological and hydrogeological knowledge. Obviously, enhanced results may be obtained, and a higher spatial resolution achieved should spatially dense and reliable data-driven calculations be employed.

Challenges and problems

Groundwater systems vulnerability assessment is a valuable tool with profound applications in environmental science. One of them could potentially be its incorporation, as a factor in the delineation of Nitrate Vulnerable Zones (NVZs), in the framework of the Nitrates Directive [67]. Thessaly as a whole is one of the first regions of the country to be declared as an NVZ in the late 1990s [76], using as key criteria the surface and groundwater concentration of nitrates, the N-input, and the geomorphological characteristics of the region. Groundwater system vulnerability consideration would add to the reliability of the

NVZ assessment and perhaps even lead to a considerably differentiated delineation pattern, accounting for the essential properties that control the potential leaching of contaminants. Therefore, such an approach would lead to a more comprehensive and meaningful delineation of NVZs and greatly assist the design of appropriate and targeted, thus efficient, measures as part of the Action Plan in the framework of the Nitrates Directive. Such an approach is directly in line with the Common Agricultural Policy, which identifies the direct relationship between agriculture and groundwater resources and the need for effective protection and preservation of the latter from agricultural inputs [77,78].

Spatially distributed flow models have been developed for the studied area and the wider Thessaly region [79–81], which provide a reliable simulation of the flow domain accounting for the key hydrodynamic evolution mechanisms. Contamination transport and even hydrogeochemical modelling coupled with such flow models would enhance understanding of evolution mechanisms and improve validation of the results. Advective and dispersive migration elements of the pollutants could be quantified and isolated, thus enabling backwards analysis of nitrates concentration distribution deduced from the monitoring exercise, accounting for reducing or oxidizing reactions that influence groundwater chemistry and thus the nitrates concentrations in the collected and analysed water samples. In this way, leaching would be the sole pathway to be reflected on the nitrates distribution, thus directly relating to the evaluation provided by RIVA, which assesses vulnerability considering the vertical pathways of potential pollutants. On a different viewpoint, such models could be coupled to RIVA to account for the advective and dispersive migration of a potential pollutant, thus contributing to a more holistic assessment of groundwater systems' vulnerability. In such an approach, the vulnerability would be calculated as a function of the examined parameters in the framework of RIVA (for leaching potential assessments) and lateral crossflows from surrounding formations or groundwater systems.

As pointed out in previous parts of the paper, seasonal variations of precipitation patterns lead to differentiated calculations of key factors of the vulnerability index. Likewise, consideration of seasonal groundwater levels as opposed to mean annual or inter-annual levels may lead to considerable differences in the calculations of the protective cap sub-factor of V factor, which has the highest weighting factor in the calculation of the vulnerability index. Previous studies have demonstrated this in similar geo-climatic environments [82,83]. In the framework of climate change, considerable variations of key climate parameters, including precipitation depth and intensity, temperature, and evapotranspiration, are anticipated, especially for the region Thessaly, which is considered amongst the vulnerable regions along with the entire Mediterranean [84–86]. These changes will potentially affect the hydrological balance of the system in terms of anticipated natural recharge and abstractions given that irrigation water needs are expected to increase under the business as usual scenario. In turn, depth to groundwater is expected to be affected. Prolonged droughts duration and frequency along with the already experienced severe floods because of high rainfall intensity directly differentiate directly the values of several sub-factors considered in the vulnerability assessment under the RIVA method.

Likewise, improving irrigation water efficiency is also expected in response to climate change and resilience development. Finally, crop distribution changes are already being discussed as an adaptation measure. These are some of the key changes expected in the future that will influence the vulnerability of the considered groundwater system. The factors above will alter the spatial distribution of vulnerability. However, some of them will increase it, whilst others decrease it. Therefore, the overall prediction of the anticipated outcome may not be safely made empirically. Still, reliable assessments of the forecasted vulnerability under climate change environment are important to augment measures designed to increase groundwater systems' resilience and overall water sector safety.

6. Conclusions

This study assessed the intrinsic vulnerability of the Tirnavos basin (Central Greece) groundwater system using the RIVA method. RIVA has the advantage that it can be applied to all types of groundwater bodies independently of the specific conditions, lithologic phases, and aquifer typology of each area. Four main factors were used to represent the natural hydrogeological conditions of the specific area: recharge, infiltration, vadose zone, and aquifer.

Modelled results were validated with ground-truth values of nitrates obtained from 43 wells and proved to be quite successful, as they presented over 80% of matching (negligible or small deviations between modelled and monitored values). The few deviations are attributed to inherent uncertainty factors, such as the interpolation of the various factors used and the lateral contamination transport, which the index and overlay methods cannot assess without the help of a spatially distributed model.

The outcomes of the RIVA application to the Tirnavos basin can be further exploited as a preliminary tool for decision making and strategic assessment. At a regional scale, the results may be further valorised for the delineation of NVZs, considered as a key target for the trade-off between sustainable agriculture and water resources protection.

Supplementary Materials: The following supporting information can be downloaded at: https://www.mdpi.com/article/10.3390/w14040534/s1, Table S1: Correlation between annual precipitation (mm/y), P values, and vulnerability classes; Table S2: Correlation between rainfall intensity (mm/d), In values, and vulnerability classes; Table S3: Correlation between irrigation dose, In values, and vulnerability class; Table S4: Classification of topographic slope (s) and correlation with vulnerability classes; Table S5: Categorization of vegetation according to CORINE land-use classification (EEA, 2020); Table S6: Calculation of F values according to soil (s) and vegetation (v) parameters and link with vulnerability classes; Table S7: Calculation of S values for the soils and correlation with the vulnerability class (USDA, 1999); Table S8: Calculation of S values for consolidated geological formations and correlation with permeability and the vulnerability class (Lewis et al., 2006); Table S9: Classification of layer reference values (ly) for representative geological formations; Table S10: "f" factor values according to the assessed fracturing or karstification degree; Table S11: Protective cover (pc) values and corresponding vulnerability characterization and V factor values; Table S12: The suggested link between hydraulic conductivity and A factor values, corresponding to specific vulnerability class.

Author Contributions: I.V. conceived the methodological approach, performed the data processing and supervised the drafting and revision of the final text. E.T., A.P. and G.S. participated pin data processing, verified and optimised the methodology, contributed to the discussion parts, and revised the final text. All authors have read and agreed to the published version of the manuscript.

Funding: This research received no external funding.

Institutional Review Board Statement: Not applicable.

Informed Consent Statement: Informed consent was obtained from all subjects involved in the study.

Data Availability Statement: Data is contained within the article.

References

1. Younger, P.L. *Groundwater in the Environment: An Introduction*; John Wiley & Sons: Hoboken, NJ, USA, 2009.
2. Wachniew, P.; Zurek, A.J.; Stumpp, C.; Gemitzi, A.; Gargini, A.; Filippini, M.; Rozanski, K.; Meeks, J.; Kværner, J.; Witczak, S. Toward operational methods for the assessment of intrinsic groundwater vulnerability: A review. *Crit. Rev. Environ. Sci. Technol.* **2016**, *46*, 827–884. [CrossRef]
3. Nageswara, R.K.; Narendra, K. Mapping and evaluation of urban sprawling in the Mehadrigedda watershed in Visakhapatnam metropolitan region using remote sensing and GIS. *Curr. Sci.* **2006**, *91*, 1552–1557.
4. Lindström, R. Groundwater Vulnerability Assessment Using Process-Based Models. Ph.D. Thesis, KTH, Stockholm, Sweden, 2005.
5. Milnes, E. Process-based groundwater salinisation risk assessment methodology: Application to the Akrotiri aquifer (Southern Cyprus). *J. Hydrol.* **2011**, *399*, 29–47. [CrossRef]

6. Popescu, I.C.; Brouyère, S.; Dassargues, A. The APSÛ method for process-based groundwater vulnerability assessment. *Hydrogeol. J.* **2019**, *27*, 2563–2579. [CrossRef]

7. Masetti, M.; Sterlacchini, S.; Ballabio, C.; Sorichetta, A.; Poli, S. Influence of threshold value in the use of statistical methods for groundwater vulnerability assessment. *Sci. Total Environ.* **2009**, *407*, 3836–3846. [CrossRef]

8. Sorichetta, A.; Masetti, M.; Ballabio, C.; Sterlacchini, S.; Beretta, G.P. Reliability of groundwater vulnerability maps obtained through statistical methods. *J. Environ. Manag.* **2011**, *92*, 1215–1224. [CrossRef]

9. Li, X.; Philp, J.; Cremades, R.; Roberts, A.; He, L.; Li, L.; Yu, Q. Agricultural vulnerability over the Chinese Loess Plateau in response to climate change: Exposure, sensitivity, and adaptive capacity. *Ambio* **2016**, *45*, 350–360. [CrossRef]

10. Gogu, R.C.; Dassargues, A. Current trends and future challenges in groundwater vulnerability assessment using overlay and index methods. *Environ. Geol.* **2000**, *39*, 549–559. [CrossRef]

11. Antonakos, A.K.; Lambrakis, N.J. Development and testing of three hybrid methods for the assessment of aquifer vulnerability to nitrates, based on the drastic model, an example from NE Korinthia, Greece. *J. Hydrol.* **2007**, *333*, 288–304. [CrossRef]

12. Pacheco, F.A.L.; Pires, L.M.G.R.; Santos, R.M.B.; Fernandes, L.S. Factor weighting in DRASTIC modeling. *Sci. Total Environ.* **2015**, *505*, 474–486. [CrossRef]

13. Boufekane, A.; Saighi, O. Application of groundwater vulnerability overlay and index methods to the Jijel plain area (Algeria). *Groundwater* **2018**, *56*, 143–156. [CrossRef] [PubMed]

14. Shirazi, S.M.; Imran, H.M.; Akib, S. GIS-based DRASTIC method for groundwater vulnerability assessment: A review. *J. Risk Res.* **2012**, *15*, 991–1011. [CrossRef]

15. Focazio, M.J. *Assessing Ground-Water Vulnerability to Contamination: Providing Scientifically Defensible Information for Decision Makers*; US Government Printing Office: Washington, DC, USA, 1984; Volume 1224.

16. Brindha, K.; Elango, L. Cross comparison of five popular groundwater pollution vulnerability index approaches. *J. Hydrol.* **2015**, *524*, 597–613. [CrossRef]

17. Aller, L.; Bennett, T.; Lehr, J.; Petty, R.J.; Hackett, G. *DRASTIC: A Standardized System for Evaluating Ground Water Pollution Potential Using Hydrogeologic Settings*; US Environmental Protection Agency: Washington, DC, USA, 1987; Volume 455.

18. Foster, S.S.D. *Fundamental Concepts in Aquifer Vulnerability, Pollution Risk and Protection Strategy*; Netherlands Organization for Applied Scientific Research: The Hague, The Netherlands, 1987; Volume 38, pp. 69–86.

19. Stempvoort, D.V.; Ewert, L.; Wassenaar, L. Aquifer vulnerability index: A GIS-compatible method for groundwater vulnerability mapping. *Can. Water Resour. J.* **1993**, *18*, 25–37. [CrossRef]

20. Navulur, K.C.S. Groundwater Vulnerability Evaluation to Nitrate Pollution on a Regional Scale Using GIS. Ph.D. Thesis, Purdue University, West Lafayette, IN, USA, 1996.

21. Doerfliger, N.; Jeannin, P.Y.; Zwahlen, F. Water vulnerability assessment in karst environments: A new method of defining protection areas using a multi-attribute approach and GIS tools (EPIK method). *Environ. Geol.* **1999**, *39*, 165–176. [CrossRef]

22. Lappas, I.; Kallioras, A.; Pliakas, F.; Rondogianni, T. Groundwater vulnerability assessment to seawater intrusion through GIS-based Galdit method. Case study: Atalanti coastal aquifer, central Greece. *Bull. Geol. Soc. Greece* **2016**, *50*, 798–807. [CrossRef]

23. Petelet-Giraud, E.; Dörfliger, N.; Crochet, P. RISKE: Méthode d'évaluation multicritère de la cartographie de la vulnérabilité des aquifères karstiques. Application aux systèmes des Fontanilles et Cent-Fonts (Hérault, Sud de la France). *Hydrogéologie (Orléans)* **2000**, *4*, 71–88.

24. Allouche, N.; Maanan, M.; Gontara, M.; Rollo, N.; Jmal, I.; Bouri, S. A global risk approach to assessing groundwater vulnerability. *Environ. Model. Softw.* **2017**, *88*, 168–182. [CrossRef]

25. Vrba, J.; Zaporozec, A. *Guidebook on Mapping Groundwater Vulnerability*; Heise: Niedersachsen, Germany, 1994.

26. Tziritis, E.; Pisinaras, V.; Panagopoulos, A.; Arampatzis, G. RIVA: A new proposed method for assessing intrinsic groundwater vulnerability. *Environ. Sci. Pollut. Res.* **2020**, *28*, 7043–7067. [CrossRef]

27. Zwahlen, F. *Vulnerability and Risk Mapping for the Protection of Carbonate (Karst) Aquifers*; Office for Official Publications of the European Communities: Luxembourg, 2003; Available online: https://www.cost.eu/publications/vulnerability-and-risk-mapping-for-the-protection-of-carbonate-karst-aquifers-final-report/(PDF) (accessed on 18 January 2022).

28. Vrouhakis, I.; Panagopoulos, A.; Stamatis, G. Current Quality and Quantity Status of Tirnavos sub-Basin Water System—Central Greece. In Proceedings of the 11th International Hydrogeological Congress of the Greece, Athens, Greece, 4–6 October 2017.

29. European Environment Agency. CORINE Land Cover, Methodology and Nomenclature. Available online: https://www.eea.europa.eu/publications/COR0-landcover (accessed on 18 January 2022).

30. Plastiras, V. *Geological Map of Greece, Larissa Sheet*; Institute of Geology and Mineral Exploitation: Athens, Greece, 1982.

31. Miggiros, G. *Geological Map of Greece, Gonnoi Sheet*; Institute of Geology and Mineral Exploitation: Athens, Greece, 1980.

32. Alexandridis, T.; Panagopoulos, A.; Galanis, G.; Alexiou, I.; Cherif, I.; Chemin, Y.; Stavrinos, E.; Bilas, G.; Zalidis, G. Combining remotely sensed surface energy fluxes and GIS analysis of groundwater parameters for irrigation assessment. *Irrig. Sci.* **2014**, *32*, 127–140. [CrossRef]

33. Panagopoulos, A. A Methodology for Groundwater Resources Management of a Typical Alluvial Aquifer System in Greece. Ph.D. Thesis, School of Earth Sciences, Faculty of Science, University of Birmingham, Birmingham, UK, 1995.

34. Vrouhakis, I.; Tziritis, E.; Panagopoulos, A.; Kulls, C.; Stamatis, G. The Use of Environmental Stable Isotopes at the Tirnavos Alluvial Basin (Central Greece). In Proceedings of the 15th International Congress of the Geological Society of Greece, Athens, Greece, 22–24 May 2019.

35. Demitrack, A. The Late Quaternary Geologic History of the Larissa Plain, Thessaly, Greece: Tectonic, Climatic, and Human Impact on the Landscape. Ph.D. Thesis, Stanford University, Stanford, CA, USA, 1986; p. 117, (Unpublished).

36. Vrouhakis, I.; Pisinaras, V.; Panagopoulos, A.; Stamatis, G. Multivariate Statistical Analyses of Groundwater Hydrochemical Data of Tirnavos Sub-basin (Central Greece). In Proceedings of the 16th International Conference on Environmental Science and Technology, Rhodes, Greece, 4–7 September 2019.

37. Hellenic National Meteorological Service (HNMS), Digital Data, 1989–2018. Available online: http://emy.gr/emy/en (accessed on 18 January 2022).

38. National Centers for Environmental Informations (NCEI), Digital Data, 1989–2018. Available online: https://www.ncei.noaa.gov/ (accessed on 18 January 2022).

39. Greek Payment Authority of Common Agricultural Policy Aid Schemes (OPEKEPE), Digital Data. 2015. Available online: https://www.opekepe.gr/en/contact-us-en (accessed on 18 January 2022).

40. Advanced Spaceborne Thermal Emission and Reflection Radiometer (ASTER) Global Digital Elevation Model Version 3 (GDEM 003), Digital Data. Available online: https://asterweb.jpl.nasa.gov/gdem.asp (accessed on 18 January 2022).

41. Toulios, M.; Katsilouli, E.; Georgiou, T.; Argyropoulos, G.; Dimogiannis, D. *Soil Study of Tirnavos Area*; National Agricultural Research Foundation, Soil Mapping and Classification Institute: Larissa, Greece, 1997.

42. Caputo, R.; Helly, B.; Pavlides, S.; Papadopoulos, G. Palaeoseismological investigation of the Tyrnavos fault (Thessaly, central Greece). *Tectonophysics* **2004**, *394*, 1–20. [CrossRef]

43. Karakostas, V.; Papazachos, C.; Papadimitriou, E.; Foumelis, M.; Kiratzi, A.; Pikridas, C.; Kostoglou, A.; Kkallas, C.; Chatzis, N.; Bitharis, S.; et al. The March 2021 Tyrnavos, central Greece, doublet (Mw6. 3 and Mw6. 0): Aftershock relocation, faulting details, coseismic slip and deformation. *Bull. Geol. Soc. Greece* **2021**, *58*, 131–178. [CrossRef]

44. Grenoble, S. *Study for the Development of Groundwater in the Thessaly Plain*; Technical Report for Groundwater in the Thessaly Plain; Land Reclamation Services of Greece: Thessaly, Greece, 1974.

45. Antonaropoulos, P.; Vainalis, D. *Hydrogeological Study of Artificial Recharge of the Karstic System of Titarisios—Pinios Rivers of Tirnavos Area*; Technical Report for Pinios Rivers of Tirnavos Area; Land Reclamation Services of Greece: Larissa, Greece, 2010.

46. Constantinides, D. Hydrodynamique d'un Systeme Aquifere Heterogene. Ph.D. Thesis, Universite de Grenoble, Saint-Martin-d'Hères, France, 1978.

47. Davis, S.N.; Turk, L.J. Best well depth in crystalline rocks. *Johns. Drillers J.* **1969**, *41*, 1–5.

48. Freeze, R.A.; Cherry, J.A. *Groundwater*; Prentice-Hall: Englewood Cliffs, NJ, USA, 1979; 604p.

49. Voudouris, K. Hydrogeology of the Environment. In *Groundwater and Environment*; Tziolas Publications: Thessaloniki, Greece, 2009; Volume 460.

50. Neuner, M.; Smith, L.; Blowes, D.W.; Sego, D.C.; Smith, L.J.; Fretz, N.; Gupton, M. The Diavik waste rock project: Water flow through mine waste rock in a permafrost terrain. *Appl. Geochem.* **2013**, *36*, 222–233. [CrossRef]

51. Owor, M.; Taylor, R.G.; Tindimugaya, C.; Mwesigwa, D. Rainfall intensity and groundwater recharge: Empirical evidence from the Upper Nile Basin. *Environ. Res. Lett.* **2019**, *4*, 035009. [CrossRef]

52. Mileham, L.; Taylor, R.G.; Todd, M.; Tindimugaya, C.; Thompson, J. The impact of climate change on groundwater recharge and runoff in a humid, equatorial catchment: Sensitivity of projections to rainfall intensity. *Hydrol. Sci. J.* **2009**, *54*, 727–738. [CrossRef]

53. Guzzetti, F.; Peruccacci, S.; Rossi, M.; Stark, C.P. The rainfall intensity–duration control of shallow landslides and debris flows: An update. *Landslides* **2008**, *5*, 3–17. [CrossRef]

54. Pavlis, M.; Cummins, E. Assessing the vulnerability of groundwater to pollution in Ireland based on the COST-620 Pan-European approach. *J. Environ. Manag.* **2014**, *133*, 162–173. [CrossRef]

55. Koutsi, R.; Stournaras, G. Groundwater Vulnerability Assessment in the Loussi Polje area, N Peloponessus: The PRESK Method. In *Advances in the Research of Aquatic Environment*; Springer: Berlin/Heidelberg, Germany, 2014; pp. 335–342. [CrossRef]

56. Kallioras, A.; Pliakas, F.; Skias, S.; Gkiougkis, I. Groundwater Vulnerability Assessment at S.W. Rhodope Aquifer System in N.E. Greece. In *Advances in the Research of Aquatic Environment*; Springer: Berlin/Heidelberg, Germany, 2011; pp. 351–358. [CrossRef]

57. Lohani, S.; Baffaut, C.; Thompson, A.L.; Aryal, N.; Bingner, R.L.; Bjorneberg, D.L.; Bosch, D.D.; Bryant, R.B.; Buda, A.; Dabney, S.M.; et al. Performance of the Soil Vulnerability Index with respect to slope, digital elevation model resolution, and hydrologic soil group. *J. Soil Water Conserv.* **2020**, *75*, 12–27. [CrossRef]

58. Descroix, L.; Viramontes, D.; Vauclin, M.; Barrios, J.G.; Esteves, M. Influence of soil surface features and vegetation on runoff and erosion in the Western Sierra Madre (Durango, Northwest Mexico). *Catena* **2001**, *43*, 115–135. [CrossRef]

59. El Kateb, H.; Zhang, H.; Zhang, P.; Mosandl, R. Soil erosion and surface runoff on different vegetation covers and slope gradients: A field experiment in Southern Shaanxi Province, China. *Catena* **2013**, *105*, 1–10. [CrossRef]

60. United States Department of Agriculture. *Soil Taxonomy a Basic System of Soil Classification for Making and Interpreting Soil Surveys*, 2nd ed.; Agriculture Handbook; United States Department of Agriculture: Washington, DC, USA, 1999; p. 436.

61. Lewis, M.A.; Cheney, C.S.; O Dochartaigh, B.E. *Guide to Permeability Indices (CR/06/160N)*; British Geological Survey: Nottingham, UK, 2006; p. 29, (Unpublished).

62. Panagopoulos, G.P.; Antonakos, A.K.; Lambrakis, N.J. Optimization of the DRASTIC method for groundwater vulnerability assessment via the use of simple statistical methods and GIS. *Hydrogeol. J.* **2006**, *14*, 894–911. [CrossRef]

63. Ravbar, N.; Goldscheider, N. Proposed methodology of vulnerability and contamination risk mapping for the protection of karst aquifers in Slovenia. *Acta Carsologica* **2007**, *36*, 3. [CrossRef]

64. Bear, J. *Dynamics of Fluids in Porous Media*; Courier Corporation: Chelmsford, MA, USA, 1988.
65. Ghazavi, R.; Ebrahimi, Z. Assessing groundwater vulnerability to contamination in an arid environment using DRASTIC and GOD models. *Int. J. Environ. Sci. Technol.* **2015**, *12*, 2909–2918. [CrossRef]
66. Zghibi, A.; Merzougui, A.; Chenini, I.; Ergaieg, K.; Zouhri, L.; Tarhouni, J. Groundwater vulnerability analysis of Tunisian coastal aquifer: An application of DRASTIC index method in GIS environment. *Groundw. Sustain. Dev.* **2016**, *2*, 169–181. [CrossRef]
67. EU Commission. Council Directive 91/676/EEC of 12 December 1991 concerning the protection of waters against pollution caused by nitrates from agricultural sources. *Off. J. Eur. Community* **1991**, *L375*, 1–8.
68. Chao, W.A.N.G.; Pei-Fang, W.A.N.G. Migration of infiltrated NH4 and NO3 in a soil and groundwater system simulated by a soil tank. *Pedosphere* **2008**, *18*, 628–637. [CrossRef]
69. Tziritis, E.; Lombardo, L. Estimation of intrinsic aquifer vulnerability with index-overlay and statistical methods: The case of eastern Kopaida, central Greece. *Appl. Water Sci.* **2017**, *7*, 2215–2229. [CrossRef]
70. Vías, J.M.; Andreo, B.; Perles, M.J.; Carrasco, F.; Vadillo, I.; Jiménez, P. Proposed method for groundwater vulnerability mapping in carbonate (karstic) aquifers: The COP method. *Hydrogeol. J.* **2006**, *14*, 912–925. [CrossRef]
71. Haidu, I.; Nistor, M.M. Groundwater vulnerability assessment in the Grand Est region, France. *Quat. Int.* **2020**, *547*, 86–100. [CrossRef]
72. Qiang, W.; Bo, L.; Yulong, C. Vulnerability assessment of groundwater inrush from underlying aquifers based on variable weight model and its application. *Water Resour. Manag.* **2016**, *30*, 3331–3345. [CrossRef]
73. Meerkhan, H.; Teixeira, J.; Espinha Marques, J.; Afonso, M.J.; Chaminé, H.I. Delineating groundwater vulnerability and protection zone mapping in fractured rock masses: Focus on the DISCO index. *Water* **2016**, *8*, 462. [CrossRef]
74. Thomas, R.; Duraisamy, V. Hydrogeological delineation of groundwater vulnerability to droughts in semi-arid areas of western Ahmednagar district. *Egypt. J. Remote Sens. Space Sci.* **2018**, *21*, 121–137. [CrossRef]
75. Kumari, R.; Datta, P.S.; Rao, M.S.; Mukherjee, S.; Azad, C. Anthropogenic perturbations induced groundwater vulnerability to pollution in the industrial Faridabad District, Haryana, India. *Environ. Earth Sci.* **2018**, *77*, 187. [CrossRef]
76. Karyotis, T.; Panagopoulos, A.; Pateras, D.; Panoras, A.; Danalatos, N.; Angelakis, C.; Kosmas, C. The Greek Action Plan for the mitigation of nitrates in water resources of the vulnerable district of Thessaly. *J. Mediterr. Ecol.* **2002**, *3*, 77–83.
77. EU Commission. Directive 2000/60/EC of the European Parliament and of the Council of 23 October 2000 establishing a framework for Community action in the field of water policy. *Off. J. Eur. Communities* **2000**, *L327*, 1–72.
78. EU Commission. CAP Reform—A Long-Term Perspective for Sustainable Agriculture. 2003. Available online: https://ec.europa.eu/commission/presscorner/detail/en/IP_03_99 (accessed on 18 January 2022).
79. Koukidou, I.; Panagopoulos, A. Application of FEFLOW for the simulation of groundwater flow at the Tirnavos (central Greece) alluvial basin aquifer system. *Bull. Geol. Soc. Greece* **2010**, *43*, 1747–1757. [CrossRef]
80. Tziatzios, G.; Sidiropoulos, P.; Vasiliades, L.; Tzabiras, J.; Papaioannou, G.; Mylopoulos, N.; Loukas, A. Effects of climate change on groundwater nitrate modelling. In Proceedings of the International Conference on Protection and Restoration of the Environment XIV, Thessaloniki, Greece, 3–6 July 2018.
81. Lyra, A.; Loukas, A.; Sidiropoulos, P.; Tziatzios, G.; Mylopoulos, N. An integrated modeling system for the evaluation of water resources in coastal agricultural watersheds: Application in Almyros Basin, Thessaly, Greece. *Water* **2021**, *13*, 268. [CrossRef]
82. Panagopoulos, A.; Domakinis, C.; Arampatzis, G.; Charoulis, A.; Vrouhakis, I.; Panoras, A. Seasonal Variations of Aquifer Intrinsic Vulnerability in an Intensively Cultivated Vulnerable Basin of Greece. In *Innovative Strategies and Policies for Soil Conservation*; Fullen, M., Famodimu, J., Karyotis, T., Noulas, C., Panagopoulos, A., Rubio, J., Gabriels, D., Eds.; Catena Verlag: Reiskirchen, Germany, 2015.
83. Luoma, S.; Okkonen, J.; Korkka-Niemi, K. Comparison of the AVI, modified SINTACS and GALDIT vulnerability methods under future climate-change scenarios for a shallow low-lying coastal aquifer in southern Finland. *Hydrogeol. J.* **2017**, *25*, 203–222. [CrossRef]
84. Panagopoulos, A.; Arampatzis, G.; Tziritis, E.; Pisinaras, V.; Herrmann, F.; Kunkel, R.; Wendland, F. Assessment of climate change impact in the hydrological regime of River Pinios basin, central Greece. *Desalination Water Treat.* **2016**, *57*, 2256–2267. [CrossRef]
85. Tziritis, E.; Pisinaras, V.; Kunkel, R.; Panagopoulos, A.; Arampatzis, G.; Wendland, F. Assessing the Potential Effects of Climate Change in the Hydrologic Budget of a Large Mediterranean Basin: The Case of River Pinios Basin, Central Greece. In Proceedings of the International Conference Adapt to Climate, Nicosia, Cyprus, 27–28 May 2014; pp. 1–13.
86. Kotsopoulos, S.; Nastos, P.; Ghionis, G.; Lazogiannis, K.; Poulos, S.; Alexiou, I.; Panagopoulos, A.; Farsirotou, E.; Alamanis, N. Evaporation and Evapotranspiration Estimates under Present and Future Climate Conditions. In Proceedings of the 12th International Conference, Protection & Restoration of the Environment—PRE12, Skiathos Island, Greece, 29 June–4 July 2014; pp. 91–97.

 *water*

Article

Hydrogeochemical Processes and Natural Background Levels of Chromium in an Ultramafic Environment. The Case Study of Vermio Mountain, Western Macedonia, Greece

Eleni Vasileiou [1], Panagiotis Papazotos [1], Dimitrios Dimitrakopoulos [2] and Maria Perraki [1,*]

[1] School of Mining and Metallurgical Engineering, Division of Geo-Sciences, National Technical University of Athens, 9 Heroon Polytechniou St., 15773 Zografou, Greece; elvas@metal.ntua.gr (E.V.); papazotos@metal.ntua.gr (P.P.)
[2] Researcher, Orologa 8, 11521 Athens, Greece; ddimitrakopoulos@gmail.com
* Correspondence: maria@metal.ntua.gr; Tel.: +30-2107722115

Abstract: The hydrogeochemical processes and natural background levels (NBLs) of chromium in the ultramafic environment of Vermio Mountain, Western Macedonia, Greece, were studied. Seventy groundwater samples were collected from 15 natural springs between 2014–2020, and an extensive set of physical and chemical parameters were determined. The ultramafic-dominated environment of western Vermio Mt. favors elevated groundwater concentrations of dissolved magnesium (Mg^{2+}), silicon (Si), nickel (Ni), and Cr in natural spring waters. Chromium was the principal environmental parameter that exhibited a wide range of concentrations, from 0.5 to 131.5 µg/L, systematically exceeding the permissible limit of 50 µg/L for drinking water. Statistical evaluation of hydrogeological, hydrochemical, and hydrological data highlighted the water-ultramafic rock process as the predominant contributor of Cr in groundwater. The NBL assessment for Cr and Cr(VI) was successfully applied to the typical ultramafic-dominated spring "Potistis" that satisfied all the methodology criteria. The NBLs of Cr and Cr(VI) were defined at 130 µg/L and 100 µg/L, respectively, revealing that a natural ultramafic-dominated environment exhibits the geochemical potential to contribute very high concentrations of geogenic Cr to groundwater. The holistic methodology, proposed herein, could be implemented in any catchment scale to assess geogenic and anthropogenic Cr-sources that degrade groundwater quality.

Keywords: chromium; ultramafic rocks; springs; water–rock interaction; natural background levels

Citation: Vasileiou, E.; Papazotos, P.; Dimitrakopoulos, D.; Perraki, M. Hydrogeochemical Processes and Natural Background Levels of Chromium in an Ultramafic Environment. The Case Study of Vermio Mountain, Western Macedonia, Greece. *Water* **2021**, *13*, 2809. https://doi.org/10.3390/w13202809

Academic Editors: Evangelos Tziritis and Andreas Panagopoulos

Received: 2 August 2021
Accepted: 2 October 2021
Published: 9 October 2021

Publisher's Note: MDPI stays neutral with regard to jurisdictional claims in published maps and institutional affiliations.

1. Introduction

Natural background levels (NBLs) are defined as "the concentration of a substance or the value of an indicator in a body of groundwater corresponding to no, or only very minor, anthropogenic alterations to undisturbed conditions" according to the Groundwater Daughter Directive (GDD) (Directive 2006/118/EC) [1]. Broadly, the term of NBLs is synonymous with the terms of environmental geochemistry "natural/geochemical background values" or "geochemical baseline" used in the past [2]. The NBLs are a set of several varying hydrogeological (i.e., the residence time of groundwater in the saturated zone, recharge by precipitation, hydraulic connection with other aquifer systems) [3–5], and hydrogeochemical (i.e., water–rock interaction, pH/redox conditions, chemical, and biological processes in the unsaturated zone) [5–7] factors. The determination of NBLs requires in-depth knowledge of geological/hydrogeochemical processes [8] and the distinguishment of natural and anthropogenic factors that affect the groundwater systems [9]. The need to separate NBLs from the anthropogenic impacts (e.g., urbanization, industrialization, agricultural activity) is frequently satisfied through statistical and pre-selection (PS) methods [10]. Such methods were applied within the EU-Specific Targeted Research Project BRIDGE (Background cRiteria for the iDentification of Groundwater thrEsholds), the objective of which was to develop a comprehensive methodology to evaluate threshold values (TVs) and NBLs of various qualitative parameters in the groundwater resources [11]. The first stage of this approach includes the PS method, which assumes that the groundwater samples represent pristine groundwater not affected by anthropogenic pressures [9]. It constitutes the most frequent method to exclude samples influenced by anthropogenic activities based on specific criteria such as concentrations of Cl^-, Na^+, NO_3^-, NH_4^+, and DO [6,10,12]. The PS method has been successfully applied to establish NBLs for different physical and chemical parameters, including EC, Cl^-, SO_4^{2-}, F^-, As, Cr, Cr(VI), Mn, Ni, Fe, and V in many European water bodies [4,5,7,13–17]. The next stage contains statistical tools such as box plots and normality tests for assessing the NBLs of the target chemical parameter. An approach that incorporates both PS and statistical methods has been performed by many researchers [7–10,14,15], providing a comprehensive methodology to boost the validity of the assessment, mainly when the geochemical and geological features are adequately considered [16]. Thus, the challenging assessment of NBLs in an environment in which the prevailing geochemical conditions favor the occurrence and mobilization of naturally occurring chemical elements could provide essential information regarding the controversial geogenic and anthropogenic inputs in a complex environmental setting.

The water–ultramafic rock interaction is of great scientific interest due to the high content of the latter in Cr (1000–3000 mg/kg), and other potentially toxic elements (PTEs) such as As, Co, Fe, Mn, and Ni compared to the Earth's crust composition [8–21] and to other rock types [22]; it constitutes the principal geogenic source of Cr in the environment [18]. Chromium is mainly hosted in spinels (e.g., chromite and magnetite) and silicates (e.g., pyroxene, serpentine, chlorite, olivine, talc). Serpentine group minerals can be highly enriched in Cr because it substitutes for magnesium (Mg) and/or iron (Fe) [18]. In the crystal lattice of most minerals, Cr occurs in the trivalent valence state [Cr(III)]. However, the geochemically immobile Cr(III) is oxidized into the mobile and toxic for the living organisms hexavalent chromium [Cr(VI)] in the presence of natural manganese oxides (MnO_2), specifically pyrolusite (b-MnO_2), in the typical range of groundwater pH (6.5–8.5) and under oxidizing redox potential (Eh) conditions [23–28]. Although an increasing number of studies focus on the occurrence and fate of Cr in the environment [29–38] only a few have systematically examined the geochemical fingerprint of water–ultramafic rock interaction in natural springs [29,33,35,39]. Typical worldwide examples of ultramafic springs with elevated groundwater concentrations of Cr(VI) have been recorded in the Province of La Spezia, Italy (up to 73 µg/L) [29], the Pollino massif, Italy (up to 30 µg/L) [40], the Gerania springs, Greece (up to 17.2 µg/L) [33], the Euboea Island, Greece (up to 37 µg/L) [41], and the Lesvos and Rhodes Islands, Greece (10–15 µg/L) [42]. The water–rock interaction constitutes a crucial and controlling factor concerning groundwater evolution. The geochemical reactions between the recharging water and the minerals of the host rocks affect the groundwater quality [43,44]. Hydrogeological and hydrogeochemical conditions such as pH, Eh, dissolved oxygen (DO), and groundwater flow path play a significant role in elevated concentrations of PTEs, including Cr, in the aquifer systems. Geochemical reactions such as ion exchange, weathering, precipitation/dissolution, and sorption process control the groundwater's composition considerably. During chemical weathering, some major ions, PTEs, and other trace elements become mobile and release from the parent rocks to the groundwater along the flow path. In addition, the mobility and solubility of these elements are controlled by water–rock contact time, Eh–pH conditions, and chemical reactions with organic matter [36]. Ionic ratios, saturation indices (SIs), and geochemical bivariate plots are usually evaluated to determine the intensity of water–rock interaction and chemical reactions [43]. Hence, the primary target of studying the mechanism of water–rock interaction is to elucidate the indissoluble association between the geological environment and the qualitative characteristics of groundwater.

In this work, we study the geochemical fingerprint of the water–ultramafic rock interaction process in the western Vermio Mt., Western Macedonia, Greece, and determine the NBLs of Cr in groundwater from natural springs. At the catchment scale of the

Sarigkiol Basin, elevated groundwater concentrations of Cr (up to ~140 µg/L) have been recorded in irrigation wells in the lowland [45]. Based on geospatial and multivariate statistical analyses of data from selected natural springs, irrigation wells, and surface waters the increased concentrations of Cr were attributed mainly to geogenic origin with the synergistic contribution of anthropogenic factors [45]. Challenged by the leaching potential of Cr of the ultramafic rocks in the area, we focus, herein, exclusively on the natural springs located in the ultramafic environment of western Vermio Mt, assessing hydrogeochemical data of a 7-year monitoring period (2014–2020). The springs are ideal for setting the NBLs at the catchment scale of the Sarigkiol Basin, because: (a) they record a strong ultramafic footprint, (b) they are located at a high altitude (>1300 m), (c) they exhibit unique worldwide high to very high concentrations of Cr (up to ~130 µg/L) [45], and (d) they are not affected by anthropogenic activities. Defining the NBLs of Cr in western Vermio Mt., will facilitate the identification of Cr origin in groundwater in the Sarigkiol Basin. This is the first systematic study of the natural springs of western Vermio Mt. and provides important hydrogeochemical data for the geogenic footprint of a natural ultramafic environment on the groundwater quality.

2. Materials and Methods

This section contains basic information about the: (i) study area, (ii) geological and hydrogeological setting, and (iii) sampling, chemical analysis procedures, and data processing.

2.1. Case Study

The present study is focused on the western Vermio Mt., located in the eastern part of the Sarigkiol Basin, Western Macedonia, Greece. The altitude of Vermio Mt is 2025 m and the average altitude of the basin is 650 m. The study area lies between the latitudes 40°25′00″ and 40°28′00″ E and the longitudes 21°56′00″ and 22°59′00″ N (Figure 1). In this area, any extensive anthropogenic activities lack except for local livestock farming and sporadically logging.

2.2. Geological and Hydrogeological Setting

The western Vermio Mt. is composed of (Figures 1 and 2) [46,47]: (a) alluvial deposits, (b) clastic conglomerates, talus cones, and breccias, (c) upper Cretaceous flysch, (d) a complex of schists and cherts formations, (e) ultramafic rocks (serpentinites and peridotites), (f) Triassic–Jurassic limestones, (g) Cretaceous limestones.

The aquifer systems in Vermio Mt. are:

(a)　The deep karstic aquifer of the Triassic-Jurassic limestones, which form the mountainous boundaries and the basement of the Sarigkiol Basin,

(b)　Perched aquifer systems that are developed in the highly fractured serpentinites of Vermio Mt. due to secondary porosity,

(c)　Small in size and capacity, karstic aquifers developed in the scattered Cretaceous limestones. There are many aquifers in which the water table varies from +700 up to +900 m. They are hydraulically connected and recharge the groundwater of the screes and talus cones in the ridges of the basin. The general flow direction of the groundwater is from the mountainous area to the center of the basin, i.e., NE–SW.

Different types of natural springs flow out in Vermio Mt. and specifically (Figure 1):

Contact springs formed where permeable formations (limestones, breccia, conglomerates) overlay formations of low permeability or impermeable (altered ultramafic rocks/ serpentinites). Contact springs studied here were: the springs S19, S10, S15 in the Agio Pnevma area, the "Potistis"-W13 spring, the spring S1 in the Agios Dimitrios area, the spring "Mouratidis"-S2, the springs S13, S14 in the Agios Panteleimonas area, and the springs S5 and S6 in the Vazelona area.

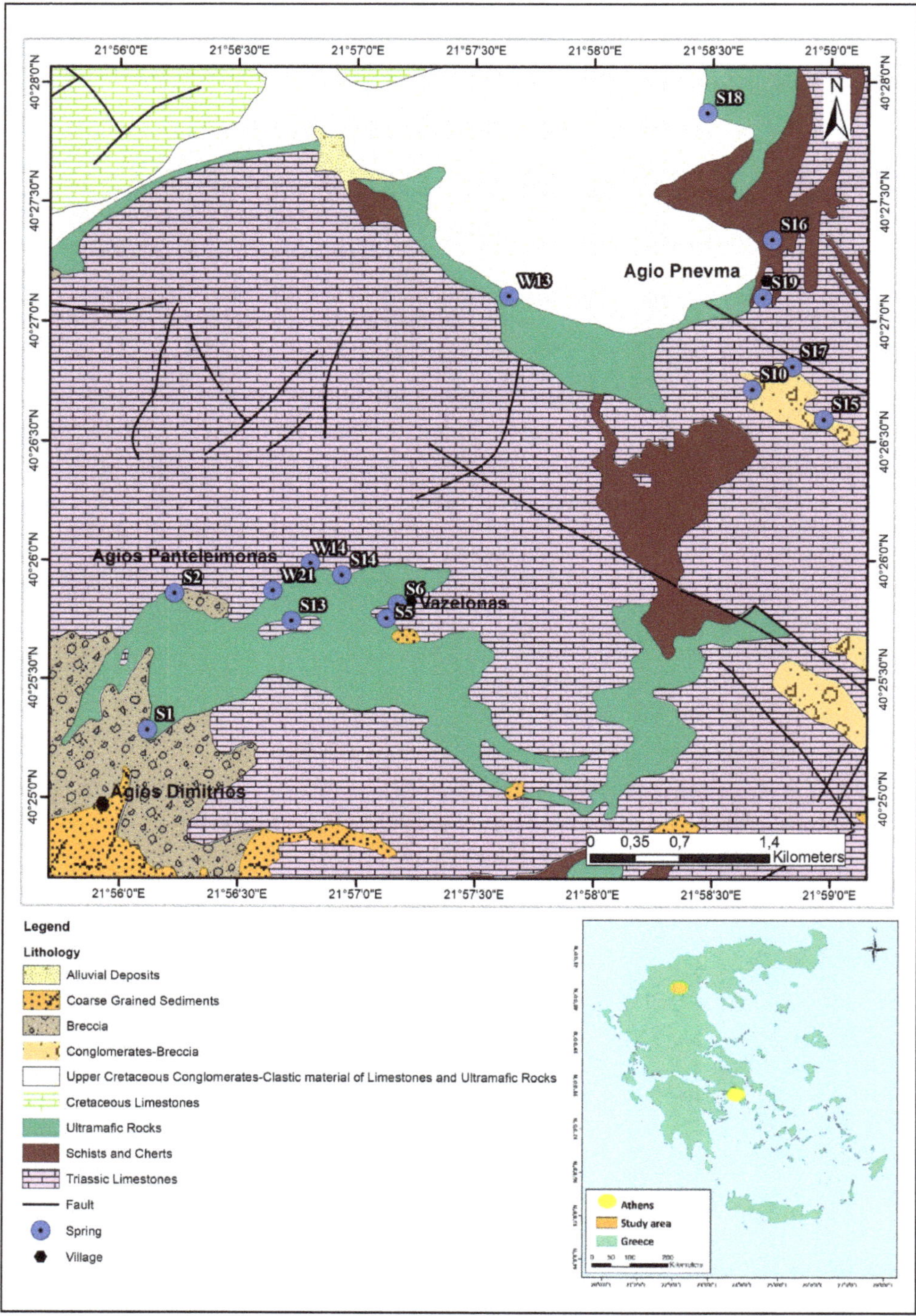

Figure 1. A simplified geological map of the western Vermio Mt.; the natural springs studied herein are marked.

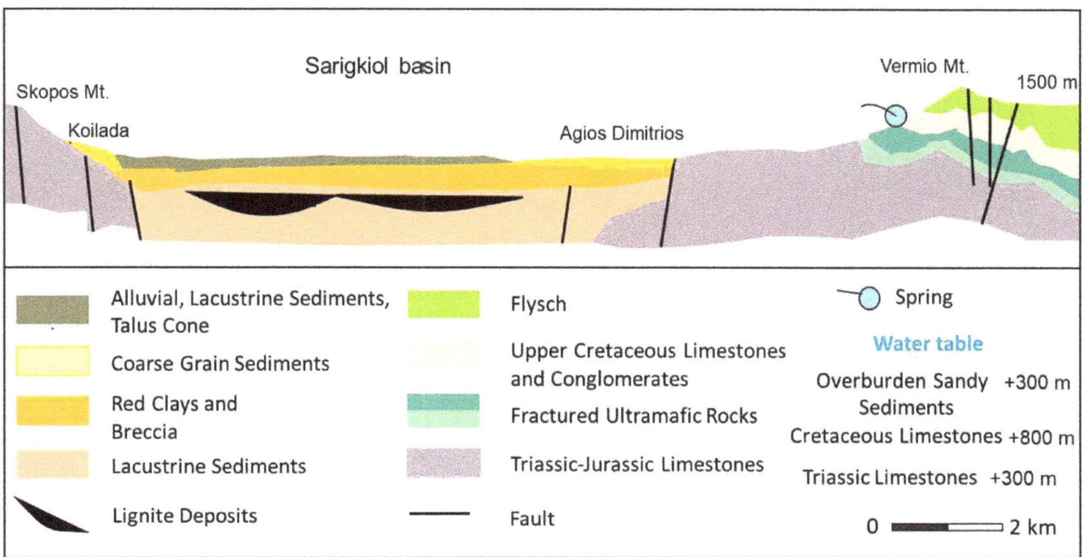

Figure 2. A simplified geological section of the western Vermio Mt.

Fault springs formed where impermeable rocks such as ultramafic rocks are in contact with an unconfined aquifer due to faulting; the springs "Elafakia"-W14 and W21 in the Agios Panteleimonas area belong to this type.

The spring "Potistis"-W13, presents great interest, because of the very high concentrations of Cr it exhibits [45]. The spring "Potistis"-W13 flows out at an elevation of 1300 m in an ultramafic environment characterized by the absence of any anthropogenic activities. It constitutes a contact-type spring in the contact of conglomerates with ultramafic clastic material and limestones and ultramafic rocks. The aquifer, which discharges via the spring, flows through a weathered zone in serpentines. The natural recharge comes mainly from the seasonal precipitations via the permeable upper unsaturated zone (conglomerates, clastic material of ultramafic rocks, and limestones). The recharge water is mainly enriched with released PTEs (mainly Cr) from the ultramafic rocks, as the rainfall infiltrates through the weathered fractured ultramafic rocks. An additional lateral recharge takes place due to secondary porosity in the fractured ultramafic rocks. The high permeability of the unsaturated zone due to the presence of conglomerates in this area facilitates the direct recharge of the aquifer in a short time. The range of discharge was calculated from 205 L/h up to 1200 L/h, with an average value of 482 L/h. In Figure 3, the simplified hydrogeological section describes the natural recharge and the operation mechanism of the spring "Potistis"-W13.

In western Vermio Mt. ultramafic rocks, mainly serpentinites, carbonates, schists and cherts occur [48]. The main mineral phases of the ultramafic rocks, depending on the degree of serpentinization, are serpentine [$(Mg, Mn, Fe, Co, Ni)_{3-x}SiO_2O_5(OH)_4$], olivine [$(Mg,Fe^{2+})_2(SiO_4)$)], pyroxene [$(Mg,Fe^{2+})(Si,Al)_2O_6$], talc [$Mg_3Si_4O_{10}(OH)_2$], chlorite [$(Mg,Fe^{2+})_5Al(Si_3Al)O_{10}(OH)_8$], tremolite [$Ca_2(Mg,Fe^{2+})_5Si_8(OH)_2O_{22}(OH)_2$], magnetite ($Fe^{2+}Fe_2^{3+}O_4$) and Cr-rich magnetite [$Fe^{2+}(Fe^{3+},Cr)_2O_4$] and chromite ($FeCr_2O_4$).

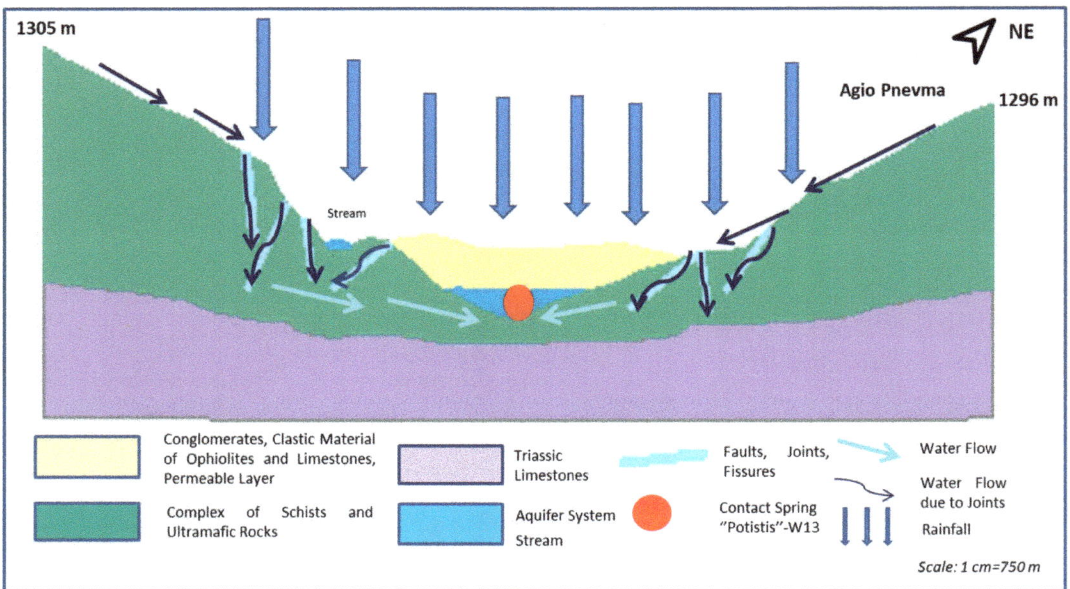

Figure 3. Simplified hydrogeological section of the natural spring "Potistis"-W13 in western Vermio Mt.

2.3. Sampling, Chemical Analyses, and Data Treatment

Springs to be systematically studied herein were selected based on their hydrogeochemical characteristics [45]. A total of 70 representative groundwater samples were collected from 15 natural springs during wet and dry periods from March 2014 to September 2020, following the groundwater sampling guidelines [49]. The 15 sampling sites (Figure 1) were classified into seven groups according to their location, lithology, and type: (i) twenty-three (23) samples were collected from the spring "Potistis"-W13, (ii) seven (7) samples from the Agio Pnevma area (S18, S16, S19, S10, S17, S15), (iii) twenty-three (23) samples from the spring "Elafakia"-W14, (iv) three (3) samples from the Agios Panteleimonas area (S13, S14, W21), (v) ten (10) samples from the spring "Mouratidis"-S2 (vi) two (2) samples (S5 and S6) from the Vazelona area, and (vii) two (2) samples from the spring S1 in the Agios Dimitrios area. Considering that the number of the water samples differs between the seven groups, each group has been treated and evaluated separately (the statistical and geochemical analysis), so the analyses are classified as reliable.

The analytical methods for the determination of physical [i.e., temperature (T), pH, oxidation-reduction potential (ORP), DO, and electrical conductivity (EC)] and chemical parameters (i.e., major ions, PTEs, and other trace elements) are provided in detail in Papazotos et al. [33]. The calculations of Eh and the total dissolved solids (TDS) values were carried out by converting ORP measurements (i.e., adding 200 mV) and the summation of major ions in each collected water sample, respectively.

AquaChem 5.0 software was used to elaborate chemical analyses, develop a Piper diagram, and calculate alkalinity. The statistical analyses of the chemical data were performed with SPSS 22.0 software.

2.4. Spearman's Rank Correlation Coefficient

Spearman's rank correlation coefficient (ρ, also signified by r_s) measures the strength and direction of association between two ranked variables, evaluating the degree of linear association or correlation between these independent variables. It presents many similarities to Pearson's coefficient except that it operates on the ranks of the data rather than the raw data [50].

The Spearman's rank correlation coefficient is calculated according to the following Equation (1) [51]:

$$r_s = \frac{6\sum_{i=1}^{n} d_i^2}{n(n^2 - 1)} \tag{1}$$

where d_i = difference in paired ranks, n = number of cases, x_i and y_i = data pair.

The formula to use when there are tied ranks is Equation (2):

$$\rho = \frac{\sum_i \left(x_i - \bar{x}\right)\left(y_i - \bar{y}\right)}{\sqrt{\sum_i \left(x_i - \bar{x}\right)^2 \sum_i \left(y_i - \bar{y}\right)^2}} \tag{2}$$

The Spearman's rank correlation coefficient, r_s, can get values from -1 to $+1$. The equation for the calculation is developed so that it gives $r_s = +1$ when the data pairs have a perfect positive correlation ($d^i = 0$) and $r_s = -1$ for the perfect negative correlation, whereas $r_s = 0$ indicates no association between ranks. The closer r_s is to zero the weaker the association between the ranks is.

The values of the correlation coefficient are classified as very strong (0.80–1), strong (0.60–0.79), moderate (0.40–0.59), weak (0.20–0.39), and very weak (0.00–0.19) [52]. The correlation coefficient is highly statistically significant, marginally statistically significant when the *p*-value is $p < 0.01$, $p < 0.05$.

2.5. Shapiro-Wilks Test

Shapiro–Wilks is a test of normality in frequentist statistics. The null hypothesis of this test is that the dataset is normally distributed. Thus, if the *p*-value is less than the chosen alpha level (0.05 in this case), then the null hypothesis is rejected and the data tested are not normally distributed. If the *p*-value is greater than the selected alpha level, then the null hypothesis cannot be rejected (Equation (3)) [53].

$$W = \frac{\left(\sum_{i=1}^{n} a_i x_{(i)}\right)^2}{\sum_{i=1}^{n} \left(x_i - \bar{x}\right)^2} \tag{3}$$

where $x_{(i)}$ is the i-th largest order statistic, x- is the sample mean, and n is the number of observations.

2.6. Quantile–Quantile Plot

The quantile-quantile (q–q) plot is a graphical tool for defining if two datasets come from populations with a common distribution [54], basically tests the conformity between the empirical distribution and the given theoretical one. On a Q–Q plot normally distributed data, the points in a Q–Q plot will fit on a straight diagonal line.

2.7. Geochemical modeling

The geochemical software PHREEQC version 3.1.2 [55] coupling with the MINTEQ database was used to calculate the saturation indices (SIs) of natural spring samples. Mineral SIs employed to define mineral dissolution and precipitation processes in the natural springs of western Vermio Mt. Saturation index is calculated by the Equation (4):

$$SI = Log\frac{IAP}{K_{sp}} = LogIAP - LogK_{sp} \tag{4}$$

where IAP = ion activity and K_{sp} = solubility product constant.

A positive SI indicates that the mineral is oversaturated or supersaturated with respect to the solution [56]; thus, the mineral could precipitate. Conversely, a negative SI indicates that the solution is undersaturated with respect to the selected mineral, suggesting that the mineral is dissolved in groundwater to reach equilibrium.

2.8. Calculation of NBLs of Cr

The assessment of NBLs for the target parameter was implemented based on the BRIDGE methodology [11]. The applied modified multi-method was separated into three steps: (a) the hydrogeochemical (bivariate plots, Piper, SI), (b) the PS method, and (c) the statistical analysis for estimating the NBLs (box plots for outliers, Q–Q (quantile–quantile) plots, and normality tests).

The applied methodology for the assessment of NBLs of Cr is described in detail in Figure 4. The pre-selection (PS) method, which is widely applied worldwide, was employed to select the suitable spring water samples for the NBLs assessment [10,16,57–60]. The PS method constitutes the methodology geochemical approach to validate the dataset according to similar geochemical characteristics and recognize the water samples that are affected by anthropogenic activities. In the first stage, the hydrochemical facies were selected based on the DO concentrations and Eh (ORP) [61]. The first dataset group included the water samples with ORP > 100 mV and DO > 3 mg/L. All water samples from the natural springs satisfied this criterion. The next criterion included consideration of redox conditions; if the prevailing conditions were oxidizing, then the concentrations of NO_3^- < 10 mg/L would be considered and if the conditions were reducing, then the NH_4^+ < 0.5 mg/L would be considered [9,61,62] to exclude the samples affected by anthropogenic activities [62,63]. Based on this criterion, the samples with NO_3^- > 10 mg/L were considered to be affected by anthropogenic activities and thus, were excluded from the new dataset. The next criterion required eight measurements per year for two years or two measurements per year for at least four years to exist for each spring [16].

2.9. Threshold Values (TVs) Derivation

The assessment of TVs was based on three scenarios [11,58] (Figure 5). The reference value was set equal to the water drinking acceptable limit (i.e., World Health Organization (WHO) guideline value).

2.10. Meteoric Genesis Index (MGI)

The meteoric genesis index (MGI) was also employed to classify the groundwater sources based on the depth of the meteoric water [64]. This index was calculated using the following Equation (5):

$$r_2 = Na^+ + K^+ - \frac{Cl^-}{SO_4^{2-}}, \text{ all concentrations are expressed in meq/L} \tag{5}$$

when r_2 < 1, the groundwater source is of deep meteoric water percolation type whereas when r_2 > 1 the groundwater is of shallow meteoric water percolation type [65].

2.11. Meteorological Data

Daily and monthly rainfall data for the period 2014–2019 were evaluated from the meteorological station in the Ermakia village (40°30′325″ N, 21°51′233″ E) which is the most representative and the nearest one, located on Vermio Mt., at an elevation of 1100 m. The annual precipitation for the period 2014–2019 was estimated at 985 mm.

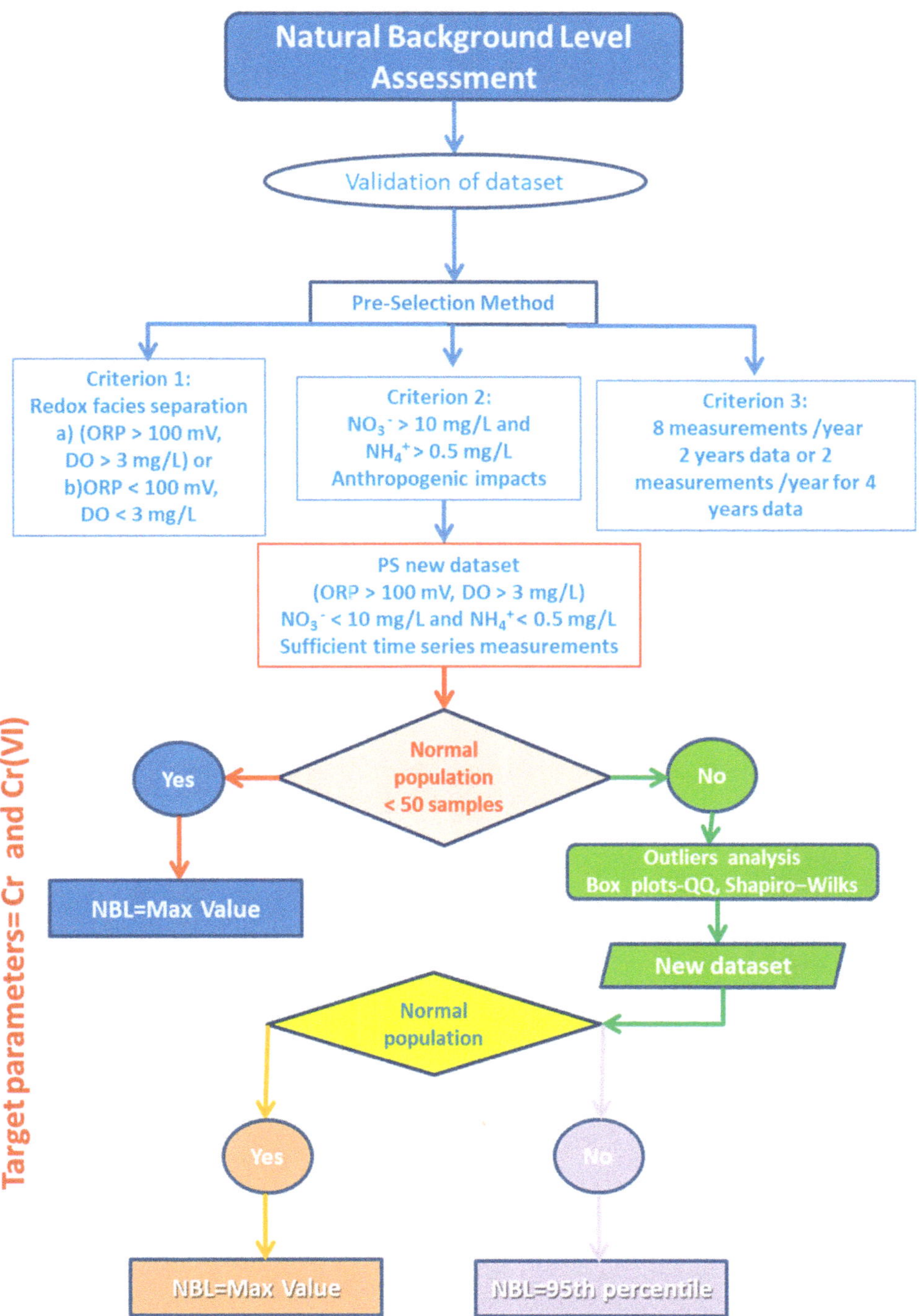

Figure 4. Flow chart of the modified conceptual model for the assessment NBLs of the target parameter.

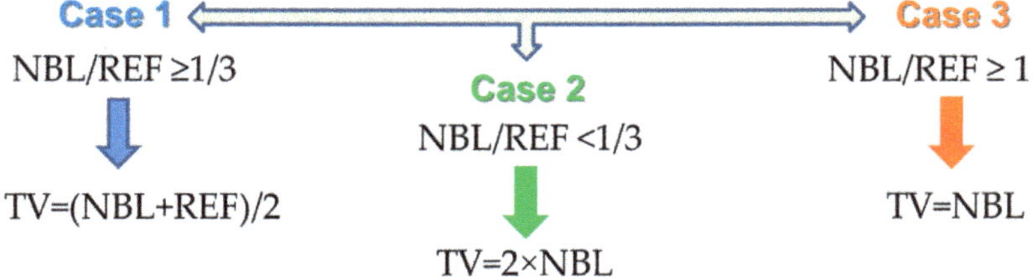

Figure 5. Flow chart for the assessment TVs for the target parameter.

3. Results

3.1. Chemometric Analysis

All chemical analyses were grouped into the above-mentioned seven categories according to their location, lithology, and type to evaluate the results. The descriptive statistics (max, min, median) of the physical and chemical parameters, for the studied spring waters from March 2014 until October 2020, are summarized in Tables 1 and 2. Some concentrations above the detection limit (DL) and below the quantification limit (QL) were not excluded because their very low concentrations do not affect the data processing of this work. All concentrations determined by ICP–MS presented a QL equal to the DL. The parameters NO_2^-, NH_4^+, PO_4^{3-}, Fe, Cd, Co were measured below the detection limit (BDL) in the majority of the samples. Silver, Au, Be, Bi, Cs, Pt, Re, Ga, Ge, Hf, Hg, In, Mo, Nb, Ta, Ti, Th, Tl, W, Zr were detected BDL in all water samples. Hence, these elements were excluded from the statistical analyses. In Table 3, the dataset for Cr-Cr(VI) and the geographical coordinates for the sampling points are given.

In the Agios Panteleimonas area, the natural spring "Elafakia"-W14 and in the Agio Pnevma area, the natural spring "Potistis"-W13 were examined separately from the other springs in these areas because of their elevated groundwater concentrations of Cr.

Box plots depicting the variation of eight groundwater quality parameters (pH, Eh, DO, EC, Ca^{2+}, Mg^{2+}, HCO_3^- and Si) are given in Figures 6 and 7. The alkalinity of the water samples was calculated in a range of 1.95×10^{-3} meq/L (S18) up to 6.71×10^{-3} meq/L "Potistis"-W13, with an average value of 4.60×10^{-3} meq/L.

The abundance of major ions varied significantly among the natural springs (Table 4). Concentrations of Cr and Cr(VI) exhibited a wide range of values (Tables 1–3, Figures 8–11). Most physical and chemical parameters (e.g., EC, NO_3^-, SO_4^{2-}, As, B, Ba, Cd, Cu, Ni, Pb, Sb, Se, Zn, etc.) in the natural springs of Vermio Mt. exhibited lower concentrations than the desirable and permissible limits for drinking water. All samples were compared with the guideline of WHO for drinking water [66]. Exceedances were recorded in the natural springs "Potistis"-W13, "Elafakia"-W14, S1-Agios Dimitrios area, and S18- Agio Pnevma area. Specifically, in the spring S18-Agio Pnevma area, concentrations of K^+ exceeded the guideline value of 12 mg/L for drinking water (WHO, [66]). Two exceedances of As, above the permissible limits of 10 µg/L, were recorded in the natural spring S1 of the Agios Dimitrios area. In the spring "Elafakia"-W14 seven water samples exceeded the WHO guideline value of 50 µg/L for Cr concentration for drinking water [66], while in the spring "Potistis"-W13 23 water samples exceeded this value. Finally, in total, three samples (one from the spring "Elafakia"-W14, and two from the spring "Potistis"-W13) exceeded the limit of 20 µg/L for the concentration of Ni for drinking water [66]. The systematic exceedances of Cr in the natural springs of Vermio Mt. were the principal reason for selecting Cr as the target parameter for NBLs assessment in this area.

Table 1. Maximum, minimum, and median values of physical and chemical parameters of natural springs in western Vermio Mt.

Parameter	Unit	QL	DL	The Agio Pnevma Area			The Agios Panteleimonas Area			"Mouratidis"			The Agios Dimitrios Area			The Vazelona Area		
				Max	Min	Median	Max	Min	Median	Max	Min	Median	Max	Min	Median	Max	Min	Median
pH	-	-	-	8.4	7.6	7.9	7.94	7.37	7.38	8.4	7.8	8.1	8.5	8.1	8.3	7.7	7.7	7.7
DO	mg/L	-	-	9.6	8.4	8.9	9.3	7.74	8.53	9.2	7.8	8.8	8.9	8.1	8.5	9.5	8.3	8.9
T	°C	-	-	16.0	10.8	14.3	18.0	13.1	13.6	25.6	9.8	13.8	24.1	8	16.1	15.8	15.0	15.4
TDS	mg/L	-	-	561.3	150.5	226.0	397.69	361.25	377.36	398	294.4	369.8	377	319.3	348.2	462.2	383.5	422.9
EC	µS/cm	10	-	593.0	293.0	460.0	505.0	448.0	494.0	520	405	426.8	456	446	451	484.0	389.0	436.5
Eh	mV	-	-	320.0	110.0	160.0	389.7	303.0	387.0	409	303.42	346.2	377	340.7	358.9	301.0	297.0	299.0
Ca^{2+}	mg/L	0.2	0.05	119.0	24.0	43.8	94.9	41.6	93.2	60.4	54.2	55.2	49.5	46.6	48.1	104.0	98.2	101.1
Mg^{2+}	mg/L	1.0	0.3	24.9	3.7	8.4	38.8	3.13	11.4	34.9	21.3	21.4	31.8	30.9	31.4	13.5	3.1	8.3
Na^+	mg/L	5.0	0.5	1.4	1.4	1.4	2.5	BDL	1.25	1.7	BDL	1.4	1.2	BDL	0.6	2.1	1.0	1.6
K^+	mg/L	0.2	0.05	33.0	0.4	0.6	1.63	0.59	1.22	1	0.3	0.3	1.8	1.5	1.7	10.8	1.4	6.1
NO_3^-	mg/L	5.0	1	BDL	BDL	BDL	BDL	BDL	BDL	9.1	8.3	8.7	BDL	BDL	BDL	1.0	BDL	BDL
Cl^-	mg/L	5.0	1	31.0	2.0	7.5	BDL	BDL	BDL	5	BDL	BDL	BDL	BDL	BDL	12.0	1.0	6.5
SO_4^{2-}	mg/L	10.0	2	31.0	10.0	20.5	19.0	12.0	19.0	20	16	16	22	BDL	11	21.0	13.0	17.0
HCO_3^-	mg/L	10.0	2	387.0	119.0	167.0	276.0	250.0	271.0	277	192	258	271	240	255.5	304.0	259.0	281.5
Al	µg/L	DL	1	8.0	2.0	3.0	1.0	1.0	1.0	1546	2	4	31	3	17	3.0	2.0	2.5
As	µg/L	DL	0.5	0.7	0.6	0.7	5.4	0.6	1.5	6.1	1.4	1.7	49.1	28.8	39	1.8	0.5	1.2
B	µg/L	DL	5	9.0	5.0	6.0	12.0	12.0	12.0	11	7	9	10	10	10	19.0	15.0	17.0
Ba	µg/L	DL	0.05	11.6	1.9	2.6	8.69	5.72	5.81	13.8	6	6.9	16.1	14.4	15.2	7.0	6.4	6.7
Br	µg/L	DL	5	25.0	8.0	13.0	14.0	11.0	12.0	16	13	14	15	9	12	15.0	13.0	14.0
Cr	µg/L	DL	0.1	47.8	0.5	3.8	18.0	1.5	1.9	38.3	10	20.4	16.6	10.9	13.8	0.8	0.5	0.7
Cr(VI)	µg/L	DL	0.1	36.7	0.5	1.8	7.2	1.0	1.0	33.9	7	16	16.5	8.7	12.6	0.5	0.1	0.3
Cu	µg/L	DL	0.1	2.8	0.9	1.2	1.2	0.6	0.7	2	0.7	1.4	1.3	1.1	1.2	2.8	1.1	2.0
Li	µg/L	DL	0.1	0.7	0.3	0.6	4.3	0.1	0.2	3.4	0.5	0.6	4.7	3.6	4.2	0.2	0.1	0.2
Mn	µg/L	DL	0.05	5.8	0.2	0.8	0.43	0.28	0.29	1.3	0.6	0.8	1.4	0.4	0.9	0.4	0.3	0.4
Ni	µg/L	DL	0.2	1.7	1.0	1.4	1.3	0.8	1.05	7.7	0.5	4.1	5.8	3.6	4.7	1.0	0.7	0.9
P	µg/L	DL	10	117	12.0	57.0	31.0	15.0	23.0	39	37	39	45	13	29	28.0	18.0	23.0
Si	µg/L	DL	40	14,327	2231	4296	24,875	3250	3441	21,350	8786	10,697	19,307	17,245	18,276	3495	3467	3481
Sr	µg/L	DL	0.01	84.0	44.5	59.8	79.03	41.76	70.15	62.9	57.5	58.4	65.2	60.7	62.9	83.2	69.7	76.5
U	µg/L	DL	0.02	0.2	BDL	0.1	7.2	0.4	0.7	0.3	0.3	0.3	0.4	0.3	0.3	0.7	0.2	0.4
V	µg/L	DL	0.2	0.7	0.2	0.3	0.08	0.08	0.08	4.6	1.3	1.5	4.7	4.3	4.5	0.8	0.5	0.7
Zn	µg/L	DL	0.5	12.7	7.5	10.4	4.3	3.4	3.9	14.8	9	13	11.9	7.3	9.6	33.1	13.6	23.4

BDL: Below the detection limit. DL: Detection limit. QL: Quantification limit.

Table 2. Maximum, minimum, and median values of physical and chemical parameters of the natural springs in western Vermio Mt.

Parameter	Unit	QL	DL	Potistis			Elafakia		
				Max	Min	Median	Max	Min	Median
pH	-	-	-	8.3	7.3	7.9	8.3	7.3	7.7
DO	mg/L	-	-	9.6	8.5	9.0	11.6	8.6	9.2
T	°C	-	-	15.2	6.2	12.3	20.5	5.8	12.5
TDS	mg/L	-	-	528.9	386.3	481.8	522.5	366.4	458.0
EC	µS/cm	10	-	620.0	374.8	574.5	718.0	357.7	546.0
Eh	mV	-	-	359.8	90.0	325.5	377.6	194.9	309.8
Ca^{2+}	mg/L	0.2	0.05	56.1	28.6	35.6	91.4	51.7	76.9
Mg^{2+}	mg/L	1.0	0.3	73.3	34.2	61.7	36.6	24.8	30.3
Na^+	mg/L	5.0	0.5	1.2	1.2	1.2	2.5	BDL	2.0
K^+	mg/L	0.2	0.05	0.5	0.1	0.3	3.1	0.6	0.7
NO_3^-	mg/L	5.0	1	BDL	BDL	BDL	1.0	BDL	BDL
Cl^-	mg/L	5.0	1	1.0	DL	1.0	8.0	BDL	2.0
SO_4^{2-}	mg/L	10.0	2	BDL	BDL	BDL	128.0	16.0	23.0
HCO_3^-	mg/L	10.0	2	409.0	298.0	382.0	369.0	250.0	318.0
Al	µg/L	DL	1	111.0	1.0	3.0	13.0	1.0	2.0
As	µg/L	DL	0.5	BDL	BDL	BDL	1.7	0.9	1.5
B	µg/L	DL	5	13.0	8.0	10.5	27.0	6.0	14.0
Ba	µg/L	DL	0.05	10.4	4.2	4.8	16.2	10.4	15.0
Br	µg/L	DL	5	18.0	10.0	14.0	28.0	18.0	21.0
Cr	µg/L	DL	0.1	131.5	39.0	103.9	57.4	26.0	47.5
Cr(VI)	µg/L	DL	0.1	100.0	39.0	90.0	51.2	18.0	41.0
Cu	µg/L	DL	0.1	4.0	0.3	0.5	6.4	0.6	1.3
Li	µg/L	DL	0.1	1.0	0.7	0.8	1.2	0.9	1.1
Mn	µg/L	DL	0.05	0.8	0.1	0.4	0.8	0.2	0.3
Ni	µg/L	DL	0.2	38.2	2.7	5.5	314.0	5.1	7.0
P	µg/L	DL	10	78.0	13.0	38.0	147.0	18.0	32.0
Si	µg/L	DL	40	22,394	15,120	20,663	17,717	13,248	14,685
Sr	µg/L	DL	0.01	40.3	32.0	35.2	90.4	58.3	74.0
U	µg/L	DL	0.02	0.1	BDL	BDL	1.1	0.6	0.8
V	µg/L	DL	0.2	1.6	0.2	0.4	1.2	0.7	1.0
Zn	µg/L	DL	0.5	37.3	4.1	5.1	74.7	4.6	8.7

BDL: Below the detection limit. DL: Detection limit. QL: Quantification limit.

Table 3. Geographical coordinates of the water sampling sites and the results of Cr-Cr(VI) for the natural springs in western Vermio Mt.

Sample ID	Latitude	Longitude	Sampling Point	Cr (µg/L)	Cr(VI) (µg/L)	Sample ID	Latitude	Longitude	Sampling Point	Cr (µg/L)	Cr(VI) (µg/L)
W13_06_2018	40°27′103″	21°57′639″	Potistis	41.6	39.2	W14_9a_2018	40°25′854″	21°56′878″	Elafakia	46.0	43.0
W13_07_2018	40°27′103″	21°57′639″	Potistis	39.0	39.0	W14_9b_2018	40°25′854″	21°56′878″	Elafakia	56.1	40.0
W13_09_2018a	40°27′103″	21°57′639″	Potistis	111.5	90.0	W14_9c_2018	40°25′854″	21°56′878″	Elafakia	56.3	40.0
W13_09_2018b	40°27′103″	21°57′639″	Potistis	109.5	90.0	W14_9d_2018	40°25′854″	21°56′878″	Elafakia	56.0	40.0
W13_09_2018c	40°27′103″	21°57′639″	Potistis	108.7	90.0	W14_9e_2018	40°25′854″	21°56′878″	Elafakia	54.5	41.0
W13_09_2018d	40°27′103″	21°57′639″	Potistis	112.6	90.0	W14_05_2019	40°25′854″	21°56′878″	Elafakia	42.5	40.0
W13_10_2018a	40°27′103″	21°57′639″	Potistis	131.5	100.0	W14_08_2019	40°25′854″	21°56′878″	Elafakia	44.8	33.0
W13_10_2018b	40°27′103″	21°57′639″	Potistis	130.2	100.0	W14_11_2019	40°25′854″	21°56′878″	Elafakia	46.3	41.0
W13_10_2018c	40°27′103″	21°57′639″	Potistis	111.9	90.0	W14_02_2020	40°25′854″	21°56′878″	Elafakia	47.5	32.0
W13_10_2018d	40°27′103″	21°57′639″	Potistis	127.8	100.0	W14_07_2020	40°25′854″	21°56′878″	Elafakia	46.6	18.0
W13_10_2018e	40°27′103″	21°57′639″	Potistis	110.7	90.0	W14_09_2020	40°25′854″	21°56′878″	Elafakia	48.0	33.0
W13_11_2018	40°27′103″	21°57′639″	Potistis	127.5	100.0	S10_11_2014	40°26′854″	21°58′711″	Agio Pnevma	47.8	36.7
W13_04_2019	40°27′103″	21°57′639″	Potistis	89.2	89.0	S15_06_2018	40°26′689″	21°58′801″	Agio Pnevma	15.9	15.2
W13_05_2019	40°27′103″	21°57′639″	Potistis	92.8	90.0	S10_06_2018	40°26′854″	21°58′711″	Agio Pnevma	18.5	15.6
W13_06_2019	40°27′103″	21°57′639″	Potistis	98.1	89.0	S16_07_2018	40°27′338″	21°58′750″	Agio Pnevma	2.4	1.8
W13_08_2019	40°27′103″	21°57′639″	Potistis	103.3	88.0	S17_07_2018	40°26′856″	21°58′757″	Agio Pnevma	3.8	1.5
W13_10_2019	40°27′103″	21°57′639″	Potistis	103.5	92.0	S18_07_2018	40°27′871″	21°58′475″	Agio Pnevma	2.2	1.6
W13_11_2019	40°27′103″	21°57′639″	Potistis	105.8	99.0	S19_07_2018	40°27′095″	21°58′711″	Agio Pnevma	0.5	0.5
W13_02_2020	40°27′103″	21°57′639″	Potistis	95.7	87.0	S2_03_2014	40°25′789″	21°56′216″	Mouratidis	38.3	33.9
W13_06_2020	40°27′103″	21°57′639″	Potistis	99.0	97.0	S2_09_2016	40°25′789″	21°56′216″	Mouratidis	17.0	13.0
W13_07_2020	40°27′103″	21°57′639″	Potistis	99.6	82.0	S2_02_2017	40°25′789″	21°56′216″	Mouratidis	10.0	7.00
W13_09_2020	40°27′103″	21°57′639″	Potistis	103.9	88.0	S2_04_2017	40°25′789″	21°56′216″	Mouratidis	15.0	12.0
W13_10_2020	40°27′103″	21°57′639″	Potistis	103.5	92.0	S2_05_2017	40°25′789″	21°56′216″	Mouratidis	28.0	23.0
W14_11_2014	40°25′854″	21°56′878″	Elafakia	53.5	51.1	S2_06_2017	40°25′789″	21°56′216″	Mouratidis	25.0	22.0
W14_07_2014	40°25′854″	21°56′878″	Elafakia	57.4	51.2	S2_07_2017	40°25′789″	21°56′216″	Mouratidis	26.0	19.0
W14_12_2015	40°25′854″	21°56′878″	Elafakia	52.5	49.1	S2_08_2017a	40°25′789″	21°56′216″	Mouratidis	12.0	8.00
W14_09_2016	40°25′854″	21°56′878″	Elafakia	26.0	23.0	S2_08_2017b	40°25′789″	21°56′216″	Mouratidis	23.7	21.4
W14_04_2017	40°25′854″	21°56′878″	Elafakia	42.0	41.0	S2_09_2017	40°25′789″	21°56′216″	Mouratidis	16.0	10.0
W14_05_2017	40°25′854″	21°56′878″	Elafakia	47.3	47.0	S1_03_2014	40°25′224″	21°55′889″	Agios Dimitrios	16.6	16.5
W14_06_2017	40°25′854″	21°56′878″	Elafakia	49.0	47.0	S1_08_2017	40°25′224″	21°55′889″	Agios Dimitrios	10.9	8.70
W14_08_2017	40°25′854″	21°56′878″	Elafakia	49.2	46.8	S13_06_2017	40°25′810″	21°56′900″	Agios Panteleimonas	1.50	1.00
W14_10_2017	40°25′854″	21°56′878″	Elafakia	45.4	35.2	S14_06_2017	40°25′801″	21°57′099″	Agios Panteleimonas	1.90	1.00
W14_07_2018	40°25′854″	21°56′878″	Elafakia	45.0	40.0	W21_08_2019	40°25′842″	21°56′856″	Agios Panteleimonas	18.0	7.2
W14_08_2018	40°25′854″	21°56′878″	Elafakia	49.2	42.0	S5_07_2014	40°25′713″	21°56′725″	Vazelonas	0.50	0.10
W14_10_2018	40°25′854″	21°56′878″	Elafakia	45.4	41.0	S6_07_2014	40°25′840″	21°56′882″	Vazelonas	0.80	0.50

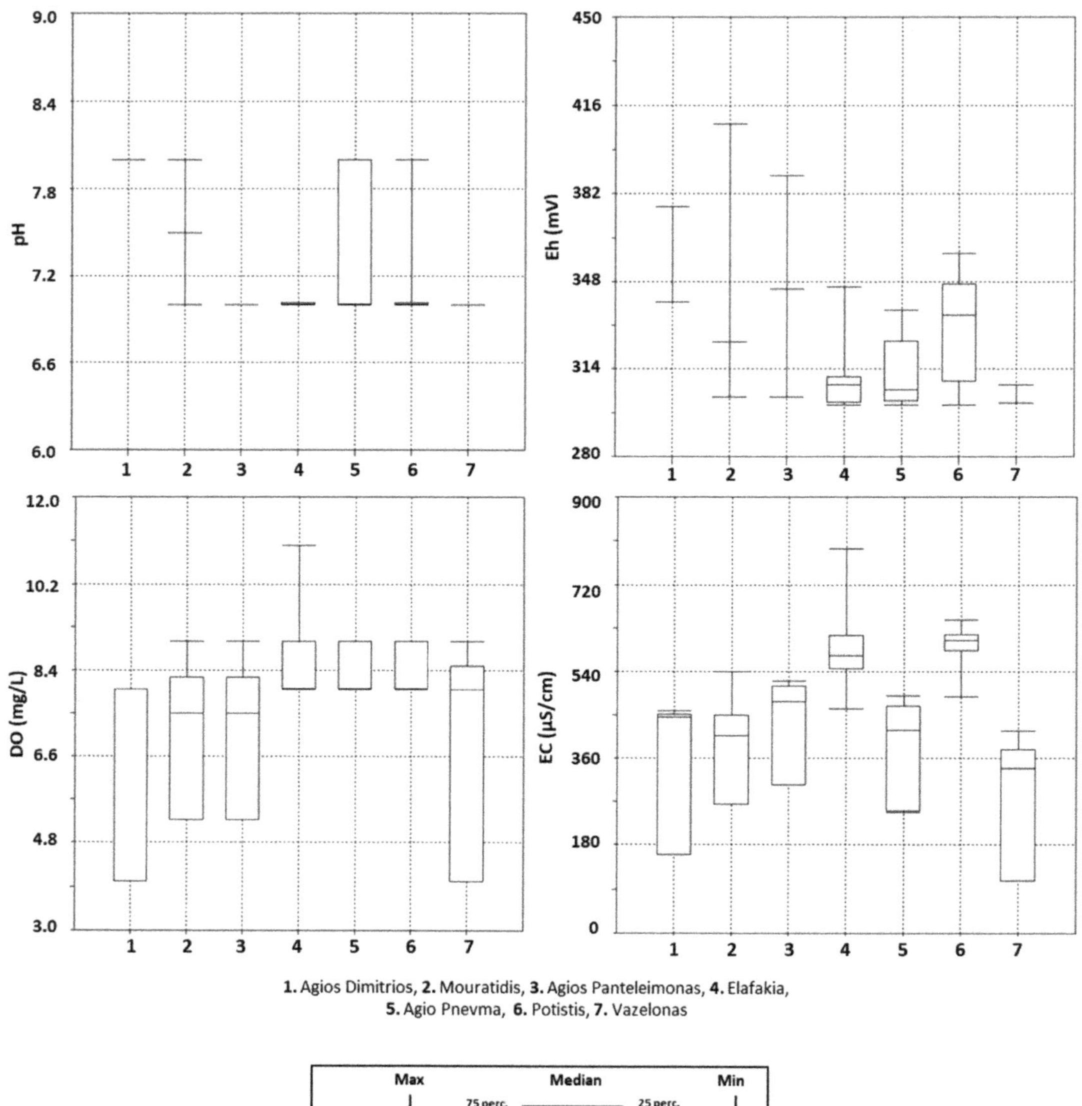

Figure 6. Box plots for the physical parameters of the natural springs of western Vermio Mt.

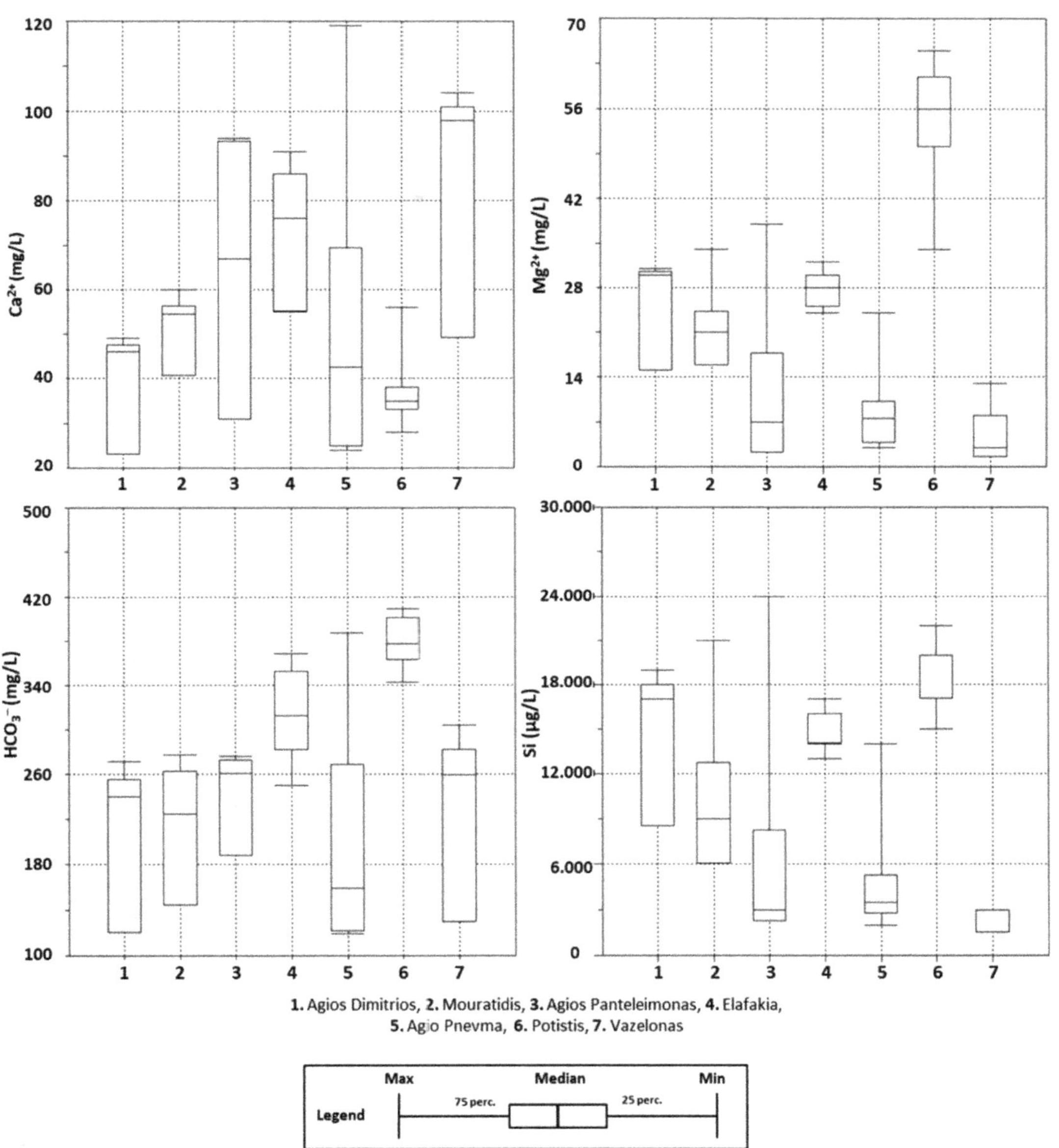

Figure 7. Box plots for the major ions and Si of the natural springs of western Vermio Mt.

Table 4. The abundance of major ions among the natural springs in western Vermio Mt.

	Area/Sampling Site	Sample ID	Cations Order	Anions Order
1	Agios Dimitrios area	S1	$Ca^{2+} > Mg^{2+} > K^+ > Na^+$	$HCO_3^- > SO_4^{2-} > Cl^- > NO_3^-$
2	Elafakia	W14	$Ca^{2+} > Mg^{2+} > Na^+ > K^+$	$HCO_3^- > SO_4^{2-} > Cl^- > NO_3^-$
3	Agios Panteleimonas area	S2, S13, S14	$Ca^{2+} > Mg^{2+} > Na^+ > K^+$	$HCO_3^- > SO_4^{2-}$
4	Potistis	W13	$Mg^{2+} > Ca^{2+} > K^+$	$HCO_3^- > SO_4^{2-} > Cl^- > NO_3^-$
5	Agio Pnevma area	S18, S16, S19, S10, S17, S15	$Ca^{2+} > Mg^{2+} > K^+ > Na^+$	$HCO_3^- > SO_4^{2-} > Cl^-$
6	Mouratidis	S2	$Ca^{2+} > Mg^{2+} > Na^+ > K^+$	$HCO_3^- > SO_4^{2-} > NO_3^- > Cl^-$
7	Vazelona area	S5, S6	$Ca^{2+} > Mg^{2+} > K^+ > Na^+$	$HCO_3^- > SO_4^{2-} > Cl^-$

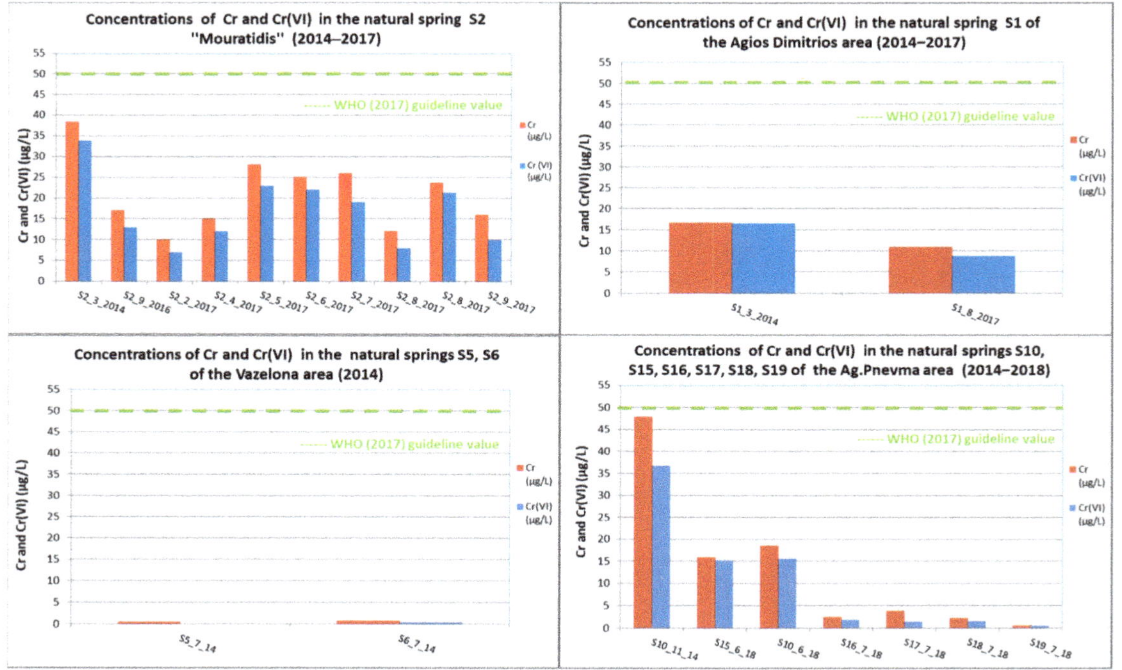

Figure 8. Concentrations of Cr and Cr(VI) in the natural springs of western Vermio Mt.

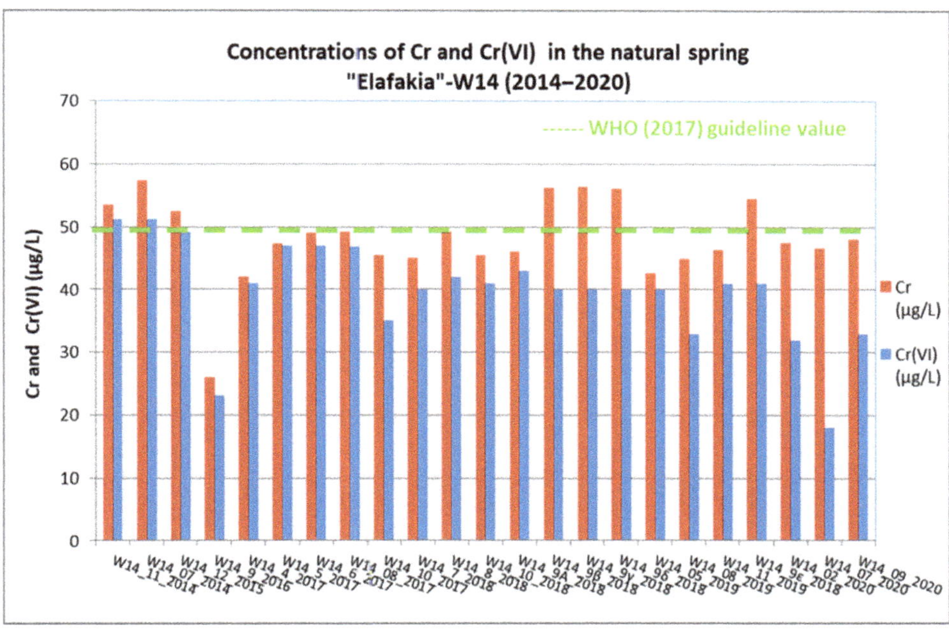

Figure 9. Concentrations of Cr and Cr(VI) in the natural spring "Elafakia"-W14 of western Vermio Mt.

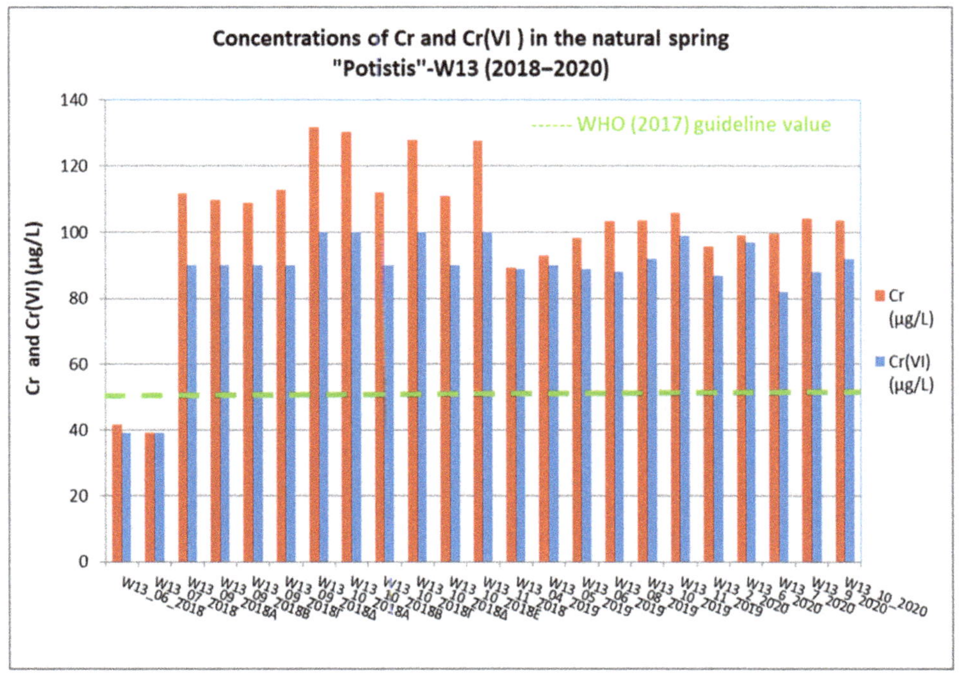

Figure 10. Concentrations of Cr and Cr(VI) in the natural spring "Potistis"-W13 of western Vermio Mt.

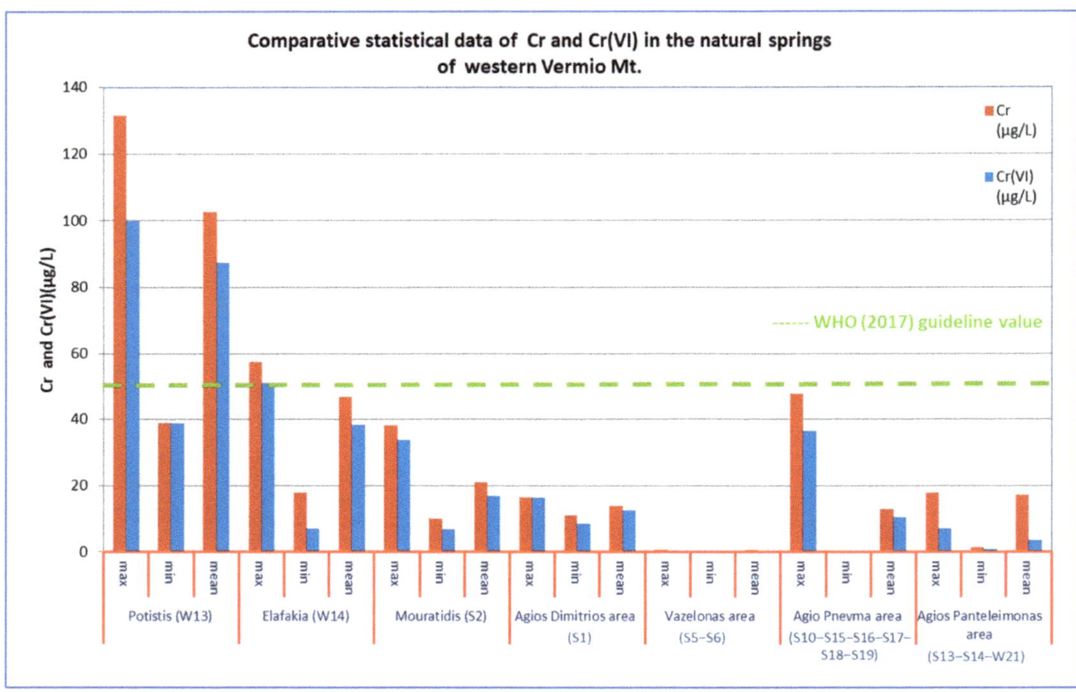

Figure 11. Statistical analyses of concentrations of Cr and Cr(VI) of the natural springs in western Vermio Mt.

3.2. Correlation Analysis of Water Samples

Spearman's rank correlation coefficient was applied to selected parameters (Table 5); the rest of the measured parameters were excluded because most of their values were BDL. The most remarkable features of the Spearman's rank correlation coefficients are the following: Cr presented a statistically significant ($p < 0.01$) very strong positive correlation coefficient with Cr(VI) ($r_s = 0.98$), strong positive correlation coefficients with Mg^{2+} ($r_s = 0.76$), Si ($r_s = 0.75$), EC ($r_s = 0.71$) and Ni (0.61), and moderate positive correlation coefficients with HCO_3^- ($r_s = 0.55$) and alkalinity ($r_s = 0.55$). Hexavalent chromium exhibited a statistically significant ($p < 0.01$) very strong positive correlation coefficient with Mg^{2+} ($r_s = 0.8$), a strong positive correlation coefficient with Si ($r_s = 0.76$) and Ni ($r_s = 0.67$), and moderate positive correlation coefficients with HCO_3^- ($r_s = 0.59$) and alkalinity ($r_s = 0.59$). Magnesium exhibited a statistically significant ($p < 0.01$) strong positive correlation coefficients with EC ($r_s = 0.68$), HCO_3^- ($r_s = 0.68$), alkalinity ($r_s = 0.68$) and Ni ($r_s = 0.60$), while very strong correlation coefficient with Si ($r_s = 0.91$). Bicarbonates presented a statistically significant ($p < 0.01$) strong positive correlation coefficient with EC ($r_s = 0.62$), and moderate positive correlation coefficients with Ni ($r_s = 0.51$), Si ($r_s = 0.48$). Sulfates had a statistically significant ($p < 0.01$) very strong positive correlation coefficient with U ($r_s = 0.82$), strong positive correlation coefficients with Sr ($r_s = 0.74$), Br ($r_s = 0.70$), Ba ($r_s = 0.63$), and Ca^{2+} ($r_s = 0.67$), and a moderate positive correlation coefficient with Na^+ ($r_s = 0.52$). Arsenic exhibited statistically significant ($p < 0.01$) strong positive correlation coefficients with U ($r_s = 0.72$), Ba ($r_s = 0.67$) and V ($r_s = 0.70$), and a moderate positive correlation coefficient with K^+ ($r_s = 0.49$).

Table 5. Spearman's rank correlation matrix of selected major and trace elements of the natural springs of western Vermio Mt.

Parameter	pH	DO	EC	Eh	Ca^{2+}	Mg^{2+}	Na^+	K^+	NO_3^-	Cl^-	SO_4^{2-}	HCO_3^-	Al	As	B	Ba	Br	Cr	Cr(VI)	Li	Mn	Ni	P	Si	Sr	U	V	Zn	Alkalinity
pH	1																												
DO	−0.11	1																											
EC	−0.15	0.21	1																										
Eh	−0.11	0.06	0.19	1																									
Ca^{2+}	−0.18	0.10	−0.20	0.15	1																								
Mg^{2+}	0.06	0.05	0.68 **	0.32	−0.42 *	1																							
Na^+	−0.17	−0.06	−0.16	−0.15	0.49 **	−0.18	1																						
K^+	0.12	−0.27	−0.28	−0.06	0.51 **	−0.511 **	0.13	1																					
NO_3^-	0.18	−0.04	−0.41 *	0.13	0.31	−0.18	0.47 **	0.03	1																				
Cl^-	−0.04	−0.15	−0.30	−0.43	0.29	−0.461 **	0.37 *	0.39 *	0.19	1																			
SO_4^{2-}	−0.06	−0.07	0.09	−0.08	0.67 **	−0.21	0.52 **	0.49 **	0.20	0.25	1																		
HCO_3^-	−0.11	0.04	0.62 **	0.14	0.00	0.68 **	−0.15	−0.31	−0.22	−0.14	0.01	1																	
Al	0.17	0.10	−0.40 *	−0.24	−0.07	−0.08	0.09	−0.03	0.18	0.28	−0.14	−0.19	1																
As	0.17	−0.20	−0.10	0.19	0.42 *	0.00	0.51 **	0.49 **	0.33	−0.04	0.60 **	−0.25	0.09	1															
B	−0.11	0.31	0.39 *	0.25	0.19	0.37 *	0.04	0.00	0.20	−0.19	0.25	0.41 *	−0.14	0.18	1														
Ba	−0.01	0.04	0.18	0.04	0.47 **	0.14	0.26	0.41 *	0.12	−0.10	0.63 **	0.18	0.06	0.67 **	0.45 **	1													
Br	−0.17	0.10	0.28	−0.22	0.33	−0.02	0.32	0.21	0.04	0.31	0.70 **	0.21	−0.05	0.20	0.26	0.45 **	1												
Cr	−0.11	0.32	0.71 **	0.19	−0.30	0.76 **	−0.09	−0.55 **	−0.19	−0.33	−0.13	0.55 **	−0.12	−0.16	0.38 *	0.07	0.17	1											
Cr(VI)	−0.14	0.32	0.73 **	0.18	−0.34	0.80 **	−0.10	−0.6 **	−0.19	−0.35 *	−0.17	0.59 **	−0.09	−0.19	0.36 *	0.06	0.17	0.98 **	1										
Li	0.17	0.04	0.42 *	0.13	−0.17	0.611 **	−0.08	0.05	−0.15	−0.32	0.23	0.22	0.11	0.48 **	0.39 *	0.63 **	0.29	0.47 **	0.48 **	1									
Mn	0.17	−0.26	−0.30	−0.44 *	−0.15	−0.19	−0.13	0.04	0.05	0.06	−0.07	−0.11	0.30	−0.04	−0.43	−0.05	−0.09	−0.41 *	−0.32	−0.14	1								
Ni	−0.27	0.30	0.60 **	0.07	−0.13	0.60 **	−0.10	−0.25	−0.25	−0.30	0.01	0.51 **	0.10	0.04	0.42 *	0.45 **	0.37 *	0.61 **	0.67 **	0.64 **	−0.10	1							
P	−0.09	−0.07	−0.27	−0.41 *	−0.08	−0.28	−0.15	−0.02	−0.03	0.10	−0.07	0.00	0.25	−0.23	−0.30	−0.13	−0.04	−0.36 *	−0.26	−0.24	0.75 **	−0.04	1						
Si	0.10	−0.07	0.58 **	0.34	−0.43 *	0.91 **	−0.20	−0.34	−0.11	−0.43 *	−0.19	0.48 **	−0.07	0.09	0.34	0.17	−0.09	0.75 **	0.76 **	0.71 **	−0.22	0.53 **	−0.33	1					
Sr	−0.13	0.01	−0.26	−0.18	0.827 **	−0.55	0.61 **	0.61 **	0.21	0.38 *	0.74 **	−0.23	0.06	0.53 **	0.12	0.56 **	0.47 **	−0.37 *	−0.43	−0.04	−0.09	−0.13	−0.07	−0.53	1				

Table 5. *Cont.*

Parameter	pH	DO	EC	Eh	Ca^{2+}	Mg^{2+}	Na$^+$	K$^+$	NO$_3^-$	Cl$^-$	SO$_4^{2-}$	HCO$_3^-$	Al	As	B	Ba	Br	Cr	Cr(VI)	Li	Mn	Ni	P	Si	Sr	U	V	Zn	Alkalinity
U	−0.14	0.12	−0.03	−0.09	0.676 **	−0.27	0.59 **	0.462 **	0.28	0.15	0.82 **	−0.15	0.05	0.72 **	0.33	0.78 **	0.58 **	−0.16	−0.18	0.29	−0.05	0.18	−0.03	−0.28	0.85 **	1			
V	0.10	0.02	−0.08	0.08	0.15	0.16	0.30	0.16	0.33	−0.11	0.41 *	−0.13	0.25	0.70 **	0.21	0.69 **	0.19	−0.10	−0.09	0.59 **	0.07	0.26	−0.07	0.22	0.24	0.54 **	1		
Zn	−0.01	−0.10	−0.50 **	−0.23	0.22	−0.33	0.24	0.18	0.52 **	0.38 *	0.04	−0.22	0.32	0.09	−0.05	0.09	−0.04	−0.23	−0.22	−0.18	0.34	−0.16	0.41 *	−0.24	0.23	0.21	0.09	1	
Alkalinity	−0.11	0.04	0.62 **	0.14	0.00	0.68 **	−0.15	−0.31	−0.22	−0.14	0.02	1 **	−0.19	−0.24	0.41 *	0.19	0.22	0.55 **	0.59 **	0.23	−0.11	0.50 **	0.00	0.48 **	−0.22	−0.14	−0.13	−0.22	1

* Correlation is significant at the 0.05 level (2-tailed). ** Correlation is significant at the 0.01 level (2-tailed).

4. Discussion

4.1. Hydrogeochemical Characterization of the Natural Springs of Western Vermio Mt. the Ultramafic Fingerprint

The dominant hydrochemical types of the studied natural springs of western Vermio Mt. were Ca-Mg-HCO$_3$ (40% of the water samples), Mg-Ca-HCO$_3$ (33% of the water samples), and Ca-HCO$_3$ (21%) (Figure 12). Other transitional water types in the study area comprised Ca-K-HCO$_3$-Cl (3%) and Ca-HCO$_3$-SO$_4$ (3%). Based on the type and the geological environment of the springs, the Ca–HCO$_3$ waters are considered to originate through the interaction of meteoric water with rocks containing Ca-bearing minerals, whereas water types enriched in Mg, were derived from the dissolution of ultramafic rocks [29]. The mixed Ca-Mg-HCO$_3$ type indicated fresh recharge waters mainly related to carbonate rocks and less to ultramafic rocks. The Mg-Ca-HCO$_3$ water type represents recharge waters related to Mg-rich rocks, suggesting the strong interaction with the ultramafic rocks of the area [67]. The springs "Potistis"-W13, S1, and W21 that belong to this type are associated with fissured aquifers in strongly serpentinised ultramafic rocks, or they are in hydraulic connection with them.

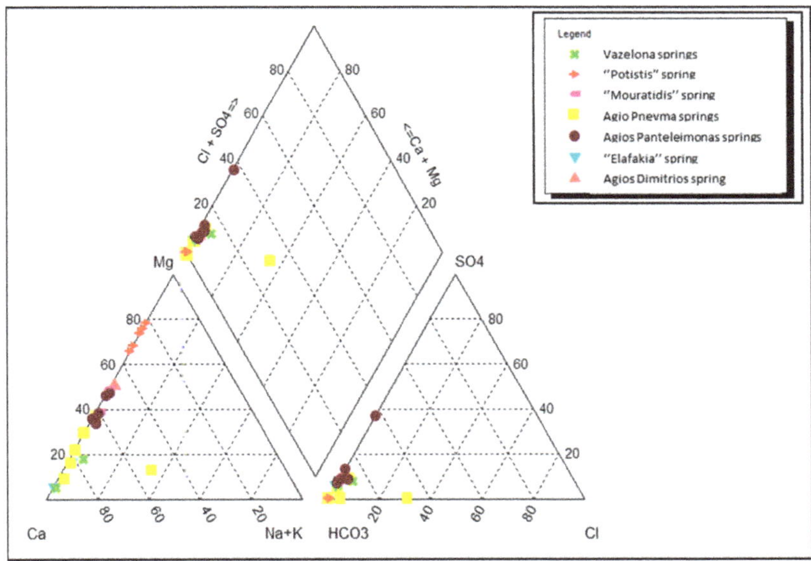

Figure 12. Piper diagram of major ion chemistry for the natural spring samples.

The water–ultramafic rock interaction typically produces Mg-HCO$_3$ water type [33,40,68] and slightly alkaline to strongly alkaline pH conditions [69] because of the absorption of dissolved CO$_2$ from atmospheric water in the serpentine and olivine according to the Equations (6) and (7) [70]:

$$Mg_3Si_2O_5(OH)_4 + 6CO_2 + 5H_2O \rightarrow 3Mg^{2+} + 6HCO_3^- + 2H_4SiO_4 \qquad (6)$$

$$Mg_2SiO_4 + 4H_2O + 4CO_2 \rightarrow 2Mg^{2+} + 4HCO_3^- + H_4SiO_4 \qquad (7)$$

The pH values that characterised the studied springs varied from 7.3 up to 8.5, indicating near-neutral to slightly alkaline conditions that are typical of groundwater interacting with ultramafic and carbonate rocks [40,71]. Redox potential conditions were oxidizing up to strong oxidizing, as indicated by Eh, ranging from 300 mV up to 410 mV. The pH and Eh conditions in the studied springs favoured the release and solubility of the Cr oxyanion in groundwater since the solubility of oxyanions such as $HCrO_4^-$, CrO_4^{2-}, $Cr_2O_7^{2-}$, $H_2AsO_4^-$, and $HAsO_4^{2-}$ is enhanced with increasing pH [72].

Various tools are usually employed to define and evaluate the water–rock interaction processes in an area [73]. Bivariate plots of major ions and ionic ratios were used to study the hydrogeochemical evolution processes in the studied springs (Figure 13). In the bivariate plot of Ca vs. Mg, the water samples were grouped into three classes based on their Ca/Mg ratio (Figure 13a). In the first class belong the water samples with a Ca/Mg ratio below the 1:3 line. This class included two seasonal water samples from the spring "Potistis"-W13 (W13_10_19 and W13_4_19) with a Mg-Ca-HCO$_3$ water type, revealing that the flow path was mainly through serpentinites. The second class contained the water samples of the natural spring S1-Agios Dimitrios area and "Potistis"-W13. They are all of Mg-Ca-HCO$_3$ type and characterized by Ca/Mg ratios plotted below the 1:1 and above the 1:3 lines; this suggests a mixture of Mg-HCO$_3$ and Ca-HCO$_3$, indicating that these waters were derived from interaction with serpentinites and Ca-rich rocks. The third class comprised the rest of natural springs, characterized by mixed water types and a Ca/Mg ratio above the 1:1 line, indicating a limited influence of serpentinites. The bivariate plot of Ca + Mg vs. HCO$_3$ (Figure 13b) suggests an excess of (Ca + Mg) over HCO$_3$ reflecting an additional non-carbonate source of Ca^{2+} and Mg^{2+} ions, such as the dissolution of silicate minerals [38,57]. Iron-Mg-silicates of ultramafic rocks, such as olivine, pyroxene, and amphibole are transformed to serpentine group minerals during the serpentinisation process. Dissolution reactions favour the Mg^{2+} and HCO$_3{}^-$ release of the Mg-rich minerals (Equations (6) and (7)) [68].

During water–rock interaction, various chemical processes (e.g., fluctuation of ionic concentrations, mobilization of the dissolved components, and change in pH) are finger-printed on the groundwater quality [74,75]. Gibbs diagrams are generally used to identify the hydrogeochemical evolution of groundwater, which involves precipitation, water–rock interaction, and evaporation–crystallization processes, based on TDS vs. Na$^+$/(Na$^+$ + Ca^{2+}), and TDS vs. Cl$^-$/(Cl$^-$+ HCO$_3{}^-$) scatter diagrams [76]. Herein, Gibbs diagrams were employed to assess hydrogeochemical processes that affect the water chemistry in the natural springs of western Vermio Mt. Figure 14 illustrates that all samples from natural springs fall into the water–rock interaction field, suggesting weathering of carbonate and silicate minerals. Although the use of Gibbs plots for groundwater has been disputed [77], the case study discussed herein exhibits none of the characteristics that could result in misuse of these plots (e.g., high SO$_4{}^{2-}$ concentrations, salinity sources, evolutionary flow paths, etc.). The implications of the Gibbs diagrams are in accordance with the calculated MGI index, according to which the waters from the natural springs are characterized as deep percolation types.

4.2. Hydrogeochemistry of Cr in Natural Ultramafic Springs

To further study the hydrogeochemistry of Cr in the studied springs, the average concentration of Cr was plotted vs. the water type of each spring (Figure 15). As shown, each water type is characterised by a wide range of concentrations of Cr, attributed to the different operation mechanisms of the spring and the weathering degree of the host geological formations. The mixed Mg-Ca-HCO$_3$ water type ranges from very high concentrations of Cr, in the spring "Potistis"-W13 (>100 µg/L), to much lower values (<20 µg/L) in the springs S1-Agios Dimitrios area and W21-Agios Panteleimonas area. The mixed Ca-Mg-HCO$_3$ water type is related to a range of concentrations of Cr from 17 to 48 µg/L. On the other hand, all springs that are characterised by a Ca-HCO$_3$ water type exhibit very low Cr concentrations (<5 µg/L) since mostly the carbonate rocks influence their hydrochemistry. In all springs, the dominant anion is HCO$_3{}^-$, the principal source of which is the dissolution of carbonate and silicate minerals [33].

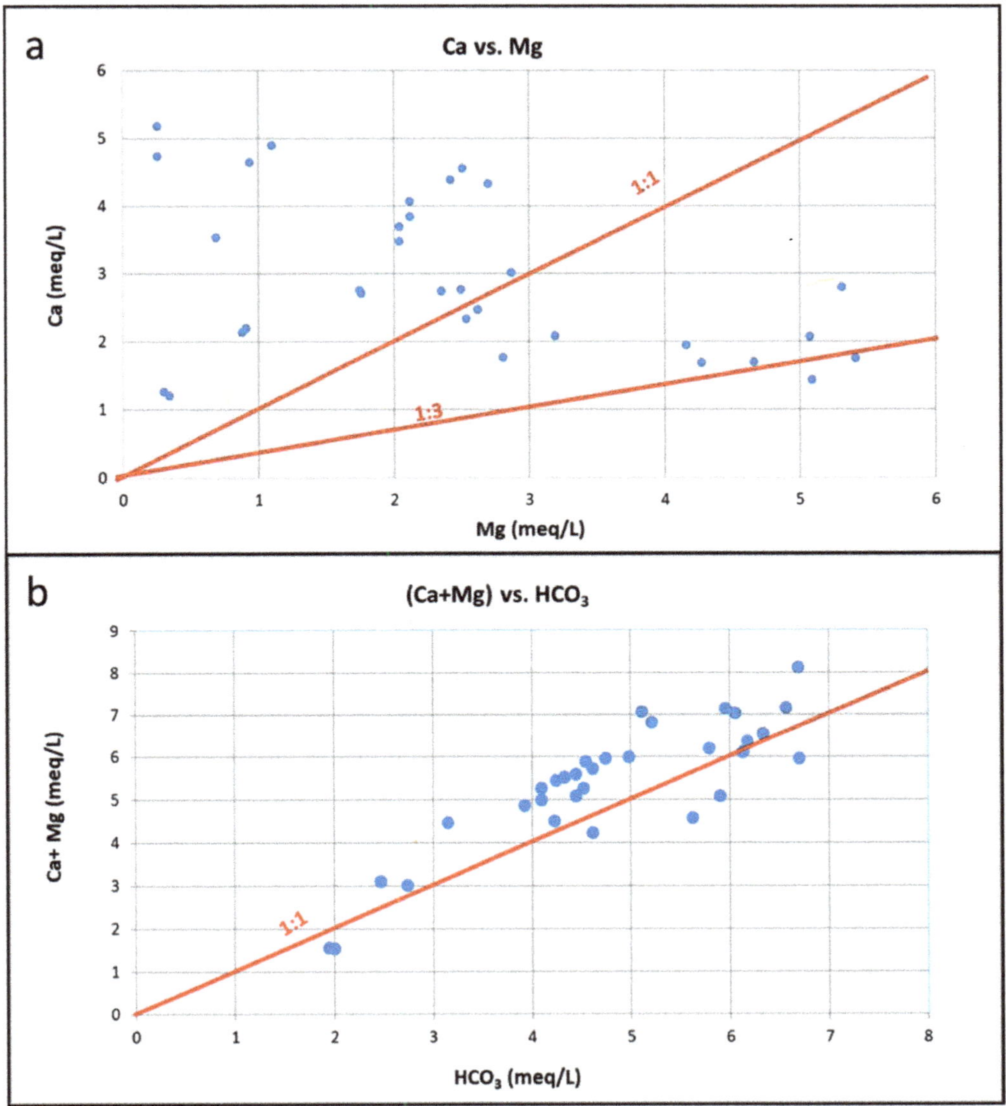

Figure 13. Bivariate plots of: (**a**) Ca vs. Mg and (**b**) (Ca+Mg) vs. HCO_3 for the natural springs of western Vermio Mt.

An interesting feature of the spring "Potistis"-W13, derived from the evaluation of hydrogeochemical, hydrological, and meteorological data, is the decrease in concentrations of Cr in a very short time after rainfall; this is further supported by the strong linear regression of Cr vs. discharge (coefficient of determination $R^2 = 0.85$) (Figures 16 and 17). Low discharge results in increased water–ultramafic rock contact time and thus, in elevated concentrations of Cr.

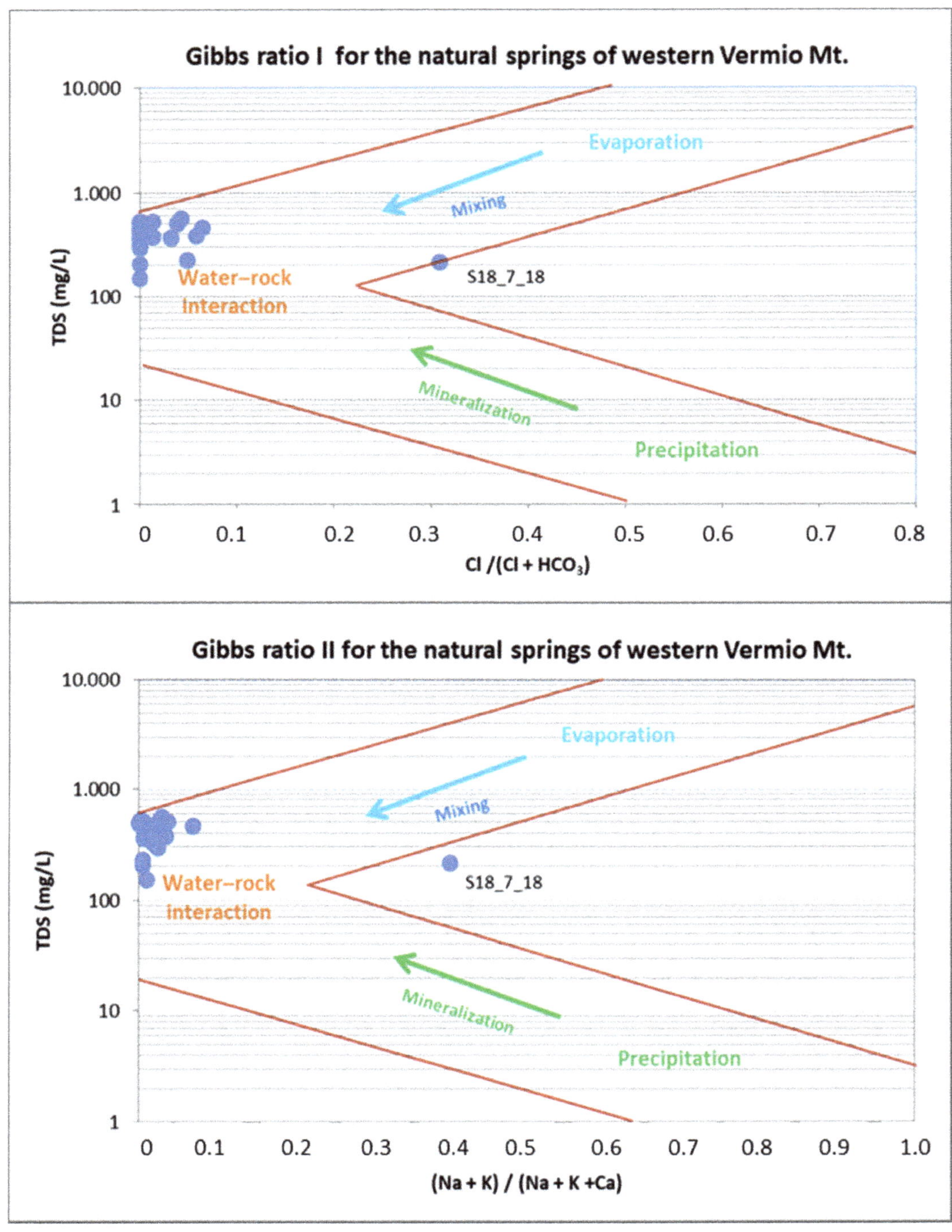

Figure 14. Gibbs diagrams of the natural springs in western Vermio Mt.

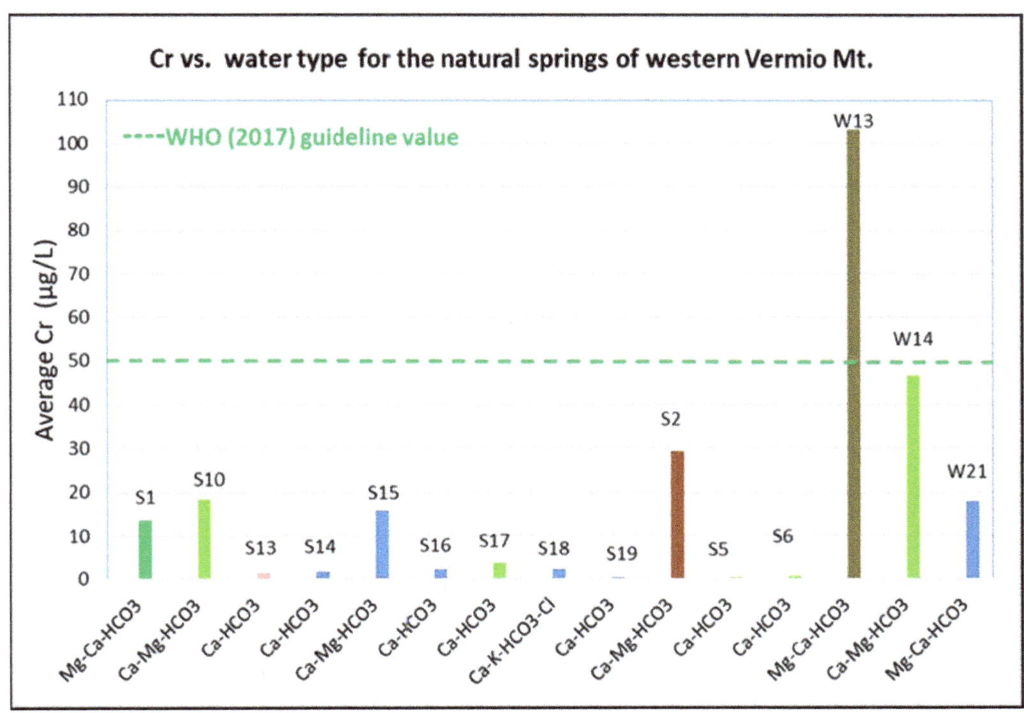

Figure 15. Average concentrations of Cr vs. water types of the natural springs in western Vermio Mt.

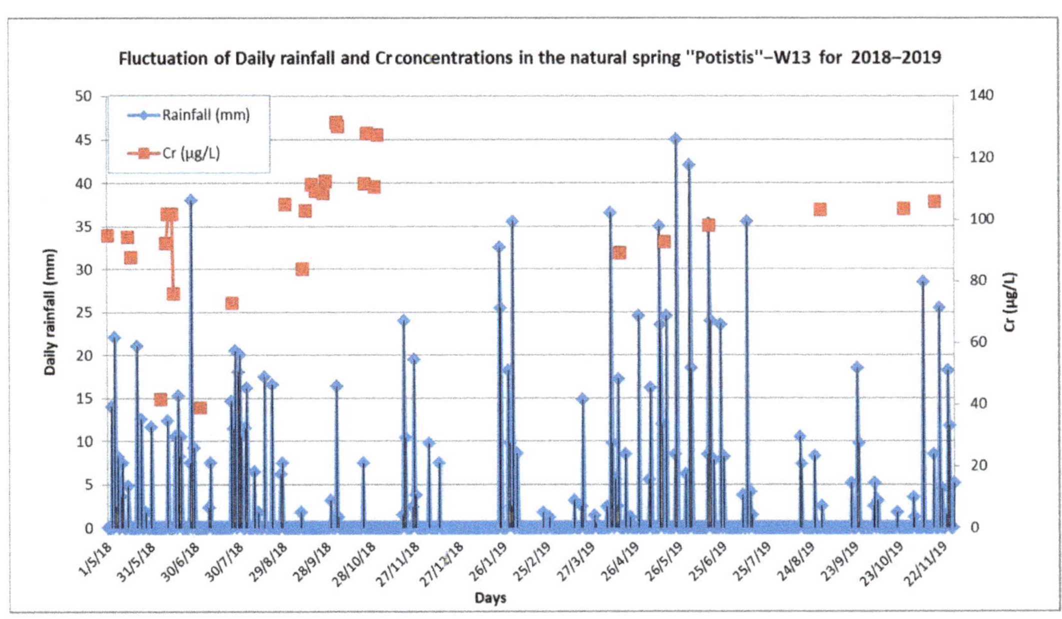

Figure 16. Fluctuation diagrams of rainfall and Cr concentrations of the natural spring "Potistis"-W13 in western Vermio Mt.

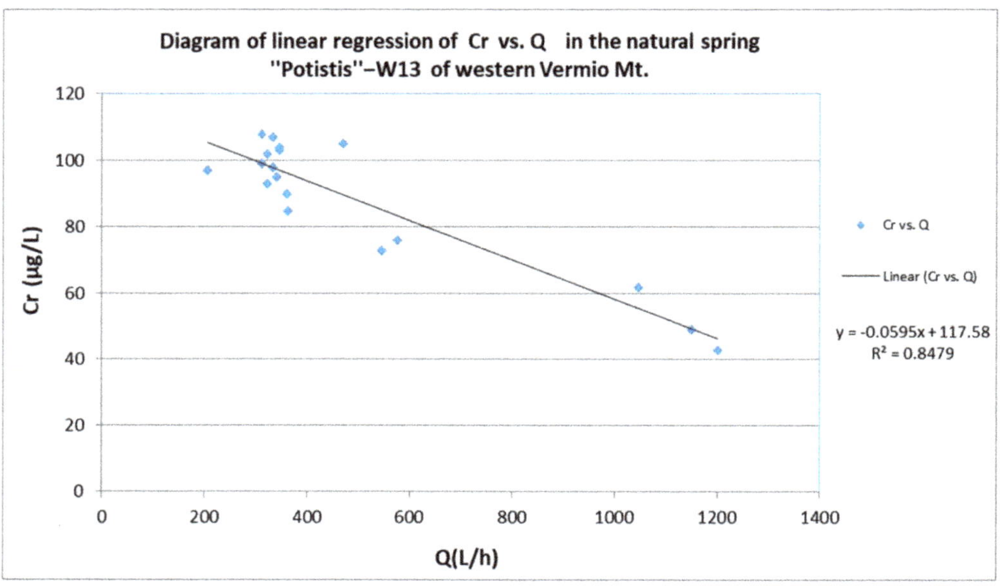

Figure 17. Plot of Cr (µg/L) vs. discharge (L/h) in the natural spring "Potistis"-W13 in western Vermio Mt.

In Figure 18a, the Mg/Si ratio vs. Cr in the springs is presented. The diagram is divided into three sub-groups according to the concentrations of Cr. The water samples with low concentrations of Cr (<30 µg/L) constituted 45.5% of the total samples, 73.3% of which exhibited a Mg/Si ratio lower than 2. Concentrations of Cr from 30 µg/L up to 50 µg/L, comprised 25% of the total water samples, 77.78% of which exhibited a Mg/Si ratio lower than 2.3; only the samples W13_6_18 and W13_7_18 which correspond to the lowest concentrations of Cr recorded in the spring "Potistis"-W13 exhibited a Mg/Si ratio higher than 2.3. Of the total water samples, 42% exceeded the permissible limit of 50 µg/L for drinking water [66], with most of them corresponding to samples from the spring "Potistis"-W13. Most samples presented a Mg/Si ratio higher than 2.3. Respective Mg/Si ratios have been reported for groundwater in other natural ultramafic environments [29,33].

The strong fingerprint of the water–rock interaction on the spring water chemistry and the geogenic origin of Cr in groundwater are indicated by the statistically significant very strong positive correlation coefficient of Cr with Si, the strong positive correlation coefficients of Cr with Mg^{2+}, EC, and Ni, and the moderate positive correlation coefficients of Cr with HCO_3 and alkalinity. Magnesium and alkalinity are two parameters usually increased with increasing degree of weathering; the latter has been reported to relate to elevated concentrations of Cr in groundwater [78]. Nickel is derived from the dissolution of Ni-bearing silicates which are released to groundwater under morphological and geochemical conditions that do not favour the occurrence of Fe-hydroxides and other secondary minerals capable of adsorbing Ni [79]. Nickel exhibited statistically significant, moderate positive correlation coefficients with Mg^{2+}, EC, and Si, further highlighting its geogenic origin. The two natural springs with high concentrations of Ni ("Potistis"-W13, "Elafakia"-W14), also exhibit high mean concentrations of dissolved Si, and are of Mg-Ca-HCO_3 and Ca-Mg-HCO_3 water type. A similar case of high concentrations of Cr and Ni in Mg-HCO_3 groundwater has been reported by Margiotta et al. [40]. Unlike the spring waters, Cr in the irrigational wells in the lowland of the Sarigkiol Basin was reported to strongly correlate with NO_3^- and P, indicating the synergistic role of the agricultural activities [45].

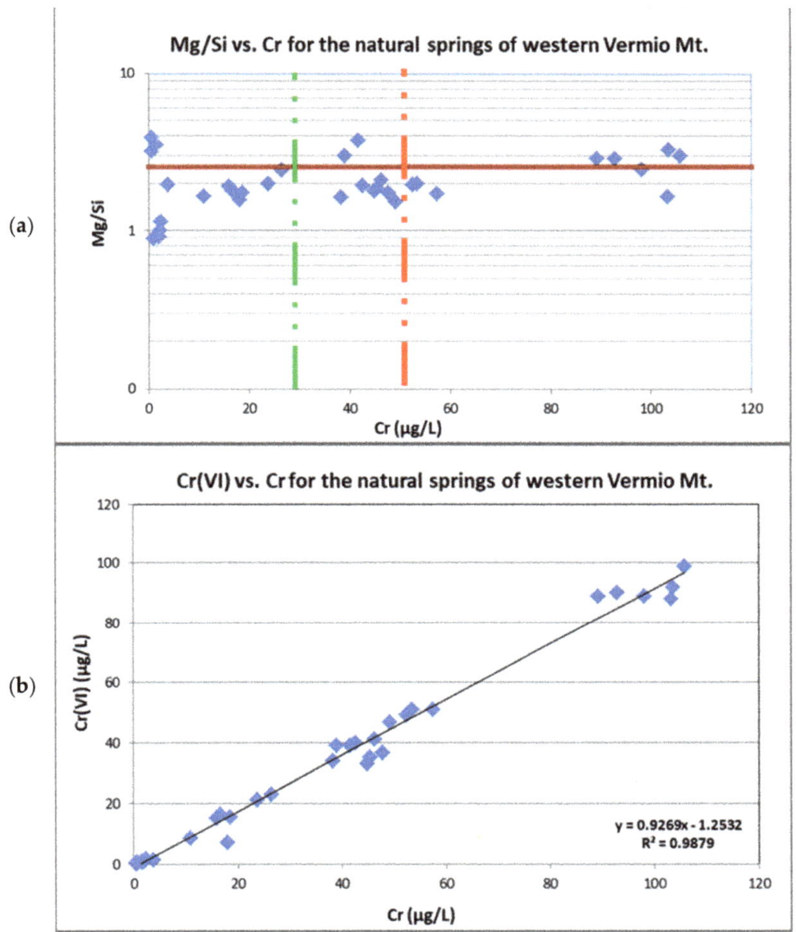

Figure 18. Plots of: (**a**) Mg/Si vs. Cr and (**b**) Cr(VI) vs. Cr (regression model) of the natural springs in western Vermio Mt.

The statistically significant, very strong positive correlation coefficients of Cr with Cr(VI) (Spearman's rank correlation coefficient r_s = 0.986) was further proven by their linear regression with a strong linear relationship (coefficient of determination R^2 = 0.99, Figure 18b). In the analysed water samples, the Cr(VI)/Cr ratio ranges from 20% up to 100%. Specifically: (a) 62.5–90.3% in the spring "Mouratidis"-S2, (b) 80–99% in the spring S1- Agios Dimitrios area, (c) 20–62.5% in the springs at the Vazelonas area, (d) 39–100% in the springs at the Agio Pnevma area, (e) 40–99% in the spring "Elafakia"-W14, (f) 76–100% in the spring "Potistis"-W13. The fluctuation in the Cr(VI)/Cr ratio depends on the prevailing geochemical conditions (redox reactions, pH), the presence of iron or manganese oxides, and competing anions in each area, suggesting that various processes take place [78]. Hexavalent chromium is the principal form of Cr in the natural water springs ("Potistis"-W13, S1-Agios Dimitrios area, "Mouratidis"-S2); several factors contribute to the high Cr(VI)/Cr ratio. Specifically, the geological environment, which is enriched in Ca and Mg-bearing minerals, enhances Cr(VI) to form complexes with Mg and Ca and inhibits Cr(VI) reduction [71]. The presence of manganese oxides enhances the Cr(III) oxidation to

Cr(VI) in ultramafic rocks, soils, and unsaturated zone releasing Cr(VI) to groundwater (Equation (8)) [24].

$$Cr(III) + 1.5\delta - MnO_{2(s)} + H_2O \rightarrow HCrO_4^- + 1.5Mn^{2+} + H^+ \tag{8}$$

The pronounced role of minerals in the concentrations of Cr in natural springs was investigated via SIs of selected mineral phases present in the study area (Figure 19). The percentage distribution of the SIs for the selected mineral phases is given in Table 6 for all collected water samples.

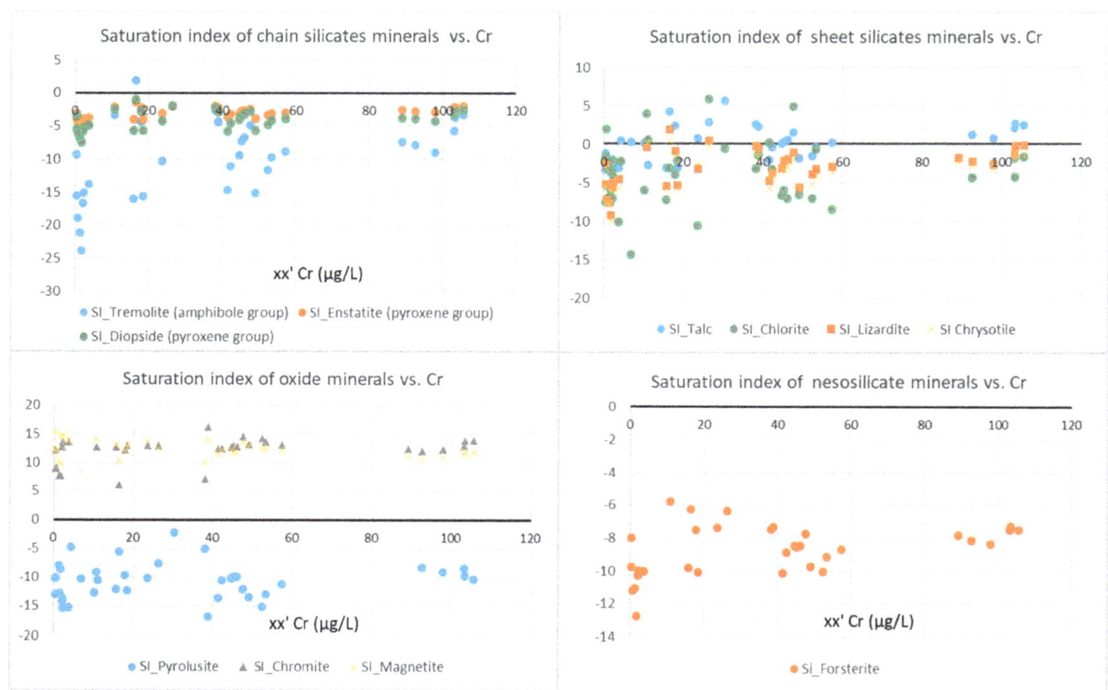

Figure 19. Plots of saturation indices vs. Cr of the natural springs in western Vermio Mt.

The water samples from the natural springs in which concentrations of Cr exceeded 50 µg/L ("Potistis"-W13 and "Elafakia"-W14) were oversaturated in: (a) the carbonate mineral calcite (100%) and (b) the oxide minerals chromite (100%) and magnetite (100%). On the other hand, they are undersaturated in: (a) the serpentine group minerals lizardite (100%) and chrysotile (100%), (b) the pyroxenes enstatite (100%) and diopside (100%), and the amphibole tremolite (100%) and (c) the olivine (100%).

In general, the mineralogical phases that appear undersaturated tend to dissolve in water. The dissolution reactions contribute major, minor elements and PTEs to the groundwater. Chromium-bearing silicate minerals (serpentine, amphibole, pyroxene, chlorite, talc) occurred mostly undersaturated in the water samples, whereas Cr-rich oxides (chromite and Cr-magnetite) were oversaturated in the water samples. Therefore, silicate minerals are the principal geogenic contributors of Cr and other major/minor elements (e.g., Mg^{2+}, Ca^{2+}, HCO_3^-, Si) and PTEs (e.g., As, Ni) to the spring waters of western Vermio Mt.

Table 6. The percentage distribution of the saturation state with respect to the selected mineral phases in the collected water samples.

Mineral Phase	Oversaturated SI > 0 (%)	Undersaturated SI < 0 (%)
Calcite	63.8	36.2
Dolomite	58.0	42.0
Magnesite	29.0	71.0
Talc	49.3	50.7
Chlorite	30.4	69.6
Tremolite	14.5	85.5
Enstatite	0.00	100
Diopside	0.00	100
Pyrolusite	0.00	100
Chromite	100	0.00
Magnetite	100	0.00
Chrysotile	14.5	85.5
Lizardite	14.5	85.5
Olivine (Forsterite)	0.00	100

4.3. NBLs of Cr in the Ultramafic Environment of Vermio Mt.

The geochemical characteristics of the natural springs, the geological environment, and the water–ultramafic rock interaction are reflected in springs' water quality. Chromium constitutes the principal environmental component in groundwater of the Sarigkiol Basin, originating primarily from geogenic and incidentally from anthropogenic sources [45]. This paper aims to assess the NBLs of Cr, which is of great interest in the catchment scale of the Sarigkiol Basin.

Based on the above-discussed hydrogeochemical data (e.g., pH, DO, Eh, Mg^{2+}, Si, Cr, alkalinity, etc.), the most representative natural springs, which flow through and interact with ultramafic rocks, are the S1-Agios Dimitrios area, "Mouratidis"-S2, "Potistis"-W13, and "Elafakia"-W14.

Take into consideration the modified methodology for assessing NBLs, the PS method was applied to create the new dataset. All samples from the natural springs satisfied the two criteria (ORP > 100 mV, DO > 3 mg/L and NO_3^- < 10 mg/L). Regarding the third criterion, the available time series of measurements, two natural springs, those of "Potistis"-W13 and "Elafakia"-W14, sufficiently satisfied this criterion. The resulting population was examined for the normality of the dataset with the Shapiro–Wilks test, a method proposed to be appropriate for a sample size less than 50 [80]. Although the number of the sampling sites is limited (2), they are considered representative because of the available time-series measurements, their hydrogeochemical characteristics, and the elevated concentrations of Cr, Si, Ni, and Mg^{2-}.

The normality test of Shapiro–Wilks showed that there was no normally distributed population of the samples, either for Cr or Cr(VI) ($p < 0.05$). The outliers were identified via Box plots to exclude these measurements in the next step until the total elimination of the outliers, and then the population of the remaining data was rechecked for its normal distribution (Figures 20 and 21). The last datasets of each parameter without outliers were double-checked for their normality with Q–Q plots and the Shapiro–Wilks test. In the spring "Elafakia"-W14, the NBLs of Cr constitutes the 95th percentile (57.24 µg/L) of the population as it was not normally distributed (Figure 20) [9]. On the other hand, Cr(VI) dataset was normally distributed, and based on the methodology, the NBL was defined to be 51.20 µg/L (NBL = the maximum value of the normally distributed dataset). Similarly, in the spring "Potistis"-W13, the NBLs of Cr is equal to the max value (130 µg/L) of the normally distributed dataset while the NBL for Cr(VI) was calculated at 100 µg/L (NBL = the 95th percentile of the dataset as non-normally distributed) (Figure 21).

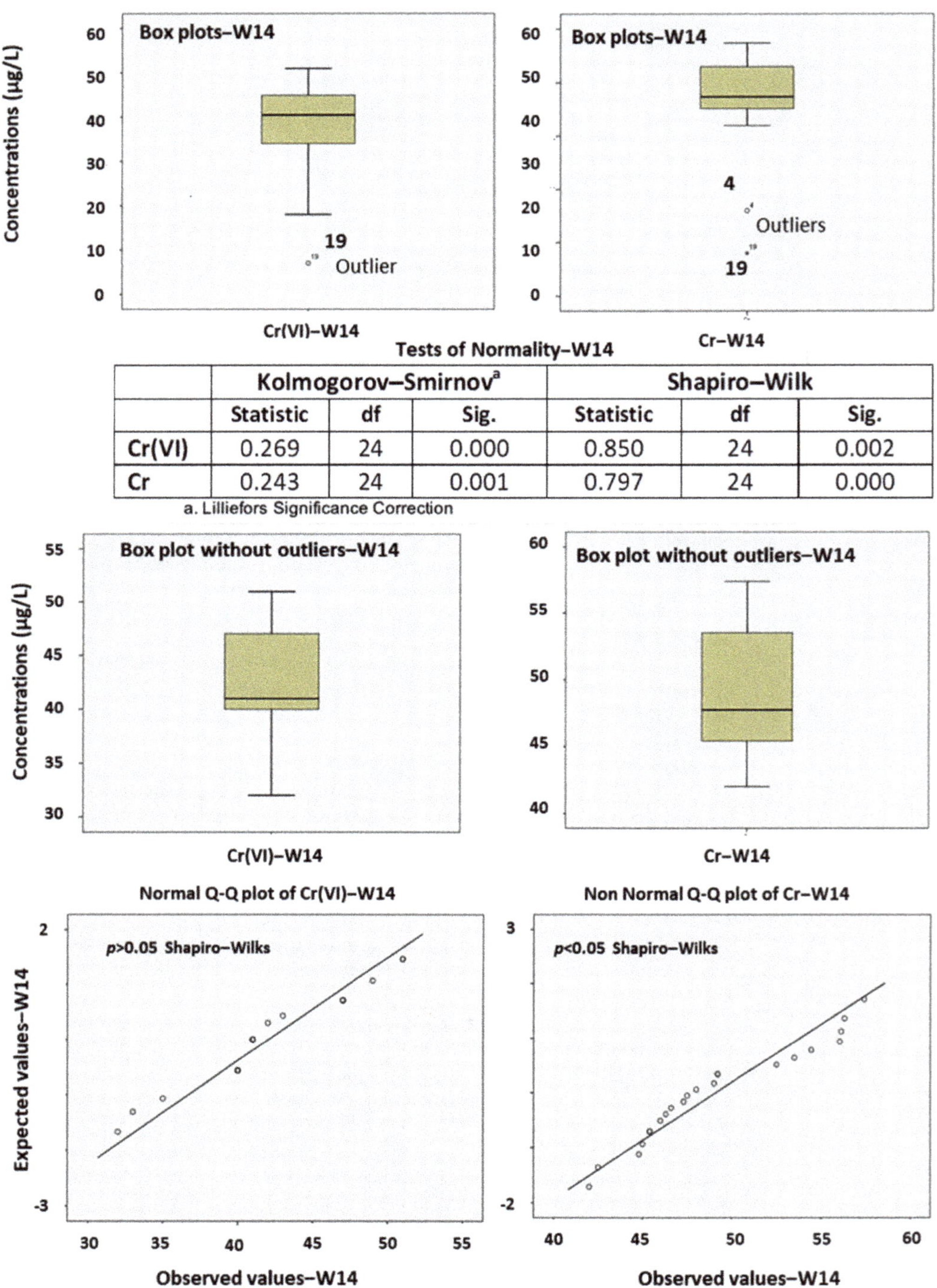

Tests of Normality–W14

	Kolmogorov–Smirnov[a]			Shapiro–Wilk		
	Statistic	df	Sig.	Statistic	df	Sig.
Cr(VI)	0.269	24	0.000	0.850	24	0.002
Cr	0.243	24	0.001	0.797	24	0.000

a. Lilliefors Significance Correction

Figure 20. Box plots and normality tests for the natural spring "Elafakia"-W14 in western Vermio Mt.

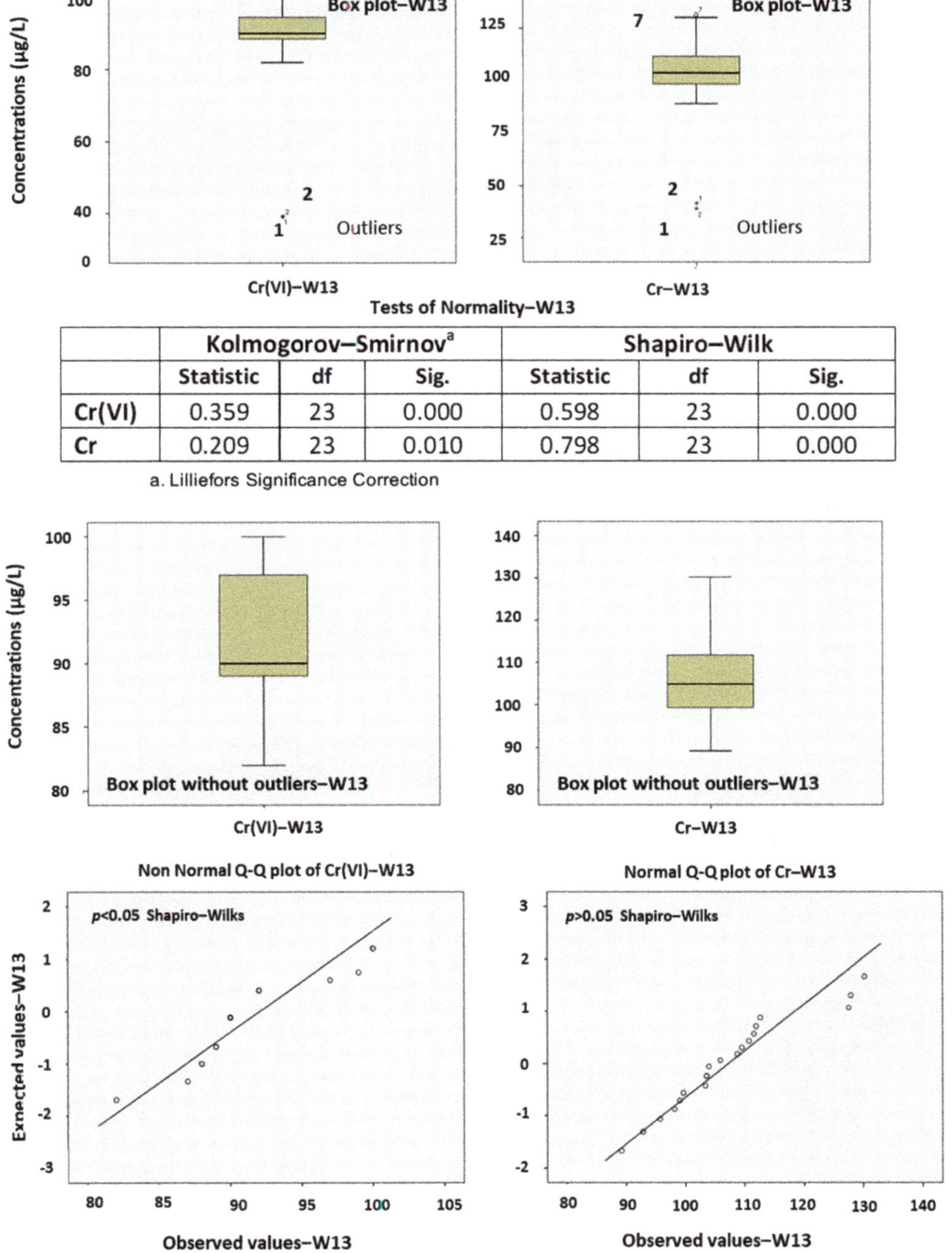

Figure 21. Box plots and normality tests for the natural spring "Potistis"-W13 in western Vermio Mt.

The estimated NBL is higher than the REF value (NBL/REF > 1) for both natural springs of Vermio. According to this suggestion, the TVs are determined as NBL for Cr in the natural springs of "Elafakia"-W14 and "Potistis"-W13. The spring "Potistis"-W13 is considered to be the most suitable one to define the NBLs in the ultramafic environment of western Vermio Mt., because its water type (Mg-Ca-HCO$_3$) indicated a strong influence by ultramafic rocks, whereas the spring "Elafakia"-W14, with a Ca-Mg-HCO$_3$ water type, was mainly influenced by carbonate formations. In Figure 22, the flow chart describes the NBLs and TVs assessment for Cr and Cr(VI), for the natural spring "Potistis"-W13.

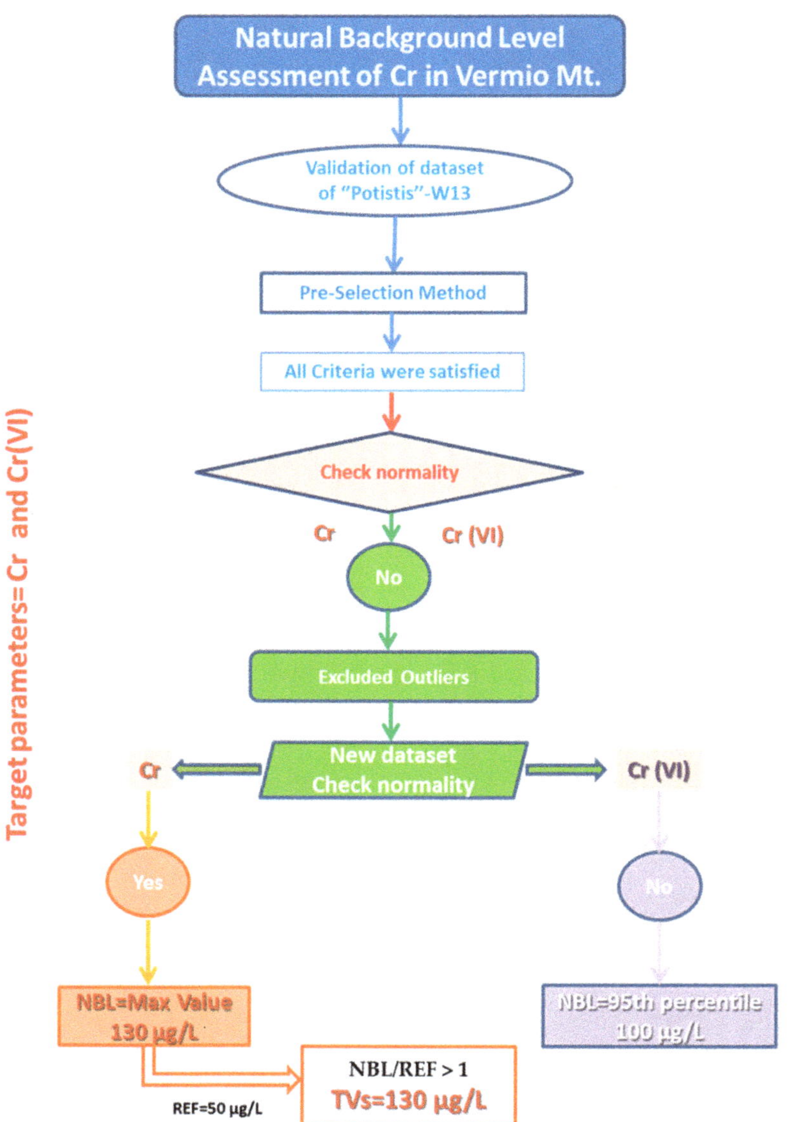

Figure 22. Flow chart of the modified conceptual model for assessing NBLs and TVs of Cr and Cr(VI) of the natural spring "Potistis"-W13, in western Vermio Mt.

The assessment of NBLs in ultramafic springs is a challenging modern methodology based on the continuous monitoring of hydrochemical parameters. Nevertheless, it is essential to mention that, since the environmental systems are complex and multicomponent, NBLs should not be treated as the absolute value above which a parameter is of anthropogenic origin; instead, NBLs constitute the minimum target value and the guide for investigating elevated groundwater concentrations of a hydrogeochemical parameter and elucidating the influence of anthropogenic factors in a study area on a larger scale (e.g., at a catchment scale). The continued monitoring of water quality parameters is likely to provide higher concentrations of the specific parameters in the future, and therefore subsequent recalculation of NBLs may lead to higher NBLs in the study area.

The high potential leaching of Cr in Vermio Mt., as derived from the above-mentioned calculated NBLs, is imprinted in the lowland of the Sarigkiol Basin [45]. The surface runoff and the discharge of springs enriched in Cr follow various flow paths via torrents or streams through the weathered mantle of ultramafic rocks in western Vermio Mt. and leach into the lowland of the Sarigkiol Basin (Figure 23). Due to the hydraulic connection between the western Vermio Mt. and the eastern part of the lowland of the Sarigkiol Basin, the defined NBLs apply to the latter, supporting the dominance of the geogenic factor in the high groundwater concentration of Cr in the Sarigkiol Basin.

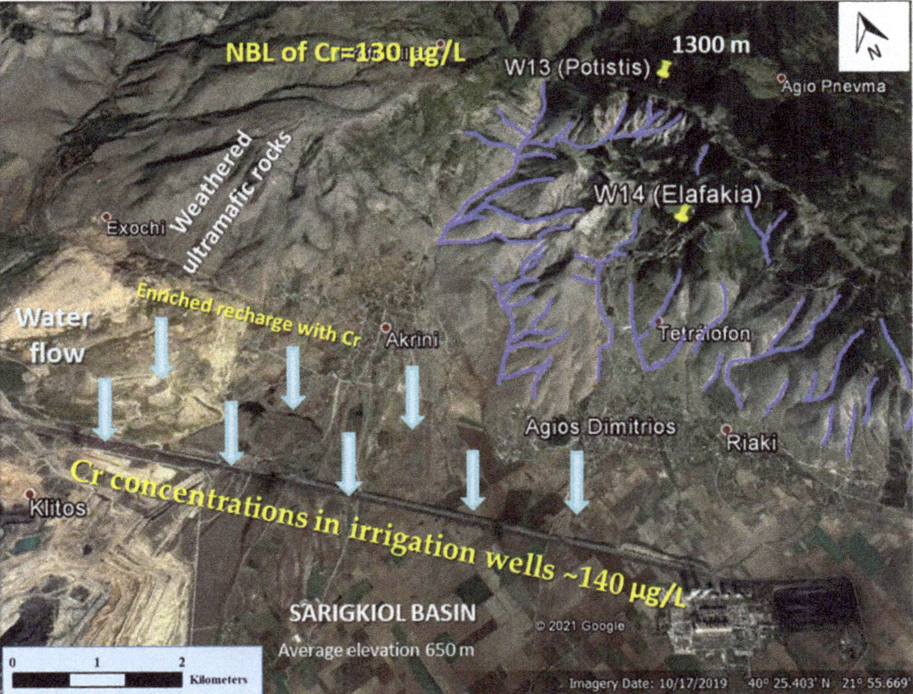

Figure 23. The flow paths and the natural recharge from western Vermio Mt. towards the eastern part of the Sarigkiol Basin (Google Earth image, 2021).

5. Conclusions

The ultramafic-dominated environment of western Vermio Mt. was fingerprinted on the groundwater chemistry and specifically on the elevated concentrations of Mg^{2+}, Si, Ni, and Cr in natural spring waters. Chromium was recognized as the principal environmental parameter in the natural spring waters of western Vermio Mt; in 42% of the studied spring water samples, the concentrations of Cr exceeded the WHO guideline value of 50 µg/L for drinking water. The geogenic origin of Cr in groundwater is recorded in the very strong positive correlation coefficients of Cr with Si, the strong positive correlation coefficients with Mg^{2+}, EC, and Ni, and the moderate positive correlation coefficients with HCO_3 and alkalinity.

The main factors that determined the concentration of Cr in the studied spring waters were:

(a) the time response of the aquifers systems to precipitations; direct infiltration on the geological formation of the aquifer results in immediate recharge of it. As a consequence, quick contaminant dilution takes place and fluctuations in Cr concentrations are observed depending on the time response of the aquifers to precipitation,

(b) the water–rock contact time; the longer the water–rock contact time is, the higher the Cr leaching is,

(c) the flow path of groundwater; a flow path through weathered ultramafic rocks results in the enrichment of groundwater in Cr,

(d) the degree of the serpentinisation of ultramafic rocks; the more serpentinised the ultramafic rocks are, the higher their leaching potential in Cr is, and

(e) the prevailing geochemical processes that favor the oxidation of Cr(III) to the soluble and mobile Cr(VI), such as alkaline pH, oxidative environment, presence of manganese oxides.

The absence of anthropic/anthropogenic activities in western Vermio Mt., the sufficient time-series data, and the hydrochemical characteristics of the studied springs allowed the assessment of NBLs of Cr by applying a multi-method approach. Considering the hydrogeological, hydrochemical, and hydrological data in western Vermio Mt. and applying the PS method, the spring "Potistis"-W13 was selected as the most representative one to define the NBLs of Cr in the area. The applied methodology is fully harmonized with the GDD and the Water Framework Directive (WFD, 2000/60/EC) [81]. The NBL of Cr was defined at 130 μg/L, and that of Cr(VI) at 100 μg/L. Based on the NBLs of Cr, TVs for Cr at a catchment scale, i.e., the Sarigkiol Basin, were defined to be equal to the NBLs. Conclusively, the ultramafic environment in western Vermio Mt. presents a high geochemical potential to dissolve and mobilize geogenic Cr.

This first systematic study of the natural springs of western Vermio Mt. provides important hydrogeochemical data for the geogenic footprint of a natural ultramafic environment on the groundwater quality. The proposed methodology could be implemented in any catchment scale aiming to distinguish between geogenic and anthropogenic groundwater deterioration and to establish new TVs considering the NBLs.

Author Contributions: Conceptualization, E.V. and M.P.; Data curation, E.V., M.P. and P.P.; Formal analysis, E.V., P.P. and M.P.; Investigation, E.V., P.P., M.P. and D.D.; Methodology, E.V., P.P. and M.P.; Project administration, M.P.; Supervision, M.P.; Visualization, E.V., P.P. and M.P.; Writing—original draft, E.V.; Writing—review and editing, E.V., P.P., M.P. and D.D. All authors have read and agreed to the summited version of the manuscript.

Funding: This research was carried out within the framework of the Research Project NTUA 623748.

Institutional Review Board Statement: Not applicable.

Informed Consent Statement: Not applicable.

Data Availability Statement: Some data is contained within the article. All hydrogeochemical data may be available for collaborative research projects by specific agreements. For information, contact maria@metal.ntua.gr.

Acknowledgments: Special thanks to G. Stamatis, Emeritus Professor in Hydrogeology (Agricultural University of Athens) and A. Dimitriadis, from Agios Dimitrios Power Plant of Public Power Corporation for their contribution to the fieldwork. A. Stamos in EAGME is acknowledged for valuable discussion on the operation mechanisms of the springs in western Vermio Mt. We would like to thank the two anonymous reviewers for their constructive com-ments and suggestions that significantly improved the quality of the paper. Special thanks are ex-pressed to the editors for their careful editorial handling.

Conflicts of Interest: The authors declare no conflict of interest.

References

1. European Union. EU Groundwater Directive 2006/118/EC. *Off. J. Eur. Union* **2006**, *L 372*, 19–31.
2. Amiri, V.; Nakhaei, M.; Lak, R.; Li, P. An integrated statistical-graphical approach for the appraisal of the natural background levels of some major ions and potentially toxic elements in the groundwater of Urmia aquifer, Iran. *Environ. Earth Sci.* **2021**, *80*, 432. [CrossRef]
3. Edmunds, W.; Shand, P.; Hart, P.; Ward, R. The natural (baseline) quality of groundwater: A UK pilot study. *Sci. Total. Environ.* **2003**, *310*, 25–35. [CrossRef]
4. Gemitzi, A. Evaluating the anthropogenic impacts on groundwaters: A methodology based on the determination of natural background levels and threshold values. *Environ. Earth Sci.* **2012**, *67*, 2223–2237. [CrossRef]
5. Preziosi, E.; Giuliano, G.; Vivona, R. Natural background levels and threshold values derivation for naturally As, V and F rich groundwater bodies: A methodological case study in Central Italy. *Environ. Earth Sci.* **2009**, *61*, 885–897. [CrossRef]
6. Urresti-Estala, B.; Carrasco-Cantos, F.; Pérez, I.V.; Gavilán, P.J. Determination of background levels on water quality of groundwater bodies: A methodological proposal applied to a Mediterranean River basin (Guadalhorce River, Málaga, Southern Spain). *J. Environ. Manag.* **2013**, *117*, 121–130. [CrossRef] [PubMed]
7. Sacchi, E.; Bergamini, M.; Lazzari, E.; Musacchio, A.; Mor, J.-R.; Pugliaro, E. Natural Background Levels of Potentially Toxic Elements in Groundwater from a Former Asbestos Mine in Serpentinite (Balangero, North Italy). *Water* **2021**, *13*, 735. [CrossRef]
8. Libera, N.D.; Fabbri, P.; Mason, L.; Piccinini, L.; Pola, M. A local natural background level concept to improve the natural background level: A case study on the drainage basin of the Venetian Lagoon in Northeastern Italy. *Environ. Earth Sci.* **2018**, *77*, 487. [CrossRef]
9. Parrone, D.; Ghergo, S.; Preziosi, E. A multi-method approach for the assessment of natural background levels in groundwater. *Sci. Total. Environ.* **2018**, *659*, 884–894. [CrossRef] [PubMed]
10. Preziosi, E.; Parrone, D.; del Bon, A.; Ghergo, S. Natural background level assessment in groundwaters: Probability plot versus pre-selection method. *J. Geochem. Explor.* **2014**, *143*, 43–53. [CrossRef]
11. Muller, D.; Blum, A.; Hart, A.; Hookey, J.; Kunkel, R.; Scheidleder, A.; Tomlin, C.; Wendland, F. *D18: Final Proposal for a Methodology to Set Up Groundwater Threshold Values in Europe*; Background Criteria for the Identification of Groundwater Thresholds; Bridge Publications: Vienna, Austria, 2006.
12. Wendland, F.; Berthold, G.; Blum, A.; Elsass, P.; Fritsche, J.-G.; Kunkel, R.; Wolter, R. Derivation of natural background levels and threshold values for groundwater bodies in the Upper Rhine Valley (France, Switzerland and Germany). *Desalination* **2008**, *226*, 160–168. [CrossRef]
13. Hinsby, K.; de Melo, M.T.C.; Dahl, M. European case studies supporting the derivation of natural background levels and groundwater threshold values for the protection of dependent ecosystems and human health. *Sci. Total Environ.* **2008**, *401*, 1–20. [CrossRef]
14. Ducci, D.; de Melo, M.T.C.; Preziosi, E.; Sellerino, M.; Parrone, D.; Ribeiro, L. Combining natural background levels (NBLs) assessment with indicator kriging analysis to improve groundwater quality data interpretation and management. *Sci. Total Environ.* **2016**, *569–570*, 569–584. [CrossRef] [PubMed]
15. Biddau, R.; Cidu, R.; Lorrai, M.; Mulas, M. Assessing background values of chloride, sulfate and fluoride in groundwater: A geochemical-statistical approach at a regional scale. *J. Geochem. Explor.* **2017**, *181*, 243–255. [CrossRef]
16. Masciale, R.; Amalfitano, S.; Frollini, E.; Ghergo, S.; Melita, M.; Parrone, D.; Preziosi, E.; Vurro, M.; Zoppini, A.; Passarella, G. Assessing Natural Background Levels in the Groundwater Bodies of the Apulia Region (Southern Italy). *Water* **2021**, *13*, 958. [CrossRef]
17. Filippini, M.; Zanotti, C.; Bonomi, T.; Sacchetti, V.; Amorosi, A.; Dinelli, E.; Rotiroti, M. Deriving Natural Background Levels of Arsenic at the Meso-Scale Using Site-Specific Datasets: An Unorthodox Method. *Water* **2021**, *13*, 452. [CrossRef]
18. Oze, C.; Fendorf, S.; Bird, D.K.; Coleman, R.G. Chromium geochemistry in serpentinized ultramafic rocks and serpentine soils from the Franciscan complex of California. *Am. J. Sci.* **2004**, *304*, 67–101. [CrossRef]
19. Tashakor, M.; Modabberi, S.; van der Ent, A.; Echevarria, G. Impacts of ultramafic outcrops in Peninsular Malaysia and Sabah on soil and water quality. *Environ. Monit. Assess.* **2018**, *190*, 333. [CrossRef] [PubMed]
20. Kelepertzis, E.; Galanos, E.; Mitsis, I. Origin, mineral speciation and geochemical baseline mapping of Ni and Cr in agricultural topsoils of Thiva Valley (central Greece). *J. Geochem. Explor.* **2013**, *125*, 56–68. [CrossRef]
21. Ryan, P.C.; Kim, J.; Wall, A.J.; Moen, J.C.; Corenthal, L.G.; Chow, D.R.; Sullivan, C.M.; Bright, K.S. Ultramafic-derived arsenic in a fractured bedrock aquifer. *Appl. Geochem.* **2011**, *26*, 444–457. [CrossRef]
22. Nriagu, J.; Nieboer, E. *Chromium in the Natural and Human Environments*; John Wiley & Sons: New York, NY, USA, 1988; Volume 20.
23. Rai, D.; Sass, B.M.; Moore, D.A. Chromium(III) hydrolysis constants and solubility of chromium(III) hydroxide. *Inorg. Chem.* **1987**, *26*, 345–349. [CrossRef]
24. Sperling, M.; Xu, S.; Welz, B. Determination of chromium(III) and chromium(VI) in water using flow injection on-line pre-concentration with selective adsorption on activated alumina and flame atomic absorption spectrometric detection. *Anal. Chem.* **1992**, *64*, 3101–3108. [CrossRef]
25. Kotaś, J.; Stasicka, Z. Chromium occurrence in the environment and methods of its speciation. *Environ. Pollut.* **2000**, *107*, 263–283. [CrossRef]

26. Berna, E.C.; Johnson, T.M.; Makdisi, R.S.; Basu, A. Cr Stable Isotopes As Indicators of Cr(VI) Reduction in Groundwater: A Detailed Time-Series Study of a Point-Source Plume. *Environ. Sci. Technol.* **2009**, *44*, 1043–1048. [CrossRef] [PubMed]

27. Johnson, C.; Xyla, A.G. The oxidation of chromium(III) to chromium(VI) on the surface of manganite (γ-MnOOH). *Geochim. Cosmochim. Acta* **1991**, *55*, 2861–2866. [CrossRef]

28. Fendorf, S.E.; Fendorf, M.; Sparks, D.L.; Gronsky, R. Inhibitory mechanisms of Cr(III) oxidation by δ-MnO2. *J. Colloid Interface Sci.* **1992**, *153*, 37–54. [CrossRef]

29. Fantoni, D.; Brozzo, G.; Canepa, M.; Cipolli, F.; Marini, L.; Ottonello, G.; Zuccolini, M. Natural hexavalent chromium in groundwaters interacting with ophiolitic rocks. *Environ. Earth Sci.* **2002**, *42*, 871–882. [CrossRef]

30. Tziritis, E.; Kelepertzis, E.; Korres, G.; Perivolaris, D.; Repani, S. Hexavalent Chromium Contamination in Groundwaters of Thiva Basin, Central Greece. *Bull. Environ. Contam. Toxicol.* **2012**, *89*, 1073–1077. [CrossRef] [PubMed]

31. Dermatas, D.; Mpouras, T.; Chrysochoou, M.; Panagiotakis, I.; Vatseris, C.; Linardos, N.; Theologou, E.; Boboti, N.; Xenidis, A.; Papassiopi, N.; et al. Origin and concentration profile of chromium in a Greek aquifer. *J. Hazard. Mater.* **2015**, *281*, 35–46. [CrossRef] [PubMed]

32. Hausladen, D.M.; Alexander-Ozinskas, A.; McClain, C.N.; Fendorf, S. Hexavalent Chromium Sources and Distribution in California Groundwater. *Environ. Sci. Technol.* **2018**, *52*, 8242–8251. [CrossRef]

33. Papazotos, P.; Vasileiou, E.; Perraki, M. Elevated groundwater concentrations of arsenic and chromium in ultramafic environments controlled by seawater intrusion, the nitrogen cycle, and anthropogenic activities: The case of the Gerania Mountains, NE Peloponnese, Greece. *Appl. Geochem.* **2020**, *121*, 104697. [CrossRef]

34. Coyte, R.; McKinley, K.; Jiang, S.; Karr, J.; Dwyer, G.S.; Keyworth, A.J.; Davis, C.C.; Kondash, A.J.; Vengosh, A. Occurrence and distribution of hexavalent chromium in groundwater from North Carolina, USA. *Sci. Total. Environ.* **2019**, *711*, 135135. [CrossRef]

35. Perraki, M.; Vasileiou, E.; Bartzas, G. Tracing the origin of chromium in groundwater: Current and new perspectives. *Curr. Opin. Environ. Sci. Heal.* **2021**, *22*, 100267. [CrossRef]

36. Vithanage, M.; Kumarathilaka, P.; Oze, C.; Karunatilake, S.; Seneviratne, M.; Hseu, Z.-Y.; Gunarathne, V.; Dassanayake, M.; Ok, Y.S.; Rinklebe, J. Occurrence and cycling of trace elements in ultramafic soils and their impacts on human health: A critical review. *Environ. Int.* **2019**, *131*, 104974. [CrossRef]

37. Liang, J.; Huang, X.; Yan, J.; Li, Y.; Zhao, Z.; Liu, Y.; Ye, J.; Wei, Y. A review of the formation of Cr(VI) via Cr(III) oxidation in soils and groundwater. *Sci. Total Environ.* **2021**, *774*, 145762. [CrossRef]

38. Papazotos, P.; Vasileiou, E.; Perraki, M. The synergistic role of agricultural activities in groundwater quality in ultramafic environments: The case of the Psachna basin, central Euboea, Greece. *Environ. Monit. Assess.* **2019**, *191*, 317. [CrossRef]

39. Oze, C.; Bird, D.K.; Fendorf, S. Genesis of hexavalent chromium from natural sources in soil and groundwater. *Proc. Natl. Acad. Sci. USA* **2007**, *104*, 6544–6549. [CrossRef] [PubMed]

40. Margiotta, S.; Mongelli, G.; Summa, V.; Paternoster, M.; Fiore, S. Trace element distribution and Cr(VI) speciation in Ca-HCO$_3$ and Mg-HCO$_3$ spring waters from the northern sector of the Pollino massif, Southern Italy. *J. Geochem. Explor.* **2012**, *115*, 1–12. [CrossRef]

41. Remoundaki, E.; Vasileiou, E.; Philippou, A.; Perraki, M.; Kousi, P.; Hatzikioseyian, A.; Stamatis, G. Groundwater Deteriora-tion: The Simultaneous Effects of Intense Agricultural Activity and Heavy Metals in Soil. *Procedia Eng.* **2016**, *162*, 545–552. [CrossRef]

42. Kaprara, E.; Kazakis, N.; Simeonidis, K.; Coles, S.; Zouboulis, A.; Samaras, P.; Mitrakas, M. Occurrence of Cr(VI) in drinking water of Greece and relation to the geological background. *J. Hazard. Mater.* **2014**, *281*, 2–11. [CrossRef]

43. Elango, L.; Kannan, R. *Chapter 11: Rock–Water Interaction and Its Control on Chemical Composition of Groundwater*; Elsevier: Amsterdam, The Netherlands, 2007; pp. 229–243. [CrossRef]

44. Sharif, M.; Davis, R.; Steele, K.; Kim, B.; Kresse, T.; Fazio, J. Inverse geochemical modeling of groundwater evolution with emphasis on arsenic in the Mississippi River Valley alluvial aquifer, Arkansas (USA). *J. Hydrol.* **2008**, *350*, 41–55. [CrossRef]

45. Vasileiou, E.; Papazotos, P.; Dimitrakopoulos, D.; Perraki, M. Expounding the origin of chromium in groundwater of the Sarigkiol Basin, Western Macedonia, Greece: A cohesive statistical approach and hydrochemical study. *Environ. Monit. Assess.* **2019**, *191*, 509. [CrossRef] [PubMed]

46. Stamos, A.; Samiotis, G.; Tsioptsias, C.; Amanatidou, E. Natural presence of hexavalent chromium in spring waters of South-West Mountain Vermion, Greece. In Proceedings of the 16th International Conference on Environmental Science and Technology (CEST 2019), Rhodes, Greece, 4–7 September 2019; p. 4.

47. Institute of Geology and Mineral Exploration of Greece. *Geological Maps of Greece, Sheet: Kozani*; Scale 1:50.000, Department of Geological Maps; Institute of Geology and Mineral Exploration of Greece: Athens, Greece, 1980.

48. Perraki, M. *Mineralogical, Petrological and Geochemical Study of Heavy Minerals with Emphasis on Chromium in the Geological Formations (Ultrabasic Rocks, Lignite, Clay Formations) and the Coal-Fired Products (Fly Ash) and the Quality of Surficial and Underground Aquifers of the Sarigkiol Basin (NW Greece)*; Technical Report; National Technical University of Athesn: Athens, Greece, 2016.

49. Nematollahi, M.J.; Ebrahimi, P.; Razmara, M.; Ghasemi, A. Hydrogeochemical investigations and groundwater quality assessment of Torbat-Zaveh plain, Khorasan Razavi, Iran. *Environ. Monit. Assess.* **2015**, *188*, 1–21. [CrossRef]

50. Esmaeili, A.; Moore, F. Hydrogeochemical assessment of groundwater in Isfahan province, Iran. *Environ. Earth Sci.* **2011**, *67*, 107–120. [CrossRef]

51. Spearman, C. The Proof and Measurement of Association between Two Things. *Am. J. Psychol.* **1904**, *15*, 72. [CrossRef]

52. Wuensch, K.L.; Evans, J.D. Straightforward Statistics for the Behavioral Sciences. *J. Am. Stat. Assoc.* **1996**, *91*, 1750. [CrossRef]

53. Gauthier, T. Detecting Trends Using Spearman's Rank Correlation Coefficient. *Environ. Forensics* **2001**, *2*, 359–362. [CrossRef]
54. Wilk, M.B.; Gnanadesikan, R. Probability plotting methods for the analysis for the analysis of data. *Biometrika* **1968**, *55*, 1–17. [CrossRef]
55. Parkhurst, D.L.; Appelo, C.A.J. *User's Guide to PHREEQC (Version 2): A Computer Program for Speciation, Batch-Reaction, One-Dimensional Transport, and Inverse Geochemical Calculations*; U.S. Geological Survey, Water Resources Investigations Report 99-4259; United States Geological Survey (USGS): Washington, DC, USA, 1999.
56. Merkel, B.J.; Planer-Friedrich, B.; Nordstrom, D.K. *Groundwater Geochemistry: A Practical Guide to Modeling of Natural and Contaminated Aquatic Systems*; Springer: Berlin, Germany, 2005.
57. Zhang, F.; Jin, Z.; Yu, J.; Zhou, Y.; Zhou, L. Hydrogeochemical processes between surface and groundwaters on the north-eastern Chinese Loess Plateau: Implications for water chemistry and environmental evolutions in semi-arid regions. *J. Geochem. Explor.* **2015**, *159*, 115–128. [CrossRef]
58. Christoforidou, P.; Panagopoulos, A.; Voudouris, K. Towards A New Procedure To Set Up Groundwater Threshold Values In Accordance With The Previsions Of The Ec Directive 2006/118: A Case Study From Achaia And Corinthia (Greece). *Bull. Geol. Soc. Greece* **2017**, *43*, 1678. [CrossRef]
59. Molinari, A.; Guadagnini, L.; Marcaccio, M.; Guadagnini, A. Natural background levels and threshold values of chemical species in three large-scale groundwater bodies in Northern Italy. *Sci. Total Environ.* **2012**, *425*, 9–19. [CrossRef] [PubMed]
60. Chidichimo, F.; de Biase, M.; Straface, S. Groundwater pollution assessment in landfill areas: Is it only about the leachate? *Waste Manag.* **2019**, *102*, 655–666. [CrossRef] [PubMed]
61. Parrone, D.; Frollini, E.; Preziosi, E.; Ghergo, S. eNaBLe, an On-Line Tool to Evaluate Natural Background Levels in Groundwater Bodies. *Water* **2020**, *13*, 74. [CrossRef]
62. Masetti, M.; Poli, S.; Sterlacchini, S.; Beretta, G.P.; Facchi, A. Spatial and statistical assessment of factors influencing nitrate contamination in groundwater. *J. Environ. Manag.* **2008**, *86*, 272–281. [CrossRef] [PubMed]
63. Menció, A.; Mas-Pla, J.; Otero, N.; Regàs, O.; Boy-Roura, M.; Puig, R.; Bach, J.; Domènech, C.; Zamorano, M.; Brusi, D.; et al. Nitrate pollution of groundwater; all right, but nothing else? *Sci. Total Environ.* **2016**, *539*, 241–251. [CrossRef] [PubMed]
64. Soltan, M.E. Evaluation Of Ground Water Quality In Dakhla Oasis (Egyptian Western Desert). *Environ. Monit. Assess.* **1999**, *57*, 157–168. [CrossRef]
65. Singh, U.V.; Abhishek, A.; Singh, K.P.; Dhakate, R.; Singh, N.P. Groundwater quality appraisal and its hydrochemical characterization in Ghaziabad (a region of indo-gangetic plain), Uttar Pradesh, India. *Appl. Water Sci.* **2013**, *4*, 145–157. [CrossRef]
66. World Health Organization. *Guidelines for Drinking-Water Quality*, 4th ed.; World Health Organization: Geneva, Switzerland, 2011.
67. Marghade, D.; Malpe, D.B.; Zade, A.B. Geochemical characterization of groundwater from northeastern part of Nagpur urban, Central India. *Environ. Earth Sci.* **2010**, *62*, 1419–1430. [CrossRef]
68. Lelli, M.; Grassi, S.; Amadori, M.; Franceschini, F. Natural Cr(VI) contamination of groundwater in the Cecina coastal area and its inner sectors (Tuscany, Italy). *Environ. Earth Sci.* **2013**, *71*, 3907–3919. [CrossRef]
69. Barnes, I.; O'neil, J.R. The relationship between fluids in some fresh alpine-type ultramafics and possible modern ser-pentinization, Western United States. *Bull. Geol. Soc. Am.* **1969**, *80*, 1947–1960. [CrossRef]
70. Cipolli, F.; Gambardella, B.; Marini, L.; Ottonello, G.; Zuccolini, M.V. Geochemistry of high-pH waters from serpentinites of the Gruppo di Voltri (Genova, Italy) and reaction path modeling of CO_2 sequestration in serpentinite aquifers. *Appl. Geochem.* **2004**, *19*, 787–802. [CrossRef]
71. Marques, J.M.; Carreira, P.M.; Carvalho, M.D.R.; Matias, M.J.; Goff, F.E.; Basto, M.J.; Graça, R.C.; Aires-Barros, L.; Rocha, L. Origins of high pH mineral waters from ultramafic rocks, Central Portugal. *Appl. Geochem.* **2008**, *23*, 3278–3289. [CrossRef]
72. Richard, F.C.; Bourg, A.C. Aqueous geochemistry of chromium: A review. *Water Res.* **1991**, *25*, 807–816. [CrossRef]
73. Zhang, B.; Zhao, D.; Zhou, P.; Qu, S.; Liao, F.; Guangcai, W. Hydrochemical Characteristics of Groundwater and Dominant Water—Rock Interactions in the Delingha. *Water* **2020**, *12*, 836. [CrossRef]
74. Redwan, M.; Moneim, A.A.A.; Amra, M.A. Effect of water–rock interaction processes on the hydrogeochemistry of ground-water west of Sohag area, Egypt. *Arab. J. Geosci.* **2016**, *9*, 111. [CrossRef]
75. Jalali, M.; Khanlari, Z.V. Cadmium Availability in Calcareous Soils of Agricultural Lands in Hamadan, Western Iran. *Soil Sediment. Contam. Int. J.* **2008**, *17*, 256–268. [CrossRef]
76. Gibbs, R.J. Mechanisms Controlling World Water Chemistry. *Science* **1970**, *170*, 1088–1090. [CrossRef] [PubMed]
77. Marandi, A.; Shand, P. Groundwater chemistry and the Gibbs Diagram. *Appl. Geochem.* **2018**, *97*, 209–212. [CrossRef]
78. McClain, C.; Maher, K. Chromium fluxes and speciation in ultramafic catchments and global rivers. *Chem. Geol.* **2016**, *426*, 135–157. [CrossRef]
79. Giammetta, R.; Telesca, A.; Mongelli, G. Serpentinites-water interaction in the S. Severino area, Lucanian Apennines, Southern Italy. *GeoActa* **2004**, *3*, 25–33.
80. Hanusz, Z.; Tarasińska, J. Normalization of the Kolmogorov–Smirnov and Shapiro–Wilk tests of normality. *Biom. Lett.* **2015**, *52*, 85–93. [CrossRef]
81. EUROPA. European Commission Water Framework Directive 2000/60/EC. *Off. J. Eur. Communities* **2000**, *L 327*, 1–73.

Article

Aquifer Parameters Estimation from Natural Groundwater Level Fluctuations at the Mexican Wine-Producing Region Guadalupe Valley, BC

Mario A. Fuentes-Arreazola [1,*], Jorge Ramírez-Hernández [2], Rogelio Vázquez-González [3], Diana Núñez [1], Alejandro Díaz-Fernández [3] and Javier González-Ramírez [3,4]

[1] Centro de Sismología y Volcanología de Occidente, Centro Universitario de la Costa, Universidad de Guadalajara, Av. Universidad, No. 203, Delegación Ixtapa, Puerto Vallarta 48280, Mexico; dianane1982@gmail.com

[2] Instituto de Ingeniería, Campus Mexicali, Universidad Autónoma de Baja California, Av. de la Normal S/N, Col. Insurgentes Este, Mexicali 21280, Mexico; jorger@uabc.edu.mx

[3] Centro de Investigación Científica y de Educación Superior de Ensenada, Departamento de Geofísica Aplicada, CICESE, Carretera Ensenada-Tijuana, No. 3918, Zona Playitas, Ensenada 22860, Mexico; vrog70@gmail.com (R.V.-G.); aldiaz@cicese.mx (A.D.-F.); Javier-Gonzalez@uan.edu.mx (J.G.-R.)

[4] Laboratorio de Oceanografía Física, Escuela Nacional de Ingeniería Pesquera, Universidad Autónoma de Nayarit, Bahía de Matanchen km. 12 Carretera a Los Cocos, San Blas 63470, Mexico

* Correspondence: marioafar@gmail.com

Citation: Fuentes-Arreazola, M.A.; Ramírez-Hernández, J.; Vázquez-González, R.; Núñez, D.; Díaz-Fernández, A.; González-Ramírez, J. Aquifer Parameters Estimation from Natural Groundwater Level Fluctuations at the Mexican Wine-Producing Region Guadalupe Valley, BC. *Water* **2021**, *13*, 2437. https://doi.org/10.3390/w13172437

Academic Editors: Evangelos Tziritis and Andreas Panagopoulos

Received: 4 June 2021
Accepted: 8 July 2021
Published: 4 September 2021

Publisher's Note: MDPI stays neutral with regard to jurisdictional claims in published maps and institutional affiliations.

Abstract: Determining hydrogeological properties of the rock materials that constitute an aquifer through stress tests or laboratory tests presents inherent complications. An alternative tool that has significant advantages is the study of the groundwater-level response as a result of the pore-pressure variation caused by the internal structure deformation of the aquifer induced by barometric pressure and solid Earth tide. The purpose of this study was to estimate the values of the physical/hydraulic properties of the geological materials that constitute the Guadalupe Valley Aquifer based on the analysis of the groundwater-level response to barometric pressure and solid Earth tide. Representative values of specific storage (1.27×10^{-6} to 2.78×10^{-6} m^{-1}), porosity (14–34%), storage coefficient (3.10×10^{-5} to 10.45×10^{-5}), transmissivity (6.67×10^{-7} to 1.29×10^{-4} m$^2 \cdot$s^{-1}), and hydraulic conductivity (2.30×10^{-3} to 2.97×10^{-1} m$\cdot$d^{-1}) were estimated. The values obtained are consistent with the type of geological materials identified in the vicinity of the analyzed wells and values reported in previous studies. This analysis represents helpful information that can be considered a framework to design and assess management strategies for groundwater resources in the overexploited Guadalupe Valley Aquifer.

Keywords: hydrogeological properties; natural groundwater fluctuations; semi-arid zones; depleting groundwater resources; Guadalupe Valley Aquifer

1. Introduction

Water supply for human consumption, agricultural, and industrial activities is a crucial topic for developing the northwest semi-arid zones of Mexico. Particularly, Guadalupe Valley in the Ensenada municipality, BC, Mexico, stands out as the region with the highest wine production in the country. The Guadalupe Valley has had groundwater exploitations as its primary source of water. However, this intense anthropogenic activity has led to a recharge-extraction deficit, resulting in an excessive decrease in groundwater levels, compromising water availability in the region [1–4].

This situation raises the need to conduct interdisciplinary studies that provide technical and scientific information to design and evaluate new water-resource-management strategies. In 2007, many institutions established an integrated-management plan for the Guadalupe Valley Aquifer [5]. One of the main problems identified was the uncertainty in the knowledge of the aquifer dynamic. Thus, the urgency to establish a hydrogeological-measurement network was pointed out. As part of the hydrogeological-monitoring recommendations, continuous monitoring wells were instrumented by using pressure transducers [6].

Groundwater records are commonly used to study storage evolution, hydraulic gradient, and define the groundwater direction flow [7]. However, it has been observed that the groundwater is sensitive to several natural phenomena (e.g., barometric pressure, earth tides, and seismic activity) [8–15]. Analyzing the groundwater response to barometric pressure and solid Earth tide constitutes an alternative, feasible, and inexpensive tool for hydrogeological parameter estimations. Especially in regions where hydraulic properties' information is insufficient or null, as a result of that, the sampling density necessary to describes it may be prohibitively expensive (e.g., drilling and testing core, and pumping tests) [7,16]. Therefore, this work aimed to estimate some hydrogeological parameters (specific storage, porosity, storage coefficient, transmissivity, and hydraulic conductivity) related to the geological materials that constitute the Guadalupe Aquifer based on the groundwater response to barometric pressure and solid Earth tide. Results of this analysis represent valuable information that can be considered as a framework to design and assess management strategies for groundwater resources in the overexploited Guadalupe Valley Aquifer.

2. Methods

2.1. Aquifer Response to Earth and Atmospheric Tides

Earth and atmospheric tides are natural phenomena throughout the Earth's crust, exerting a uniformly distributed surface load that causes a subsurface strain constituted of superimposed signals of various frequencies and amplitudes. Aquifer formations experiment compression and extension at their inner structure as a result of the strain induced. Part of the strain is absorbed by the soil grains, and the rest is transmitted to the water contained in the porous medium modifying the pore pressure, so water level fluctuates, and their amplitude is modulated by geologic materials hydraulic properties that constitute the aquifer [8,10,11,13]. Earth and atmospheric tides utilization is a feasible and inexpensive alternative tool for hydrogeological parameters estimations [14–17]. The latter was based on the premise that only three variables are required to compute values for some aquifer parameters: (i) computed strain-tensor associated to Earth tides, (ii) measured barometric pressure, and (iii) recorded groundwater heads.

2.2. Groundwater-Level Response to Atmospheric Pressure

Groundwater table (WL) variations show an inverse and proportional correlation to barometric pressure (BP) fluctuations. WL variations are related to BP changes through barometric efficiency (BE), which can be obtained according to Rasmussen and Crawford [18] as follows:

$$BE = -\frac{WL}{BP} \tag{1}$$

BP fluctuations generate an evenly distributed strain field on the Earth's surface. This latter causes elastic deformation from the rock materials that constitute the aquifer, and it is also transmitted to the fluid contained into the porous medium [10]. If the aquifer formation presents high transmissivity or specific yield, a drained condition is favored (i.e., mass transfer through flow). Thus, a groundwater response to BP may not be observed [19]. However, it is a common practice to consider the lateral flow negligible as a result of the vast lateral extension of the aquifer formation and the almost uniform effect of atmospheric load on the ground surface [20].

WL variations within the borehole can be conceptualized as aquifer pore-pressure changes, except in wells that are open to the atmosphere, on which the BP also exerts even pressure to the water surface [17]. Thus, a lag in WL response is often produced due to the air contained in the vadose zone. This lag causes pressure differences between the aquifer and borehole, propitiating in- and out-flows, resulting in WL variations [21].

Several methods have been developed for barometric efficiency estimation. Some of them consider independence on the frequency domain of WL–BP and calculated it by using linear regression techniques [18,22–24]. In contrast, other methods consider dependency on the frequency domain of WL–BP through transfer functions. There also assesses the simultaneous effect of the solid Earth tide [20,25–27].

2.3. Groundwater-Level Response to Solid Earth Tide

Solid Earth tide (SET) corresponds to small periodic variations in the Earth's shape as a result of expansion and compression forces generated by the gravitational attraction of celestial bodies, mainly the Moon and Sun. These gravitational forces are balanced by pore-pressure changes in an aquifer that generate WL variations within boreholes drilled typically in confined and semi-confined aquifers [10,19]. Pore pressure (PP) is related to the vertical stress (σ_{zz}) associated with SET through the tidal efficiency (γ_e), which is obtained according to Jacob [8] as follows:

$$\gamma_e = -\frac{PP}{\sigma_{zz}} \tag{2}$$

Strictly speaking, the fluid contained in the porous medium responds to a three-dimensional strain tensor (ε_v). Nevertheless, considering the induced deformation associated with SET and tectonic activity, the ε_v is not well-known as a priori [17,20]. Moreover, the ε_{zz} value measured on the terrain surface is approximately equal to the value of a horizontal strain component but with an opposite sign [20]. Therefore, it is more appropriate to analyze the WL response to an areal tidal strain (ε_A) defined by Rojstaczer and Agnew [13] and calculated as follows:

$$\varepsilon_A = \varepsilon_{xx} + \varepsilon_{yy} \tag{3}$$

Because ε_{zz} has an opposite sign, ε_A value is higher than ε_v. Thus, the WL response may be lower when ε_v is used rather than ε_A.

The strain tensor associated with the SET can be estimated from the theoretical gravitational potential, W_2 [28]. This differs from the measured gravitational potential due to geological and topographical local discontinuity effects [28–30]. Geologic and topo-graphic influence is complicated to define a priori. Thus, in the absence of strain measurements, the use of the theoretical gravitational strain is appropriate [20].

2.4. Aquifer Parameters Estimation

Jacob [8] derived a mathematical expressions that relate BE and γ_e with the elastic properties of the rock materials that constitute the aquifer, and can be written as follows:

$$BE = \frac{1}{1 + \frac{\beta_k}{\varphi \beta_W}} \tag{4}$$

$$\gamma_e = \frac{\frac{\beta_k}{\varphi \beta_W}}{1 + \frac{\beta_k}{\varphi \beta_W}} \tag{5}$$

where β_k is the rock matrix compressibility, φ corresponds to porosity, and β_W is water compressibility. If both expressions are added, the result is unity. Therefore, in calculating any of the previous parameters, it is possible to define the other one (BE = $1 - \gamma_e$).

In case that the compressibility of the rock materials that constitute the aquifer is not considered, it is possible to estimate values of specific storage and porosity based on the WL response as a result of the SET and BP effects. WL variations produced by aquifer dilatation related to the SET is a function of specific storage (S_S) of the rock materials. Bredehoeft [10] indicated that S_S could be calculated from the water-table fluctuations record (dh), and assuming a characteristic value of Poisson's ratio (v) in undrained conditions. Van der Kamp and Gale [11] derived an expression to estimate S_S as follows:

$$S_S = -\left[\left(1 - \frac{K_K}{K_U}\right) \cdot \left(\frac{1 - 2v}{1 - v}\right) \cdot \left(\frac{2h - 6l}{Er \cdot g}\right)\right] \frac{dW_2}{dh} \tag{6}$$

where K_U is the rock matrix compressibility under undrained conditions, h = 0.6031 and l = 0.0839 are the Love numbers [31]; Er corresponds to the Earth's radius, and g indicates gravity acceleration.

The relationship between W_2 and dh is equivalent to the relation among the amplitude of the dominant harmonic components of W_2 denoted as (A_2 (τ, θ)), and the amplitude of the dh (A_{dh} (τ)) at the same period (τ). Merrit [14] proposed that the derivatives (dW_2 and dh) can be approximated by a finite differential scheme; thus, Equation (6) can be written as follows:

$$S_S = -\left[\left(1 - \frac{K_K}{K_U}\right) \cdot \left(\frac{1 - 2v}{1 - v}\right) \cdot \left(\frac{2h - 6l}{Er \cdot g}\right)\right] \frac{A_2(\tau, \theta)}{A_{dh}(\tau)} \tag{7}$$

where $A_2(\tau, \theta)$ is calculated as follows:

$$A_2(\tau, \theta) = g \cdot K_m \cdot b \cdot f(\theta), \tag{8}$$

The general lunar coefficient (K_m), the particular amplitude factor (b) for each harmonic component with a period (τ), and the latitude function (f (θ)) values were determined by Merrit [14].

The classic method to study W_2 is to represent it through a finite set of harmonic functions, sinus, and cosines. Each k-tidal harmonic component has a particular frequency (f_{Tk}), amplitude (A_{Tk}), and phase angle (Φ_{Tk}) [32]. Amplitude (A_{dhk}) and phase-angle (Φ_{dhk}) estimations from the WL variations at the exact frequencies of the harmonic components of W_2 are calculated from the regression coefficients (a_{dhk} and b_{dhk}) obtained as follows [17]:

$$A_{dhk} = \sqrt{a_{dhk}^2 + b_{dhk}^2} \tag{9}$$

$$\Phi_{dhk} = \tan^{-1}\left(-\frac{b_{dhk}}{a_{dhk}}\right) \tag{10}$$

Similarly, A_{Tk} and Φ_{Tk} are computed from the theoretical strain-tensor associated to Earth tides, using Equations (9) and (10). Thus, areal strain sensitivity (A_{SK}) is calculated based on A_{dhk} and A_{Tk} according to Rojstaczer and Agnew [13]:

$$A_{Sk} = -\frac{WL}{\varepsilon_A} = \frac{A_{dhk}}{A_{Tk}} \approx \frac{A_2(\tau, \theta)}{A_{dhk}(\tau)} \tag{11}$$

In case the rock materials that constitute the aquifer are incompressible, the volume aquifer changes as a result of the deformation induced by SET could be approximated as a variation on the porosity [33]. This assumption is appropriate for most of the aquifers studied in hydrogeology; the exceptions are aquifers related to low-porosity rocks [10]. Thus, the porosity can be estimated according to Merrit [14] as follows:

$$\varphi = \frac{BE \cdot S_S}{\beta_W \cdot g \cdot \rho}, \tag{12}$$

where ρ is the fluid density, and it is a function of its temperature.

Cooper et al. [9] demonstrated that the WL harmonic response depends on transmissivity (T), storage coefficient (Sc), periodicity of disturbance (τ), radius of the well casing (RWC), and radius of the well screened (RWS). A set of dimensionless parameters that relate these hydraulic rock properties and borehole characteristics was derived by Hsieh et al. [34]:

$$\frac{T \cdot \tau}{R_{WC}^2} \tag{13}$$

$$\frac{S_C \cdot R_{WS}^2}{R_{WC}^2}, \tag{14}$$

Graphs of the amplitude ratio and phase shift as a function of Equation (13) for selected values of the parameters in Equation (14) were prepared by Hsieh et al. Figures 2 and 3 [34].

Values of T can be estimated if the phase shift and an order of magnitude estimate of storage coefficient are known [34]. The phase shift of the kth-tidal harmonic component (η_k) is determined by Hsieh et al. [34] as follows:

$$\eta_k = \phi_{dhk} - \phi_{Tk} \tag{15}$$

3. Study Area and Database

3.1. Study Area

The Guadalupe Valley (GV) is located in the Guadalupe River basin (GRB), northwest region of Baja California, Mexico. The runoff in the basin originates in the Sierra Juarez and flows in a NE–SW direction trough the Ojos Negros-Real del Castillo, Guadalupe, and La Mision Valleys up to its discharge in the Pacific Ocean (Figure 1).

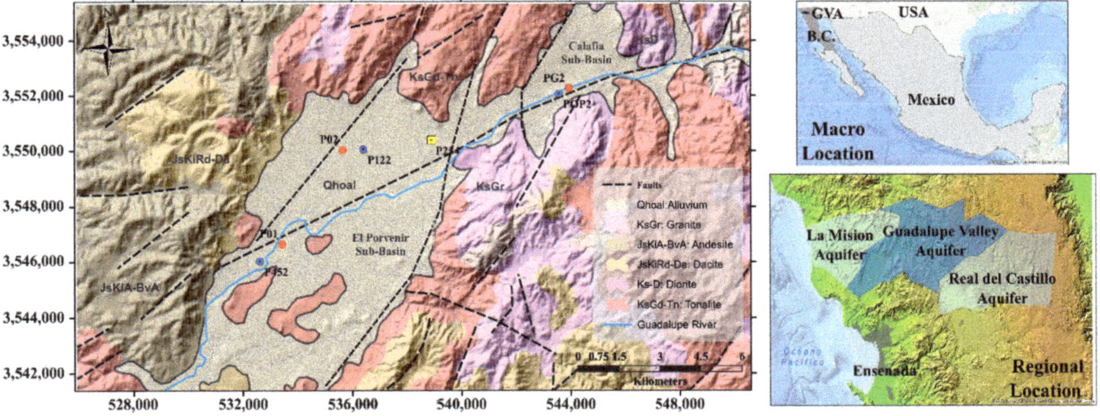

Figure 1. Macro and regional location of the Guadalupe Valley Aquifer. Monitoring wells (P452, P122, and POP2) instrumented are shown. Reference wells (P01, P02, and PG02) included as a part of the study (lithology column availability) are shown. The location of the barometer (P254) is also shown. Geological features are illustrated and described at inset legend [35,36].

The region's climate is characterized by a moderate semi-arid Mediterranean climate, according to the Kopen classification [37]. Mean monthly temperatures vary from 0.6 to 30 °C [38]. Rainfall events are generally intense, and mean annual precipitation may range from 12 to over 750 mm [39]. As a result, streamflow is highly seasonal, with most of the winter precipitation deriving streamflow from December through February and corresponds to the major source of recharge of the Guadalupe aquifer [40].

From a geological perspective, in GV, several tectonic processes originated two sub-basins: Calafia and El Porvenir (Figure 1). These are aligned to a NE–SW direction and eventually were filled with unconsolidated material from erosion, transport, and sedimentation processes. Intrusive and extrusive igneous rocks delimit the valley and constitute the hydraulic basement of the aquifer. Granodiorite, Tonalite, and Granite rocks from the Upper Cretaceous dominate in the north, east, and south regions. Meanwhile, Rio-dacite and Andesite rocks from the Upper Jurassic prevail in the west zone. Quaternary unconsolidated alluvial deposits constitute the Guadalupe Valley Aquifer (GVA) [3].

In a hydrogeological setting, the GVA is considered as a heterogeneous, unconfined aquifer formation made of three principal hydrogeological units of variable thickness: (i) highly permeable unit (alluvium, gravel, sand, and silt); (ii) semi-permeable unit (gravel, sand, and clay); and (iii) low permeability unit (igneous basement). These hydrogeological units are present in both sub-basins and constitute the main groundwater reservoir in the GVA (Figure 2a). The El Porvenir sub-basin (EPSB) prevails in the southern region of the GV and varies in depth from 70 to 100 m. The Calafia sub-basin (CSB) dominates the northeastern zone of the GV, and its depth varies from 300 to 350 m. The aquifer recharge is based on two dominant processes: (i) horizontal recharge, as a result of superficial and subterranean Guadalupe River flows; and (ii) vertical recharge, associated to percolation of precipitation and agricultural-irrigation excess [2,4–6].

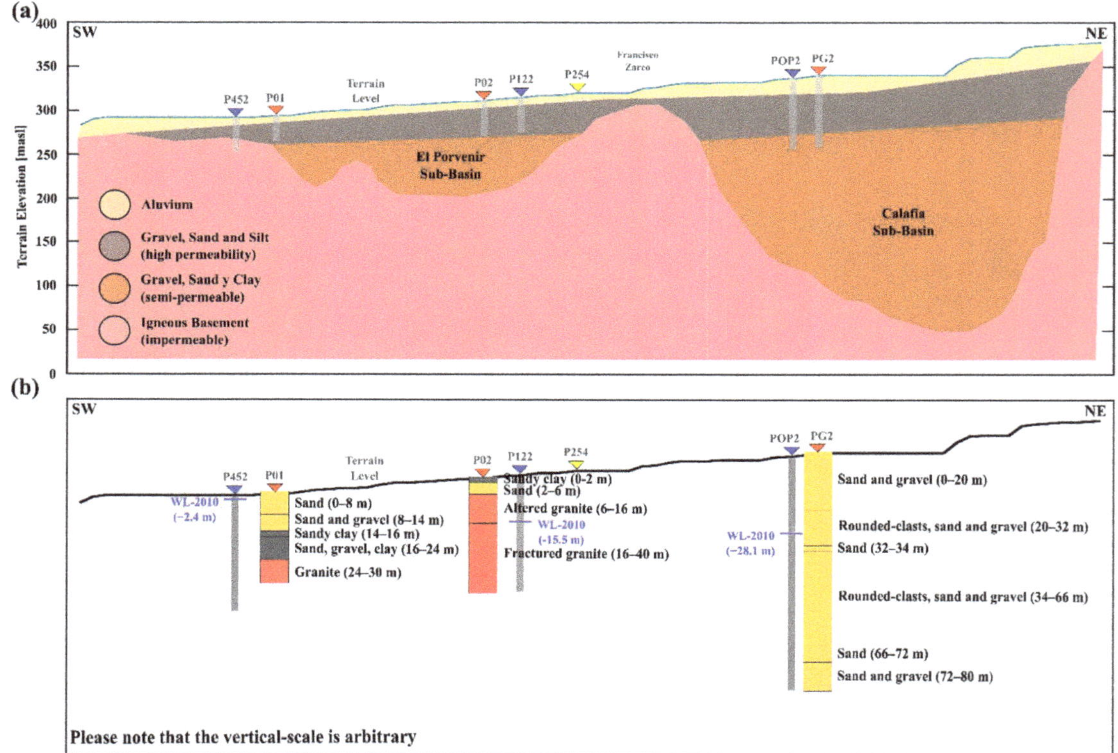

Figure 2. (**a**) Schematic hydrogeological conceptualization of the Guadalupe Valley Aquifer, after Reference [5]. The approximate location of the monitoring wells included as a part of this study is shown. (**b**) Simplified stratigraphic columns of the rock units drilled by the reference wells (P01, P02, and PG2) [2]. The water level measured in 2010 is indicated.

The available information regarding the hydraulic properties of the rock materials that constitute the GVA is limited. Only two studies based on pumping tests have been carried out to estimate hydrogeological properties in the GVA. Andrade-Borbolla [1] determined that, the transmissivity values vary from 0.34×10^{-3} to 52.40×10^{-3} m$^2 \cdot$s^{-1}, prevailing higher values of 1.00×10^{-3} m$^2 \cdot$s^{-1}. CNA [4] estimated transmissivity values ranging from 0.04×10^{-3} to 60.00×10^{-3} m$^2 \cdot$s^{-1}, hydraulic conductivity values between 0.05 to 64.80 m$\cdot$d^{-1}; and mean values of storage coefficient of 0.00005 and specific yield of 0.065.

Some previous studies have aimed at determining the spatial distribution of the groundwater-table elevation, for which, hydrogeological properties values have been proposed to control the adjusting between field measurements and modeled water-table elevations. Campos-Gaytan and Kretzschmar [41] developed a groundwater-flow regional model based on historical water-level measurements for GVA. Moreover, hydraulic conductivity values for the rock materials that filled the sub-basins were estimated based on the misfit of the measurement and modeled water-table elevation. A typical value of hydraulic conductivity for EPSB and CSB was calculated as 5.47 $m \cdot d^{-1}$. The exception to this was the southwestern region of EPSB, where the characteristic value determined was 68.49 $m \cdot d^{-1}$. Hydraulic conductivity and storage coefficient values ranging from 2.00 to 8.00 $m \cdot d^{-1}$, and 0.10 to 0.28, respectively; were used to simulate the groundwater-table response to extraordinary rainfall events within GV by [3].

On the other hand, Del Toro-Guerrero et al. [40] conducted the water-balance estimation in El Mogor sub-basin that derives in GV. As a part of the characterization activities, 48 soil samples from the vadose zone were analyzed to define its grain size. As a result, porosity values ranging from 26 to 38%, and hydraulic conductivity values ranging from 0.50 to 31.85 $m \cdot d^{-1}$ were calculated by using the Vukovic–Soro and Kozeny–Carman empirical equations proposed by [42]. Molina-Navarro et al. [38] and Montecelos-Zamora [43] modeled the Global warming hydrogeological impact on the northern zone of the Guadalupe River, using a SWAT model. As a result of the simulation, typical values of hydraulic conductivity, ranging from 2.14 to 2.71 $m \cdot d^{-1}$, were calculated.

3.2. Data

Vázquez-González et al. [6] established a groundwater monitoring network in the GVA. The monitoring wells were instrumented by using ten pressure-transducers of semi-continuous records (Solinst Levelogger and Solinst Barologger). González-Ramírez and Vázquez-González [3] reported that the monitoring network consisted of up to 17 observation wells, but the pressure-transducers were installed on the monitoring wells during different periods. In 2012, the monitoring network only had eight instrumented wells, five of them located in the CSB, and three in the EPSB. For this study, as a result of the database continuity inspection, only a relatively short period (1 June 2010 to 31 January 2011) of simultaneous record on three monitoring-wells was identified. The instrumented wells were POP2 (SW region of CSB), P122 (N region of EPSB), and P452 (SW region of EPSB). Additionally, during the same period, well P254 (NE region of EPSB) was instrumented to record barometric pressure. The location of the monitoring wells is illustrated in Figure 1. Groundwater-table level and barometric-pressure time-series recorded in each previously mentioned monitoring wells are shown in Figure 3a. Some design characteristics of the monitoring-wells considered in this study are presented in Table 1. The three monitoring wells were drilled into Quaternary alluvial deposits. Unfortunately, there is no information regarding the drilled lithological column. Nevertheless, establishing a correlation with near wells described by Campos-Gaytan [2] is feasible (Figure 2b and Table 2).

Table 1. Summary of the characteristics of the three monitoring wells studied. Nomenclature: well head elevation (W_{HE}, meters above sea level (masl)), borehole depth (B_D), water-table elevation (W_{TE}), radius well-casing (R_{WC}), radius well-screened (R_{WS}), and saturated thickness (B).

| Well ID | Coordinates [1] | | W_{HE} | B_D | W_{TE} | R_{WC}/R_{WS} | B |
	Longitude X (m)	Latitude Y (m)	(masl)	(m)	(msnmm)	(m)	(m)
P452	532,619	3,546,036	301.80	40.00	299.40	0.33/0.10	37.60
P122	536,402	3,550,067	323.07	40.00	307.50	0.30/0.10	24.43
POP2	543,576	3,552,069	345.40	80.00	317.30	0.10/0.10	51.90

[1] Projected coordinates, Universal Transversal Mercator UTM. Datum: World Geodetic Systems, year 1984, WGS-84.

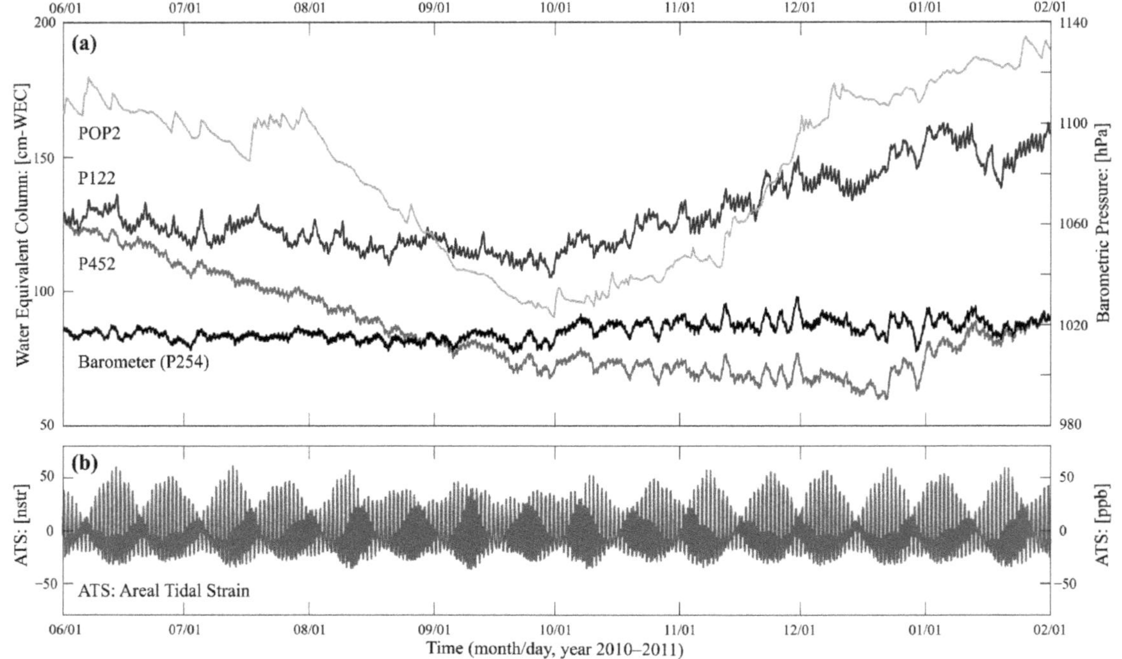

Figure 3. (**a**) Records of water level (P452, P122, and POP2) and barometric pressure (P254) in terms of centimeters of Water Equivalent Column (cm-WEC) and hectopascals. (**b**) Calculated tidal strain (P254) expressed as nanostrain (1 nstr = 1 ppb).

Table 2. Summary of the characteristics of the reference wells included as a part of this study.

Well ID	Name	Location	Depth (m)	Lithology Material (Interval)
P01	Porvenir-1	~1000 m NE-direction from P452	28.57	1. Sand (0 to 8 m) 2. Sand and gravel (8 to 14 m) 3. Sandy clay (14 to 16 m) 4. Sand, gravel, clay (16 to 24 m) 5. Granite (24 to 30 m)
P02	Porvenir-2	~650 m W-direction from P122	41.70	1. Sandy clay (0 to 2 m) 2. Sand (2 to 6 m) 3. Altered granite (6 to 16) 4. Fractured granite (16 to 40 m)
PG2	Guadalupe-2	~400 m NE-direction from POP2	83.90	1. Alternating layers of sand and gravel. The igneous basement was not drilled.

3.3. Data Processing

The theoretical gravitational potential (W_2) and its strain tensor (ε_A) were calculated at each well location, using the SPOTL package ver. 3.3.0.2 [44,45]. The geologic–topographic discontinuities and oceanic tide influence were not considered. The WL, BP, and ε_A time-series were processed and analyzed by using a set of MatLab codes written particularly for this study. The recorded time-series were detrended by using polynomial functions to represent it in a stationary fashion. A third-degree polynomial better reproduces the influence of annual and semi-annual cycles. Using the characteristic polynomial equation, the very low frequency effect was calculated and removed from the measured time-series. From the detrended data, BE was calculated with the method proposed by Rahi [24].

This technique estimates BE only considering BP perturbations and filtering the areal strain effect.

The periodic fluctuations in the time-series were identified utilizing the Discrete Fourier-Transform technique. The WL response to areal strain was analyzed from the discrete amplitude spectra. The high-frequency WL variations (>3.00 cycles per day, cpd) were removed by using a low-pass filter (Chebyshev-I, frequency-cut = 3.00 cpd). Then the low-frequency WL fluctuations (<0.50 cpd) were eliminated applying a high-pass filter (Chebyshev-I, frequency-cut = 0.50 cpd).

Amplitudes (A_{dhk}) and phase angle (Φ_{dhk}) values were determined at the exact frequencies of the tidal harmonic components, using the t-tide code [46], and applying Equations (9) and (10). Similarly, A_{Tk} and Φ_{Tk} were calculated. Areal strain sensitivity was calculated based on A_{dhk} and A_{Tk}, using Equation (11). Moreover, the phase shift was estimated based on Φ_{dhk} and Φ_{Tk}, using Equation (15). Therefore, the transmissivity magnitude order was estimated by utilizing Equation (13) and considering the values of R_{WC} and R_{WS} reported in Table 1.

Based on the A_{SK} estimates, the specific storage was calculated by using Equation (7). For this, gravitational acceleration at a GV representative latitude was calculated as $g = 9.795$ [m·s^{-2}]. Moreover, the Earth's radius of 6,371,000 m and Poisson's ratio equal to 0.25 [47] were assumed. Using estimations of BE and S_S, porosity values were calculated applying Equation (12). For this $\beta_W = 4.40 \times 10^{-10}$ [Pa^{-1}] and $\rho = 998.20$ [kg·m^{-3}] were used.

The saturated thickness (B) was determined relating the well head elevation (W_{HE}), borehole depth (B_D), and water-table elevation (W_{TE}) reported in Table 1. From B, approximation of the storage coefficient was conducted based on the relation ($S_C = S_S \cdot B$). Similarly, the hydraulic conductivity magnitude-order was estimated from the relation ($K = T \cdot B^{-1}$).

4. Results and Discussion

This study synthesized methods for estimating hydraulic aquifer properties from water-level fluctuations measured in a set of monitoring wells at Guadalupe Valley, Mexico. While this analysis was limited to the response of the well-aquifer system to deformation induced by Earth and atmospheric tides, similar methods are available to study water-level fluctuations due to other naturally occurring stresses, such as seismic events (e.g., see Reference [9]). The major simplifying assumption is that solids grains are incompressible. In addition, the primary uncertain source was the use of the tidal strain derived from the theoretical tidal potential. Nonetheless, the methods described in this work showed to be capable of providing reasonable aquifer properties estimates.

An example of the theoretical areal tidal strain calculated at well P254 reported in nanostrain units (1 nstr = 1 ppb) is shown in Figure 3b. The discrete amplitude spectra calculated for WL variations observed in wells P452, P122, and POP2 are shown in Figure 4a. Moreover, the spectra associated with the recorded BP and theoretical areal strain calculated in well P254, are shown in Figure 4b,c, respectively. The dominant harmonic components in the ε_A were five, two of them are diurnal (O1, Lunar; K1, Lunar-Solar) and three are semi-diurnal (N2, Lunar; M2, Solar; and S2, Lunar). Its period and nomenclature also are indicated in inset Figure 4c. The amplitude of the tidal harmonic components calculated in the three monitoring wells was comparable. On the reference well P254, amplitudes estimated were O1 = 5.3, K1 = 7.1, N2 = 1.7, M2 = 8.2, S2 = 4.2 nstr. These last harmonic components are responsible for 95% of tidal potential and play an essential role in hydrogeological studies [10,48].

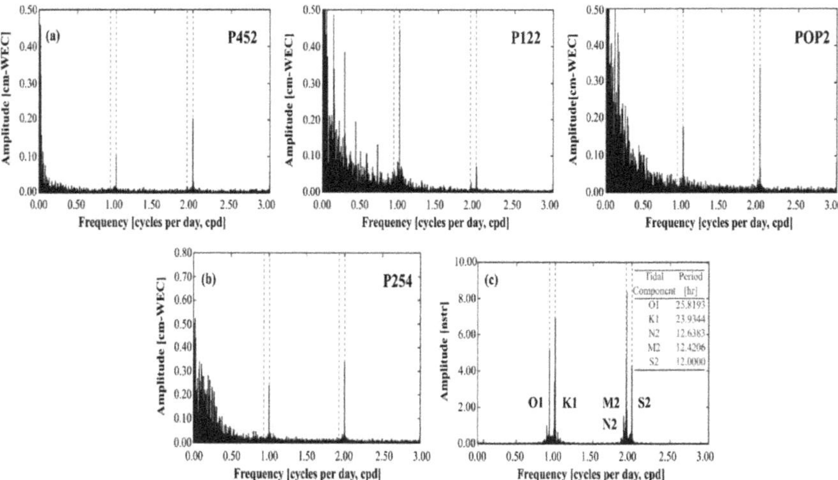

Figure 4. Discrete amplitude spectra for water levels in wells P452, P122, and POP2 (**a**); barometric pressure expressed as centimeters of Water Equivalent Column (cm-WEC) (**b**); and the calculated areal tidal strain (**c**). The dominant Earth tides' frequencies are indicated in the spectra.

BP periodic fluctuations in well P254 were diurnal and semi-diurnal. Its values were K1 = 2.6 and S2 = 3.4 mm-WEC (mm of Water Equivalent Column). These fluctuations are generally associated with the warming and cooling processes of the air column as a result of solar radiation. Moreover, diurnal and semi-diurnal periodic fluctuations in the WL spectra were identified. Semi-diurnal variations were dominant in wells P452 and POP2 (S2 = 2.0 and S2 = 3.3 mm-WEC, respectively), while diurnal fluctuations were dominant in well P122 (K1 = 4.4 mm-WEC).

The tidal harmonic components (K1 and S2) were also identified in the BP spectra. Therefore, the WL response analysis at these specific frequencies is challenging since both phenomena simultaneously influence the well-aquifer system. Moreover, the WL amplitude in the N2 frequency component is typically smaller compared with the other dominant components. As a result, the signal ratio is low and is often discarded since it becomes a relevant error source in the analysis [17]. Based on the above, K1, S2, and N2 harmonic components have been ignored in the WL response analysis. Consequently, only O1 and M2 harmonic components were used to estimate hydrogeological properties of the rock material in the vicinity of the studied monitoring wells.

The highest amplitude in the WL spectra for the O1 and M2 harmonic components was identified in well P122 (0.6 and 0.3 mm-WEC, respectively). While in wells P452 and POP2, amplitudes were lower than 0.1 mm-WEC. Bredehoeft [10] and Weeks [19] mentioned that it is unusual to identify the O1 and M2 harmonic components in the WL variations from wells drilled on unconfined aquifers, which is the typical conceptualization of the GVA. However, Rahi and Halihan [21] indicated that if O1 and M2 signatures are present in the WL spectra, it may be related to a lag of fluctuations K1 and S2 as a result of passing through the vadose zone, suggesting conditions of a semi-confined aquifer.

Estimation of A_{dhk} and Φ_{dhk} at the specific frequencies of the O1 and M2 harmonic components was carried out from the regression coefficients a_{dhk} and b_{dhk}, using Equations (9) and (10). The highest amplitudes were determined in well P122 (O1 = 1.06 mm-WEC and M2 = 0.59 mm-WEC). These last values were approximately two times the observed value on the amplitude spectra. The amplitudes determined from wells P452 and POP2 were minor relative to well P122 and are shown in Table 3. Similarly, values of A_{Tk} and Φ_{Tk} at the exact frequencies of O1 and M2 were calculated. The amplitude value for O1 was 10.67 nstr and 19.88 nstr for M2. These last two values are nearly two times the observed value on the amplitude spectra. The underestimated amplitude from the frequency spectra may be related to digital filtering and Discrete Fourier-Transform inherent problems, for example, the aliasing.

Table 3. Summary of results of the regression analysis, areal strain sensitivity, and barometric efficiency.

Well ID	$A_{dhk} \cdot (10^{-1})$ (mm) O1/M2	Φ_{dhk} (°) O1/M2	A_{Tk} (nstr) O1/M2	Φ_{Tk} (°) O1/M2	η_k (°) O1/M2	$A_{Sk} \cdot (10^{-2})$ (mm·nstr^{-1}) O1/M2	BE (%)
P-452	1.61/4.24	−68/−83	10.67/19.89	−81/−44	13/−39	1.51/2.13	41.46
P-122	10.57/5.98	55/−35	10.67/19.88	89/48	−34/−83	9.90/3.01	48.32
POP-2	1.21/2.66	66/78	10.68/19.87	−78/16	−12/62	1.13/1.34	79.79

Areal strain sensitivity values were calculated from the amplitudes and phase angles determined of the WL variations and the theoretical areal strain (Table 3). Likewise, the phase shift values were also calculated and reported (Table 3). The highest value of A_{SK} was calculated in well P122 for the harmonic component O1 = 9.90×10^{-2} mm·nstr^{-1}; this value was approximately three times the value determined for M2. Added to this, a negative phase shift was determined in well P122 (η_k-O1 = $-34°$, and η_k-M2 = $-83°$). These last results indicate that the WL variations are produced as a result of the areal tidal strain effect. In wells, P452 and POP2 values of areal strain sensitivities ranging between 1.13×10^{-2} to 2.13×10^{-2} mm·nstr^{-1} of A_{SK} were calculated. WL variations as a result of the areal tidal strain effect were determined in wells P452 (M2, harmonic component) and POP2 (O1 harmonic component). In contrast, a positive phase shift was determined for the O1 harmonic component in well P452 and for the M2 component in well POP2. In previous studies, the positive phase shift has been related to the borehole storage effect and water diffusion processes [18,25,26,49].

Barometric efficiency values were calculated and reported for each monitoring well (Table 3). In wells P452 and P122 located on the EPSB, the estimated BE values were similar, 41.46% and 48.32%, respectively. In contrast, for well POP2 located on the CSB, the value calculated was 79.79%, this value is higher in relation to those calculated for the monitoring-wells on EPSB. BE is related to the rock materials that constitute the aquifer and is also an indicator of the confinement conditions. The BE represents the fraction of induced stress held by the rock materials; the remaining fraction is transmitted to the fluid [50]. Typically, a BE value of zero implies that the pressure perturbation is entirely held by the fluid contained in the porous media. While a unit BE value signifies that the pressure perturbation is held by the grains of the rock materials. Based on the above, the rock materials (gravel, sand, clay, and altered/fractured granite) that characterize the EPSB hold up 40–50% of the pressure perturbation related to BP fluctuations. While the sand/gravel alternating layers of the CSB support almost 80% of the stress related to BP fluctuations, this last hydraulic behavior may be explained if the presence of clays (high compressibility) on the EPSB is considered. It is contrasting with the relatively low compressibility of the sand and gravel that constitutes the CSB.

Added to this, in an ideal unconfined aquifer with shallow water table, the BE value should be zero. Instead, when the water table is relatively deep or the rock material generates confinement conditions, the BE value increases. Based on this, the BE values determined could suggest that in the vicinity of the analyzed wells, the aquifer is semi-confined. Moreover, the observed O1 and M2 harmonic components in the WL spectra support that locally the GVA is a semiconfined formation. This result is surprising and contrasts with the typical conceptualization of the GVA [2,6,51]. Nonetheless, the characteristics of rock materials that constitute the aquifer and the water-table relative depth may justify the local semi-confined hydraulic behavior of the GVA. Despite this latter, regionally the GVA is a unique unconfined aquifer formation.

Estimations of S_S in the GVA has not been obtained because previously conducted studies have considered an unconfined aquifer, where the specific yield is much higher than S_S. Nevertheless, the results of this study suggest local semi-confined behavior in the GVA. The determined S_S values ranged from 1.27×10^{-6} to 2.78×10^{-6} m^{-1} (Table 4). The lowest value was calculated for well P122, while the highest value was estimated for well P452. The comparison between the S_S estimations and the expected values as a function of the rock-materials type is shown in Figure 5. In general, the S_S estimations were two orders of magnitude lower than the expected values related to the rock materials that dominate the lithologic column of wells P01, P02, and PG2 reported by Campos-Gaytán [2]. However, these stratigraphic columns also showed the presence of clay lens (P01 and P02), granite (P01), and altered/fractured granite (P02). Based on these last rock materials, the calculated S_S values for wells P452 and P122 are slightly in agreement with the expected S_S values (Figure 5). S_S estimations for well POP2 showed relevant discrepancies concerning the expected S_S values as a function of the rock materials observed in the lithologic column of well PG2 (sand and gravel). Nonetheless, shallow clay layers have been interpreted on recent electromagnetic surveys (TEMs) conducted in the CSB [52]. This last geological feature may explain the calculated S_S values in well POP2 and support the GVA local semi-confined hydraulic behavior deduced.

Porosity values were calculated based on estimations of S_S and BE. The estimated porosity values ranged from 14 to 34% (Table 4). The lowest value was calculated for well P122, located in the EPSB in which a shallow hydraulic basement has been reported. The highest porosity value was estimated for well POP2 situated in the CSB and is consistent with the expected value associated with the rock materials that constitute the PG2 reference stratigraphic column. Furthermore, calculated porosity values are comparable with the porosity values (26–38%) determined in El Mogor, GVA's tributary sub-basin [40]. Additionally, estimated porosity values are consistent with the expected porosity values reported in the classic hydrogeological literature [50,53]. For practical purposes and in the absence of local determinations, a representative porosity value for the rock materials in the CSB is 30%, 20% for EPSB, and 25% for the GVA.

Table 4. Summary of estimations of hydrogeological parameters for the rock materials that constitute the Guadalupe Valley Aquifer.

Well ID	S_S (10^{-6}) (m^{-1}) O1/M2	φ (%) O1/M2	S_C (10^{-5}) O1/M2	T (10^{-6}) (m$^2 \cdot$s^{-1}) O1/M2	K (10^{-2}) (m$\cdot$d^{-1}) O1/M2
P-452	2.78/1.93	26.88/18.60	10.45/7.26	129.46/74.28	29.75/17.07
P-122	1.27/2.53	14.22/28.46	3.10/6.18	32.05/0.66	11.33/0.23
POP-2	1.83/1.52	33.95/28.27	9.49/7.88	12.38/1.99	2.06/0.33

Nomenclature: specific storage, S_S; porosity, φ; storage coefficient, S_C; transmissivity, T; hydraulic conductivity, K.

Storage coefficient values were calculated based on the estimations of S_S and B. The estimated S_C values ranging from 3.10×10^{-5} to 10.45×10^{-5} (Table 4). The lowest S_C value was calculated for well P122, and the highest S_C value was in well P452; both wells are in the EPSB. The S_C value estimated for well POP2 situated in the CSB was lower than the calculated value for well P452. The estimated S_C values were up to four orders of magnitude lower in comparison to those used in the water-table simulations by González-Ramírez and Vázquez-González [3]. Nevertheless, estimated S_C values are similar in the order of magnitude (10^{-5}) with those determined through pumping tests by CNA [4].

Transmissivity values were calculated based on estimations of η_k, the order of magnitude of S_C, and Figure 2 from Hsieh et al. [34]. The estimated T-values were ranging from 6.67×10^{-7} to 1.29×10^{-4} m$^2 \cdot$s^{-1} (Table 4). The lowest T-value was calculated for well P122, and the highest T-value in well P452, both wells are in the EPSB. Estimated T-values are comparable with those (3.40×10^{-4} to 52.40×10^{-3} m$^2 \cdot$s^{-1}) determined by Andrade-Borbolla [1], and to those (4.00×10^{-5} a 60.00×10^{-3} m$^2 \cdot$s^{-1}) calculated by CNA [4]. Finally, hydraulic conductivity values were calculated from estimations of T and B. The estimated K-values ranged between 2.30×10^{-3} to 2.97×10^{-1} m$\cdot$d^{-1} (Table 4). The highest K-value was calculated for well P452 located at the SW of EPSB; a similar hydraulic behavior was described by Campos-Gaytán and Kretzschmar [41]. The lowest K-value was calculated for well POP2 located on the CSB. In general, the estimated K-values were up to two or four orders of magnitude lower in comparison (Figure 6) to those determined from water-table elevation modeling [3,38,41,43]. In contrast, the estimated K-values are comparable with those reported in the classic hydrogeological literature [50,53]. Moreover, the estimated K-values are pretty similar to those determined from the soil grain-size analysis by Del Toro-Guerrero et al. [40] and with those calculated from pumping tests in wells of the GVA by CNA [4].

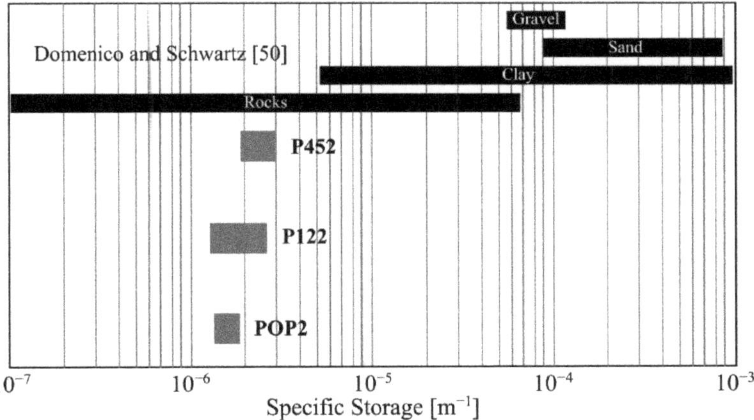

Figure 5. Comparison between the estimated specific storage values (P452, P122, and POP2) and the expected values as a function of the rock-material type.

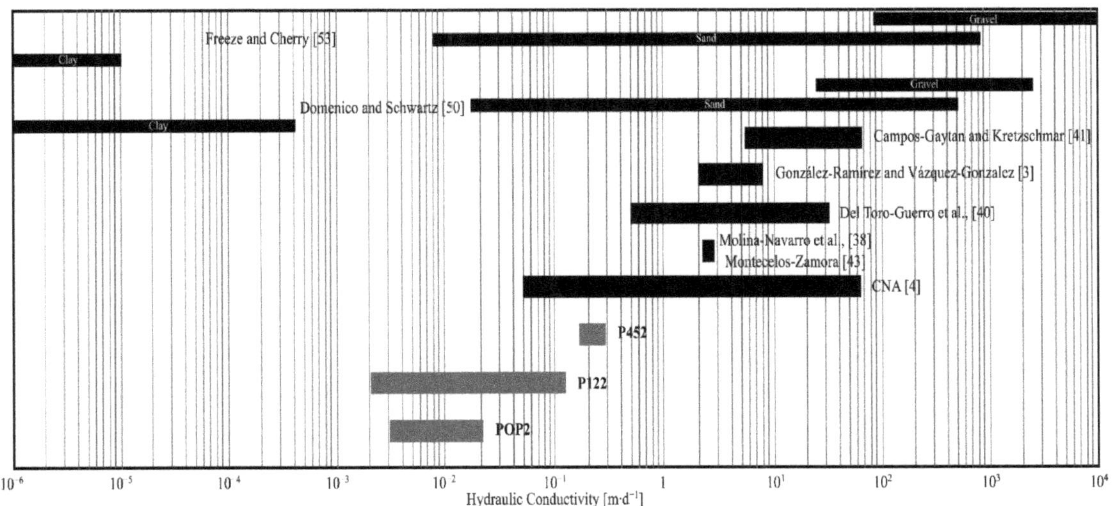

Figure 6. Comparison between the estimated hydraulic-conductivity values (P452, P122, and POP2) and the expected values as a function of the rock-material type, and determined values from previous studies.

5. Conclusions

Based on the analysis of the groundwater response in three monitoring wells to barometric pressure and solid Earth tide, we determined crucial information about the hydrogeological properties of the rock materials that constitute the Guadalupe Valley Aquifer. In particular, representative values of specific storage (1.27×10^{-6} to 2.78×10^{-6} m^{-1}), porosity (14–34%), storage coefficient (3.10×10^{-5} to 10.45×10^{-5}), transmissivity (6.67×10^{-7} to 1.29×10^{-4} m$^2 \cdot$s^{-1}), and hydraulic conductivity (2.30×10^{-3} to 2.97×10^{-1} m$\cdot$d^{-1}) were calculated. These results were consistent with previous determinations. Moreover, based on our literature review, the calculated specific storage values correspond to the first estimations reported in the Guadalupe Valley Aquifer.

About the hydraulic behavior of the rock materials as a result of the induced stress tensor related to perturbation of barometric pressure and areal tidal strain, the results suggested local semi-confined conditions of the aquifer formation. This behavior differed from the typical conceptualization of the Guadalupe Valley Aquifer. Nevertheless, the observed clay-lens in lithologic columns, the interpreted electrical-resistivity models, and the storage coefficient values determined from pumping tests, corroborated the local conditions of semi-confinement identified in this study.

The main sources of uncertainty of the estimations correspond to using the theoretical areal strain and the assumed saturated thickness. Nonetheless, the estimated hydrogeological values showed consistency with those expected for the rock-materials types reported in the classic literature. In addition, a notable similarity was defined between the estimated values and those calculated directly from aquifer stress tests. In the absence of hydrogeological information, the estimated parameters of this study may be considered as a benchmark and used to design and assess management strategies for the groundwater in the Guadalupe Valley Aquifer.

Future research should be focused on integrating water-level records from a broader set of monitoring wells to extend the hydrogeological characterization of the Guadalupe Valley Aquifer. Moreover, the investigation should explore hydrogeologic–poroelastic relationships to determine geomechanical properties associated with the rock materials that constitute the Guadalupe Valley Aquifer.

Author Contributions: Conceptualization, M.A.F.-A., J.R.-H. and R.V.-G.; methodology, M.A.F.-A. and J.R.-H.; software, M.A.F.-A.; validation, M.A.F.-A. and J.R.-H.; formal analysis, M.A.F.-A. and J.R.-H.; investigation, M.A.F.-A., R.V.-G. and A.D.-F.; resources, M.A.F.-A., R.V.-G., D.N. and J.G.-R.; data curation, M.A.F.-A., A.D.-F. and J.G.-R.; writing—original draft preparation, M.A.F.-A., J.R.-H., R.V.-G. and J.G.-R.; writing—review and editing, M.A.F.-A., J.R.-H. and D.N.; visualization, M.A.F.-A.; funding acquisition, R.V.-G. All authors have read and agreed to the summited version of the manuscript.

Funding: This research received no external funding. The APC was partially funded by the Guadalajara University Project 258044 (Programa de Difusión de los Resultados de Investigación del Centro Universitario de la Costa de la Universidad de Guadalajara, Proyecto 258044).

Data Availability Statement: The groundwater-level and barometric data may be available for collaborative research projects by specific agreements. For information, contact marioafar@gmail.com.

Acknowledgments: The authors express their gratitude to the staff of the Comité Técnico de Aguas Subterráneas del Valle de Guadalupe for providing the water-level and barometric-pressure recorded data of the wells analyzed in this study. The authors gratefully acknowledge Geraldine Castillo-Martínez for her valuable English language edition. Thanks are extended to the reviewers and Editor for useful comments that improved the manuscript.

Conflicts of Interest: The authors declare no conflict of interest. The funders had no role in the design of the study; in the collection, analyses, or interpretation of data; in the writing of the manuscript; or in the decision to publish the results.

References

1. Andrade-Borbolla, M. *Actualización hidrogeológica del Valle de Guadalupe, Municipio de Ensenada, Baja California*; Grupo Agroindustrial Valle de Guadalupe: Ensenada, Mexico, 1997; p. 66. (In Spanish)
2. Campos-Gaytán, J.R. Simulación Del Flujo de Agua Subterránea en El Acuífero Del Valle de Guadalupe. Ph.D. Thesis, Centro de Investigación Científica y de Educación Superior de Ensenada, Ensenada, Mexico, 2008; 240p. (In Spanish). Available online: https://biblioteca.cicese.mx/catalogo/tesis/ficha.php?id=17841 (accessed on 14 May 2021).
3. Gonzalez-Ramirez, J.; González, R.V. Modeling of the water table level response due to extraordinary precipitation events: The case of the guadalupe valley aquifer. *Int. J. Geosci.* **2013**, *4*, 950–958. [CrossRef]
4. CNA; Comisión Nacional del Agua. *Actualización de la Disponibilidad Media Anual de Agua en El Acuífero Guadalupe (0207)*; Comisión Nacional del Agua: Mexico, Mexico, 2018; 28p, (In Spanish). Available online: https://sigagis.conagua.gob.mx/gas1/Edos_Acuiferos_18/BajaCalifornia/DR_0207.pdf (accessed on 14 May 2021).
5. Ramírez-Hernández, J.; Carreón, C.D.; Campbell, R.H.; Palacios, B.R.; Leyva, C.O.; Ruíz, M.L.; Vázquez, G.R.; Rousseau, F.P.; Campos, G.R.; Mendoza, E.L.; et al. *Informe Final. Plan de Manejo Integrado de las Aguas Subterráneas en el Acuífero de Guadalupe, Estado de Baja California. Tomo I. Reporte Interno. Elaborado por la*; Convenio: SGT-OCPBC-BC-07-GAS-001; Universidad Autónoma de Baja California para la Comisión Nacional del Agua, Organismo de Cuenca Península de Baja California, Dirección Técnica: Ensenada, Mexico, 2007. (In Spanish)
6. Vázquez-González, R.; Romo-Jones, J.M.; Kretzschmar, T. *Estudio Técnico Para El Manejo Integral Del Agua en El Valle de Guadalupe*; División de Ciencias de la Tierra, Centro de Investigación Científica y de Educación Superior de Ensenada: Baja California, Mexico, 2007; p. 53. (In Spanish)
7. Fuentes-Arreazola, M.A. Estimación de Parámetros Geohidrológicos, Poroelásticos Y Geomecánicos Con Base en El análisis de Las Variaciones Del Nivel Del Agua Subterránea en Pozos de Monitoreo en El Valle de Mexicali. Ph.D. Thesis, Centro de Investigación Científica y de Educación Superior de Ensenada, Baja California, Mexico, 2018; p. 127. (In Spanish). Available online: https://biblioteca.cicese.mx/catalogo/tesis/ficha.php?id=25077 (accessed on 14 May 2021).
8. Jacob, C.E. Flow of groundwater. In *Engineering Hydraulic*; Rouse, H., Ed.; John Wiley & Son, Inc.: New York, NY, USA, 1950; pp. 321–386.
9. Cooper, H.H.; Bredehoeft, J.D.; Papadopulos, I.S.; Bennett, R.R. The response of well-aquifer systems to seismic waves. *J. Geophys. Res. Space Phys.* **1965**, *70*, 3915–3926. [CrossRef]
10. Bredehoeft, J.D. Response of well-aquifer systems to Earth tides. *J. Geophys. Res. Space Phys.* **1967**, *72*, 3075–3087. [CrossRef]
11. Van Der Kamp, G.; Gale, J.E. Theory of earth tide and barometric effects in porous formations with compressible grains. *Water Resour. Res.* **1983**, *19*, 538–544. [CrossRef]
12. Roeloffs, E.A. Hydrologic precursors to earthquakes: A review. *Pure Appl. Geophys.* **1988**, *126*, 177–209. [CrossRef]
13. Rojstaczer, S.; Agnew, D. The influence of formation material properties on the response of water levels in wells to Earth tides and atmospheric loading. *J. Geophys. Res. Space Phys.* **1989**, *94*, 12403–12411. [CrossRef]
14. Merritt, M.L. *Estimating Hydraulic Properties of the Floridan Aquifer System by Analysis of Earth-Tide, Ocean-Tide, And Barometric Effects, Collier and Hendry Counties, Florida*; US Geological Survey: Reston, VA, USA, 2004; p. 80.
15. Fuentes-Arreazola, M.A.; Vázquez-González, R. Estimation of some geohydrological properties in a set of monitoring wells in Mexicali Valley, B.C., México. *Revista Ingeniería del Agua* **2016**, *20*, 87. [CrossRef]
16. Fuentes-Arreazola, M.A.; Ramírez-Hernández, J.; Vázquez-González, R. Hydrogeological properties estimation from groundwater level natural fluctuations analysis as a low-cost tool for the Mexicali Valley aquifer. *Water* **2018**, *10*, 586. [CrossRef]
17. Cutillo, P.A.; Bredehoeft, J.D. Estimating aquifer properties from the water level response to earth tides. *Ground Water* **2010**, *49*, 600–610. [CrossRef] [PubMed]
18. Rasmussen, T.; Crawford, L.A. Identifying and removing barometric pressure effects in confined and unconfined aquifers. *Ground Water* **1997**, *35*, 502–511. [CrossRef]
19. Weeks, P.E. Barometric fluctuations in wells tapping deep unconfined aquifers. *Water Resour. Res.* **1979**, *15*, 1167–1176. [CrossRef]
20. Galloway, D.; Rojstaczer, S. Analysis of the frequency response of water levels in wells to earth tides and atmospheric loading. In *Proceedings of the Fourth Canadian/American Conference in Hydrogeology*; Canada National Ground Water Association: Banff, AB, Canada, 1988; pp. 100–113.
21. Rahi, K.A.; Halihan, T. Identifying aquifer type in fractured rock aquifers using harmonic analysis. *Ground Water* **2013**, *51*, 76–82. [CrossRef]
22. Clark, W.E. Computing the barometric efficiency of a well. *J. Hydraul. Div.* **1967**, *93*, 93–98. [CrossRef]
23. Toll, N.J.; Rasmussen, T. Removal of barometric pressure effects and earth tides from observed water levels. *Ground Water* **2007**, *45*, 101–105. [CrossRef] [PubMed]
24. Rahi, K.A. Estimating the Hydraulic Parameters of the Arbuckle-Simpson Aquifer by Analysis of Naturally-Induced Stresses. Ph.D. Dissertation, Oklahoma State University, Oklahoma, OK, USA, 2010; p. 168. Available online: https://core.ac.uk/download/pdf/215232243.pdf (accessed on 14 May 2021).
25. Rojstaczer, S. Intermediate period response of water levels in wells to crustal strain: Sensitivity and noise level. *J. Geophys. Res. Space Phys.* **1988**, *93*, 13619–13634. [CrossRef]
26. Rojstaczer, S. Determination of fluid flow properties from the response of water levels in wells to atmospheric loading. *Water Resour. Res.* **1988**, *24*, 1927–1938. [CrossRef]

27. Lai, G.; Ge, H.; Wang, W. Transfer functions of the well-aquifer systems response to atmospheric loading and Earth tide from low to high-frequency band. *J. Geophys. Res. Solid Earth* **2013**, *118*, 1904–1924. [CrossRef]

28. Harrison, J.C. *New Computer Programs for the Calculation of Earth Tides*; Cooperative Institute for Research in Environmental Sciences: Colorado, CO, USA, 1971; p. 30.

29. Berger, J.; Beaumont, C. An analysis of tidal strain observation from the United States of America II: The inhomoge-neous tide. *Bull. Seismol. Soc. Am.* **1976**, *66*, 1821–1846.

30. Harrison, J.C. Cavity and topographic effects in tilt and strain measurement. *J. Geophys. Res. Space Phys.* **1976**, *81*, 319–328. [CrossRef]

31. Agnew, D.C. Earth tides. In *Treatise on Geophysics and Geodesy*; Herring, T.A., Ed.; Elsevier: New York, NY, USA, 2007; pp. 163–195.

32. Doodson, A.T.; Warburg, H.D. *Admiralty Manual of Tides*; Her Majesty´s Stationary Office: London, UK, 1941; p. 270.

33. Jacob, C.E. On the flow of water in an elastic artesian aquifer. *Trans. Am. Geophys. Union* **1940**, *21*, 574–586. [CrossRef]

34. Hsieh, P.A.; Bredehoeft, J.D.; Farr, J.M. Determination of aquifer transmissivity from Earth tide analysis. *Water Resour. Res.* **1987**, *23*, 1824–1832. [CrossRef]

35. INEGI. Datos de geología del Instituto Nacional de Estadística y Geografía. 2018. Available online: http://gaia.inegi.org.mx/mdm6/ (accessed on 14 May 2021).

36. SGM. Sistema GEOINFOMEX del Servicio Geológico Mexicano. 2018. Available online: https://www.sgm.gob.mx/GeoInfoMexGobMx (accessed on 14 May 2021).

37. García, E. *Modificaciones Al Sistema de Clasificación Climática de Köpen Para Adaptarlo a Las Condiciones de la República Mexicana*; Instituto de Geografía, Universidad Nacional Autónoma de México: Ciudad Universitaria, Mexico, 1981; p. 97, (In Spanish). Available online: http://www.publicaciones.igg.unam.mx/index.php/ig/catalog/view/83/82/251-1 (accessed on 14 May 2021).

38. Molina-Navarro, E.; Hallack-Alegría, M.; Martínez-Pérez, S.; Ramírez-Hernández, J.; Moctezuma, A.M.; Sastre-Merlín, A. Hydrological modeling and climate change impacts in an agricultural semiarid region. Case study: Guadalupe River basin, Mexico. *Agric. Water Manag.* **2016**, *175*, 29–42. [CrossRef]

39. Hallack-Alegría, M.; Ramírez-Hernández, J.; Watkins, D.W. ENSO-conditioned rainfall drought frequency analysis in northwest Baja California, Mexico. *Int. J. Clim.* **2011**, *32*, 831–842. [CrossRef]

40. Del Toro-Guerrero, F.J.; Kretzschmar, T.; Hinojosa-Corona, A. Estimación del balance hídrico en una Cuenca semiárida, El Mogor, Baja California, México. *Technol. Cienc. Agua* **2014**, *5*, 69–81.

41. Campos-Gaytán, J.R.; Kretzschmar, T. Numerical understanding of regional scale water table behavior in the Guada-lupe Valley aquifer, Baja California, Mexico. *Hydrol. Earth Syst. Sci. Discuss.* **2006**, *3*, 707–730. [CrossRef]

42. Odong, J. Evaluation of empirical formulae for determination of hydraulic conductivity based on grain-size analysis. *J. Am. Sci.* **2007**, *3*, 54–60.

43. Montecelos-Zamora, Y. Modelación Del Efecto de la Variación Climática en El Balance Hídrico en Dos Cuencas (México Y Cuba) Bajo Un Escenario de Cambio Climático. Ph.D. Thesis, Centro de Investigación Científica y de Educación Superior de Ensenada: Baja California, Mexico, 2018; p. 100. (In Spanish). Available online: https://biblioteca.cicese.mx/catalogo/tesis/ficha.php?id=25188 (accessed on 14 May 2021).

44. Berger, J.; Farrell, W.; Harrison, J.C.; Levine, J.; Agnew, D.C. *ERTID 1: A Program for Calculation of Solid Earth Tides*; Technical Report; Scripps Institution of Oceanography: La Jolla, CA, USA, 1987; p. 20.

45. Agnew, D.C. *SPOTL: Some Programs for Ocean-Tides Loading*; Technical Report; Scripps Institution of Oceanography: La Jolla, CA, USA, 2012; p. 30. Available online: https://igppweb.ucsd.edu/~{}agnew/Spotl/spotlmain.html (accessed on 14 May 2021).

46. Pawlowicz, R.; Beardsley, B.; Lentz, S. Classical tidal harmonic analysis including error estimates in MATLAB using T_TIDE. *Comput. Geosci.* **2002**, *28*, 929–937. [CrossRef]

47. Gercek, H. Poisson's ratio values for rocks. *Int. J. Rock Mech. Min. Sci.* **2007**, *44*, 1–13. [CrossRef]

48. Melchior, P. Earth tides. In *Research in Geophysics*; Odishaw, H., Ed.; Massachusetts Institute of Technology Press: Cambridge, MA, USA, 1964; pp. 183–193.

49. Rojstaczer, S.; Riley, F.S. Response of the water level in a well to Earth tides and atmospheric loading under unconfined conditions. *Water Resour. Res.* **1990**, *26*, 1803–1817. [CrossRef]

50. Domenico, P.A.; Schwartz, F.W. Groundwater movement, hydraulic conductivity and permeability of geological material. In *Physical and Chemical Hydrogeology*, 2nd ed.; John Wiley & Sons, Inc.: New York, NY, USA, 1997; pp. 33–54.

51. Hernández-Rosas, M.; Mejía-Vázquez, R. *Relación de las Aguas Superficiales y Subterráneas del Acuífero BC-07*; Gerencia Regional de la Península de Baja California, Subgerencia Regional Técnica, Technical Report; Comisión Nacional del Agua: Valle de Guadalupe, Mexico, 2003; p. 13. (In Spanish)

52. Monge-Cerda, F.E. Detección Del Nivel Freático Con Métodos Geoeléctricos: El Caso Del Valle de Guadalupe. Master's Thesis, Centro de Investigación Científica y de Educación Superior de Ensenada, B.C. Baja California, Mexico, 2020; p. 97. (In Spanish). Available online: https://biblioteca.cicese.mx/catalogo/tesis/ficha.php?id=25639 (accessed on 14 May 2021).

53. Freeze, R.A.; Cherry, J.A. Physical properties and principles. In *Groundwater*; Prentice-Hall, Inc.: Hoboken, NJ, USA, 1979; pp. 14–79.

Article

Spatial Prediction of Groundwater Potentiality in Large Semi-Arid and Karstic Mountainous Region Using Machine Learning Models

Mustapha Namous [1], Mohammed Hssaisoune [2,3], Biswajeet Pradhan [4,5,*], Chang-Wook Lee [6,*], Abdullah Alamri [7], Abdenbi Elaloui [8], Mohamed Edahbi [9], Samira Krimissa [1], Hasna Eloudi [2], Mustapha Ouayah [1], Hicham Elhimer [10] and Tarik Tagma [11]

1   Laboratory of Biotechnology and Sustainable Development of Natural Resources, Polydisciplinary Faculty, Sultan Moulay Slimane University, Mghila B.P. 592, Beni Mellal 23000, Morocco; mustapha.namous@usms.ma (M.N.); samira_krimissa@yahoo.fr (S.K.); mustaphaouayah@gmail.com (M.O.)

2   Applied Geology and Geoenvironment Laboratory, Faculty of Sciences, Ibn Zohr University, Agadir 80000, Morocco; m.hssaisoune@uiz.ac.ma (M.H.); hasnaeloudi@gmail.com (H.E.)

3   Faculty of Applied Sciences, Ibn Zohr University, B. O. 6146, Ait Melloul 86153, Morocco

4   Centre for Advanced Modelling and Geospatial Information Systems (CAMGIS), Faculty of Engineering and Information Technology, University of Technology Sydney, Sydney, NSW 2007, Australia

5   Earth Observation Center, Institute of Climate Change, Universiti Kebangsaan Malaysia (UKM), Bangi 43600, Selangor, Malaysia

6   Division of Science Education, Kangwon National University, Chuncheon-si 24341, Gangwon-do, Korea

7   Department of Geology & Geophysics, College of Science, King Saud University, P.O. Box 2455, Riyadh 11451, Saudi Arabia; amsamri@ksu.edu.sa

8   Water and Remote Sensing Team (GEVARET), Faculty of Sciences and Techniques, Sultan Moulay Slimane University, Beni Mellal 23000, Morocco; a.elaloui@usms.ma

9   Higher School of Technology of Fkih Ben Salah, Sultan Moulay Slimane University, Beni Mellal 23000, Morocco; edahbimohamed@gmail.com

10   Laboratory of Geostructures, Geomaterials and Water Resources, Faculty of Sciences Semlalia, Cadi Ayyad University, Marrakesh 44000, Morocco; elhimer_h@yahoo.fr

11   Laboratoire Multidisciplinaire de Recherche et d'Innovation (LAMRI), Equipe Ingénierie des Ressources Naturelles et Impacts Environnementaux (IRNIE), Polydisciplinary Faculty of Khouribga, Sultan Moulay Slimane University, Khouribga 25000, Morocco; tariktagma@usms.ma

*   Correspondence: Biswajeet.Pradhan@uts.edu.au (B.P.); cwlee@kangwon.ac.kr (C.-W.L.)

Citation: Namous, M.; Hssaisoune, M.; Pradhan, B.; Lee, C.-W.; Alamri, A.; Elaloui, A.; Edahbi, M.; Krimissa, S.; Eloudi, H.; Ouayah, M.; et al. Spatial Prediction of Groundwater Potentiality in Large Semi-Arid and Karstic Mountainous Region Using Machine Learning Models. *Water* **2021**, *13*, 2273. https://doi.org/10.3390/w13162273

Academic Editors: Evangelos Tziritis and Andreas Panagopoulos

Received: 13 June 2021
Accepted: 16 August 2021
Published: 19 August 2021

Publisher's Note: MDPI stays neutral with regard to jurisdictional claims in published maps and institutional affiliations.

Abstract: The drinking and irrigation water scarcity is a major global issue, particularly in arid and semi-arid zones. In rural areas, groundwater could be used as an alternative and additional water supply source in order to reduce human suffering in terms of water scarcity. In this context, the purpose of the present study is to facilitate groundwater potentiality mapping via spatial-modelling techniques, individual and ensemble machine-learning models. Random forest (RF), logistic regression (LR), decision tree (DT) and artificial neural networks (ANNs) are the main algorithms used in this study. The preparation of groundwater potentiality maps was assembled into 11 ensembles of models. Overall, about 374 groundwater springs was identified and inventoried in the mountain area. The spring inventory data was randomly divided into training (75%) and testing (25%) datasets. Twenty-four groundwater influencing factors (GIFs) were selected based on a multicollinearity test and the information gain calculation. The results of the groundwater potentiality mapping were validated using statistical measures and the receiver operating characteristic curve (ROC) method. Finally, a ranking of the 15 models was achieved with the prioritization rank method using the compound factor (CF) method. The ensembles of models are the most stable and suitable for groundwater potentiality mapping in mountainous aquifers compared to individual models based on success and prediction rate. The most efficient model using the area under the curve validation method is the RF-LR-DT-ANN ensemble of models. Moreover, the results of the prioritization rank indicate that the best models are the RF-DT and RF-LR-DT ensembles of models.

Keywords: drinking and irrigation water scarcity; groundwater potential mapping; machine learning; remote sensing; GIS; karstic mountainous aquifers; Morocco

1. Introduction

Mountainous areas cover more than 20% of the Earth's land surface where 25% of the global population lives [1]. The mountains areas are well known to provide 50% of freshwater [2] compared to the other critical resources (i.e., food and wood). These areas constitute the main recharge areas of several porous and continuous aquifers in downstream lowland regions [3–5]. Nevertheless, understanding the details of groundwater functioning in a mountain massif requires a comprehensive knowledge of the most semi-arid areas [6].

Furthermore, mountainous areas assume the deep dynamics of groundwater, denying us direct access to groundwater outcrops that aid in hydrogeological exploration. An additional complication arises because mountainous regions are frequently fractured (e.g., the Atlas Mountains) and contain a discontinuous aquifer (e.g., a karstic aquifer). The groundwater potential in mountainous aquifers is governed by several parameters (i.e., lithology, geomorphology, topography, secondary porosity, geological structures, fracture density, permeability, drainage pattern and density, groundwater recharge, piezometric level, slope, land use/cover and climatic conditions, and their interrelationships) [7].

Overall, there is insufficient geodatabase related to groundwater, notably in fractured and karstic bedrock aquifers [8]. The knowledge gaps in term of geodatabase make the development of numerical models difficult and consequently understand the aquifer functioning in mountainous areas [9]. In order to resolve this issue, various statistical models and machine learning algorithms have been employed for groundwater potential modelling and mapping using inventories of springs for dependent variables (i.e., binary logistic regression (LR) [10], certainty factor (CF) [11], weights-of-evidence (WE) [12], artificial neural networks (ANNs), random forest (RF), support vector machines (SVMs), naïve Bayes (NB) and decision tree (DT) [13]). Generally, machine learning methods have shown more robustness and stability during modelling, and thus, they have been popular and cost-effective in predicting groundwater potentiality. Furthermore, the application of machine learning algorithms remains more known in the prediction of natural disasters such as landslides, floods [14] and gully erosion. Recently, several researchers have tested machine learning ensemble models to improve the performance of prediction. Furthermore, other machine learning ensemble methods have been tested to delineate groundwater potential zones based on spring or well location inventories [15]. The purpose of this research is to apply four models individually (random forest (RF), logistic regression (LR), decision tree (DT) and artificial neural network (ANN)) and test different possible combinations using two, three and four models in each ensemble to produce groundwater potential maps in a large mountainous area.

In the semi-arid Oum Er-Rbia catchment located in the central part of Morocco, water resources are threatened by climate change. As a result, the economy and population requirements will be increased in the future [16–18]. Groundwater resources are derived from two system types: (1) the multi-layered system of the Tadla plains; and (2) the karst aquifer of the Atlas Mountains. The first system is overexploited and polluted by anthropogenic activities [5,19,20]. The second, located in the High Atlas Mountains of Beni Mellal, is characterized by the emergence of several springs (more than 370). The high flow rate of these springs indicates the presence of important groundwater reserves in this mountainous area. Several studies [21–25] using chemical and isotopic tools have been conducted to determine the water quality, hydrodynamic functioning and recharge area. In addition, the groundwater potential mapping of this aquifer has not yet worked out. In fact, mapping potential groundwater areas allow to determine zoning areas, permitting the identification of new sites that can provide a water supply for both drinking and irrigation. The results of this paper can be used by water managers and stakeholders to find potential water supplies and manage the water resources which are more vulnerable to climate change and anthropogenic impacts.

The main objectives of this study are the following: (1) Evaluate the performance of individual and ensemble models in the prediction of groundwater potentiality in karstic mountainous areas; (2) compare the performance, robustness and stability of these models using several statistical and validation techniques; (3) test the importance of using a maximum of groundwater influencing factors and (4) produce reliable maps of the spatial distribution of groundwater potential in the study area.

2. Study Area

2.1. Geographic and Climatic Context

The study area is located in the Oum Er Rbia catchment, specifically in the High Atlas Mountains of Beni Mellal. It is bounded by the Tadla plain in the North and in the West, by the El Aabid River in the South and by the Oum Er-Rbia River in the East (Figure 1).

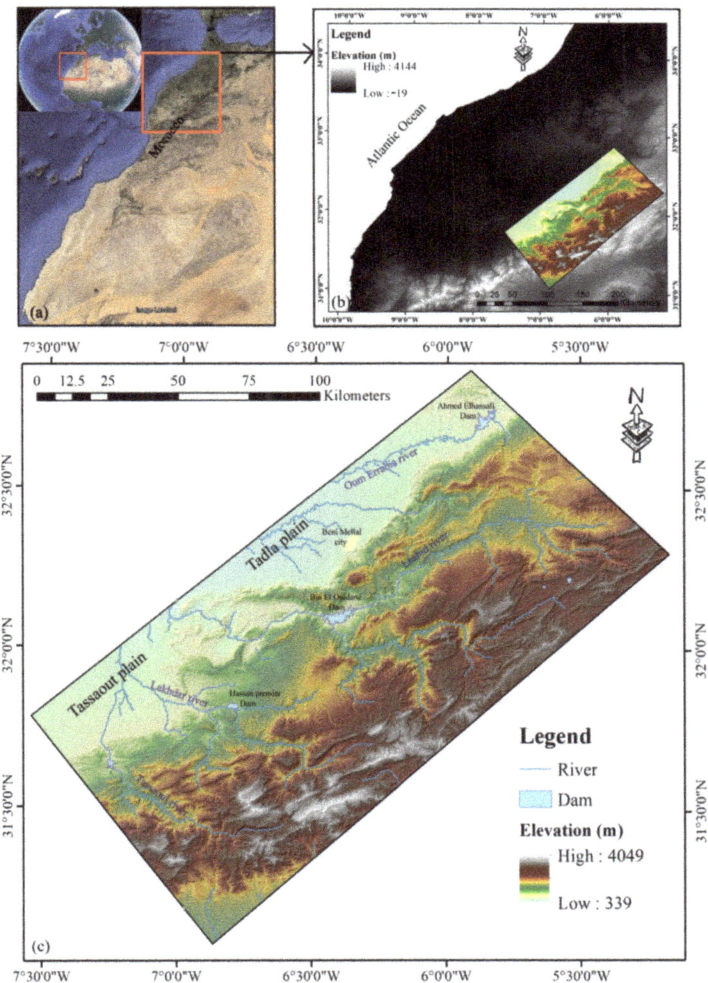

Figure 1. Geographical situation of study area at (**a**) national scale, (**b**) regional scale and (**c**) digital elevation model showing altitudes variability of the study area.

The climate ranges from semi-arid at the borders to sub-humid at high elevations, with a dominance in both cases of two distinct wet and dry seasons [18]. This Mediterranean climate is characterized by an annual rainfall varied between 300 to 750 mm and poorly distributed throughout the year. The average annual temperature is 18 °C (with peak periods of over 40 °C in August and 0 to 4 °C in January). The snow appears from 900 m of altitude and the prevailing wind is the Chergui in the summer period.

2.2. Geological and Hydrogeological Setting

The Atlas of Beni Mellal is composed of a Liassic and Middle Jurassic massive dolomitic and/or calcareous facies [5]. The other formations are composed of (1) Cenomanian conglomerates and sandstones interbedded with clays, gypsum marls and limestones, (2) limestones and karstic dolomites interlayered with marl horizons of the Turonian, (3) marls and limestones with evaporitic characters of the Senonian and (4) a complex of limestones, marls and phosphate sandstones forming the phosphate series (Figure 2) [26]. The geological structure of the basin implies a continuation of the north Atlas thrust fault toward the Tadla plain [5,27] (Figure 2).

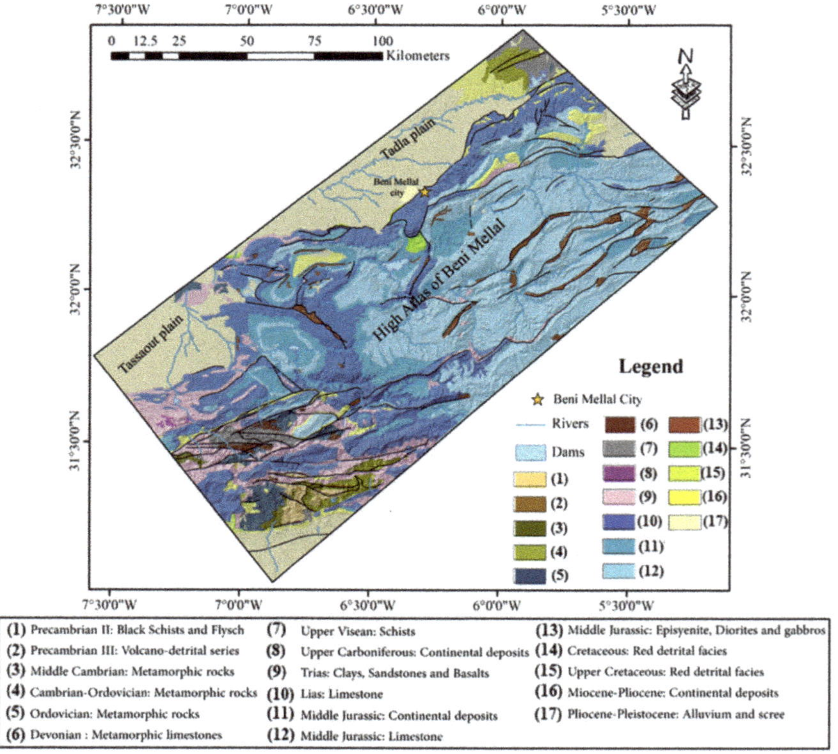

(1) Precambrian II: Black Schists and Flysch	**(7)** Upper Visean: Schists	**(13)** Middle Jurassic: Episyenite, Diorites and gabbros
(2) Precambrian III: Volcano-detrital series	**(8)** Upper Carboniferous: Continental deposits	**(14)** Cretaceous: Red detrital facies
(3) Middle Cambrian: Metamorphic rocks	**(9)** Trias: Clays, Sandstones and Basalts	**(15)** Upper Cretaceous: Red detrital facies
(4) Cambrian-Ordovician: Metamorphic rocks	**(10)** Lias: Limestone	**(16)** Miocene-Pliocene: Continental deposits
(5) Ordovician: Metamorphic rocks	**(11)** Middle Jurassic: Continental deposits	**(17)** Pliocene-Pleistocene: Alluvium and scree
(6) Devonian : Metamorphic limestones	**(12)** Middle Jurassic: Limestone	

Figure 2. Geological setting of the study area.

In the study area, the groundwater resources are derived from two system types [5,22]: (1) the shallow and deep aquifers of the Tadla plain and (2) the karst aquifer of the Atlas Mountains. The first system is composed of four aquifers: (1) Mio-Plio-Quaternary, (2) Eocene, (3) Senonian and (4) Turonianwhich is the main productive aquifer in the region and separated by impermeable or semi-permeable horizons. The groundwater of the second system (the subject of this study) is contained in the aquifer calcareous rocks—of Liasic age, which are mainly karstic system which possess high storage and conductive capacities.

These formations are exposed on the mountains and favor water percolation (rainwater and snowmelt), which constitutes the natural replenishment of the aquifer [22,28].

From the areas of recharge in the mountains, the groundwater flows in the aquifer to the points of discharge outlets: natural springs, underground seepage and pumping wells in the plain area [5]. The major karst outlets are located along the northern High Atlas accident of the High Atlas Mountains from Timoulilt in the southwest to Zaouit Cheikh in the northeast; the most important is Ain Asserdoune, with an average flow rate of 700 L/s. It is important to note that, in this environment, most karst eminences are probably underground and are therefore not visible, particularly at the mountain's transition of the Tadla plain.

3. Materials and Methods

A spring can be defined as a window through which groundwater flows from an aquifer to the Earth's surface [29]. Based on this characteristic, the emergence of springs reflects the groundwater potentiality. To assess the relationship between source occurrence and factors controlling groundwater flow, the groundwater potential mapping (GPM) a tool has been used to provide spatial information [29].

The methodology of this study is summarized in the flowchart (Figure 3). The main steps are as follows: (1) preparation of the data for modelling (preparation of the spring inventory map and preparation of the conditioning factors datasets). Two methods were applied for the factors selection which contribute to springs emergence (*IG*) and variance inflation factor (*VIF*). (2) A frequency ratio method was applied to determine the spatial relationships between spring occurrence and its predisposing factors. (3) RF, LR, DT and ANN models were applied for mapping groundwater potential; then, different ensemble of models were tested in order to find the best rate of prediction, in addition to the production of various groundwater potentiality maps. (4) Several statistical parameters were applied to test the results of the models application, and a general comparison was carried out based on a compound factor (*CF*) method and priority rank (*PR*). Geographic Information System environments and statistics software were used during the current study for database preparation and groundwater potential mapping, and R packages for machine learning algorithm modelling were also used (randomForest, C50, neuralnet and calibrateBinary). Table 1 highlights the spatial datasets used in this study.

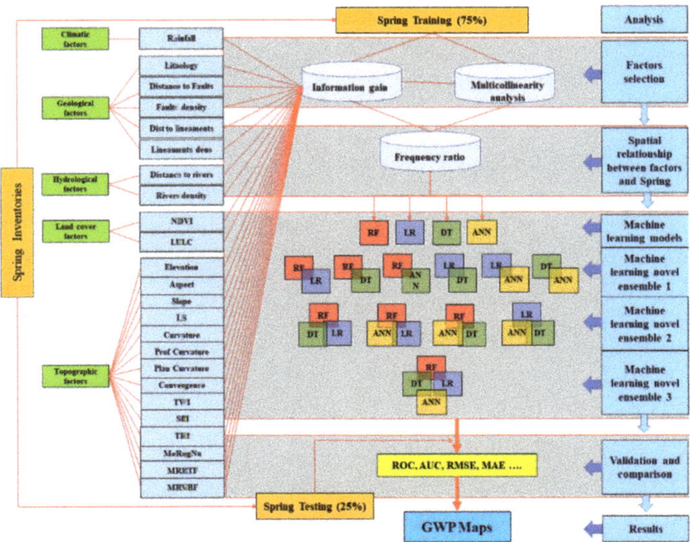

Figure 3. Methodology used in this research.

Table 1. Spatial database of the study area.

Factors	Data Layers	Data Provider
Spring inventory		Previous studies Field investigation Topographic maps
Topographic factors	Elevation Aspect Slope Curvature Profile curvature Plan curvature Convergence TWI SPI TRI MeRugNu MRRTF MRVBF LS	SRTM-DEM from (http://gdex.cr.usgs.gov/gdex/ (accessed on 10 January 2020)) pixel size of 30 m × 30 m.
Geologic factors	Lithology Distance to Faults Faults Density Distance to lineaments Lineaments density	Geological map of Morocco at the scale 1:500,000 Geological map of Morocco at the scale 1:500,000 LANDSAT satellite image at 30 m from (https://earthexplorer.usgs.gov/ (accessed on 10 January 2020))
Hydrologic factors	Distance to rivers Rivers density	DEM at 30 m
Land cover factors	NDVI LULC	LANDSAT satellite image at 30 m from (https://earthexplorer.usgs.gov/ (accessed on 10 January 2020))
Climatic factors	Rainfall	Climatic stations data from hydraulic basin agency of Oum Erabia Tropical Rainfall Measuring Mission

3.1. Groundwater Springs Inventory (SI)

The spring inventory map was developed using an extensive field data. A total of 374 springs were identified, where 280 springs (75%) have been randomly selected for the training dataset. The remaining springs (25%) were used for the validation dataset (Figure 4). The springs' discharge values vary between 0.1 and 1450 L/s.

For spatial modelling, several researchers have recommended the use of equal proportions of spring and non-spring pixels, while, others have suggested to use a high number of non-spring compared to the number of spring pixels when the study area is generally large and it cannot ensure a good spatial representativeness of the springs. Due to the large size of the study area, a high number of non-spring pixels were selected for this study. As a result, randomly mapped of 840 non-spring points for training data and 282 non-spring points for testing data were tested (three times the number of spring pixels) (Figure 4).

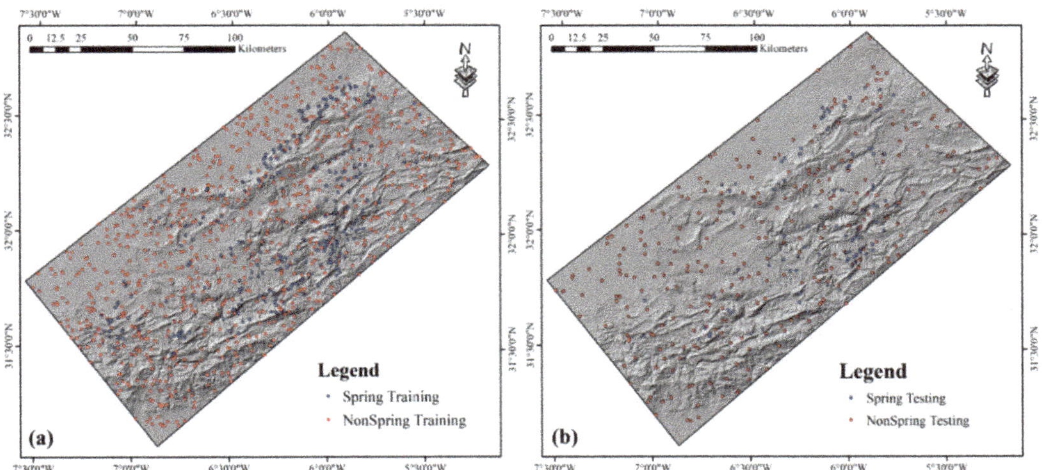

Figure 4. Location of springs and non-springs in the study area, (**a**) Training datasets and (**b**) Testing datasets.

3.2. Groundwater Influencing Factors (GIFs)

The selection of the groundwater potentiality influencing factors is very challenging due to the complexity of the groundwater functioning phenomenon. Moreover, this choice is very difficult since there are no exact standard norms. For the present study, our challenging aim was to combine many factors as possible that may have an influence on the groundwater potentiality. Consequently, we have prepared a total of 24 geological, hydrological, climatic, topographic and land cover/use factors (Figure 5).

3.2.1. Climatic Factors

Climate is a key factor directly involved in groundwater availability. In that sense, rainfall permits and directly encourages the recharge of aquifers. Annual precipitation data were obtained from the Tropical Rainfall Measuring Mission (TRMM) between 1998 and 2016 (validation of estimated TRMM rainfall data by [30]). According to the rainfall map produced, the annual average rainfall varies between 119 and 889 mm/year in the study area. The most significant values are located in the northern part, while in the south, the precipitation decreases intensely (Figure 5u).

3.2.2. Hydrological Factors

The hydrological factors chosen are the distance from rivers and the density of rivers. The distance to rivers was calculated by the Euclidean distance method in ArcGIS environment for the purpose of determining the distance of the spring from the drainage system (Figure 5n), while river density helps us recognize the spatial distribution of streams in the study area (Figure 5o). The maps show that the distances to the rivers vary between 0 and 5077 m, and that the rivers are more concentrated in the northwest part (plain area) than in the southeast part (mountainous area).

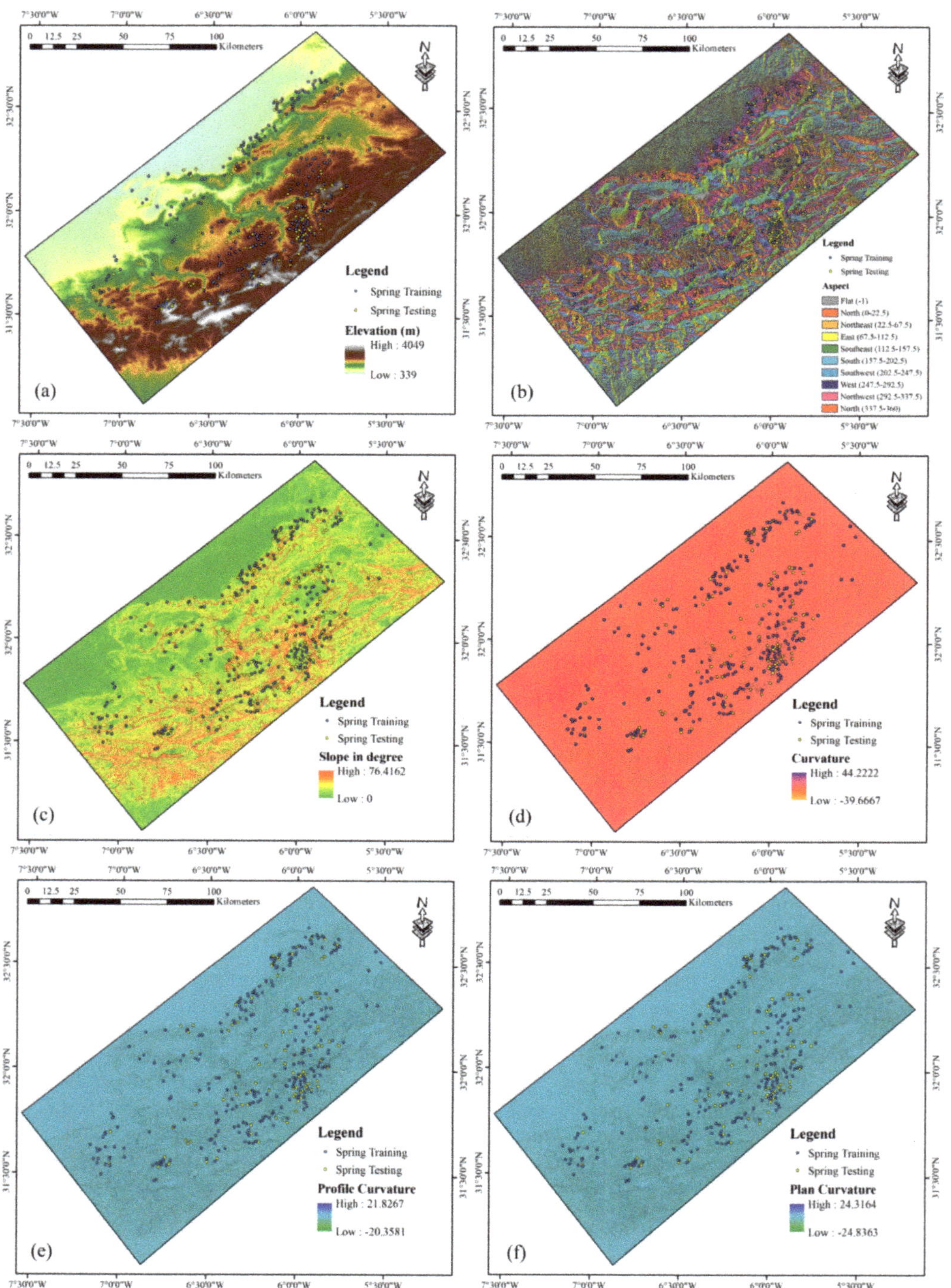

Figure 5. *Cont.*

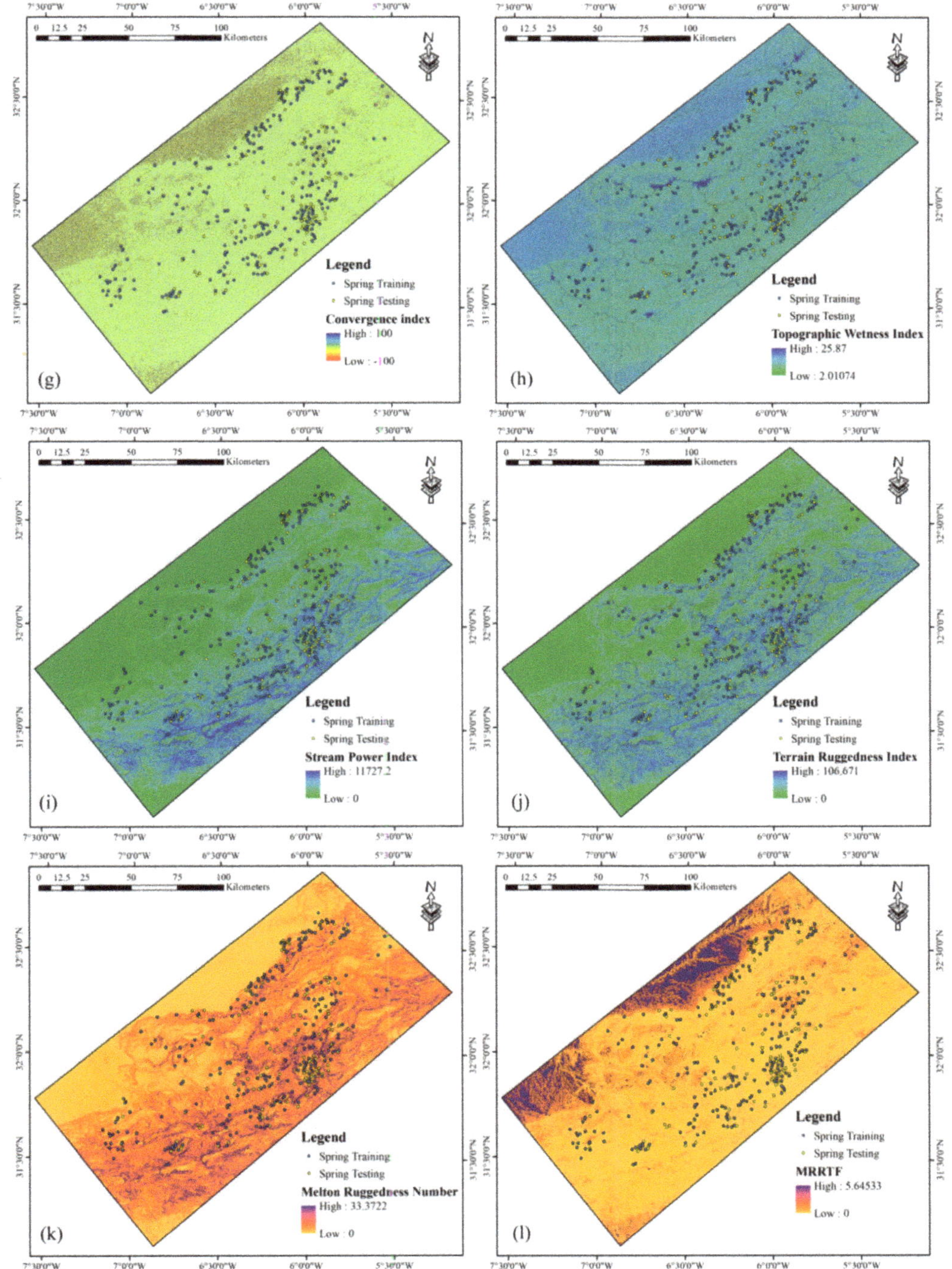

Figure 5. *Cont.*

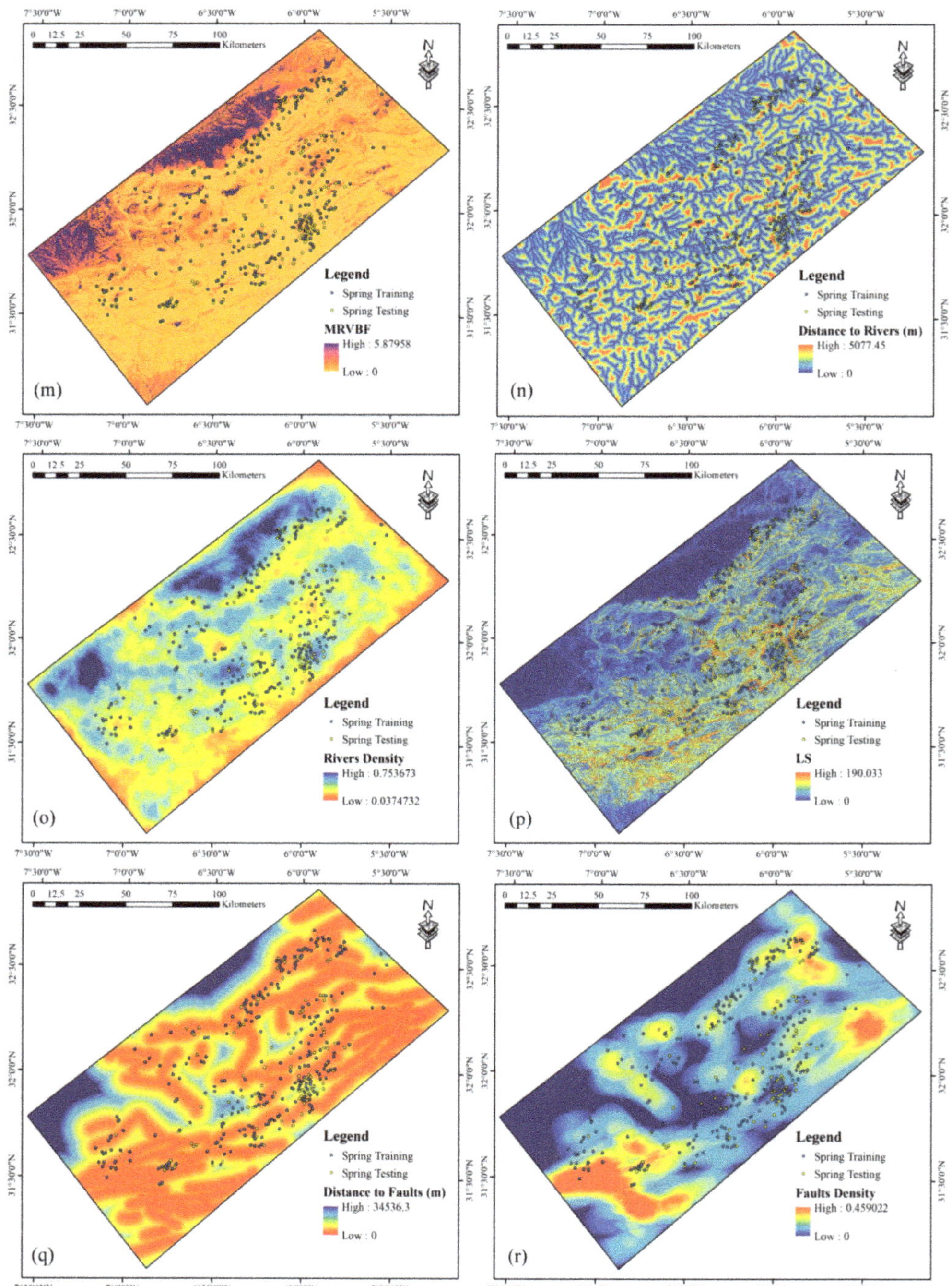

Figure 5. *Cont.*

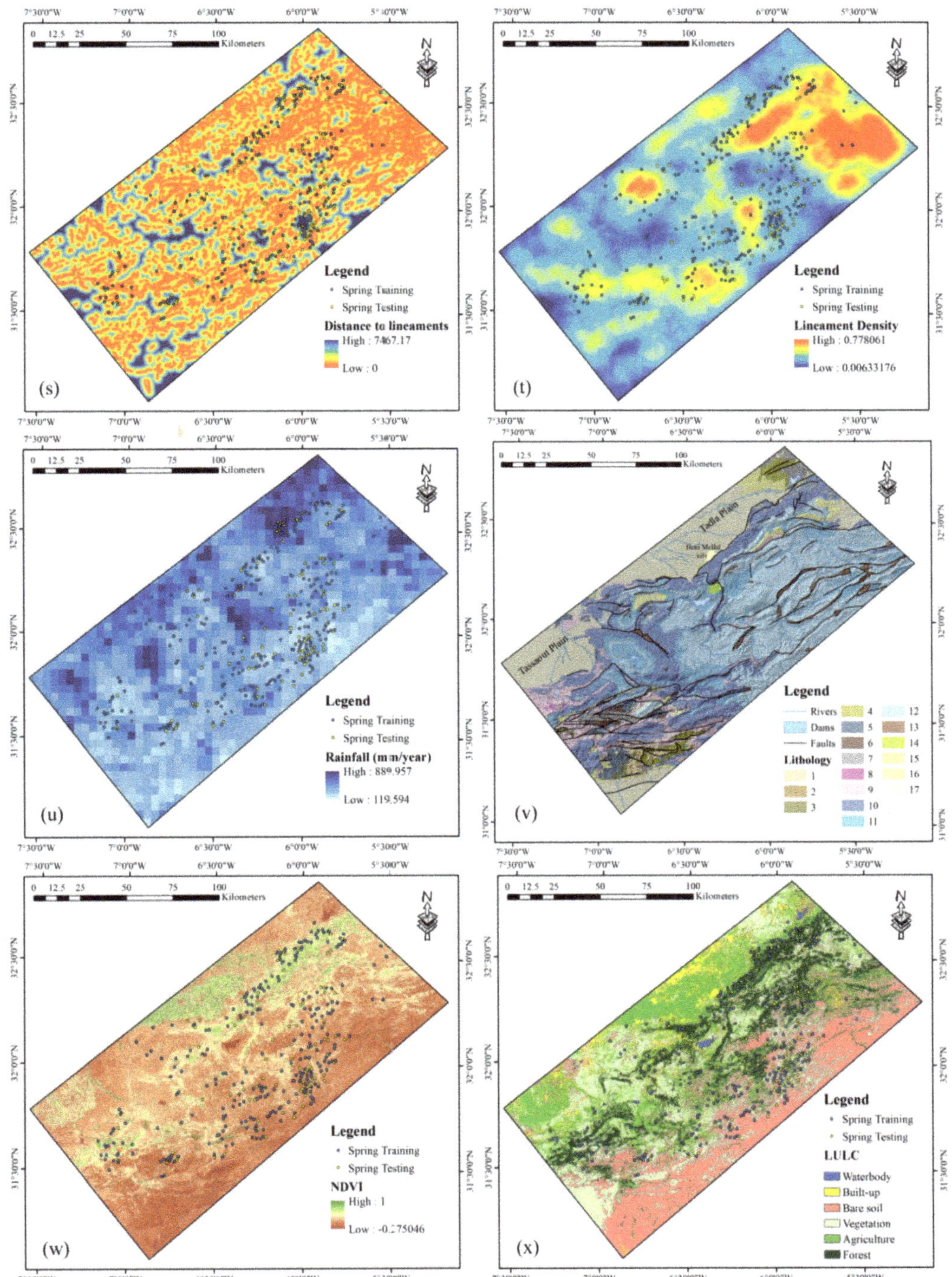

Figure 5. Groundwater influencing factors considered in the present study. (**a**) Elevation, (**b**) Aspect, (**c**) Slope, (**d**) Curvature, (**e**) Profile Curvature, (**f**) Plane Curvature, (**g**) Convergence, (**h**) TWI, (**i**) SPI, (**g**) TRI, (**k**) MeRugNu, (**l**) MRRTF, (**m**) MRVBF, (**n**) Distance to Rivers, (**o**) Density of Rivers, (**p**) LS, (**q**) Distance to Faults, (**r**) Density of Faults, (**s**) Distance to lineament, (**t**) Density of lineament, (**u**) Rainfall, (**v**) Lithology, (**w**) NDVI, (**x**) LULC.

3.2.3. Geological Factors

There are many geological factors, and they play an essential role in the formation, availability and recharge of groundwater. The main geological factor is lithology, since rock types determine aquifer formation and its continuous recharge by controlling the permeability and water circulation [31]. The 1/500,000 geological map of Morocco is used to digitalize the main lithological units in this study; the results are presented in Figures 2 and 5v.

In addition, the faults influence the presence/recharge of groundwater and the emergence of springs (secondary permeability of rocks). In our case, the Beni Mellal High Atlas is largely fractured by a dominant northeast–southwest oriented fault network, which gives great importance to the investigation of the fault-groundwater potentiality spatial relationship. Thus, two fault factor maps are produced, one representing the distance to faults and the other represent the fault density (Figure 5q,r, respectively).

The lineament is very important in the third type of environment, as it will give an idea of the spatial distribution and density of fractures in the karst landscape. The fractures frequently participate in groundwater recharge and spring emergence [32]. In this study, lineaments were detected through the interpretation of Landsat OLI imagery; the maps of distance to lineaments and lineament density are shown in Figure 5s,t.

3.2.4. Topographic Factors

Topographic factors play an essential role in controlling hydrological conditions, such as the flow of groundwater and soil moisture. In this study, 14 topographic factors were used: elevation, aspect, slope, curvature, profile curvature, plan curvature, convergence, topographic wetness index (TWI), sediment power index (SPI), terrain ruggedness index (TRI), Melton ruggedness number (MeRugNu), multi-resolution ridge top flatness (MRRTF), multi-resolution valley bottom flatness (MRVBF) and slope length (LS). These topographic factors are shown in Figure 5.

The elevation and slope factors generally negatively control the groundwater potential where in flat areas and low elevation the rainwater has much more time to infiltrate and recharge groundwater [33]. For the aspect factor, the exposure of the slopes favors more water infiltration on slopes exposed to humid winds and protected from solar radiation [15]. In this study, the potential of groundwater will be higher on the north- and northwest-facing slopes. The curvature influences groundwater recharge. The same influence may also be related to the LS factor. The MRRTF and MRVBF indicate the flatness and size of valley bottoms; low values show a smaller possibility of the existence of a groundwater aquifer, and high values indicate the potential zones [15]. In this study, the highest values for these two parameters are calculated in the plains bordering the Atlas Mountains towards the North. The TWI factor shows the influence of topography on runoff generation and flow accumulation; a high groundwater potentiality is favored when TWI values increase. The SPI expresses the erosive influence and power of water flow [34], and the TRI indicates the difference in elevation between adjacent cells of a digital elevation grid [35].

3.2.5. Land Use/Cover Factors

Land use/cover factors (LULC) strongly influence hydrological processes, such as infiltration, evapotranspiration and surface runoff, and they consequently play a significant role in groundwater potentiality. Two factors in relation to land cover have been prepared: land use/land cover (LU/LC) and normalized difference vegetation index (NDVI).

The LULC map of the study area was prepared using supervised classification and a maximum likelihood algorithm in ArcGIS environment from three merged images of the Landsat Operational Land Imager (OLI); their dates of acquisition are July 9 and 18, 2019. From the same merged images, we calculate the NDVI to determine the density of the vegetation. The LULC map contained six different classes: Waterbody, Built-up, Bare soil, Vegetation, Agriculture and Forest (Figure 5x). The NDVI map is shown in Figure 5w.

3.3. Groundwater Influencing Factors (GIFs) Analysis

3.3.1. Multicollinearity Analysis and Confusion Matrix

Multicollinearity analysis was used in statistics to detect the linearity between the conditioning factors of a given phenomenon, and detect and quantify information redundancies between the parameters that may have a negative impact on the model performance. Multicollinearity refers to the non-independence of conditioning factors that may occur in datasets. It is widely used in the prediction of several phenomena, such as landslides, gully erosion and groundwater potentiality. In this study, the multicollinearity for the groundwater influencing factors was identified using confusion matrix, tolerances and variable inflation factor (VIF) methods, according to Equations (1) and (2):

$$Tolerance = 1 - R_j^2 \tag{1}$$

$$VIF = \left[\frac{1}{Tolerance} \right] \tag{2}$$

where:

R_j^2 is the coefficient of determination.

When $VIF \geq 10$, there are linear relationships between conditioning factors.

3.3.2. Selection of Groundwater Influencing Factors

The ability to estimate groundwater potentiality depends on the factors introduced into the model. Indeed, some factors can decrease this capacity. As a result, a preliminary selection of the factors is necessary. To meet that condition, we used the information gain method to select GIFs. The information gain (IG) value for a groundwater influencing factor Xi and a class Y is calculated using Equations (3)–(5):

$$IG(Y, X_i) = H(Y) - H(Y_i|L_i) \tag{3}$$

where:

$$H(Y) = - \sum_i P(Y_i) \, Log_2(P(Y_i)) \tag{4}$$

$$H(Y_i|L_i) = - \sum_i P(Y_i) \sum_j P(Y_i|L_i) Log_2((P(Y_i|L_i)) \tag{5}$$

where:

$H(Y)$ is the entropy value of Y_i;

$H(Y_i|L_i)$ is the entropy of Y after associating the values of the landslide conditioning factor L_i;

$P(Y_i)$ is the prior probability of the out-class Y;

$P(Y_i|L_i)$ are the posterior probabilities of Y given the values of the conditioning factor L_i.

Factors that have a negative IG are considered to have no effect on groundwater potential, and therefore, they will be eliminated from the analysis.

3.3.3. Weight of the Groundwater Influencing Factors

In order to assign a weight of each class of factor before the modelling phase, several researchers recommend the use of the frequency ration (FR) method. The FR method helps to determine the spatial relationship between the predisposition factors and the dependent factor [36]. Each factor is segmented into several classes; Fr index value is calculated for each class of factors using the following equation (Equation (6)):

$$Fr = \frac{PSi}{PDi} = \frac{\left(\frac{NSi}{NSt} \right) \times 100}{\left(\frac{NAi}{NAt} \right) \times 100} \tag{6}$$

where:

PSi denotes the percentage of spring pixels for each class *i* of influencing factors, relative to the total number of spring pixels in the study area;

PDi is the percentage of each class *i* of influencing factors, relative to the total area;

NSi is the number of spring pixels in a thematic class *i*;

NSt is the number of pixels of all springs;

NAi is the total number of pixels in a thematic class *i*;

NAt is the total number of all pixels.

The results obtained represent the correlation between each class of influencing factors and the groundwater spring areas. The final step is the standardization of the FR to give equal importance to the different factors. The method used is to arrange the values of FR between 0.01 and 0.99 by the max-min normalization method according to Equation (7):

$$FRN = \frac{FR - Max(FR)}{Max(FR) - Min(FR)} \times (0.99 - 0.01) + 0.01 \tag{7}$$

where:

FRN is the normalized *FR* matrix;

FR is the original data matrix.

3.4. Methods

3.4.1. Random Forest (RF) Model

Random forest model has been developed based on the classification and regression trees (CARTs) [37]. The objective of the method is to evaluate the relationships between all factors: the springs are dependent factors and the GIFs are independent factors. The purpose is to identify the most appropriate model to build the GPM map and determine the weight of each factor. The approach is based on the creation of several decision trees, and for each tree, there will be a random selection of a set of predictive factors that will be used at each node to improve the prediction [38]. Consequently, all trees' prediction results are averaged to build the final set of model predictions [39], and the data that are not involved in the analysis are defined as the out-of-bag (OOB) error.

The basic parameters to build the RF model are mtry and ntree. mtry designates the number of factors to be considered in each tree-building process, and ntree designates the number of trees. The advantage of the RF model is that mtry and ntree can be changed and varied to test different possibilities to choose the best performance pathways and the minimum OOB error. In addition, the RF method also allows to classify the factors according to their importance. The calculation of the weights is done by measuring the mean decrease in prediction accuracy.

3.4.2. Logistic Regression (LR) Model

Logistic regression is a machine learning method developed to solve classification problems. The LR predictive analysis algorithm accepts continuous or discrete variables for model input, and it does not require that they have a normal distribution [40]. The approach is based on the concept of using probability to determine the relationship between independent factors (GIFs) and the dependent factor (SI). To achieve this objective, the dependent factor is coded as 1's and 0's (binary variable), where important groundwater potential is coded as a 1 and weak groundwater potential is coded as a 0. Nevertheless, independent factors can be continuous or categorical. In this study, we chose to classify all GIFs into numerical values representing their weights based on the frequency ratio method.

The groundwater probability potentiality is calculated according to Equations (8) and (9):

$$P = \frac{1}{1 + e^{-z}} \tag{8}$$

$$Z = \beta_0 + \beta_1 \, x_1 + \beta_2 \, x_2 + \ldots + \beta_n \, x_n \tag{9}$$

where:

P is the probability;

Z is the linear combination of the independent variables;

β_0 is the intercept of the model;

$\beta_1, \beta_2 \ldots \beta_n$ are the coefficients of the logistic regression model;

$x_1, x_2 \ldots x_n$ are the independent variables;

n is the number of independent variables.

3.4.3. Decision Tree (DT) Model (C 5.0)

A decision tree is used to classify future observations based on an already classified data set. The base of the tree corresponds to a root. Then, a series of branches whose intersections are called nodes end in leaves that each correspond to one of the classes to be predicted. Each node of the decision tree makes a binary decision that separates one class, or several classes, from the other classes [41]. In this study, we have chosen to use the C 5.0 classification algorithm, which is more efficient than its predecessor C 4.5 and offers similar results with smaller decision trees [42]. The algorithm uses the adaptive boosting method to improve the model accuracy, and it is based on the concept of entropy. A calculation of the information gain of the variables is carried out beforehand to classify them according to the maximum values; this helps to eliminate the leaves of null or weak values, which improves the classification accuracy [39,43].

3.4.4. Artificial Neural Network (ANN) Model

In this study, the multi-layer perceptron (MLP) architecture was chosen, which contains three layers connected by several neurons: the input layer, the hidden layer and the output. For the input layer, which has one input and several output pathways for each neuron, each node is connected with the different determining factors (GIFs). Hidden nodes, where there are several inputs and output connections for each neuron, use weighted connections to learn and process the problem; weights can take positive or negative values. Usually, for the modelling phase, the ANN method starts with the adjustment of the weights of the different connections between neurons during the training phase; then, the output prediction stage is based on the constructed models [44]. In the ANN method, the input (xi) and output (yi) layers can be expressed by the following equations (Equations (10) and (11)):

$$net = \sum_{i=0}^{n} wixi \tag{10}$$

$$yi = f(net) \tag{11}$$

where:

xi are the inputs;

wi are the corresponding weights;

yi is the output.

3.5. Ensembles of Models

To improve the performance and accuracy of model prediction, ensembles of models have been used and tested by many researchers. They have confirmed their effectiveness and efficiency in landslide prediction and soil erosion assessment. However, those who have tested the method for assessing groundwater potential remain limited. The choice of ensemble in this study was based on a weighted aggregation of the individual RF, LR, DT and ANN models to determine the best possible combination. Three different combinations were tested: two models, three models and four models. The equation used is:

$$EM = \frac{\sum_{i=1}^{n}(AUCSi \times Mi)}{\sum_{i=1}^{n} AUCS} \tag{12}$$

where:

EM is the ensemble of models;
AUCSi is the area under the success rate curve for the model Mi;
Mi is the individual model.

3.6. Performance Metrics and Comparison

Validation of the results in modelling is an essential step to confirm the validity of the results and the performance of the models. However, when the database partition changes, it is important to assess the stability of the computations. To do this work, we have examined the success and prediction rate gains of the following sample divisions: 25/75%, 50/50% and 75/25%. Then, statistical metrics and the area under the receiver operating characteristic curve (AUC) were used for the testing dataset to evaluate the performance of the RF, LR, DT and ANN models and the different ensembles.

3.6.1. Statistical Metrics

The validation approach is based on the calculation of four parameters, True Positive (TP), True Negative (TN), False Positive (FP) and False Negative (FN). Their determination is based on the calculation of spring pixels which are correctly or incorrectly classified as springs in the training and testing datasets. The sensitivity is the proportion of spring pixels that are correctly classified as spring occurrences, while the specificity is the proportion of the non-spring pixels that are correctly classified as non-spring [45]. In addition, other parameters have been calculated to improve the comparison between the models: accuracy, precision, FP-Rate, MCC, RMSE, MAE and the Kappa index. Higher values of sensitivity, specificity, accuracy, precision, FP-Rate and MCC indicate better performance of a model, especially if the RMSE and MAE values are close to 0. A Kappa index value of 1 indicates a perfect model, whereas -1 represents a non-reliable model. All the equations used in the calculation of these parameters are written below:

$$Sensitivity = \frac{TP}{TP + FN} \tag{13}$$

$$Specificity = \frac{TN}{FP + TN} \tag{14}$$

$$Accuracy = \frac{TN + TP}{TP + FP + TN + TP} \tag{15}$$

$$Precision = \frac{TP}{TP + FP} \tag{16}$$

$$FPRate = \frac{FP}{FP + TN} \tag{17}$$

$$MCC = \frac{TP \times TN - FP \times FN}{\sqrt{(TP + FP)(TP + FN)(TN + FP)(TN + FN)}} \tag{18}$$

$$Kappa = \frac{Accuracy - B}{1 - B} \tag{19}$$

where:

$$B = \frac{(TP + FN)(TP + FP) + (FP + TN)(FN + TN)}{\sqrt{TP + TN + FN + FP}} \tag{20}$$

$$RMSE = \sqrt{\frac{1}{n} \sum_{i=1}^{n} (X_P - X_A)2} \tag{21}$$

$$MAE = \frac{1}{n} \sum_{i=1}^{n} |(X_P - X_A)| \tag{22}$$

3.6.2. ROC Curve

In terms of the excellence and the performance of machine learning models, the ROC curve represents the most useful way to validate the results [45]. For this method, a comparison was done between the groundwater potentiality map, the training, and validation of spring inventory maps. The receiver operating characteristics curve is a graphical representation that plots the true positive percentage in the y-axis and the cumulative false positive percentage in the x-axis [46]. Finally, the area under the curve was calculated (AUC) from the ROC curve, and the precision of the model was evaluated. The area under the ROC curve varies between 0 and 1; it can be categorized as low (0.5–0.6), medium (0.6–0.7), good (0.7–0.8), very good (0.8–0.9) and excellent (0.9–1.0) [47,48].

3.6.3. Model Prioritization Using Compound Factor

The compound factor (CF) method was used to rank the different models and to compare their performance and accuracy. The method is based first of all on the ranking of all models and ensembles of models with respect to AUC values and statistical metrics. Then, to find the best fit model for producing GPMs, the CF—based prioritization was calculated in terms of the accuracy, precision, specificity, sensitivity, FP-Rate, MCC, Kappa index, AUC, MAE and RMSE values among the 15 models. The CF calculation is performed according to the equation below (Equation (23)):

$$CF = \frac{1}{n} \sum_{i=1}^{n} R \tag{23}$$

where:
R is the variable rank;
n is the number of variables.

4. Results

4.1. GIF Selection and Analysis

After the first step of the analysis, which included the realization of an inventory map of springs and non-springs that constituted the basic document to start the modelling phase, an analysis of the influencing factors was undertaken to select the most useful GIFs and eliminate those that have no effect or those that have a multicollinearity.

Primarily, the multicollinearity analyses of the 24 groundwater influencing factors show that tolerance values vary between 0.231 for the TRI factor to 0.983 for the aspect factor. In the same way, the VIF values fluctuate between 1.017 for the aspect factor and 4.323 as the maximum value for the TRI factor (Table 2). These results are acceptable: the tolerance values are greater than 0.1 and the VIF values are less than 10, which confirms that all of the selected GIFs have no multicollinearity. However, the confusion matrix diagram results indicate a linear relationship between several variables, including TRI (0.75), slope (0.74), LS (0.75), faults density (0.69), distance to faults (0.69), LULC (0.49) and NDVI (0.49). To avoid data redundancy, we conducted many experiments, including the elimination of redundant factors, by testing the success and prediction rates for all models. Results show that there is no significant impact on the rates of learning and prediction of the various models except DT model, where the prediction rate has been decreased significantly from 0.738 to 0.609. These findings indicate that the redundancy of data from some factors has only a little impact on performance, indicating that all factors must be taken into account in this analysis.

Afterwards, the results of the analysis using the information gain method show that the lithology, fault density and distance to faults factors have the highest values (0.073, 0.030 and 0.029, respectively), followed by the rainfall (0.025), MRRTF (0.020), MRVBF (0.019), elevation (0.017), and lineament density (0.016) (Table 2 and Figure 6). The minimum IG values were calculated for the distance to rivers and MeRugNu factors (0.003 and 0.002, respectively). However, the 24 GIFs had a positive information gain, and for that reason, all of them were included in this analysis.

Table 2. Multicollinearity diagnosis, average information gain and parametric statistics (LR model) for the groundwater influencing factors.

	Information Gain	*Collinearity Statistics*		*LR Model*
Influencing Factors	**Average Merit**	**Tolerance**	*VIF*	β
Elevation	0.017	0.530	1.886	0.967
Aspect	0.008	0.983	1.017	2.758
Slope	0.012	0.411	2.434	0.021
Curvature	0.004	0.667	1.499	−0.751
Profile Curvature	0.003	0.696	1.436	−0.116
Plan Curvature	0.006	0.773	1.294	0.389
Convergence	0.008	0.754	1.327	−2.774
TWI	0.004	0.907	1.103	−0.081
SPI	0.011	0.471	2.122	0.522
TRI	0.009	0.231	4.323	−0.965
MeRugNu	0.002	0.707	1.414	−0.947
MRRTF	0.020	0.615	1.625	1.687
MRVBF	0.020	0.546	1.832	1.232
Distance to Rivers	0.002	0.977	1.024	1.971
Density of Rivers	0.019	0.899	1.112	2.136
LS	0.010	0.330	3.034	0.749
Distance to Faults	0.029	0.407	2.456	0.510
Density of Faults	0.030	0.503	1.988	1.380
Distance to lineament	0.006	0.975	1.026	1.962
Density of lineament	0.016	0.949	1.054	2.979
Rainfall	0.025	0.829	1.207	1.836
Lithology	0.073	0.741	1.350	1.519
NDVI	0.013	0.681	1.469	2.313
LULC	0.008	0.681	1.469	−0.540
Constant				−12.0977

The rank of the weights of the different GIF classes by the FR analysis shows a positive correlation between the spring potentiality and high values of TWI, since the class 15.39–25.87 holds the highest FR weight (1.811). This class is closely followed by class 10 of lithology relating to Lias limestones, with a value of 1.808. Even if the slope factor has a very low GI value, its class 34.16–76.42 shows a high FR value of 1.792, followed by the class 0.99–24.32 of the curvature factor and the class 19.66–106.67 of the TRI factor. The values of FR which follow are on the order of 1.696, 1.592, 1.583, 1.565 and 1.546, characterizing the classes 3484.68–7467.17 (distance to lineaments factor), Forest (LULC factor), 13.41–26.83 (LS factor), 0.21–0.30 (NDVI factor) and 1379.67–2299.45 (SPI factor), respectively. Finally, the lowest (null) FR values are calculated for the classes 2.30–3.45 (MRRTF), 3.45–5.65 (MRRTF), 4.04–5.88 (MRVBF), 26.83–190.03 (LS), 18,419.36–34,536.29 (distance to faults), (1, 2, 3, 4, 5, 6, 8, 14 and 15) lithology and waterbody (LULC), indicating their minimal effect on groundwater potentiality.

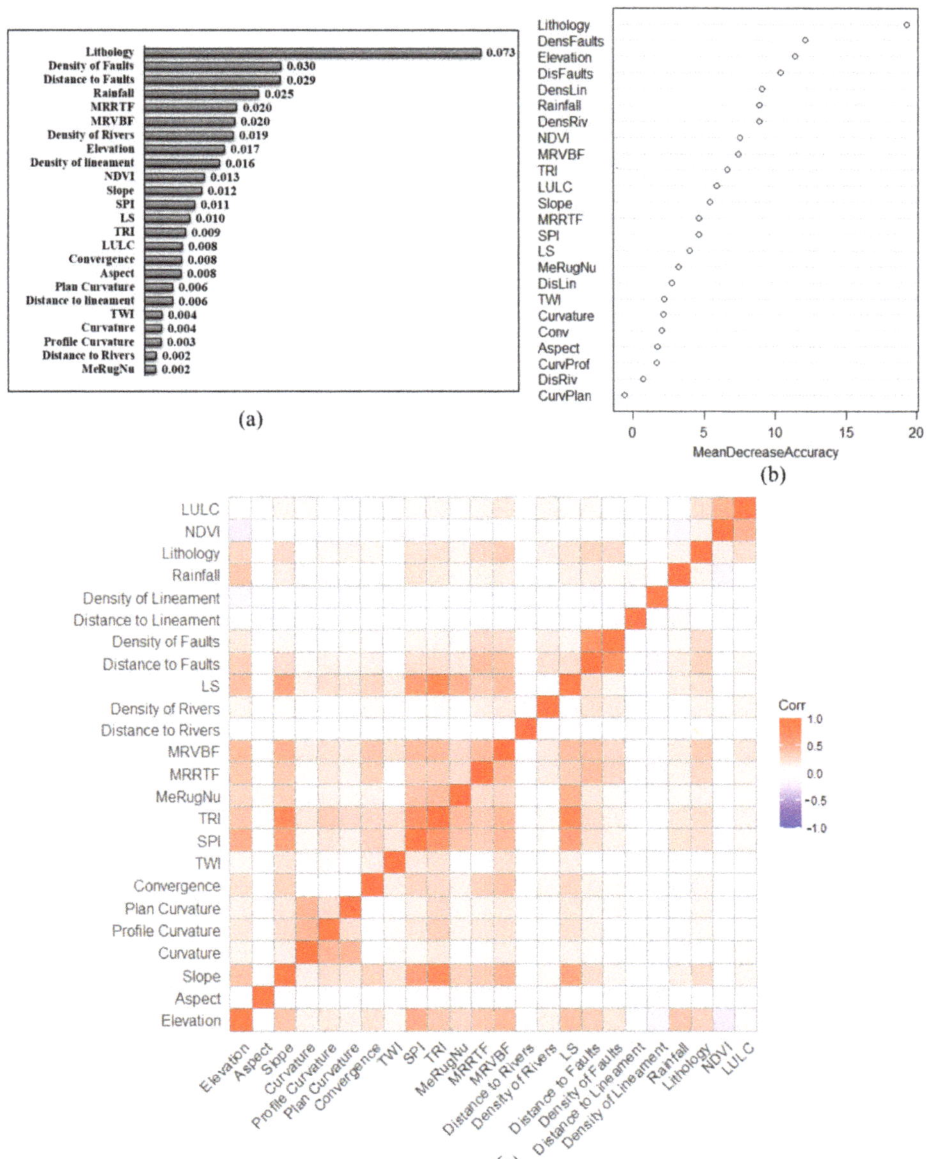

Figure 6. Predictive capabilities using information gain method: (**a**) importance of GIFs derived from Random Forest model; and (**b**) correlation matrix of the twenty-four GIFs (**c**).

Following the developed methodology, the random forest method was employed for estimating the importance of the GPM related factors that were selected by the IG method (Figure 6). The values of RF ranged between −0.598 and 19.313, with the lowest value corresponding to the variable plane curvature and the highest to variable lithology. In decreasing order of RF importance value, after the lithology factor, we have fault density, elevation, distance to faults, lineament density, rainfall and river density, with values of 12.187, 11.419, 10.446, 9.095, 8.921 and 8.913, respectively. The factors of least importance are the distance to rivers, plane curvature, profile curvature, aspect, convergence and curvature.

4.2. Models Building and Hyperparameters Tuning

This is a critical step in modeling using machine learning algorithms. To do this, we randomly subdivided the data Training (75%) and Testing (25%) dataset, then the cross-validation method is applied on the Training data to minimize the splitting error. The findings show that the RF model has an average accuracy of 0.71 and a kappa value of 0.16. The optimization of the RF parameters was applied using a random search and based on the OBB error rate (Figure 7a). Then, accuracy was used to select the optimal model using the largest value (Table 3). The final value used for the RF model was mtry = 13. These results confirm that the subdivision of the data does not have a great impact on accuracy, which allowed us to apply the other models without fearing a degradation of the accuracy for the different partitions. For the ANN model, after several tests, the optimum number of hidden layers is equal 2 (Figure 7b).

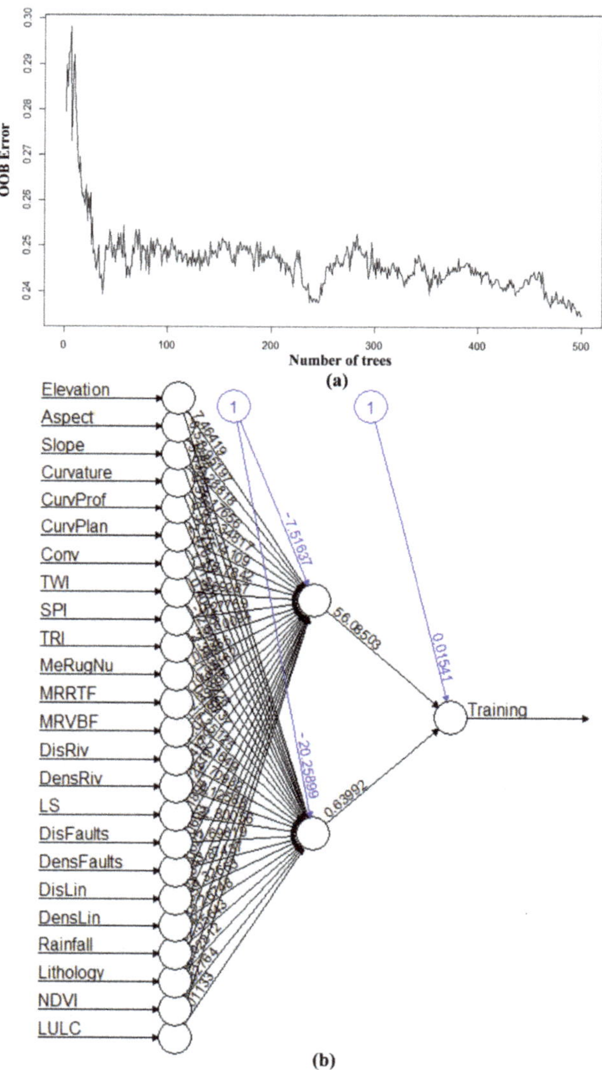

Figure 7. Number of trees optimized based on OOB estimates of the error rate in RF model (**a**) and diagram of ANN model (**b**).

Table 3. Resampling results across tuning parameters using RF model. Accuracy was used to select the optimal model using the largest value. The final value used for the model was mtry = 13.

	Accuracy	Kappa	Resample
1	0.7321429	0.21052632	Fold02
2	0.6785714	0.10000000	Fold01
3	0.7017544	0.17637059	Fold04
4	0.7079646	0.12196845	Fold03
5	0.6902655	0.09309791	Fold06
6	0.6696429	−0.01369863	Fold05
7	0.7232143	0.19480519	Fold08
8	0.7678571	0.31578947	Fold07
9	0.7142857	0.20000000	Fold10
10	0.7232143	0.25301205	Fold09

4.3. Groundwater Potential Mapping

The main objective of this study is to produce groundwater potentiality maps (GPMs) using individual and ensemble machine learning models. Fifteen GPMs were produced based on the application of four models, both individually and in ensembles: random forest, logistic regression, decision tree and artificial neural network. The GPMs produced were divided into five classes based on Jenk's natural breaks classification method (calibration results). The five classes are very low, low, moderate, high and very high. The results are shown in Figure 8. Accordingly, the spatial distribution are: (1) very low and low classes which are dominant; (2) the areas with the highest potential are located in the middle of the study area, and in the northern parts, especially in the mountains–plains transition zone; and (3) the regions with the least potential are those located in the southern, western and eastern parts of the study area.

For the individual models four groundwater potential maps are shown in Figure 8. The rate of springs in each class showed that all of the training and validation springs were identified in the very high class for the RF, DT and ANN models, with a maximum training value for the RF GPM (100% of springs in the high class) and a maximum validation value identified for the ANN GPM (46.81%). In the case of the GPM based on the LR model (Figure 8b), the high class makes up the major part of the training (35.00%) and validation (32.98%) springs, followed by the moderate class (25.36% for training and 26.60% for validation).

Following the completion of the GPMs for the different models individually, the different possible combinations were tested. The first sets of models tested involved two models, so six ensembles were considered: RF-LR, RF-DT, RF-ANN, LR-DT, LR-ANN and DT-ANN. The GPMs that were derived and the statistical results of the spatial distribution of the different classes are shown in Figure 8. On first examination, it is clear that the very low and low classes cover the major part of the territory; the percentages of the very low class vary from 47.92% for the DT-ANN ensemble (the maximum value) to 25.01% for RF-LR (the minimum value). For the low classes, the values vary from 32.23% (RF-DT) to 15.70% (DT-ANN). On the other hand, the area covered by the high and very high classes represent the lowest percentages for all model ensembles. From the point of view of the spring percentage, it has been observed that the training springs largely cover the high and very high classes. Indeed, except for the LR-ANN ensemble, where the high class holds the major part of the training springs (33.93%), all of the other ensembles show very high class which contains the most important percentages of training springs: 92.14%, 86.07%, 93.21%, 50.36% and 67.14% for RF-LR, RF-DT, RF-ANN, LR-DT and DT-ANN, respectively. To finish, the highest percentage of validation springs are located at the very high-class level, with 30.85% for RF-DT, 41.49% for RF-ANN, 26.60% for LR-DT, 29.79% for LR-ANN and 35.11% for DT-ANN. Only for the RF-LR ensemble does the moderate class contain the majority of the validation springs (32.98%), followed by the very high class (31.91%).

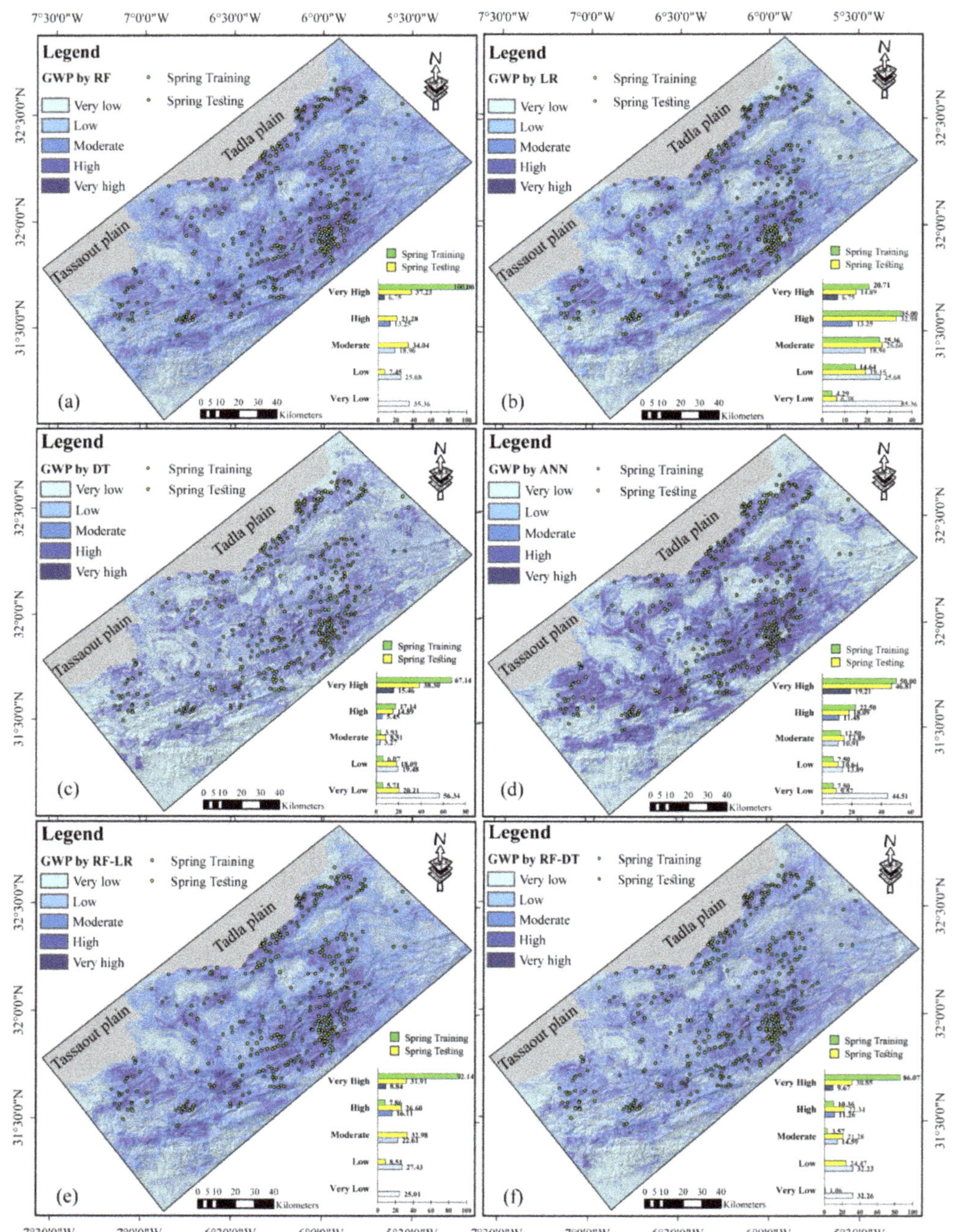

Figure 8. *Cont.*

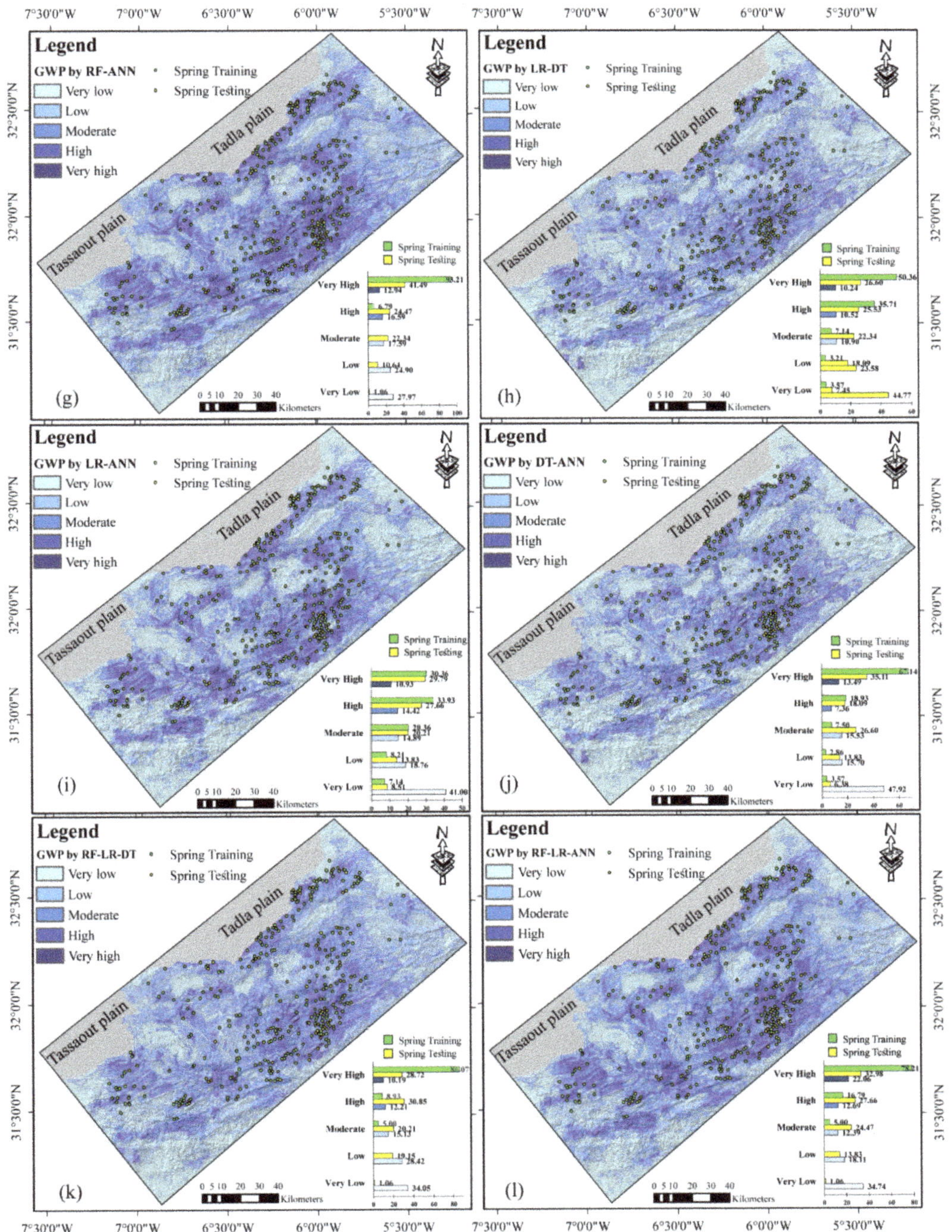

Figure 8. *Cont.*

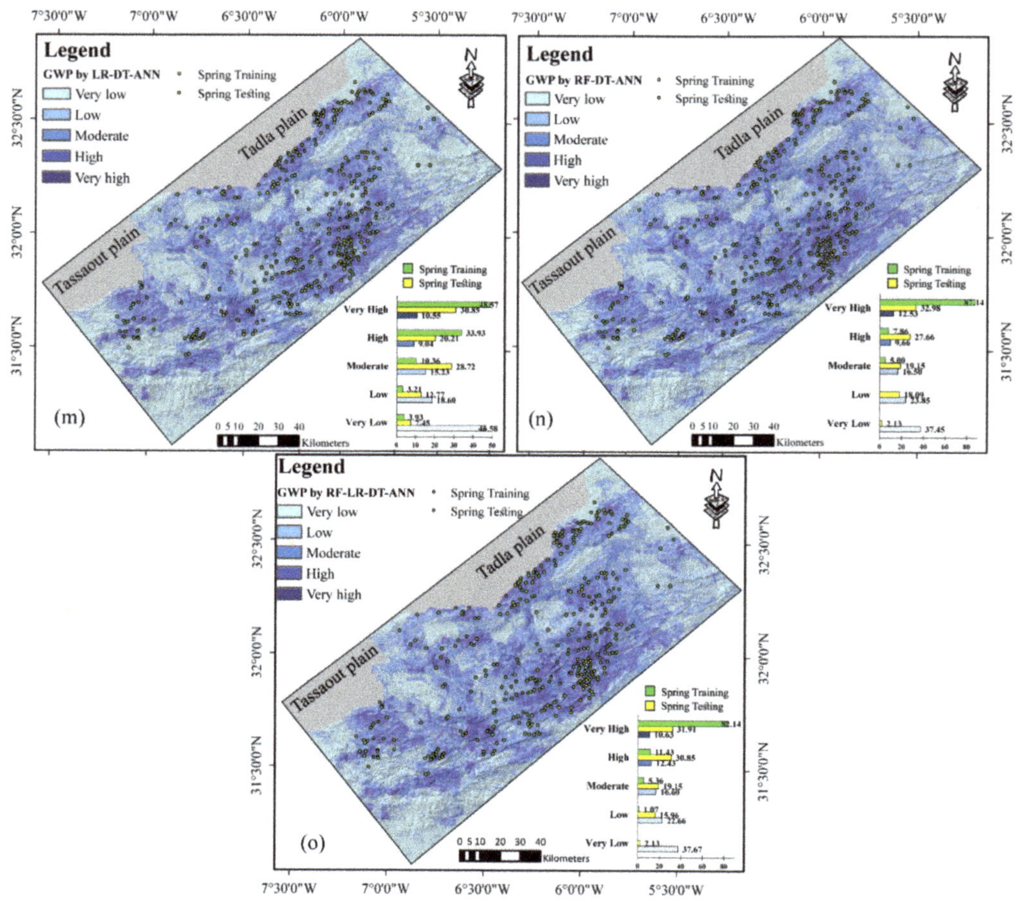

Figure 8. GSPM obtained from RF (**a**), LR (**b**), DT (**c**), ANN (**d**), RF-LR (**e**), RF-DT (**f**), RF-ANN (**g**), LR-DT (**h**), LR-ANN (**i**), DT-ANN (**j**), RF-LR-DT (**k**), RF-LR-ANN (**l**), RF-DT-ANN (**m**), LR-DT-ANN (**n**) and RF-LR-DT-ANN (**o**) models.

Combining three models at the same time has allowed to produce four new GPMs: RF-LR-DT, RF-LR-ANN, RF-DT-ANN and LR-DT-ANN. The results and the spatial distribution of the different classes of potentiality are presented in Figure 8. Generally, the percentages covered by the different potentiality classes' decrease from the very low class to the very high class. From the point of view of the spatial distribution of the springs, all ensembles of three models show that the training springs are mostly located in the very high class, with values of 87.14%, 86.07%, 78.21% and 48.57% for the RF-DT-ANN, RF-LR-DT, RF-LR-ANN and LR-DT-ANN ensembles, respectively. For the validation springs, the very high classes of the RF-LR-ANN and RF-DT-ANN ensembles contain the highest percentage (32.98%), followed by the LR-DT-ANN ensemble (30.85%) and the high class (30.85%) of the RF-LR-DT ensemble.

Finally, a set of all four models was used to produce one more GPM. The map produced and the statistics relating to the spaces occupied by the different potentiality classes are shown in Figure 8. Given the spatial distribution, very low is the dominant class, with a percentage of 37.67%, followed by the low class (22.66%), the moderate class (16.60%), the high class (12.43%) and finally, the very high class (10.63%). Inversely, the percentage of training and validation springs increases from the very low to the very high classes.

4.4. Performance Metrics and Comparison

In this part, a comparison was realized to test values of success and prediction rate 25/75%, 50/50% and 75/25% (Table 4). The table reveals a considerable stability in the overall performance in all tests (25%, 50% and 75%) except the prediction rate of LR and DT, where a slight instability was observed. For other individual or ensemble models, the success and prediction rates increase generally from 25% to 75% for the sample.

Table 4. Gain in success and prediction rate using the partition Training/Testing progresses from 25/75% to 75/25% with redundant factors elimination.

Models	25/75% of Overall Sample		50/50% of Overall Sample		75/25% Of Overall Sample		Elimination of Redundant Factors	
	Success Rate	Prediction Rate	Success Rate	Prediction Rate	Success Rate	Prediction Rate	Success Rate	Prediction Rate
RF	1.000	0.719	1.000	0.767	1.000	0.786	1.000	0.780
LR	0.729	0.780	0.730	0.755	0.784	0.744	0.781	0.746
DT	0.939	0.612	0.911	0.596	0.964	0.738	0.946	0.609
ANN	0.788	0.693	0.776	0.743	0.784	0.744	0.782	0.743
RF-LR	0.999	0.773	0.999	0.775	0.999	0.779	0.999	0.773
RF-DT	0.996	0.722	0.997	0.742	0.998	0.787	0.997	0.725
RF-ANN	0.999	0.722	1.000	0.760	1.000	0.775	1.000	0.770
LR-DT	0.920	0.759	0.904	0.739	0.949	0.779	0.931	0.702
LR-ANN	0.783	0.753	0.796	0.750	0.794	0.749	0.784	0.746
DT-ANN	0.922	0.697	0.919	0.714	0.954	0.771	0.933	0.700
RF-LR-DT	0.995	0.757	0.995	0.758	0.996	0.790	0.995	0.732
RF-LR-ANN	0.985	0.757	0.981	0.766	0.983	0.772	0.980	0.765
RF-DT-ANN	0.995	0.723	0.995	0.748	0.997	0.789	0.996	0.730
LR-DT-ANN	0.915	0.745	0.905	0.739	0.943	0.780	0.920	0.713
RF-LR-DT-ANN	0.987	0.750	0.984	0.756	0.989	0.791	0.985	0.738

The success rate of all samples is stable. In order to verify the obtained results, the GPMs and the spring inventory locations were compared (Figure 8). For the four individual models, we see that the majority of the validation and training springs fall into the high and very high susceptibility classes. In addition, the very low susceptibility class either has very weak or no spring occurrence in all GPMs. Even better, for the model sets, the results show an increase in the percentages of validation and training springs in high and very high classes, especially for RF-LR, DT-ANN and RF-ANN. It is clear from these results that the field-recorded spring locations have a better fit with the RF, DT, ANN, RF-LR, DT-ANN and RF-ANN maps than with the other GPMs.

The results of validation techniques and accuracy prioritization based on the training datasets are shown in Table 5, and the results based on the testing datasets are shown in Table 6. Ten parameters have been calculated to improve the comparison between the models: accuracy, sensitivity, specificity, precision, FP-Rate, MCC, KAPPA, AUC, RMSE and MAE. For the training datasets, the accuracy ranges from 0.754 to 1.000; the highest value is identified for the RF model, and the lowest is from the LR-ANN and ANN models. Like accuracy, the maximum sensitivity value (1.000) was found for the RF individual model and the minimum for the LR model (0.257). Specificity varies between 0.797 (ANN model) to 1.000 (RF and RF-ANN models). For precision, the values range from 0.311 (ANN) to 1.000 (RF). The FP-Rate calculation indicates that the best result is always for the RF model (0.334), and the worst result is for LR (0.086). In additon, for the other parameters, RF is better, with values of 1.000 for Kappa and AUC. The minimum values are Kappa = 0.233 (LR-ANN) and AUC = 0.784 (LR and ANN models). For the reliability, which has been assessed by applying the MAE and RMSE methods, the training results indicate minimum values for the RF model (0.000) and maximum values for LR (0.137) and ANN (0.496).

Table 5. Results of validation techniques and accuracy prioritization based on training datasets.

	Training																		Rank Total		
	Ac	R	Sen	R	Sp	R	Pr	R	FPR	R	MCC	R	Ka'1	R	AUC	R	MAE	R	RMSE	R	RT	CF	PR
RF	1.000	1	1.000	1	1.000	1	1.000	1	0.334	1	1.000	1	1.000	1	1.000	1	0.000	1	0.000	1	10	1	1
LR	0.765	13	0.257	15	0.935	13	0.567	13	0.086	15	0.262	13	0.234	14	0.784	14	0.137	15	0.485	13	138	13.8	13
DT	0.942	8	0.861	5	0.969	11	0.903	12	0.287	5	0.843	8	0.843	8	0.964	9	0.012	2	0.241	9	77	7.7	7
ANN	0.754	15	0.512	13	0.797	15	0.311	15	0.092	13	0.256	15	0.244	13	0.784	14	0.098	13	0.496	15	141	14.1	15
RF-LR	0.956	6	0.825	7	1.000	1	1.000	1	0.275	7	0.883	6	0.876	6	0.999	3	0.044	10	0.209	7	54	5.4	6
RF-DT	0.968	2	0.886	2	0.995	6	0.984	6	0.295	2	0.914	2	0.911	2	0.998	4	0.025	3	0.179	2	31	3.1	2
RF-ANN	0.963	5	0.850	6	1.000	1	1.000	1	0.283	6	0.900	5	0.895	5	1.000	1	0.038	7	0.194	5	42	4.2	4
LR-DT	0.929	9	0.793	9	0.974	9	0.910	9	0.264	10	0.804	9	0.801	9	0.949	11	0.038	8	0.194	6	89	8.9	9
LR-ANN	0.763	14	0.264	14	0.929	14	0.552	14	0.088	14	0.257	14	0.233	15	0.794	13	0.130	14	0.487	14	140	14	14
DT-ANN	0.929	10	0.793	10	0.974	9	0.910	9	0.264	10	0.804	10	0.801	10	0.954	10	0.032	6	0.267	10	94	9.4	10
RF-LR-DT	0.967	3	0.882	3	0.995	6	0.984	6	0.294	3	0.911	3	0.909	3	0.996	6	0.026	4	0.182	3	40	4	3
RF-LR-ANN	0.910	11	0.646	11	0.998	4	0.989	5	0.215	11	0.754	11	0.728	11	0.983	8	0.087	12	0.300	11	95	9.5	11
RF-DT-ANN	0.966	4	0.871	4	0.998	4	0.992	4	0.290	4	0.909	4	0.906	4	0.997	5	0.030	5	0.184	4	42	4.2	5
LR-DT-ANN	0.893	12	0.639	12	0.977	8	0.904	11	0.213	12	0.700	12	0.683	12	0.943	12	0.073	11	0.327	12	114	11.4	12
RF-LR-DT-ANN	0.950	7	0.814	8	0.955	12	0.983	8	0.271	8	0.865	7	0.859	7	0.989	7	0.043	9	0.224	8	81	8.1	8

Table 6. Results of validation techniques and accuracy prioritization based on testing datase.

	Testing																			Rank Total		
	Ac	R	Sen	R	Sp	R	Pr	R	FPR	R	MCC	R	Ka	R	AUC	R	MAE	R	RMSE	R	RT	CF	PR
RF	0.753	7	0.213	14	0.933	5	0.513	6	0.071	14	0.207	15	0.181	15	0.786	5	0.141	9	0.497	8	98	9.8	15
LR	0.756	5	0.223	11	0.933	6	0.525	5	0.074	11	0.220	13	0.193	14	0.744	13	0.143	10	0.494	6	94	9.4	14
DT	0.737	11	0.511	2	0.813	13	0.475	8	0.170	1	0.316	1	0.316	2	0.738	15	0.019	3	0.512	12	68	6.8	3
ANN	0.756	6	0.520	1	0.792	15	0.277	15	0.080	9	0.245	8	0.227	7	0.744	14	0.117	8	0.494	6	89	8.9	11
RF-LR	0.761	3	0.234	10	0.936	4	0.550	4	0.078	10	0.240	10	0.211	10	0.779	7	0.143	11	0.489	3	72	7.2	6
RF-DT	0.737	12	0.468	4	0.827	11	0.473	9	0.155	3	0.296	3	0.296	4	0.787	4	0.003	1	0.512	12	63	6.3	1
RF-ANN	0.761	4	0.213	14	0.943	2	0.556	3	0.071	15	0.230	11	0.197	12	0.775	9	0.154	13	0.489	3	86	8.6	9
LR-DT	0.735	13	0.500	3	0.813	14	0.470	11	0.166	2	0.306	2	0.306	3	0.779	7	0.154	14	0.489	3	72	7.2	7
LR-ANN	0.764	2	0.223	11	0.943	3	0.568	2	0.074	12	0.243	9	0.209	11	0.749	12	0.151	12	0.486	2	76	7.6	8
DT-ANN	0.729	15	0.436	5	0.827	12	0.456	14	0.145	4	0.267	4	0.267	5	0.771	11	0.010	2	0.520	15	87	8.7	10
RF-LR-DT	0.748	8	0.351	7	0.880	8	0.493	7	0.117	6	0.261	5	0.255	6	0.790	2	0.072	6	0.502	9	64	6.4	2
RF-LR-ANN	0.767	1	0.223	11	0.947	1	0.583	1	0.074	13	0.251	6	0.215	9	0.772	10	0.154	15	0.483	1	68	6.8	4
RF-DT-ANN	0.735	14	0.340	8	0.866	9	0.457	13	0.113	7	0.229	12	0.225	8	0.789	3	0.064	5	0.515	14	93	9.3	13
LR-DT-ANN	0.740	9	0.362	6	0.866	10	0.472	10	0.120	5	0.250	7	0.339	1	0.780	6	0.058	4	0.510	10	68	6.8	5
RF-LR-DT-ANN	0.740	10	0.277	9	0.894	7	0.464	12	0.092	8	0.208	14	0.197	13	0.791	1	0.101	7	0.510	10	91	9.1	12

In the case of testing datasets (Table 6), validation results indicate that the RF-LR-ANN ensemble is the best performing model in terms of accuracy (0.767), specificity (0.947), precision (0.583) and RMSE (0.483). For sensitivity, the maximum value is identified for the ANN model (0.520), and for the FP-Rate and MCC parameters, the best results were calculated for DT (FPR = 0.170, MCC = 0.316). The calculated Kappa index values show a maximum value of 0.339 for the ensemble of three LR-DT-ANN models, and the minimum value of the MAE parameter was calculated for the RF-DT ensemble model (MAE = 0.003). Finally, the most efficient model in terms of AUC was the ensemble of four models, RF-LR-DT-ANN (AUC = 0.791).

Additionally, the estimation of prediction capability for the fifteen models is obtained by comparing the spring training and validation inventories with the GPMs. Then, the rate curves were created (ROC), and the areas under each curve (AUCs) were calculated (Figure 9). For training datasets, the prediction-rate curve showed that the maximum AUC values were 1.000, 0.999 and 0.998 for the RF and RF-ANN, RF-LR and RF-DT models, respectively. Moreover, in the prediction-rate curve obtained by comparing the spring validation data with GPMs, it was observed that all models present tolerable performance for groundwater potentiality mapping (AUC > 0.7). The RF-LR-DT-ANN ensemble model achieved the best performance (AUC = 0.791), followed by the RF-LR-DT ensemble model (AUC = 0.790), the RF-DT-ANN model (AUC = 0.789) and the RF-DT model (AUC = 0.787).

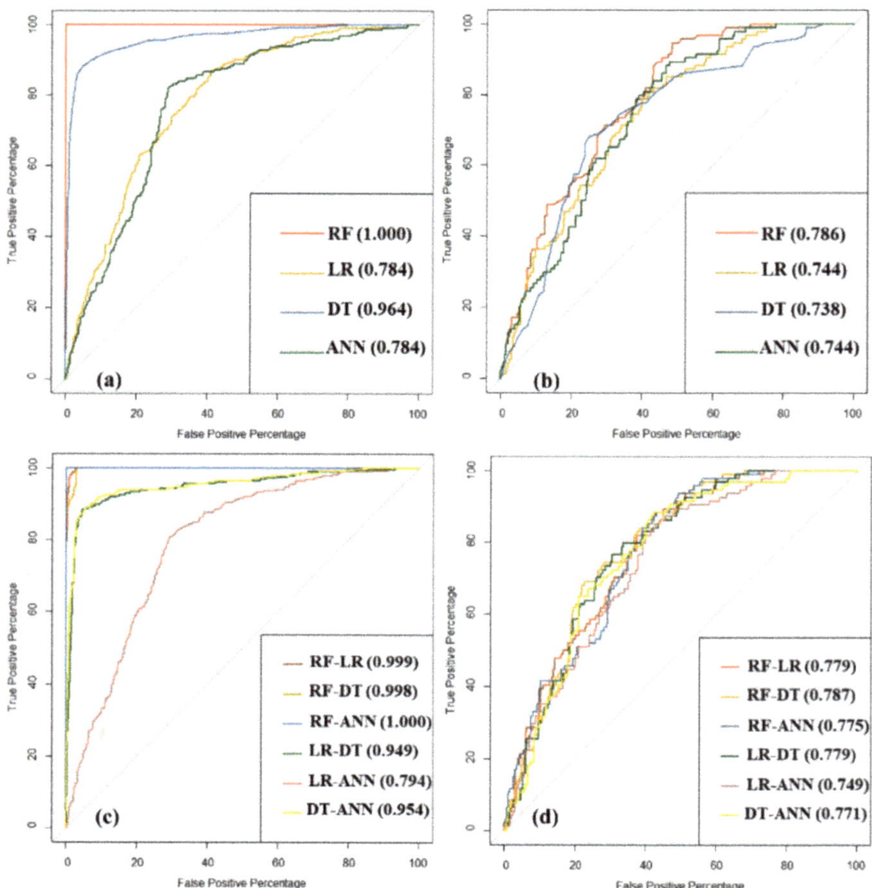

Figure 9. *Cont.*

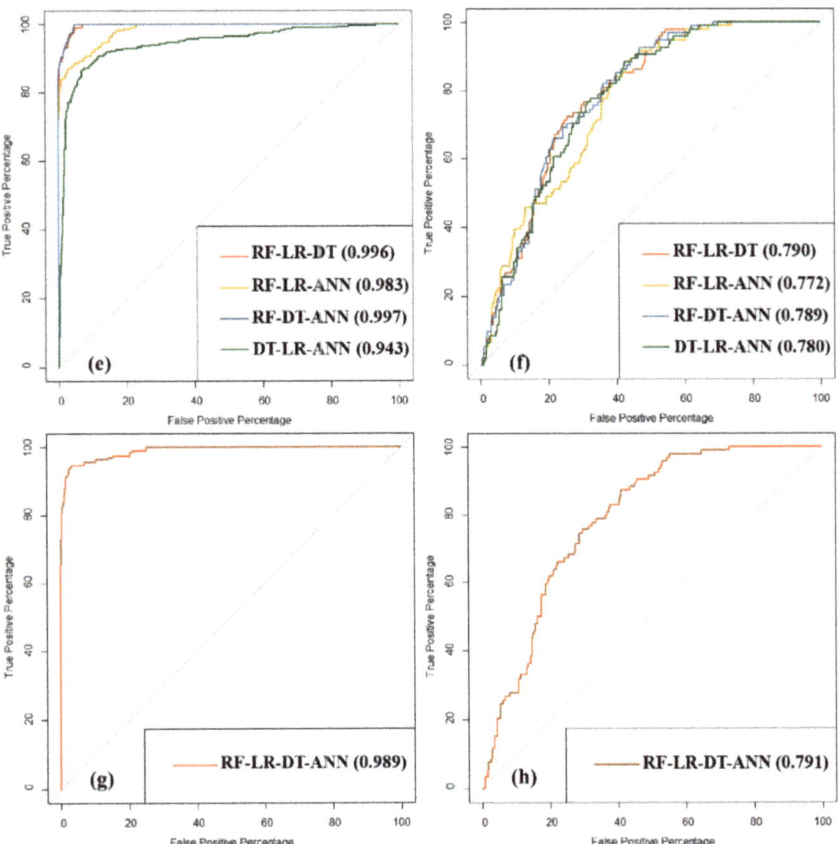

Figure 9. ROC curve: (**a**) success rate of individual models, (**b**) prediction rate of individual models, (**c**) success rate of ensemble of two models, (**d**) prediction rate of ensemble of two models, (**e**) success rate of ensemble of three models, (**f**) prediction rate of ensemble of three models, (**g**) success rate of ensemble of four models and (**h**) prediction rate of ensemble of four models.

Furthermore, we have calculated the prioritization rank based on all evaluation criteria; the results are shown in Tables 5 and 6 and Figure 10. The prioritization results by compound factor (CF) analysis using the training GPMs of all models found that the RF and RF-DT models had the best success rates; it ranked RF at 1 and RF-DT at 2. Moreover, for prediction aptitude using testing GPMs, the prioritization analysis indicates that the best models are RF-DT and RF-LR-DT. Taking into account the results in terms of success and prediction rate, the two best models for mapping the groundwater potential in our mountainous study area are RF-DT, followed by RF-LR-DT.

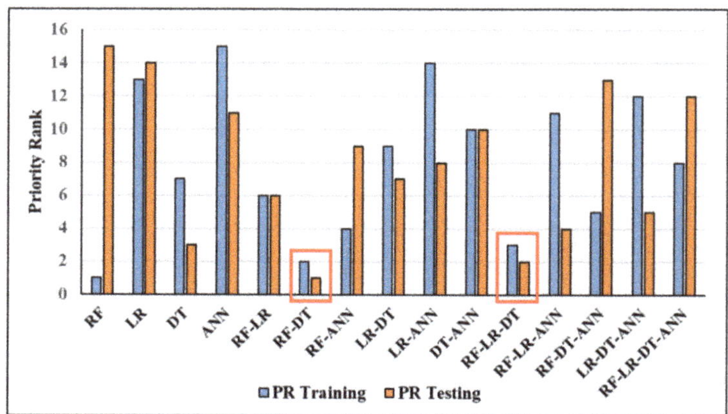

Figure 10. Prioritization results of all models using compound factor (CF) analysis used in the training and testing processes.

5. Discussion

The discussion will focus on three main points: (1) the analysis and selection of GIFs; (2) role of factors in the occurrence of karst springs, and (3) the performance of individual and ensemble machine learning algorithms in GWP mapping in large-scale areas.

5.1. GIF Selection and Analysis

According to IG analysis, lithology, fault density, distance to faults and rainfall were identified as factors that were highly predictive of the presence of groundwater; moreover, MeRugNu, distance to rivers, profile curvature and curvature were identified as the least predictive. This seems logical given the very important role of the rock type in the genesis and recharge of aquifers, in addition to the structuring factors (faults and fractures) that facilitate recharge and the emergence of springs at the same time. According to FR analysis results, groundwater potentiality is more likely to be found within Liasic limestone areas, in areas with a high wetness index and within steep slopes. These last two factors underline the importance of topographic control on hydrological processes and on GPMs in mountain areas in particular. The results, showing the importance of high TWI values in the prediction of groundwater potential.

As in the IG factor rankings, geological factors were classified as the best predictor variables of GPM in our study according to the RF importance value. The importance given to the elevation factor in this study may be related to the fact that the lower elevation zones correspond to the Tadla and Tassaout plains, where the emergence of springs is very rare due to their low slopes (<10%). Elevation is a key factor in the production of groundwater potential maps in our region and in other mountainous regions around the world. The rainfall factor also remains a very good predictor, since it is the source of aquifer recharge in addition to the snow cover that characterizes these mountainous areas, especially during the winter and spring seasons [28], which is in line with several previous studies [49,50]. Another factor that presents a robustness of prediction under the RF importance calculation is the river density, since rivers represent an important source of exchange between aquifer systems and surface hydrology. The importance of the predictors such as rainfall and river density in GPM has been seen in other semi-arid regions around the world [11,51].

According to the results obtained from this study on GIF selection and analysis, it is recommended to take into account many factors as possible in the analysis and in the modelling processes, especially in regional studies where GIFs vary spatially and influence groundwater potentiality from one location to another.

5.2. Role of Factors in the Occurrence of Karst Springs

Geological factors were identified as the top predictor variables of GPM based on the RF importance value and IG results. Certainly, lithology is the most relevant factor, followed by fault density and distance to faults. Indeed, the karstic environment is largely composed of the Lias limestones [22] which explains the importance of faults and lineaments in the karstification process. These findings are in accordance with results reported by previous studies in similar geological contexts, which detect higher groundwater potentiality in limestone-dominated karstic areas with severe fracturing [52].

According to the results of this study in the Beni Mellal, Atlas Mountains, the factors controlling the karst springs development can be divided into four categories: lithology, geological structure, topography and climate condition: The liasic and Middle Jurassic limestone contains 85% of the inventoried springs in the research region. The remaining springs are found in Middle Jurassic and Pliocene–Pleistocene continental and alluvium deposits. This confirms the important role of carbonated lithology in the emergence of these springs by the effect of underground dissolution by rainwater. In addition, near to the faults the probability of emerging sources increases, the same result is observed in areas where the density of the faults is important. Regarding the topographic factors, the altitude controls the most of this phenomenon, which mainly explains the emergence of most springs in the transition area between the mountains and the Tadla plain (Dir). Finally, the non-homogeneous appearance of the springs in this large mountain region appears strongly controlled by the spatial variation of the precipitation, this will be mainly related to the importance of the rains on the northern slopes compared to those of the south.

5.3. Machine Learning Algorithm Performance

Generally, the use of the RF, LR, DT and ANN individual learning machine models has given good results, even though there is little stability between the success and the prediction rate of some models (e.g., RF and DT models). Indeed, the RF and DT models have a very high success rate (AUC = 1.000 for RF and AUC = 0.964 for DT), but the prediction rate has decreased significantly for both models (AUC = 0.786 for RF and AUC = 0.738 for DT).

In order to improve the performance and prediction rates of the models, combinations of these four models were used. Different combinations were tested, and the prioritization rank method was used to select the best models for groundwater potentiality mapping in mountainous areas by integrating the maximum number of influencing factors in the analysis. Thus, a significant difference between the individual models and ensembles of models based on the predictive performance can be perceived. Indeed, the average prediction rate using AUC values showed an interesting progression. The average prediction rate for the four individual models was 0.753; then, it increased to 0.773 for the ensembles of two models. Then, it increased to 0.783 on average for the ensembles of three models, and finally, the AUC recorded its maximum value, which was reached by the ensemble based on all four models, RF-LR-DT-ANN (0.791). This is confirmed by the compound factor method, which has allowed us to draw up a general ranking of individual models and ensembles of models using several statistics metrics applied to the training and validation datasets. The results of the prioritization clearly indicate that the ensembles of models improve performance and reduce some errors related to data preparation or modelling processes. The best set of models for groundwater potentiality mapping in our study are the RF-DT and RF-LR-DT ensembles. This is also the situation in several studies that compare the predictive performance of either optimized, hybrid or ensemble models. This is also confirmed by previous studies [53] that indicated that hybrid models show better accuracies than individual models. However, in our study, while the majority of the model ensembles performed better, others were marked by a decrease in performance: for example, LR-ANN's success rate and RF-DT-ANN's prediction rate. This requires us to make the maximum number of groupings possible in order to select the best ensemble of models.

6. Conclusions

In the present study, a wide variety of methodologies were applied based on GIS, remote sensing and the use of individual and ensemble machine learning algorithms to evaluate GP in large-scale mountainous areas. In addition, the novel aspect of this study is that we have tried to integrate many of groundwater potentiality influencing variables as possible, which include geological, topographical, hydrological, climatic and land cover factors; 24 factors have been considered. Then, after a test of multicollinearity and an information gain calculation, all of the factors were retained to produce groundwater potentiality maps. In the same way, the importance of GIFs has been estimated from the random forest method. Lithology represents the factor that most influences groundwater potentiality in our karstic zone, followed by tectonic factors (faults and lineaments) and a climatic factor (rainfall). Alternatively, several machine learning algorithms were used for GP mapping. The RF, LR, DT and ANN models were chosen due to their satisfactory results in other regions of the world. The application of individual models indicates that RF represented the best model in terms of success and prediction rate. To improve the performance and robustness of the prediction, ensemble models based on combinations of the RF, LR, DT and ANN models were developed to investigate their capability to predict groundwater potentiality in our large-scale mountainous area. According to the results, the ensembles of models showed better performance, especially from a prediction rate point of view. The AUC recorded an interesting progression: the maximum value was calculated for the RF-LR-DT-ANN ensemble model, with AUC = 0.791. Furthermore, to test the performance and reliability of different models and ensembles of models, several statistics metrics were applied, and a prioritization rank was carried out based on the compound factor method. Consequently, the best results were obtained for the RF-DT and RF-LR-DT ensembles. Finally, the methodology developed in this study may be useful for detecting groundwater potential zones, especially in mountainous areas with difficult access and where the application of geophysical methods of exploration remains costly and difficult to initiate for very large areas.

Author Contributions: Conceptualization, M.N., M.H., B.P., A.A., C.-W.L. and T.T.; methodology, M.N., M.H., B.P., A.E. and S.K.; software, M.N., B.P. and M.E.; validation, M.N. and H.E. (Hicham Elhimer); formal analysis, M.N., A.E., M.E., S.K. and H.E. (Hicham Elhimer); investigation, M.N., M.H., M.E. and T.T.; resources, M.H., H.E. (Hasna Eloudi) and M.O.; data curation, M.N., M.H., A.E., M.E., S.K., M.O. and H.E. (Hicham Elhimer); writing—original draft preparation, M.N., M.H., B.P., C.-W.L., S.K., H.E. (Hasna Eloudi), M.O., and T.T.; writing—review and editing, M.N., M.H., B.P., A.A., C.-W.L., and T.T.; visualization, A.A., A.E., M.E, H.E. (Hicham Elhimer) and H.E. (Hasna Eloudi); supervision, B.P., A.A. and T.T.; project administration, B.P., A.A. and T.T.; funding acquisition, B.P., C.-W.L. and A.A.; All authors have read and agreed to the published version of the manuscript.

Funding: This research is supported by the Centre for Advanced Modelling and Geospatial Information Systems (CAMGIS), University of Technology Sydney (UTS). This research was also supported by grants from the National Research Foundation of Korea, provided by the Korea government (No. 2019R1A2C1085686) and King Saud University, Riyadh, Saudi Arabia (grant number RSP-2021/14).

Institutional Review Board Statement: Not applicable.

Informed Consent Statement: Not applicable.

Data Availability Statement: The data presented in this study are available on request from the corresponding author.

Acknowledgments: The authors express their gratitude to the Hydraulic Basin Agency of Haouz and Oum Er-Rabia for their help and collaboration.

Conflicts of Interest: The authors declare no conflict of interest.

References

1. Price, M.F.; Byers, A.C.; Friend, D.A.; Kohler, T.; Price, L.W. (Eds.) *Mountain Geography: Physical and Human Dimensions*; University of California Press: Berkeley, CA, USA, 2013.
2. Kohler, T.; Giger, M.; Hurni, H.; Ott, C.; Wiesmann, U.; Von Dach, S.W.; Maselli, D. Mountains and Climate Change: A Global Concern. *Mt. Res. Dev.* **2010**, *30*, 53–55. [CrossRef]
3. FAO. *State of the World's Forests 2003*; FAO: Rome, Italy, 2003.
4. Bouchaou, L.; Michelot, J.; Qurtobi, M.; Zine, N.; Gaye, C.; Aggarwal, P.; Marah, H.; Zerouali, A.; Taleb, H.; Vengosh, A. Origin and residence time of groundwater in the Tadla basin (Morocco) using multiple isotopic and geochemical tools. *J. Hydrol.* **2009**, *379*, 323–338. [CrossRef]
5. Bouchaou, L.; Michelot, J.; Vengosh, A.; Hsissou, Y.; Qurtobi, M.; Gaye, C.; Bullen, T.; Zuppi, G. Application of multiple isotopic and geochemical tracers for investigation of recharge, salinization, and residence time of water in the Souss–Massa aquifer, southwest of Morocco. *J. Hydrol.* **2008**, *352*, 267–287. [CrossRef]
6. Bouimouass, H.; Fakir, Y.; Tweed, S.; Leblanc, M. Groundwater recharge sources in semiarid irrigated mountain fronts. *Hydrol. Process.* **2020**, *34*, 1598–1615. [CrossRef]
7. Rathay, S.; Allen, D.; Kirste, D. Response of a fractured bedrock aquifer to recharge from heavy rainfall events. *J. Hydrol.* **2018**, *561*, 1048–1062. [CrossRef]
8. Voeckler, H.; Allen, D.M. Estimating regional-scale fractured bedrock hydraulic conductivity using discrete fracture network (DFN) modeling. *Hydrogeol. J.* **2012**, *20*, 1081–1100. [CrossRef]
9. Shenga, Z.D.; Baroková, D.; Šoltész, A. Numerical modeling of groundwater to assess the impact of proposed railway construction on groundwater regime. *Pollack Period.* **2018**, *13*, 187–196. [CrossRef]
10. Ozdemir, A. Using a binary logistic regression method and GIS for evaluating and mapping the groundwater spring potential in the Sultan Mountains (Aksehir, Turkey). *J. Hydrol.* **2011**, *405*, 123–136. [CrossRef]
11. Razandi, Y.; Pourghasemi, H.R.; Neisani, N.S.; Rahmati, O. Application of analytical hierarchy process, frequency ratio, and certainty factor models for groundwater potential mapping using GIS. *Earth Sci. Inform.* **2015**, *8*, 867–883. [CrossRef]
12. Madani, A.; Niyazi, B. Groundwater potential mapping using remote sensing techniques and weights of evidence GIS model: A case study from Wadi Yalamlam basin, Makkah Province, Western Saudi Arabia. *Environ. Earth Sci.* **2015**, *74*, 5129–5142. [CrossRef]
13. Golkarian, A.; Naghibi, S.A.; Kalantar, B.; Pradhan, B. Groundwater potential mapping using C5.0, random forest, and multivariate adaptive regression spline models in GIS. *Environ. Monit. Assess.* **2018**, *190*, 149. [CrossRef] [PubMed]
14. Ganguly, K.K.; Nahar, N.; Hossain, B.M.M. A machine learning-based prediction and analysis of flood affected households: A case study of floods in Bangladesh. *Int. J. Disaster Risk Reduct.* **2019**, *34*, 283–294. [CrossRef]
15. Rizeei, H.M.; Pradhan, B.; Saharkhiz, M.A.; Lee, S. Groundwater aquifer potential modeling using an ensemble multi-adoptive boosting logistic regression technique. *J. Hydrol.* **2019**, *579*, 124172. [CrossRef]
16. Hssaisoune, M.; Bouchaou, L.; Sifeddine, A.; Bouimetarhan, I.; Chehbouni, A. Moroccan Groundwater Resources and Evolution with Global Climate Changes. *Geosciences* **2020**, *10*, 81. [CrossRef]
17. Milewski, A.; Seyoum, W.M.; Elkadiri, R.; Durham, M. Multi-Scale Hydrologic Sensitivity to Climatic and Anthropogenic Changes in Northern Morocco. *Geosciences* **2019**, *10*, 13. [CrossRef]
18. Ouatiki, H.; Boudhar, A.; Ouhinou, A.; Arioua, A.; Hssaisoune, M.; Bouamri, H.; Benabdelouahab, T. Trend analysis of rainfall and drought over the Oum Er-Rbia River Basin in Morocco during 1970–2010. *Arab. J. Geosci.* **2019**, *12*, 128. [CrossRef]
19. Barakat, A.; Hilali, A.; El Baghdadi, M.; Touhami, F. Assessment of shallow groundwater quality and its suitability for drinking purpose near the Béni-Mellal wastewater treatment lagoon (Morocco). *Hum. Ecol. Risk Assess. Int. J.* **2020**, *26*, 1476–1495. [CrossRef]
20. Barakat, A.; Mouhtarim, G.; Saji, R.; Touhami, F. Health risk assessment of nitrates in the groundwater of Beni Amir irrigated perimeter, Tadla plain, Morocco. *Hum. Ecol. Risk Assess. Int. J.* **2019**, *26*, 1864–1878. [CrossRef]
21. Barakat, A.; Meddah, R.; Afdali, M.; Touhami, F. Physicochemical and microbial assessment of spring water quality for drinking supply in Piedmont of Béni-Mellal Atlas (Morocco). *Phys. Chem. Earth Parts A/B/C* **2018**, *104*, 39–46. [CrossRef]
22. Bouchaou, L.; Chauve, P.; Mudry, J.; Mania, J.; Hsissou, Y. Structure et fonctionnement d'un hydrosystème karstique de montagne sous climat semi-aride: Cas de l'Atlas de Beni-Mellal (Maroc). *J. Afr. Earth Sci.* **1997**, *25*, 225–236. [CrossRef]
23. Bouchaou, L.; Michelot, J.L. Contribution Des Isotopes à l'étude de La Recharge Des Aquifères de La Région de Béni Mellal (Tadla, Maroc). *IAHS-AISH Publ.* **1997**, *244*, 37–44.
24. Ettazarini, S. Incidences of water-rock interaction on natural resources characters, Oum Er-Rabia Basin (Morocco). *Environ. Geol.* **2004**, *47*, 69–75. [CrossRef]
25. Ettazarini, S. Groundwater pollution risk mapping for the Eocene aquifer of the Oum Er-Rabia basin, Morocco. *Environ. Geol.* **2006**, *51*, 341–347. [CrossRef]
26. Archambault, C.; Combe, M.; Ruhard, J.P. The Phosphates Plateau. Water Resources. *Notes Mém. Sér. Géol. Rabat* **1975**, *231*, 239–258.
27. Brede, R. Structural aspects of the Middle and the High Atlas (Morocco) phenomena and causalities. *Geologische Rundschau* **1992**, *81*, 171–184. [CrossRef]

28. Boudhar, A.; Ouatiki, H.; Bouamri, H.; Lebrini, Y.; Karaoui, I.; Hssaisoune, M.; Arioua, A.; Benabdelouahab, T. Hydrological Response to Snow Cover Changes Using Remote Sensing over the Oum Er Rbia Upstream Basin, Morocco. In *Mapping and Spatial Analysis of Socio-Economic and Environmental Indicators for Sustainable Development*; Springer: Cham, Switzerland, 2020; pp. 95–102.

29. Kordestani, M.D.; Naghibi, S.A.; Hashemi, H.; Ahmadi, K.; Kalantar, B.; Pradhan, B. Groundwater potential mapping using a novel data-mining ensemble model. *Hydrogeol. J.* **2019**, *27*, 211–224. [CrossRef]

30. Ouatiki, H.; Boudhar, A.; Tramblay, Y.; Jarlan, L.; Benabdelouhab, T.; Hanich, L.; El Meslouhi, M.R.; Chehbouni, A. Evaluation of TRMM 3B42 V7 Rainfall Product over the Oum Er Rbia Watershed in Morocco. *Climate* **2017**, *5*, 1. [CrossRef]

31. Shaban, A.; Khawlie, M.; Abdallah, C. Use of remote sensing and GIS to determine recharge potential zones: The case of Occidental Lebanon. *Hydrogeol. J.* **2005**, *14*, 433–443. [CrossRef]

32. Yeh, H.-F.; Cheng, Y.-S.; Lin, H.-I.; Lee, C.-H. Mapping groundwater recharge potential zone using a GIS approach in Hualian River, Taiwan. *Sustain. Environ. Res.* **2016**, *26*, 33–43. [CrossRef]

33. Botzen, W.J.W.; Aerts, J.C.J.H.; Bergh, J.V.D. Individual preferences for reducing flood risk to near zero through elevation. *Mitig. Adapt. Strat. Glob. Chang.* **2013**, *18*, 229–244. [CrossRef]

34. Gokceoglu, C.; Sonmez, H.; Nefeslioglu, H.A.; Duman, T.Y.; Çan, T. The 17 March 2005 Kuzulu landslide (Sivas, Turkey) and landslide-susceptibility map of its near vicinity. *Eng. Geol.* **2005**, *81*, 65–83. [CrossRef]

35. Riley, S.J.; DeGloria, S.D.; Elliot, R. Index that quantifies topographic heterogeneity. *Intermt. J. Sci.* **1999**, *5*, 23–27.

36. Juliev, M.; Mergili, M.; Mondal, I.; Nurtaev, B.; Pulatov, A.; Hübl, J. Comparative analysis of statistical methods for landslide susceptibility mapping in the Bostanlik District, Uzbekistan. *Sci. Total Environ.* **2019**, *653*, 801–814. [CrossRef]

37. Breiman, L. Random forests. *Mach. Learn.* **2001**, *45*, 5–32. [CrossRef]

38. Knoll, L.; Breuer, L.; Bach, M. Large scale prediction of groundwater nitrate concentrations from spatial data using machine learning. *Sci. Total Environ.* **2019**, *668*, 1317–1327. [CrossRef] [PubMed]

39. Kuhn, M.; Johnson, K. *Applied Predictive Modeling*; Springer: New York, NY, USA, 2013.

40. Bourenane, H.; Guettouche, M.S.; Bouhadad, Y.; Braham, M. Landslide hazard mapping in the Constantine city, Northeast Algeria using frequency ratio, weighting factor, logistic regression, weights of evidence, and analytical hierarchy process methods. *Arab. J. Geosci.* **2016**, *9*, 1–24. [CrossRef]

41. Saito, H.; Nakayama, D.; Matsuyama, H. Comparison of landslide susceptibility based on a decision-tree model and actual landslide occurrence: The Akaishi Mountains, Japan. *Geomorphology* **2009**, *109*, 108–121. [CrossRef]

42. Tseng, C.-J.; Lu, C.-J.; Chang, C.-C.; Chen, G.-D.; Cheewakriangkrai, C. Integration of data mining classification techniques and ensemble learning to identify risk factors and diagnose ovarian cancer recurrence. *Artif. Intell. Med.* **2017**, *78*, 47–54. [CrossRef]

43. Quinlan, R.C. *4.5: Programs for Machine Learning*; Morgan Kaufmann Publishers Inc.: San Francisco, CA, USA, 1993.

44. Kawabata, D.; Bandibas, J. Landslide susceptibility mapping using geological data, a DEM from ASTER images and an Artificial Neural Network (ANN). *Geomorphology* **2009**, *113*, 97–109. [CrossRef]

45. Song, Y.; Niu, R.; Xu, S.; Ye, R.; Peng, L.; Guo, T.; Li, S.; Chen, T. Landslide Susceptibility Mapping Based on Weighted Gradient Boosting Decision Tree in Wanzhou Section of the Three Gorges Reservoir Area (China). *ISPRS Int. J. Geo-Inf.* **2018**, *8*, 4. [CrossRef]

46. Demir, G.; Aytekin, M.; Akgun, A. Landslide susceptibility mapping by frequency ratio and logistic regression methods: An example from Niksar–Resadiye (Tokat, Turkey). *Arab. J. Geosci.* **2014**, *8*, 1801–1812. [CrossRef]

47. Yesilnacar, E.; Topal, T. Landslide susceptibility mapping: A comparison of logistic regression and neural networks methods in a medium scale study, Hendek region (Turkey). *Eng. Geol.* **2005**, *79*, 251–266. [CrossRef]

48. Fawcett, T. An introduction to ROC analysis. *Pattern Recognit. Lett.* **2006**, *27*, 861–874. [CrossRef]

49. Avand, M.; Janizadeh, S.; Bui, D.T.; Pham, V.H.; Ngo, P.T.T.; Nhu, V.-H. A tree-based intelligence ensemble approach for spatial prediction of potential groundwater. *Int. J. Digit. Earth* **2020**, *13*, 1408–1429. [CrossRef]

50. Nguyen, P.T.; Ha, D.H.; Avand, M.; Jaafari, A.; Nguyen, H.D.; Al-Ansari, N.; Van Phong, T.; Sharma, R.; Kumar, R.; Van Le, H.; et al. Soft Computing Ensemble Models Based on Logistic Regression for Groundwater Potential Mapping. *Appl. Sci.* **2020**, *10*, 2469. [CrossRef]

51. Nampak, H.; Pradhan, B.; Manap, M.A. Application of GIS based data driven evidential belief function model to predict groundwater potential zonation. *J. Hydrol.* **2014**, *513*, 283–300. [CrossRef]

52. Karami, G.H.; Bagheri, R.; Rahimi, F. Determining the groundwater potential recharge zone and karst springs catchment area: Saldoran region, western Iran. *Hydrogeol. J.* **2016**, *24*, 1981–1992. [CrossRef]

53. Tiwari, M.K.; Chatterjee, C. Development of an accurate and reliable hourly flood forecasting model using wavelet–bootstrap–ANN (WBANN) hybrid approach. *J. Hydrol.* **2010**, *394*, 458–470. [CrossRef]

Freshwater–Saltwater Interactions in a Multilayer Coastal Aquifer (Ostia Antica Archaeological Park, Central ITALY)

Margherita Bonamico, Paola Tuccimei *, Lucia Mastrorillo and Roberto Mazza

Dipartimento di Scienze, Università "Roma Tre", 00146 Roma, Italy; bonamicomargherita@gmail.com (M.B.); lucia.mastrorillo@uniroma3.it (L.M.); roberto.mazza@uniroma3.it (R.M.)
* Correspondence: paola.tuccimei@uniroma3.it

Abstract: An integrated research approach consisting of hydrogeologic and geochemical methods was applied to a coastal aquifer in the Ostia Antica archaeological park, Roma, Italy, to describe freshwater–saltwater interactions. The archaeological park of Ostia Antica is located on the left bank of the Tevere River delta which developed on a morphologically depressed area. The water monitoring program included the installation of multiparametric probes in some wells inside the archaeological area, with continuous measurement of temperature, electrical conductivity, and water table level. Field surveys, water sampling, and major elements and bromide analyses were carried out on a seasonal basis in 2016. In order to understand the detailed stratigraphic setting of the area, three surface boreholes were accomplished. Two distinct circulations were identified during the dry season, with local interaction in the rainy period: an upper one within the archaeological cover, less saline and with recharge inland; and a deeper one in the alluvial materials of Tevere River, affected by salinization. Oxygen and carbon isotopic signature of calcite in the sediments extracted from the boreholes, along with major elements and Br concentration, allowed us to recognize the sources of salinity (mainly, local interaction with Roman salt pans and agricultural practices) and the processes of gas–water–rock interaction occurring in the area. All these inferences were confirmed and strengthened by PCA analysis of physicochemical data of groundwater.

Keywords: freshwater–saltwater interactions; multilayer coastal aquifer; hydro-geochemistry; Tevere River delta; Ostia Antica archaeological park

Citation: Bonamico, M.; Tuccimei, P.; Mastrorillo, L.; Mazza, R. Freshwater–Saltwater Interactions in a Multilayer Coastal Aquifer (Ostia Antica Archaeological Park, Central ITALY). *Water* **2021**, *13*, 1866. https://doi.org/10.3390/w13131866

Academic Editors: Evangelos Tziritis and Andreas Panagopoulos

Received: 1 June 2021
Accepted: 1 July 2021
Published: 4 July 2021

Publisher's Note: MDPI stays neutral with regard to jurisdictional claims in published maps and institutional affiliations.

1. Introduction

Salinization of fresh groundwater is a global issue and a major threat to sustainable groundwater resources [1]. It is mainly caused by evaporite dissolution [2], fossil seawater [3] and seawater intrusion [4]. Seawater intrusion is defined as the mass transport of saline waters into zones previously occupied by fresher waters [5] due to natural processes or human activities.

Worldwide, aquifers in low-lying coastal areas are threatened by saltwater occurrence, as a result of small head gradients, high groundwater abstraction rates, and drain management of the landscape [6]. Urban and industrial development and the expansion of irrigated agriculture have led to a drastic increase in the exploitation of groundwater resources. The over-exploitation of coastal aquifers has caused a seawater intrusion and has seriously degraded groundwater quality [1]. In order to assess the influence of seawater on a coastal aquifer, it is essential to elucidate the source(s) of salinity and to understand the hydraulic and hydrogeochemical conditions [7].

In coastal environments, rivers are preferential way to lead seawater inland through salt-wedge intrusion [8]. Some authors [9] reported how anthropogenic land subsidence, land reclamation drainage system, and groundwater pumping in coastal areas of North Queensland, Australia, have an impact on dynamics of seawater intrusion in the phreatic aquifer. Another study on the salinization of a coastal aquifer in South Korea [10] showed that water hydrochemistry is controlled by several intermixed processes, such as seawater mixing, anthropogenic contamination, and water–rock interaction. A comprehensive review of groundwater salinization processes in coastal areas of the Mediterranean region was recently published [11].

Lately, a significant amount of literature has focused on subsurface water exchange between land and sea, both in coastal areas and marine environments. This issue was addressed using either physical approaches such as hydrogeological analysis, seepage meters, and geophysical methods, or employing chemical methods such as terrestrial water quality analysis, marine chemistry, and nutrients [12]. Modelling approaches were also applied [12]. Many papers were devoted to sub-seafloor freshened groundwater research and to application of the modern 3D shallow seismic technology to detect offshore freshened groundwater systems, for example in the Atlantic Ocean [13,14] and in the Baltic Sea [15–17].

No similar studies are available for the submerged Tevere River Delta (TRD) in central Italy, which is the area investigated in this work, and proximate offshore areas. Existing research made use of deep seismic technology to reconstruct the crustal structure and the complex geological settings of the Central Mediterranean Sea around the Italian Peninsula, e.g. ref. [18] or were focused on the reconstruction of the sedimentary succession of the emerged Tevere River Delta [19]. Furthermore, the coastal plain of the Tevere River is not among the hydrogeological complexes of the Tyrrhenian Sea with highest discharge. Most relevant water resources (in the order of some hundreds of Mm^3/year) are available for the Volturno Plain, Pontina Plain, Solofrana–Sarno Plains and Sele Plain [20].

This study focuses on the freshwater–saltwater interactions in a multilayer coastal aquifer of Roma, hosted in the Tevere River Delta depositional sequence. Recent studies [21] suggest that the Roman coastal aquifer reaches electrical conductivity (EC) values up to 5000 μS/cm and that groundwater salinization could be related to a combination of land use and historical development of the TRD, rather than to seawater. In the Roman coastal aquifer, EC shows a wide variability from area to area, due to agriculture practices which employ brackish water for irrigation and for the presence of the ancient salt pans [22]. Moreover, the salinization of groundwater on the left bank of Tevere River could be related to the salt-wedge intrusion along the river course up to a distance of 8 km from the mouth [23] and to a lateral inflow of the river into the aquifer, due to different hydraulic heads and triggered by the drainage system [21].

Finally, in the dune part of the Roman coastal aquifer covered by *Pinus pinea* forest, the chemistry of groundwater is dependent on the barrier-effect accomplished by canopies to the wind-transported sea salt aerosol, periodically discharged into the aquifer by rain-falls [24]. The deposition of sea salt aerosol mainly occurs in spring and summer when the winds blow from the west at speeds above 4 m/s [25].

The goal of this paper is the study of the groundwater of the Ostia Antica archaeo-logical park, which is a part of the vast Roman coastal aquifer (Roma, central Italy) using an integrated hydrogeological and hydrogeochemical approach. In particular, we aim to deepen the knowledge of freshwater–saltwater interactions and identify the sources of local salinity. Gas–water–rock interaction processes were considered to justify groundwater composition and understand the mixing dynamics of the aquifer with the Tevere River and the marine water.

2. Study Area

The study area, located SW of Roma (Italy), is bounded to the S by the Castel Fusano Reserve, to the NNE by the Tevere River, and to the E by the Tyrrhenian Sea. It includes the archaeological park of Ostia Antica, with an extension of 1.5 km^2, and the Castle of Julius II to the NNE (Figure 1).

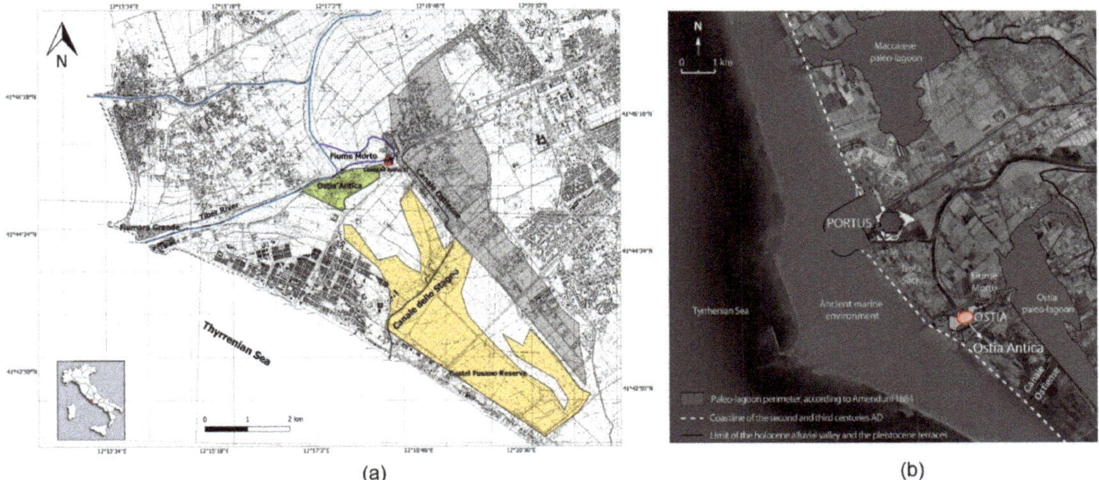

Figure 1. (**a**) Study area features: Ostia Antica archaeological park (green area), the old river meander called "Fiume Morto" with the Castle of Julius II (red dot), Tevere River (blue line), reclamation channels (thick black line), Roman salt pans (grey area), and the Castel Fusano reserve (yellow area). (**b**) Localization of the Ostia Antica archaeological park and the old meander called "Fiume Morto" with respect to paleo-lagoon areas of Ostia and Maccarese (modified from [26]).

The area, with an elevation of approximately 2–4 m a.s.l., has a typically Mediterranean climate with a warm and dry period from mid-May to mid-August. Meteorological data (period 1971–2000) show that the average annual rainfall is 741 mm, distributed over 72 days, with a minimum in summer and a maximum in autumn. The month of January has the coldest temperatures with an average of 8.6 °C, while August is the warmest month with an average temperature of 24.1 °C [27].

The study site develops on the left bank of the Tevere River on a morphologically flat environment. The general evolution of the TRD after the Last Glacial Maximum was mainly driven by the post-glacial uplift of the sea level (between 18,000 and 6000 years ago), and by the variations in river sediment discharge in the last 6000 years [19]. The shoreline progressively moved away from the coastal basins that developed into brackish ponds, currently reclaimed. The basin located on the left bank of the river, called Ostia paleo-lagoon, developed behind the current Ostia Antica archaeological park, extending to the S for at least 6 km, and communicating with the sea through the present Canale dello Stagno channel (Figure 1a). The basin of Ostia and the one located on the right bank, the basin of Maccarese (Figure 1b) have been exploited as salt pans from Roman to more recent times.

Different phases of progradation and erosion of the delta alternated. During the 15th century, coinciding with the four historically documented major floods of the Tevere River (AD 1530, 1557, 1598, and 1606), a rapid progradation of TRD took place. During the disastrous flood of 1557 AD, the Tevere River changed course, leaving a meander in the area of Ostia Antica, which later became a lake and then an isolated marshy area, now known as "Fiume Morto" (Figure 1). Towards the end of the 19th century, a law on reclamation was approved and an important drainage system was built [28]. The reclamation system consists of a network of channels that collect rainwater and discharge groundwater from a pumping station into the sea, with the function of preventing the flooding of low ground and keeping groundwater below the ground surface.

The inland delta consists of transitional mid-littoral deposits, such as sands, silty-sands, and clays interbedded with gravels, with volcanoclastic grains from the erosion of Colli Albani volcanic rocks. They were deposited during several Pleistocene transgressive cycles, with an overall thickness of 50 m.

The seaside area of the delta includes dune cordons, grown parallel to the coast during the progradation stages of the last 2500 years. It is characterized by Holocene deposits, mostly composed of coastal and eolian sands, with a lateral transition to the 10–30 m thick alluvial sediments of Tevere River [29].

A fresh-brackish water multilayer aquifer is hosted in the Pleistocene and Holocene sediments of the TRD left bank. The Plio-Pleistocene clayey bedrock acts as a regional basal aquiclude (Figure 2). Throughout the aquifer, the preferential direction of groundwater flow is towards the sea, with the exception of the area surrounding Ostia Antica, where the reclamation pumping system controls the local base level of groundwater circulation. The main direction of groundwater flow in the archaeological park is from E to W, towards an area of induced lowering of the water table up to −2 m a.s.l. [21,30,31].

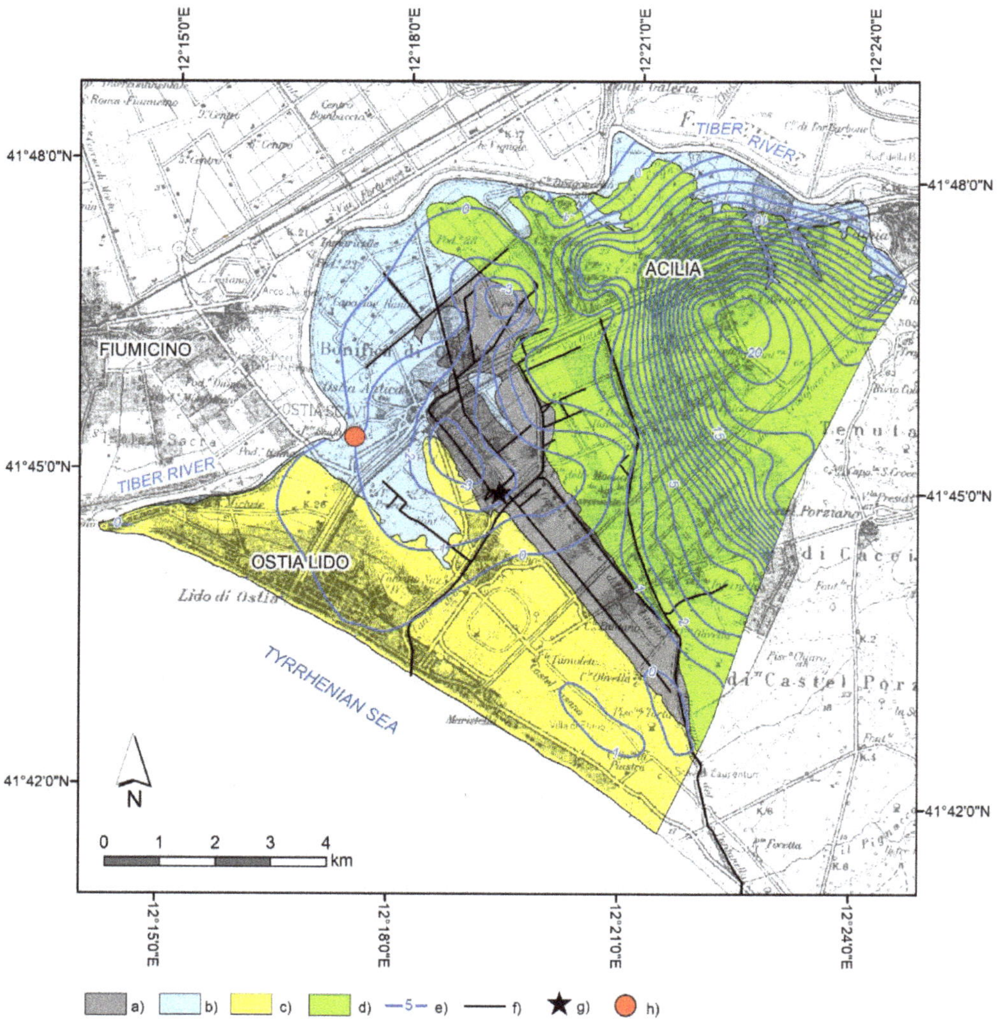

Figure 2. Hydrogeological settings of the left bank of the Tevere River Delta. Legend: (**a**) swamp deposits HOLOCENE; (**b**) sandy, silty and clayey alluvial deposits HOLOCENE; (**c**) sandy beach deposits HOLOCENE; (**d**) heterogeneous clastic deposits (sandy-silt and clay deposits interbedded with gravels) PLEISTOCENE; (**e**) groundwater contour lines (1 m interval); (**f**) reclamation channels; (**g**) Ostia pumping station; (**h**) study area (modified from [32]).

3. Material and Methods

Groundwater was collected from wells located inside the archaeological area (P1, P2, P8, P9, and P10) near the old meander (P6 and P7) and in the castle of Julius II (P4); the Tevere River was sampled near the archaeological park (P5) (Figure 3a,b).

Figure 3. (**a**) Location of sampling wells (red points) and drillings (yellow points). (**b**) Some monitored wells: P1, P2, P6, and P7. For geographic coordinates in Figure 3a, the reader is referred to Figures 1 and 2.

The field surveys were carried out on a seasonal basis in the months of April, June, October, and December of 2016 in order to obtain representative data of the dry and wet seasons. Periodical field surveys consisted of manually measuring the static level, pH, temperature, and electrical conductivity of groundwater and taking water samples. Groundwater level and electrical conductivity (EC) were monitored continuously since 2014 at 6 h intervals at wells P1, P2, and P3. The 2016 rainfall data measured by the Ostia pluviometric station near the study area were also acquired.

It is worth noting that the well bottoms are placed below the sea level, and vary from a maximum of about -2.5 m a.s.l. in P4 and a minimum of -0.17 m a.s.l. for well P10.

Temperature, pH, and EC were measured using a portable multiparameter probe. Samples were collected and stored in clean PET bottles, rinsed thrice with native water, before sampling. An aliquot for cation analysis was filtered (0.45 μm Millipore filters) and acidified with concentrated HNO_3 (68%) before storage. In order to avoid chemical reactions, samples were stored at a temperature of 4 °C.

Alkalinity was measured in the field by titration with 0.02 N HCl, Ca^{2+} and Mg^{2+} were determined using the EDTA titration method, Na^+ and K^+ by flame emission photometric method, Cl^- and Br^- by potentiometric methods with ion-selective electrodes, and SO_4^{2-} by a colorimetric method using turbidimetric techniques [33]. The precision of photometric, potentiometric, and spectrometric measurements was always better than 0.5%. The accuracy of measurements, checked against standard reference material was found to be generally within 4%.

In June 2016 three 3–5 m deep drillings (p1, p3, and p6) were performed next to wells P1, P3, and P6, employing a two-man auger Stihl BT 360 motor drill (Figure 3a). A small amount of sediment samples collected along the three drillings were dried at 100 °C and made to react with 1N HCl to check the presence of calcite cement observed in the sedimentary sequence.

Saturation index for calcite of groundwater was modelled with PHREEQC for Windows (version 2.18.00), a hydrogeochemical transport software developed by [34].

Base EXchange index (BEX), sensu [35], was calculated for groundwater average composition, as reported below:

$$[Cl^- - (Na^+ + K^+)]/Cl^- \tag{1}$$

where ion concentration is expressed in meq/L.

Carbon and oxygen isotope composition of calcite cements were determined at the IGAG, CNR laboratory (Montelibretti, Roma, Italy) using a Mat 252V mass spectrometer, manufactured by Finnigan (Bremen, Germany) according to the procedure described in [36].

OriginPro 9.0 (ADALTA) software was used to apply principal component analysis (PCA) to water chemistry data in order to transform a large set of variables into a smaller one that still contains most of the information and investigate the main geochemical trends. Eleven variables (chemical parameters) were chosen for the multivariate analysis: HCO_3, Cl, SO_4, Ca, Mg, Na, K, Br, Water table elevation, Br/Cl, and EC.

4. Results

4.1. Groundwater Data

Groundwater physicochemical data are shown in Table S1 (Supplementary Material). Lowest piezometric levels were recorded in October 2016 and highest in December 2016; in all survey campaigns the minimum static level was found in wells P6 and P4, at $-1/-1.4$ m a.s.l. and -1.3 m a.s.l., respectively. Wells P6 and P7, located outside the archaeological area, generally had a static level below sea level. Tevere River had a constant level of +0.01 m a.s.l. in April, June, and October 2016, while in December 2016 the level of the river was lower (-0.31 m a.s.l.). Finally, wells P2, P8, P9, and P10 showed groundwater table fluctuations between -0.5 and +0.4 m a.s.l.

The average temperature of groundwater was 16.8 °C, while the temperature of the Tevere River ranged from 15.1 (April 2016) to 19.3 °C (June 2016). Groundwater and Tevere River generally showed neutral and slightly basic pH values, except for the June survey when waters were slightly acidic (6.6 on average). EC values range from 404 (well P10 in December 2016) to 4820 (well P6 in December 2016) µS/cm. Two main groups of groundwater may be recognized: P2, P3, P4, P8, P9, and P10 with EC from 400 to 1450 µs/cm, and P1, P6, and P7 with EC between 1350 and 4800 µS/cm.

Figure 4 shows the static levels recorded in continuum in wells P1, P2, and P3 and the daily rainfall value measured in Figure 4a. In Figure 4b the EC values recorded in P1, P2, and P3 were reported. It can be observed that the water table of P1 ranged between approximately -0.7 m a.s.l. in summer/autumn and +0.3 m a.s.l. in winter; well P2 had values on average higher than P1 with a static level fluctuation between -0.55 m a.s.l. (in September–October 2016) and about +0.3 a.s.l. (in December 2016). Well P3 showed a maximum static level of +0.4 m a.s.l in December 2016 and a minimum value of -0.4 m a.s.l. in September–October 2016.

4.2. Stratigraphy

The stratigraphic sequence unveiled by the drillings consists of 2–2.3 m of anthropic backfill (pozzolanaceous and/or alluvial material with layers filled with calcite cements), and sandy or silty-clayey deposits up to a depth of about -3.7 m below ground level, where a grey clayey layer was identified. The water table was crossed by p6 drilling at -1 m a.s.l (-3.65 m below ground level). All sediment samples taken at different depths from the three drillings reacted with HCl, confirming the presence of calcite cement, very frequent in alluvial and pozzolanic materials. Oxygen and carbon isotopic compositions of calcite cements in the sedimentary sequence of p1 and p6 drillings are different. Calcite in p1 is characterized by $\delta^{18}O$ and $\delta^{13}C$ values of -5.06 and $-5.50‰$ (VPDB), respectively, whereas p6 sample, has a more negative isotopic signature, specifically of −8.96 and −12.88‰ (VPDB) for oxygen and carbon.

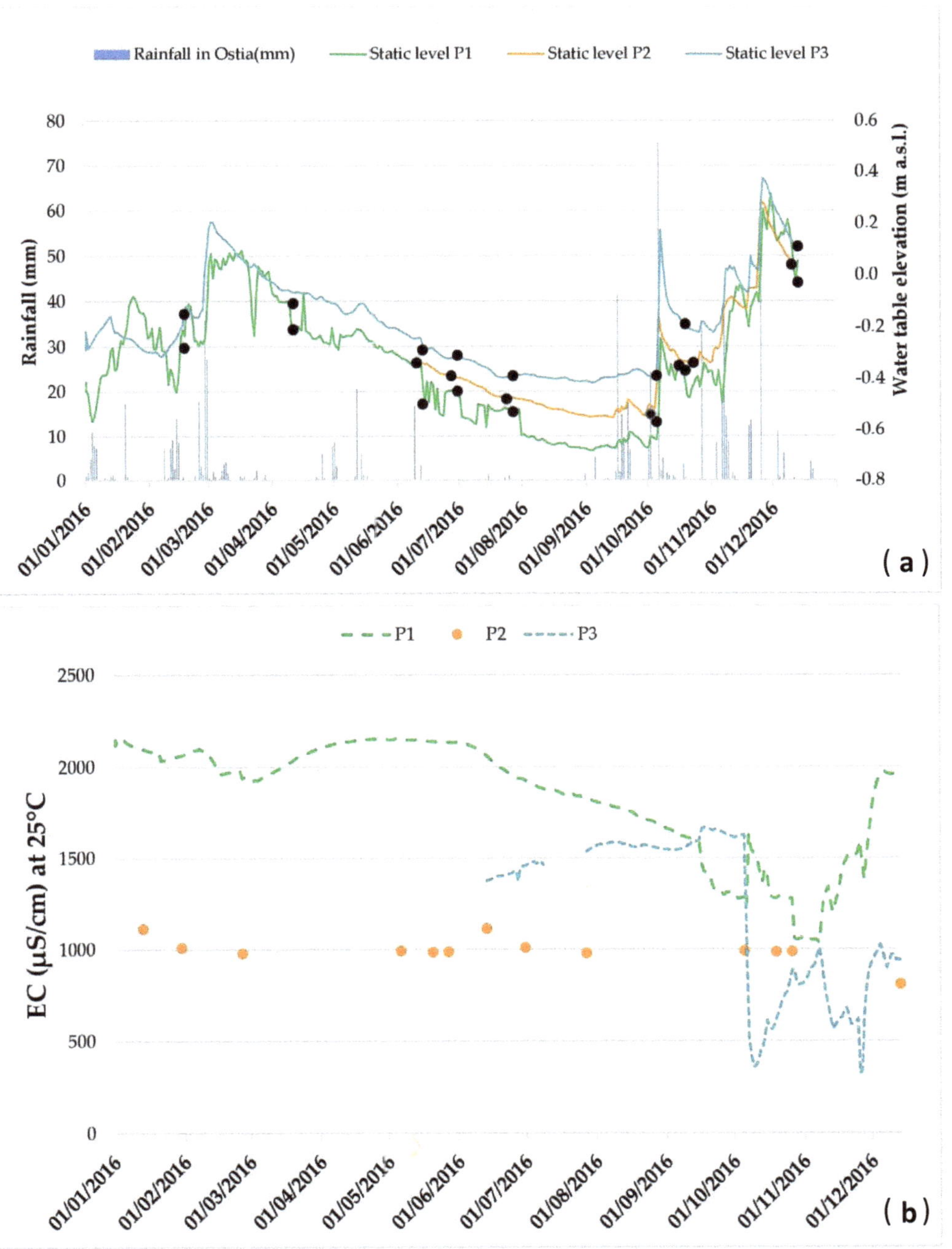

Figure 4. (**a**) Rainfall amount and water table level recorded at wells P1, P2, and P3 during the year 2016. Manual water level measurements (black dots) are superimposed on the continuous water level recorded at wells P1, P2, and P3. (**b**) Electrical conductivity (EC) of wells P1, P2 (discontinuous manual measurements), and P3 during the year 2016.

4.3. Hydrochemistry

Major ions and Br^- concentrations of groundwater are reported in Table S2 (Supplementary Material). Main cations and anions abundances are shown in Figure 5 to recognize the hydrochemical facies. Groundwater from wells P2, P4, and P9 are alkaline bicarbonate, with data points located at the boundary with the calcium-magnesium bicarbonate facies, therefore rich in Ca^{2+} (and less in Mg^{2+}), as well as in alkalis (Figure 5). Groundwater from P3, P8, and P10 belongs to the calcium-magnesium bicarbonate facies, enriched in calcium but also in alkalis (mainly Na^+). All these samples belong to the shallower circulation.

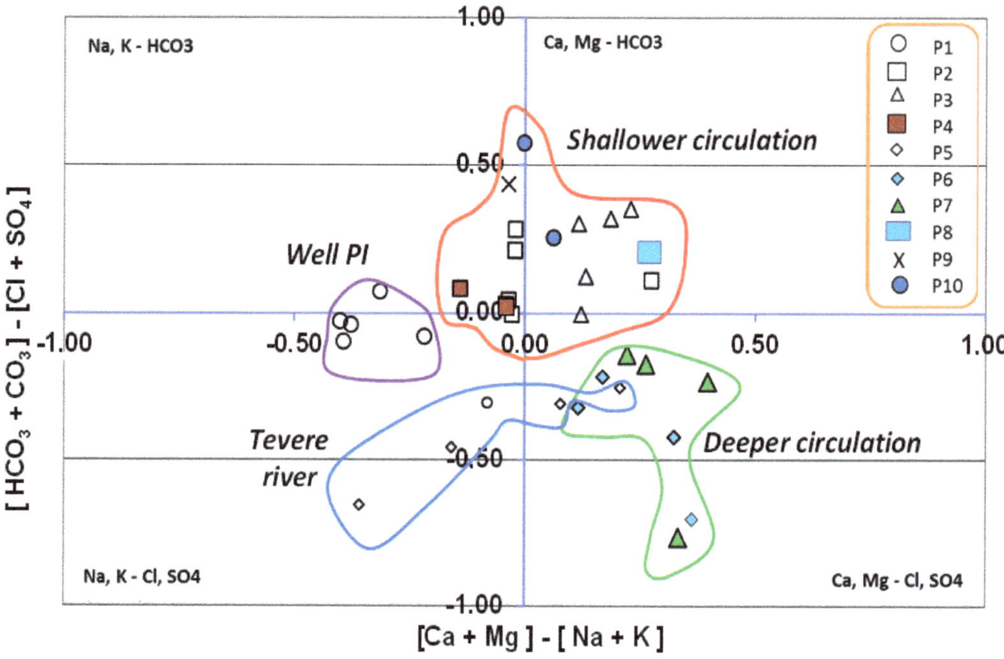

Figure 5. Chada diagram with composition of groundwater sampled during the year 2016.

P6 and P7 groundwater falls in the calcium-magnesium chloride-sulphate field of Figure 5 and are referred to the deeper circulation. The Tevere River (P5) is a mixing between alkaline chloride-sulphate and calcium-magnesium chloride-sulphate waters. Finally, P1 is an alkaline chloride-sulphate water, with an enrichment in bicarbonates and Mg^{2+}. Average Cl/Br (molar) ratio of the wells P2, P3, P8, P9, and P10 is 180; Tevere River (P5) and P1 are characterized by ratios of 260 and 185, respectively; wells P6 and P7 have an average Cl/Br (molar) ratio of 400.

These considerations are supported by the PCA analysis, as discussed in the following section.

5. Discussion

5.1. Groundwater Data

Interpretation of the groundwater levels data suggests the presence of two possible circulations. An upper circulation was identified in the archaeological park and a lower one in the abandoned meander of the Tevere River area; the two circulations have the same flow direction (WSW–ENE) and base level (the Primary Collector channel, at −2 m a.s.l), but are characterized by different values of salt concentration (EC values). The shallowest aquifer, hosted in the pozzolanic and sandy materials in the archaeological area, displays the highest water table elevation (0.3–0.2 m a.s.l.). The deepest circulation, flowing in the alluvial deposits, has the highest groundwater level near the Tevere River bank, between -0.1 and -0.6 m a s.l. The aquiclude interposed between the two circulations probably consists of a low permeability sandy clayey layer. This layer was probably intersected by two out of three drillings from about -1.00 to -1.40 m a.s.l.

In proximity of well P1 the two circulations seem to be hydraulically connected during the rainy period (in December 2016), when the levels of the water table correspond, while they are well distinguished during the dry season (in October 2016). This is confirmed by the physicochemical analyses of groundwater from well P1 which displays a constant composition throughout the year, compared with other samples with medium-high EC values whose EC and hydrochemical facies change greatly depending on the season. No information is available on the lateral continuity of the deepest circulation, so it is not possible to predict if it is present below the shallowest one. Current hydrogeochemical data seem to suggest that the two circulations are not completely isolated during the dry season when groundwater static levels are different.

The coastal aquifer is recharged mainly in autumn and winter when rainfalls are concentrated (Figure 4a), as shown by the sudden rise of piezometric levels in all monitoring points (P1, P2, and P3). Water-table fluctuations are about 1 m between the dry season and the rainy one.

The rise of groundwater levels recorded in October–December 2016 in monitoring wells P1, P2, and P3 is directly correlated with a general decrease in the concentration of dissolved salts. In the period January–March 2016 when rainfalls were modest, only a slight decrease in EC values was measured for P1 and P3 groundwater, with values of approximately 1500–2000 μS/cm. Finally, no continuous record of EC values was available for well P2, but using existing discontinuous data it can be observed that EC values ranged between 800 and 1110 μS /cm (Figure 4b).

An inverse correlation between EC and groundwater levels was observed in the three monitored wells, where a relative increase in groundwater level corresponded to a decrease in water salinity. This is particularly evident for well P1 where the abundant rainfalls of October 2016 produced an increase of about 0.5 m in the piezometric level and a decrease in EC values from 1620 to 1000 μS/cm (Figure 4).

Another correlation can be underlined between EC values and well bottom elevations. Groundwater with higher average EC values (1350–4800 μS/cm) were sampled from wells P1, P6, and P7 whose bottom depth is lower than -1 m a.s.l., while samples with average lower EC values (400–1450 μS/cm) were collected from wells P2, P3, P8, P9, and P10 with a depth higher than -1 m a.s.l. It is worth noting that P6 and P7 wells are located near the abandoned meander of Tevere River and are linked with the deepest circulation, while the second group of wells are placed in the archaeological park where the shallowest circulation has been identified. Finally, P1 in the archeologic park represents the connection between the two circulations in the rainy period.

Groundwater average concentration of major elements resulting from the four surveys and average composition of sea water are plotted on the Schoeller diagrams, where relative ratios among ions are displayed allowing a direct comparison of waters' chemical composition (Figure 6a,b).

Samples from wells P2, P3, P4, P8, P9, and P10 (Figure 6a) are enriched in bicarbonates, in alkalis and in calcium. In particular, bicarbonate and alkalis are the dominant ions in P2, P4, and P9 waters, with high Ca/Mg and Cl/SO$_4$ ratios (with the exception of well P4). P3, P8, and P10 are enriched in bicarbonates and calcium, with alkalis concentration comparable to that of calcium. The Cl/SO$_4$ ratio distinguishes between P8 and P10 groundwater on one side and P3 on the other. Chlorides are the dominant ion in the former wells, while sulphates are enriched in the latter one. None of the waters under investigation have sections of the curve equivalent to those of sea water. Finally, P1 groundwater is characterized by a low Ca/Mg ratio, whereas the Tevere River (P5) water is relatively enriched in calcium compared with magnesium and in chlorides (Figure 6b).

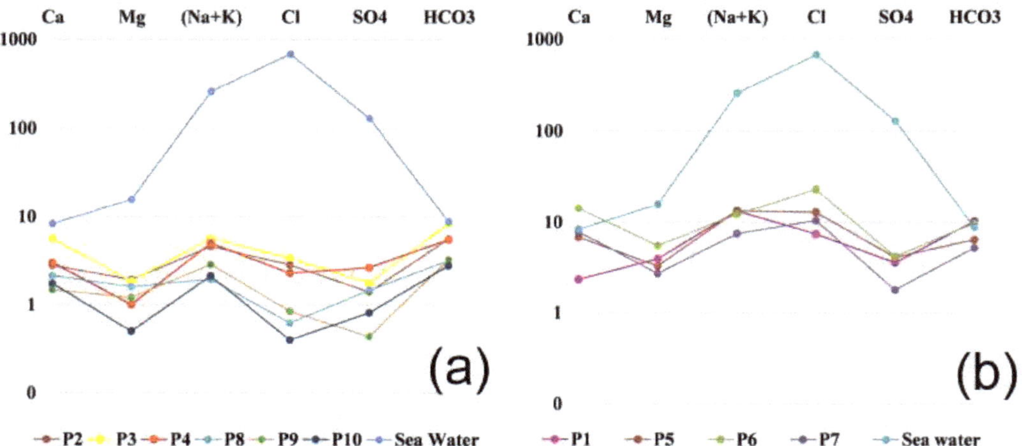

Figure 6. Schoeller diagrams with major elements ratios of groundwater (mg/L) from wells P2, P3, P4, P8, P9, and P10 (**a**) and P1, P5, P6, and P7 (**b**). Sea water composition is reported to better evaluate freshwater–saltwater interactions.

5.2. Water–Rock Interaction

Calcium bicarbonate groundwater prevails in the archaeological park, characterizing wells P3, P8, and P10. Pozzolanic materials outcropping in the area contains minerals deriving from the erosion of Colli Albani volcano products. Groundwater produces leaching and incongruent dissolution of leucite and pyroxene, very abundant in these deposits, with release of soluble cations and formation of alteration minerals such as zeolites or clay minerals [37]. Possible reactions affecting leucite and diopside, which justify groundwater composition, are reported below:

$$2\,KAlSi_2O_6\ (leucite) + 2\,H_2CO_3 + H_2O \rightarrow 2\,K^+ + 2\,HCO_3^- + 2\,SiO_2 + Al_2Si_2O_5(OH)_4\ (caolinite)$$

$$CaMgSi_2O_6\ (diopside) + 4\,CO_2 + 6\,H_2O \rightarrow Ca^{2+} + Mg^{2+} + 2\,H_4SiO_4 + 4\,HCO_3^-$$

Alkaline bicarbonate groundwater of wells P2, P4, and P9 are linked to the presence of dissolved CO_2 (enrichment in bicarbonates) and to the incongruent dissolution and cationic exchange phenomena (enrichment in alkalis). Negatively charged surfaces of clay minerals, very abundant in alluvial and delta environments, may adsorb and exchange cations, according to reactions such as:

$$Ca^{2+} + 2\,Na\text{-}X_2 \leftrightarrow Ca\text{-}X_2 + 2\,Na^+,$$

when fresh and/or recharge water rich in calcium interacts with minerals which had previously adsorbed sodium from marine waters. This interaction produces the release of sodium from the minerals and the simultaneous absorption of calcium onto the mineral surface, joined to a dilution of chlorides concentration by fresher water. Alkaline bicarbonate waters are the result of this process, but could also be related to a local mixing between groundwater and Tevere River waters, where a wind-induced salt-wedge intrusion in the river mouth produces EC values up to 2400 µS/cm [23].

These phenomena could also be invoked to justify the composition of alkaline chloride-sulphate water of the Tevere River and groundwater from well P1. This well is located in a morphologically depressed area located in proximity of the river at −2/−3 m a.s.l., where the river could locally infiltrate into the ground, recharging the aquifer. In addition to that, groundwater from well P1 is characterized by a strong depletion in calcium (Figure 6b), which is compatible with the precipitation of calcium carbonate, as shown by calcite minerals in the stratigraphic column.

Calcium chloride-sulphate composition characterizes groundwater from wells P6 and P7. This composition could be due to a cationic exchange reaction, such as that reported below:

$$Na^+ + \frac{1}{2} Ca\text{-}X_2 \leftrightarrow Na\text{-}X + \frac{1}{2} Ca^{2+}$$

An input of salt water into the aquifer generates the release of calcium and the adsorption of sodium on the mineral surfaces, with a contemporaneous enrichment of groundwater with chlorides. The source of salts could be the ancient salt pans of Ostia, near the abandoned meander; actually, P6 and P7 water are characterized by high values of EC, respectively, up to 4800 and 2600 µS/cm. Finally, irrigation of the fields for agricultural practices could introduce salt water into the aquifer.

5.3. Sources of Salinity

In order to classify groundwater and identify the sources of salinity, we employed the Cl/Br ratio [38], the BEX index (Table S3, Supplementary Material), sensu [35], and the chloride content of groundwater.

First of all, it is possible to plot EC values versus Cl/Br ratios (Figure 7). P6 and P7 groundwater is characterized by high EC and Cl/Br ratios, where chlorides are abundant compared with bromides and the total ion contents. Cl/Br ratios of wells P6 and P7 approach more than other samples the value of sea water, according to the presence of the ancient salt pans and the anthropogenic effect due to agricultural practice (see field 5a in Figure 7). Moreover, P5 shows intermediate EC and high, but extremely variable, Cl/Br ratio. P1 groundwater, with EC comparable to that of the Tevere River, stands out for lower Cl/Br ratios, along a hypothetical mixing line between the shallowest and the deepest circulations. Finally, wells P2, P3, P8, P9, and P10 exhibit low EC and variable Cl/Br ratios; they are plotted adjacent to the field of "recharge in inland areas" (field 2b in Figure 7).

Secondly, the BEX index of groundwater, sensu [35], was used to elucidate salinization phenomena (Table S3). This index indicates whether an aquifer was affected mainly by salinization or freshening phenomena; negative values of the BEX index were calculated for wells P1, P2, P3, P4, P8, P9, and P10, indicating a relative abundance of alkalis compared with chlorides, justified by a prevalent freshening and meteoric recharge of the aquifer, in agreement with Cl/Br suggestion. On the contrary, BEX index is positive for wells P6 and P7 and for the Tevere River (P5), demonstrating the input of salt water into the aquifer, as confirmed by high values of EC and Cl/Br ratio.

5.4. Precipitation of CaCO₃ from the Sampled Water

5.4. Precipitation of $CaCO_3$ from the Sampled Water

Average saturation index for calcite of wells P1, P5, P6, P7, and P10 was positive (Table S3), indicating that groundwater composition is compatible with calcite cements permeating the sediments extracted from the drillings, while moderately negative values of wells P2 and P3 suggest a minor tendency to dissolve the mineral.

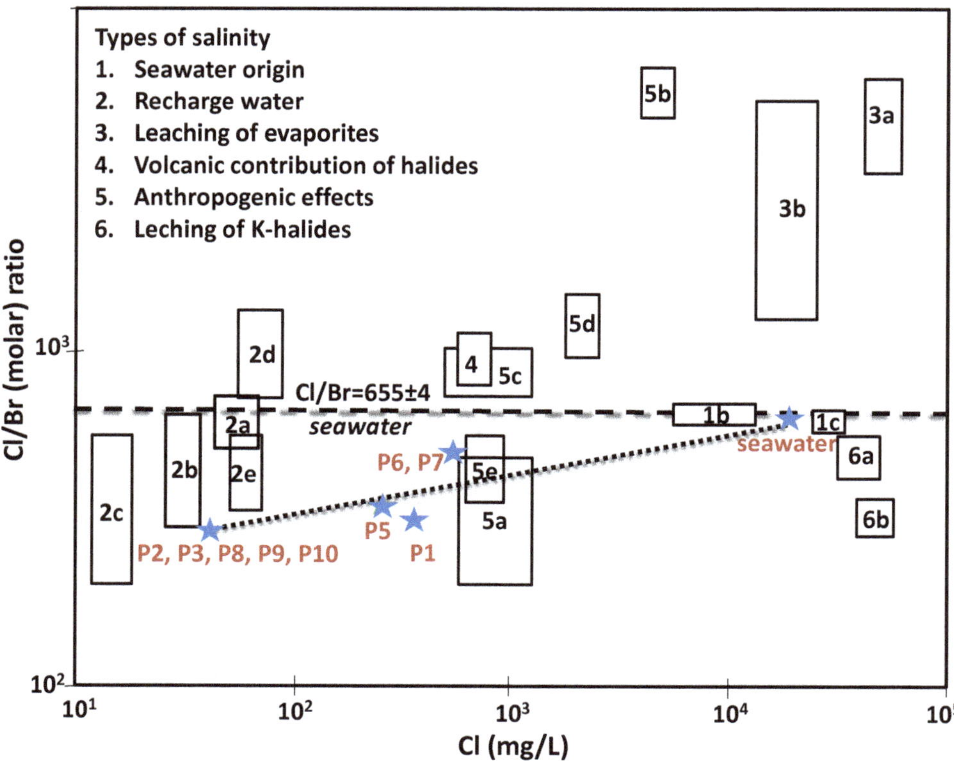

Figure 7. Cl/Br (molar) ratio versus Cl concentration of groundwater in the Ostia Antica area (Roma, central Italy) plotted over the salinity classification fields set from 24 selected aquifers of Spain and Portugal (modified from [39]). Stars stand for the average composition of well from the 2016 survey. 1b—seawater intrusion; 1c—seawater brines; 2a—recharge in coastal areas; 2b—recharge in inland areas; 2c—recharge in high altitude/continental areas; 2d—recharge in coastal arid climate; 2e—recharge in coastal polluted areas; 3a—leaching of natural halite; 3b—leaching of gypsum containing halite; 4—volcanic contribution of halides; 5a—agricultural pollution; 5b—leaching of industrial halite; 5c—leaching of garbage and solid water; 5d—urban wastewater; 5e—septic waste; 6a—leaching of carnalite; 6b—leaching of sylvite.

Since the occurrence of carbonate cement in the sedimentary sequence could provide information on the sources of carbon dioxide dissolved in groundwater, carbon ad oxygen isotopic analyses were carried out on calcite samples from drillings p1 and p6, located in proximity of wells P1 and P6, respectively (see Figure 3 for location). The isotopic signature of the two samples is plotted over a $\delta^{18}O$ vs. $\delta^{13}C$ diagram, where the composition of other carbonates (speleothems, travertines, and calcite encrustations) in the Roma area (Italy) are reported for comparison (Figure 8) [40–42]. The cement in p1 sequence, characterized by a $\delta^{13}C$ of −5.5‰ and $\delta^{18}O$ of −5.06‰ VPDB, falls very close to speleothems formed in artificial caves excavated in ignimbrites from the Colli Albani volcano, such as in a cellar in Ariccia town or in the "Acqua Vergine" and "Antonianiano" aqueducts in Roma [42]. This finding is sound because pozzolanic materials were largely used in the construction activities of ancient Romans and are abundantly present in the archaeological sequence of Ostia Antica park. Moreover, the carbon composition of p1 calcite corresponds to that of other carbonates in artificial conduits developed in the ignimbrites of Colli Albani volcano (emissary of Castiglione crater and "Anagnina" mushroom farm) [43] and in the pre-Roman remains of the ancient Portuense road at Ponte Galeria, where deep-seated fluids rich in CO_2 upsurged, forming pools in the valley floor and depositing carbonate layers, a few kilometres NE of the road [40]. Such fluids continued to deposit calcite in the Middle Age and very scarcely today, with an isotopic composition progressively more negative and now compatible with the decomposition of organic materials in the prograding river delta. The isotopic composition of calcites in p6 drilling (δ^{13}C of −12.2‰ and δ^{18}O of −9.1‰) is a mixing between current calcite at Ponte Galeria area and p1 calcite levels, suggesting a significant contribution of organic CO_2 in the peaty area where the old abandoned meander of the Tevere River is placed.

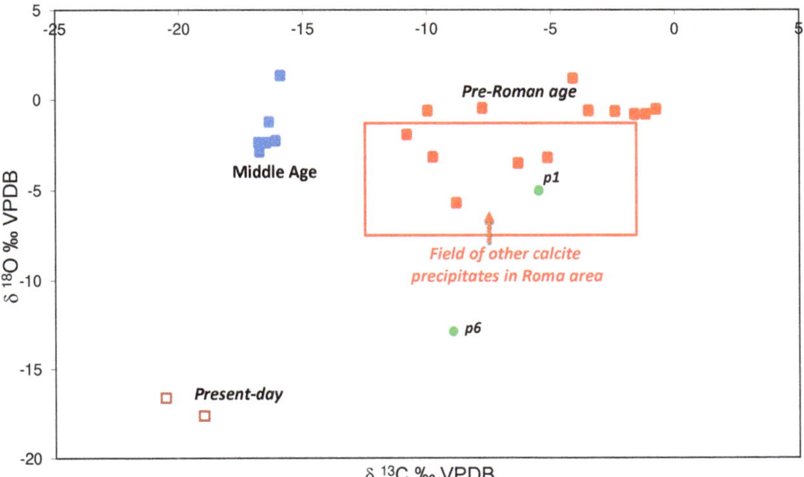

Figure 8. δ^{13}C versus δ^{18}O of calcite precipitates (p1 and p6) in Ostia Antica stratigraphic sequence compared with other calcite in sediments, aqueducts, and caves of the Roma area (Italy). Squares represent calcites layers of different ages in the pre-Roman remains of the ancient Portuense road ([40,42,43] and Tuccimei, unpublished data).

The change in isotopic composition of Ponte Galeria calcites is in agreement with the TRD progradation in the last 2000 years and with the deposition of sediments rich in organic matter where decomposition phenomena produce carbon dioxide with a more negative signature. Calcite cements from p6 drilling at Ostia Antica archaeological park follow this environmental evolution, even if they are partly influenced by the deep CO_2 upsurge. This isotopic composition (frankly, less negative) is dominant in p1 cements, located in proximity of a fault which borders the Primary Collector channel, where high concentrations of dissolved CO_2 were reported [41]. This tectonic feature likely favors the rise of deep CO_2, as in other areas of Roma [42], enhancing calcite deposition.

PCA analysis of the physicochemical data of groundwater (HCO_3, Cl, SO_4, Ca, Mg, Na, K, Br, Water table elevation, Br/Cl, and EC) supports the main results of our research. Only the Eigenvalues that have a value greater than 1 were arbitrarily selected as they are more significant and explain 76% of the total variance: respectively 60.27% and 15.67%.

PC1 was mainly correlated positively to EC, content of Br, Cl, Mg, Ca, and SO_4 and Br /Cl ratio, and negatively to groundwater level, while factor 2 has strong positive weighs on K and HCO_3-, and less on SO_4 and is negatively correlated to Br/ Cl ratio and content of Ca and Cl. The coefficients of PC1 and PC2 for the 11 variables are reported in Table S4 (Supplementary Material). Accordingly, factor 1 accounts for freshwater–saltwater interactions, while factor 2 mostly reflects gas–water–rock interaction processes. The bi-plot of PC1 and PC2 (Figure 9) distinguishes the two circulations very well, evidencing the role of well P1 as the mixing point between the two circulations in the rainy period.

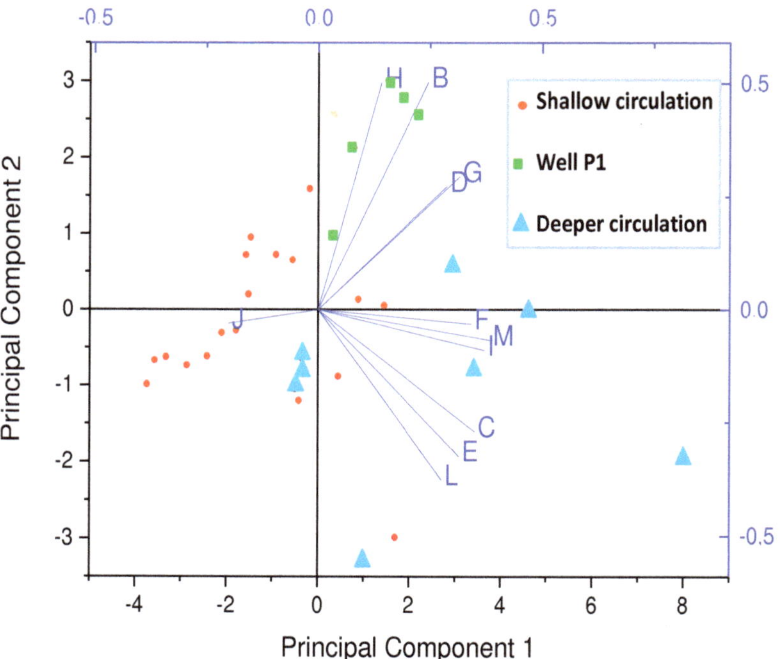

Figure 9. Bi-plot of principal components PC1 and PC2 of physicochemical data of groundwater in the area of Ostia Antica archaeological park. Eleven variables were selected: B. HCO_3; C. Cl; D. SO_4; E. Ca; F. Mg; G. Na; H. K; I. Br; J. Water table level; L. Br/Cl; M. EC. Shallow circulation is referred to wells P2, P3, P4, P8, P9, and P10; deeper circulation is linked to wells P6 and P7.

6. Conclusions

The purpose of this work was to study the processes of freshwater–saltwater interactions in a multilayer coastal aquifer (Ostia Antica archaeological park) using a hydrogeochemical approach.

Two different groundwater circulations were identified during the dry season, a shallower one in the archaeological park and a deeper one in the abandoned meander of Tevere River. The two circulations are separated by a partially impermeable sandy clay layer, intersected by the drillings from about -1.00 to -1.40 m a.s.l. During the rainy season, the two circulations merge at well P1, but no further information is available for the nearby area. The recharge area of both circulations is located in the south-western sector of the study site, with a flow in WSW–ENE direction towards the Primary Collector channel.

This model of groundwater circulation was confirmed by hydrochemical data and PCA analysis. The upper circulation, typically alkaline-bicarbonate and calcium-bicarbonate, is characterized by moderately high EC values (about 850 μS/cm) and relatively low chloride contents and Cl/Br ratios attesting recharge inland. Negative BEX values strengthen the prevalence of freshening over salinization phenomena.

The deepest circulation hosted in the alluvial materials of the Tevere River is characterized by calcium chloride-sulphate and alkaline chloride-sulphate facies, with high average EC values (about 2600 μS/cm). Chloride concentration and Cl/Br ratios indicate that the sources of salinity are due to a local interaction of groundwater with the Roman salt pans and to agricultural practices. Positive values of BEX indices support the occurrence of salinization processes. Carbon and oxygen isotopic compositions point to a river delta environment where degradation of organic matters takes place releasing CO_2 with a strong negative isotopic signature, very well matched with the nature of alluvial sediments hosting this circulation.

Hydrochemical data of well P1 demonstrate the mixing between the two circulations in the rainy period. Intermediate values of EC, chlorides, and Cl/Br ratios are in agreement with this scenario. Isotopic data are compatible with a circulation within the archaeological cover containing volcanic minerals and clearly record that calcite cements precipitated by groundwater rich in CO_2 of deep provenance, likely rising along a fault that borders the Primary Collector channel. Finally, a salt-wedge intrusion along the Tevere River was demonstrated, especially during the spring and summer.

Supplementary Materials: The following are available online at https://www.mdpi.com/article/10.3390/w13131866/s1, Table S1: Temperature, Electrical Conductivity (EC), pH and water table elevation of groundwater, Table S2: Major ions and bromide concentration (meq/L) in groundwater of Ostia Antica archeological park. Cl/Br ratio is expressed as molar ratio to fit Figure 7, modified from [30], Table S3: Average BEX of groundwater, sensu [27], sampled in the year 2016. Four groups of samples representing waters with homogenous and similar compositions were selected: the shallowest (wells P2, P3, P8, P9, and P10) and the deepest (wells P6 and P7) circulations, the Tevere River (P5), and well P1. The table also reports average saturation index (SI) for calcite, Table S4: Eigenvectors extracted from the correlation matrix of 11 physicochemical variables of groundwater. Data are available in Tables S1 and S2.

Author Contributions: Conceptualization, M.B., P.T., L.M. and R.M.; methodology, M.B., P.T., L.M. and R.M.; validation, M.B., P.T., L.M. and R.M.; writing—original draft preparation, M.B. and P.T. writing—review and editing, M.B., P.T., L.M. and R.M.; supervision, P.T., L.M. and R.M.; funding acquisition, R.M., P.T. and L.M. All authors have read and agreed to the published version of the manuscript.

Funding: This research was funded by Ministero dei Beni e delle Attività Culturali e del Turismo, Sovrintendenza Speciale per i Beni Archeologici di Roma. No Grant Number available. The grant to Department of Science, Roma Tre University (MIUR—Italy Dipartimento di Eccellenza, Articolo 1, Commi 314–337 Legge 232/2016) is also gratefully acknowledged.

Institutional Review Board Statement: Not applicable.

Informed Consent Statement: Informed consent was obtained from all subjects involved in the study.

Data Availability Statement: Data is contained within the article or supplementary material.

Acknowledgments: Authors wish to thank Renato Matteucci and Carlo Rosa for stimulating discussion. Moreover, special thanks to Carlo Rosa for providing us with the two-man auger Stihl BT 360 motor drill used for the three 3–5 m drillings.

Conflicts of Interest: The authors declare no conflict of interest.

References

1. Trabelsi, R.; Abid, K.; Zouari, K.; Yahyaoui, H. Groundwater salinization processes in shallow coastal aquifer of Djeffara plain of Medenine, Southeastern Tunisia. *Environ. Earth Sci.* **2002**, *66*, 641–653. [CrossRef]
2. Ahmed, M.; Samie, S.; Badawy, H. Factors controlling mechanisms of groundwater salinization and hydrogeochemical processes in the quaternary aquifer of the Eastern Nile Delta, Egypt. *Environ. Earth Sci.* **2013**, *68*, 69–394. [CrossRef]
3. Farid, I.; Trabelsi, R.; Zouari, K.; Abid, K.; Ayachi, M. Hydrogeochemical processes affecting groundwater in an irrigated land in Central Tunisia. *Environ. Earth Sci.* **2013**, *68*, 1215–1231. [CrossRef]
4. Tran, L.T.; Larsen, F.; Pham, N.Q.; Christiansen, A.V.; Nghi, T.; Vu, H.V.; Tran, L.V.; Hoang, H.V.; Hinsby, K. Origin and extent of fresh groundwater, salty paleowaters and recent saltwater intrusions in Red River flood plain aquifers, Vietnam. *Hydrogeol. J.* **2012**, *20*, 1295–1313. [CrossRef]
5. Bear, J.; Cheng, A.H.; Sorek, S.; Ouazar, D.; Herrera, D.H.I. *Seawater Intrusion in Coastal Aquifers-Concepts, Methods and Practices*; Kluwer Academic Publishers: Dordrecht, The Netherlands, 1999.
6. Meyer, R.; Engesgaard, P.; Sonnenborg, T.O. Origin and Dynamics of Saltwater Intrusion in a Regional Aquifer: Combining 3-D Saltwater Modeling with Geophysical and Geochemical Data. *Water Resour. Res.* **2019**, *55*, 1792–1813. [CrossRef]
7. Mahesha, A.; Nagaraja, S.H. Effect of natural recharge on sea water intrusion in coastal aquifers. *J. Hydrol.* **1996**, *174*, 211–220. [CrossRef]

8. Custodio, E.; Llamas, M.R. Rapporti tra le acque dolci e salate nelle regioni costiere. In *Idrologia Sotterranea*; Dario Flaccovio: Palermo, Italy, 2005; Volume 2, pp. 1248–1274.

9. Narayan, K.A.; Schleeberger, C.C.; Bristow, K.L. Modelling seawater intrusion in the Burdekin Delta Irrigation Area, North Queensland, Australia. *Agric. Water Manag.* **2007**, *89*, 217–228. [CrossRef]

10. Park, S.C.; Yun, S.; Chae, G.T.; Yoo, I.S.; Shin, K.S.; Heo, C.H.; Lee, S.K. Regional hydrochemical study on salinization of coastal aquifers, western coastal area of Korea. *J. Hydrol.* **2005**, *313*, 182–194. [CrossRef]

11. Mastrocicco, M.; Colombani, N. The Issue of Groundwater Salinization in Coastal Areas of the Mediterranean Region: A Review. *Water* **2021**, *13*, 90. [CrossRef]

12. Langevin, C.; Sanford, W.; Polemio, M.; Povinec, P. Background and summary. A new focus on groundwater–seawater interations. Proceedings of Symposium HS1001 at IUGG2007, Perugia, Italy, 10 July 2007. *IAHS Publ.* **2007**, *312*, 3–10.

13. Gustafson, C.; Key, K.; Evans, R.L. Aquifer systems extending far offshore on the U.S. Atlantic margin. *Sci. Rep.* **2019**, *9*, 8709. [CrossRef]

14. Micallef, A.; Person, M.; Haroon, A.; Weymer, B.A.; Jegen, M.; Schwalenberg, K.; Faghih, Z.; Duan, S.; Cohen, D.; Mountjoy, J.J.; et al. 3D characterisation and quantification of an offshore freshened groundwater system in the Canterbury Bight. *Nat. Commun.* **2020**, *11*, 1372. [CrossRef] [PubMed]

15. Idczak, J.; Brodecka-Goluch, A.; Lukawska-Matuszewska, K.; Graca, B.; Gorska, N.; Klusek, Z.; Pezacki, P.D.; Bolałek, J. A geophysical, geochemical and microbiological study of a newly discovered pockmark with active gas seepage and submarine groundwater discharge (MET1-BH, central Gulf of Gdansk, southern Baltic Sea). *Sci. Total Environ.* **2020**, *742*, 140306. [CrossRef]

16. Jakobsson, M.; O'Regan, M.; Morth, C.M.; Stranne, C.; Weidner, E.; Hansson, J.; Gyllencreutz, R.; Humborg, C.; Elfwing, T.; Norkko, A.; et al. Potential links between Baltic Sea submarine terraces and groundwater seeping. *Earth Surf. Dyn.* **2020**, *8*, 1–15. [CrossRef]

17. Pydyn, A.; Popek, M.; Kubacka, M.; Janowski, Ł. Exploration and reconstruction of a medieval harbour using hydroacoustics, 3-D shallow seismic and underwater photogrammetry: A case study from Puck, southern Baltic Sea. *Archaeol. Prospect.* **2021**. [CrossRef]

18. Finetti, I.R. The CROP profiles across the Mediterranean Sea (CROP MARE I and II). *Mem. Descr. Carta Geol. It.* **2003**, *62*, 171–184.

19. Bellotti, P.; Davoli, L.; Terragoni, C. L'evoluzione del litorale tiberino negli ultimi 3000 anni sotto le forzanti naturali e antropiche. *Stud. Costieri* **2014**, *22*, 33–43.

20. EASAC–European Academies Science Advisory Council. *Groundwater in the Southern Member States of the European Union: An Assessment of Current Knowledge and Future Prospects*; German Academy of Sciences Leopoldina: Halle, Germany, 2010.

21. Mastrorillo, L.; Mazza, R.; Manca, F.; Tuccimei, P. Evidences of different salinization sources in the roman coastal aquifer (Central Italy). *J. Coast. Conserv.* **2016**, *20*, 423–441. [CrossRef]

22. Capelli, G.; Mazza, R.; Papiccio, C. Saline intrusion in the Tiber Delta Geology, hydrology and hydrogeology of the coastal plain of the roman sector. *Giorn. Geol. Appl.* **2007**, *5*, 13–28.

23. Manca, F.; Capelli, G.; La Vigna, F.; Mazza, R.; Pascarella, A. Wind-induced salt-wedge intrusion in the Tiber river mouth (Rome–Central Italy). *Environ. Earth Sci.* **2014**, *72*, 1083–1095. [CrossRef]

24. Tuccimei, P.; D'Angelantonio, M.; Manetti, M.C.; Cutini, A.; Amorini, E.; Capelli, G. The chemistry of precipitation and groundwater in a coastal Pinus Pinea forest (Castel Fusano area, Central Italy) and its relation to stand and canopy structure. In *Horizons in Earth Science Research*; Szigethy, B.V., Ed.; Nova Science Publishers, Inc.: Suite N Hauppauge, NY, USA, 2011; Volume 4, pp. 1–17.

25. Manca, F.; Capelli, G.; Tuccimei, P. Sea salt aerosol groundwater salinization in the Litorale Romano Natural Reserve (Rome, Central Italy). *Environ. Earth Sci.* **2015**, *73*, 4179–4190. [CrossRef]

26. Goiran, J.P.; Salomon, F.; Mazzini, I.; Bravard, J.P.; Pleuger, E.; Vittori, C.; Boetto, G.; Christiansen, J.; Arnaud, P.; Pellegrino, A.; et al. Geoarcheology confirms location of the ancient harbour basin of Ostia (Italy). *J. Archaeol. Sci.* **2013**, *41*, 89–398.

27. Aeronautica Militare—Servizio Meteorologico. *Atlante Climatico d'Italia 1971–2000*; Centro Nazionale di Meteorologia e climatologia aereonautica; Pratica di Mare: Rome, Italy, 2009; Volume 3.

28. Bellotti, P.; Chiocci, F.L.; Milli, S.; Tortora, P.; Valeri, P. Sequence stratigraphy and depositional setting of the Tiber delta: Integration of high-resolution seismics, well logs, and archeological data. *J. Sediment Res. Sect. Stratigr. Glob. Stud.* **1994**, *64*, 416–432.

29. Funiciello, R.; Giordano, G.; Mattei, M. *Carta Geologica del Comune di Roma Scala 1:50.000*; S.EL.CA.: Firenze, Italy, 2008.

30. Mazza, R.; La Vigna, F.; Capelli, G.; Dimasi, M.; Mancini, M.; Mastrorillo, L. Idrogeologia del territorio di Roma. *Acque Sotter. It. J. Groundw.* **2016**, *4*, 19–30.

31. Mastrorillo, L.; Mazza, R.; Tuccimei, P.; Rosa, C.; Matteucci, R. Groundwater monitoring in the archaeological site of Ostia Antica (Rome, Italy): First results. *Acque Sotter. It. J. Groundw.* **2016**, *5*, 35–42. [CrossRef]

32. Mastrorillo, L.; Mazza, R.; Viaroli, S. Recharge process of a dune aquifer (Roman coast, Italy). *Acque Sotter. It. J. Groundw.* **2018**, *356*, 6–19. [CrossRef]

33. Petitta, M.; Primavera, P.; Tuccimei, P.; Aravena., R. Interaction between deep and shallow groundwater systems in areas affected by Quaternary tectonics (Central Italy): A geochemical and isotope approach. *Environ. Earth Sci.* **2011**, *63*, 11–30. [CrossRef]

34. Packhurst, D.L.; Appelo, C.A.J. Description of input and examples for PHREEQC version 3. A computer program for speciation, batch-reaction, one-dimensional transport, and inverse geochemical calculations. In *Department of the Interrior U.S. Geological Survey, Groundwater Book 6, Modeling Techniques*; A43 US: Denver, CO, USA, 2013; Volume 43.

35. Schoeller, H. Les échanges de base dans les eaux souterraines; trois exemples en Tunisie. *Bull. Soc. Geol. Fr.* **1934**, *4*, 389–420.
36. Brilli, M.; Giustini, F.; Kadioglu, M. Black limestone used in antiquity: Recognizing the limestone of Teos. *Archaeometry* **2019**, *61*, 282–295. [CrossRef]
37. Gaeta, M.; Freda, C.; Christensen, J.N.; Dallai, L.; Marra, F.; Karner, D.B.; Scarlato, P. Time-dependent geochemistry of clinopyroxene from the Alban Hills (Central Italy): Clues to the source and evolution of ultrapotassic magmas. *Lithos* **2006**, *86*, 330–346. [CrossRef]
38. Sánchez-Martos, F.; Pulido-Bosch, A.; Molina-Sánchez, L.; Vallejos-Izquierdo, A. Identification of the origin of salinization in groundwater using minor ions (Lower Andarax, Southeast Spain). *Sci. Total Environ.* **2002**, *297*, 43–58. [CrossRef]
39. Naily, W.; Naily, S. Cl/Br Ratio to Determine Groundwater Quality. In Proceedings of the IOP Conference Series: Earth and Environmental Science, Yogyakarta, Indonesia, 20–24 August 2018.
40. Tuccimei, P.; Soligo, M.; Arnoldus-Huyzendveld, A.; Morelli, C.; Carbonara, A.; Tedeschi, M.; Giordano, G. Datazione U/Th di Depositi Carbonatici Intercalati ai Resti della Via Portuense Antica (Ponte Galeria, Roma): Attribuzione Storico-Archeologica della Strada e Documentazione Cronologica Dell'attività Idrotermale del Fondovalle Tiberino. Available online: www.fastionline. docs/FOLDER-it-2007-97.pdf (accessed on 12 December 2007).
41. Ciotoli, G.; Etiope, G.; Marra, F.; Florindo, F.; Giraudi, C.; Ruggiero, L. Tiber Delta CO_2-CH_4 degassing: A possible hybrid, tectonically active Sediment-Hosted Geothermal System near Rome. *J Geophys. Res. Solid Earth* **2015**, *121*, 48–69. [CrossRef]
42. Tuccimei, P.; Giordano, G.; Tedeschi, M. CO_2 release variations during the last 2000 years at the Colli Albani volcano (Roma, Italy) from speleothems studies. *Earth Planet. Sci. Lett.* **2006**, *243*, 449–462. [CrossRef]
43. Rustico, L.; Buonfiglio, M.; Zanzi, G.L.; Brilli, M.; Soligo, M.; Tuccimei, P. Relazioni su scavi, trovamenti, restauri in Roma e Suburbio 2017–2019. Regione XII. Viale Guido Baccelli, largo Enzo Fioritto, viale Giotto. Rinvenimento di una diramazione dell'Acquedotto Antoniniano. *Bull. Comm. Archeo!. Comunale Roma* **2019**, *120*, 359–364.

Article

Development of Seawater Intrusion Vulnerability Assessment for Averaged Seasonality of Using Modified GALDIT Method

Il Hwan Kim, Il-Moon Chung and Sun Woo Chang *

Department of Land, Water and Environmental Research, Korea Institute of Civil Engineering and Building Technology, Goyang 10223, Korea; kimilhwan@kict.re.kr (I.H.K.); imchung@kict.re.kr (I.-M.C.)
* Correspondence: chang@kict.re.kr; Tel.: +82-31-910-0278

Abstract: Climate change and anthropogenic activities are necessitating accurate diagnoses of seawater intrusion (SWI) to ensure the sustainable utilization of groundwater resources in coastal areas. Here, vulnerability to SWI was assessed by classifying the existing GALDIT into static parameters (groundwater occurrence (G), aquifer hydraulic conductivity (A), and distance from shore (D)) and dynamic parameters (height to groundwater-level above sea-level (L), impact of existing status of seawater intrusion (I), and aquifer thickness (T)). When assessing the vulnerability of SWI based on observational data (2010–2019), 10-year-averaged data of each month is used for GALDIT dynamic parameter for representing the seasonal characteristics of local water cycles. In addition, the parameter L is indicated by the data observed at the sea-level station adjacent to the groundwater level station. The existing GALDIT method has a range of scores that can be divided into quartiles to express the observed values. To sensitively reflect monthly changes in values, the range of scores is divided into deciles. The calculated GALDIT index showed that the most vulnerable month is September, due to relatively low groundwater level. The proposed method can be used to apply countermeasures to vulnerable coastal areas and build water resources management plan considering vulnerable seasons.

Keywords: GALDIT; monthly vulnerability; seawater intrusion (SWI); vulnerability assessment; effective weight; densely populated area

Citation: Kim, I.H.; Chung, I.-M.; Chang, S.W. Development of Seawater Intrusion Vulnerability Assessment for Averaged Seasonality of Using Modified GALDIT Method. *Water* **2021**, *13*, 1820. https://doi.org/10.3390/w13131820

Academic Editors: Evangelos Tziritis and Andreas Panagopoulos

Received: 21 May 2021
Accepted: 24 June 2021
Published: 30 June 2021

Publisher's Note: MDPI stays neutral with regard to jurisdictional claims in published maps and institutional affiliations.

1. Introduction

Coastal areas host large populations of people owing to their prosperity. Since the 20th century, 21 megalopolises in the coastal areas have grown rapidly to achieve a population of more than eight million, and more than a third of the global population resides within 100 km of the shore [1]. With the increasing area affected by seawater intrusion (SWI) in coastal areas, the available amount of water resources is decreasing due to the aquifer salinization. Furthermore, changes caused by the salinization of coastal aquifers, such as limitations in the cultivation environment of agricultural and marine products, are damaging economic activities [2,3].

The land-use changes due to industrial development increase surface runoff and decrease recharge of the groundwater system. Furthermore, climate change increases rainfall intensity due to change in the rainfall pattern. As the number of days without rain rises, the amount of water resources discharged to the surface increases, while that of recharge to the aquifer decreases. Consequently, groundwater resources gradually decrease [4–8]. The continuous rise in sea levels accelerates the increase in the SWI range with respect to freshwater body, seawater–freshwater interface and mixing zone [9]. According to the analysis method of Ghyben–Herzberg, the effect of a 1 m rise in the sea level on the freshwater aquifer corresponds to 40 m of freshwater thickness [10–12]. Sherif and Singh [13] claimed that when the sea level rises by 0.5 m, the effect of SWI reaches up to 9 km from the shore. The imbalance between the inflow and outflow from the aquifer can cause a faster drop in the freshwater groundwater level in areas with a larger pumping water quantity [14]. SWI accelerates due to the extensive use of groundwater in coastal areas, and the resulting effects by the excessive pumping of groundwater are being actively researched [15–21]. To efficiently establish response measures to SWI damage, one must select an area of most active SWI damage and choose response measures in line with the regional characteristics. Research on seawater intrusion in coastal aquifers through the use of monitoring data, assessing groundwater quantity and quality by using modeling, and improving management approaches is being actively conducted [22]. One diagnostic method is the SWI assessment for a coastal groundwater aquifer. The general vulnerability assessment method for groundwater resources involves overlaying thematic maps linked with the scored geographic information system (GIS) data, using the overlaying technique and assessing vulnerability according to the value [23]. For the vulnerability assessment, the range of fixed scores is classified and presented under subjective judgment, depending on the values and types of factors associated with groundwater resources [24,25]. The vulnerability of groundwater resources is defined as their sensitivity to human activities and natural phenomena, and the recharge required to maintain groundwater resources and the possibility of the spread of pollutants by potential pollution sources have likewise been defined [26]. Representative vulnerability parameters for the potential pollution of groundwater resources include DRASTIC ([27]; Depth to groundwater, net Recharge, Aquifer media, Soil media, Topography, Impact of the vadose zone, and hydraulic Conductivity), and SINTACS (depth to the groundwater table (S), effective infiltration (I), unsaturated zone attenuation capacity (N), soil attenuation capacity (T), hydrogeological characteristics of the aquifer (A), hydraulic conductivity (C), and topographical slope (S)) [28]. To consider the effect of coastal aquifers on SWI, GALDIT ([29]; groundwater occurrence (G), aquifer hydraulic conductivity (A), height of groundwater level above the sea (L), distance from the shore (D), impact of the existing status of SWI (I), and saturated thickness of the aquifer (T)) was developed as a representative vulnerability assessment method. GALDIT is highly reliable in assessing seawater intrusion vulnerability in coastal aquifers [30]. Recently, the assessment method of the GALDIT index has been modified for the range of the existing score and weight [31,32]. The parameter replacement of GALDIT factors and the improvement of data interpolation methods have been researched, as well [33–35].

Several previous studies on SWI in South Korea addressed the inflow of seawater into the aquifer, using the seawater monitoring network (SIMN) that was built at the national level [36]. Numerous studies on SWI have also been conducted on Jeju Island in South Korea [37]. A vulnerability assessment for SWI, using GALDIT for Jeju Island, was conducted for the first time in South Korea [38]. Recently, studies on seawater intrusion in the inland areas of Korea are incomplete compared with those in the island areas, but studies on the coastal areas of the west coast have started. For example, Kim and Yang [39] prioritized three SWI response measures for SWI-vulnerable areas when climate change was applied using the multi-criteria decision-making (MCDM) method. Chun et al. [40] conducted a two-dimensional numerical analysis of the effects of SWI on coastal areas according to different climate change scenarios.

Studies on SWI can set different time scales according to the study objectives. For example, studies on the mid-to-long-term effects of SWI, such as climate change, use time scales of ten to several hundred years [16,18–20]. In contrast, studies on short repetitive variation characteristics, such as tidal effects, conduct hourly analyses use small time scales [41–43]. In the past, techniques such as vulnerability assessments used representative values obtained through statistical tests of longer-term data [44,45]. To establish response measures to SWI, the flow characteristics according to the periods of saltwater and freshwater groundwater resources must also be considered. The assessment of flow characteristics for groundwater resources consists of factors for the spatial distribution and temporal changes in the groundwater level recorded in a time series [46,47].

Therefore, the objective of this study was to develop a method to assess SWI vulnerability for averaged seasonality based on the original GLADIT. GALDIT is the most representative SWI assessment method, and the reliability of the method is guaranteed through continuous research. Modified GALDIT methods have been applied to regions with various hydrogeological properties. In this study, the modified GALDIT method for assessing the averaged seasonality is used to intuitively and easily express monthly data with various fluctuations.

Data on SWI of coastal aquifers over the last 10 years were collected to analyze the monthly variations. The monthly vulnerability changes of the SWI were analyzed by classifying the collected data into monthly means. The GALDIT method, which is the most representative SWI assessment method, was used. We attempted to indicate spatiotemporally vulnerable areas and periods by classifying the six parameters of GALDIT into parameters that change monthly (L, I, and T) and parameters that change little over time (G, A, and D).

2. Materials and Methods

2.1. Data Collection for Monthly GALDIT Assessment

GALDIT is a diagnostic method based on the index and ranking that evaluates the vulnerability of coastal aquifers, using six parameters, considering groundwater occurrence (G), aquifer hydraulic conductivity (A), distance from the shore (D), height of groundwater level above sea level (L), impact of the existing status of seawater intrusion (I), and saturated thickness of the aquifer (T) to examine the physical effects of coastal aquifers on the SWI. We developed a monthly assessment method for seawater intrusion vulnerability, using GALDIT parameters. Data were collected from the National Groundwater Information Center [48], the National Geographic Information Institute [49], and the Korea Hydrographic and Oceanographic Agency [50]. The groundwater level, drill log, and groundwater survey report data were collected from the National Groundwater Information Center. Digital elevation maps (DEM) and topographic map data were collected from the National Geographic Information Institute. Tidal and other data were collected from the Korea Hydrographic and Oceanographic Agency.

For a monthly vulnerability assessment of the SWI, the parameters that were relatively static and those that changed temporally were classified as (i) static parameters and (ii) dynamic parameters, respectively. For the static parameter group, G, A, and D were selected. The aquifer type and A are regarded relatively static parameters in the absence of human activities. D changes when the sea level rises in the long term; however, this was excluded in this study considering that the monthly change of the coastline due to the sea level is insignificant compared to the reference value. With regard to D, the data observed to date were used without considering the future rises in the sea level. Fluctuations in D were extracted from the average distances obtained in the last 10 years.

For the input data of the parameters that change with time, we employed L, I, and T. The data collected for dynamic parameters included groundwater level, seawater level, and electrical conductivity, and the average data for 10 years per month were extracted. As shown in Figure 1, monthly data for sea level and precipitation data around the groundwater level station are collected. As an example, observation data from 2010 to 2019 were used at Gimpo Walgot groundwater level station, Ganghwa Bridge seawater level station, and Gimpo precipitation station. Observed data were averaged for 10 years each month. The groundwater and seawater level stations are located in the Gimpo area, north of the study area. The Ganghwa Bridge seawater level station is located about 3.2 km away from the Gimpo Walgot groundwater level station. The groundwater level is on average 6.59 m above the reference sea level for the last 10 years. The highest was 6.77 m in July, and the lowest was 6.51 m in March. The electrical conductivity showed an average of 370.71 s/cm over the last 10 years, the highest at 387.55 s/cm in December and lowest at 359.53 s/cm in June. The sea level is located 3.32 m above the reference sea level for the last 10 years, highest being 3.58 m in August and lowest being 3.14 m in January. When showing the distribution of dynamic parameters around Gimpo Walgot, L is the most vulnerable at 2.97 m in August, where the difference between the groundwater level and seawater level is the smallest. Parameter I appear to be the most vulnerable in December, where the electrical conductivity is the highest, and T is the most vulnerable in July, where the groundwater level is highest. The seasonality of the dynamic parameter appears differently depending on the complex characteristics of the observed area. The parameter L was calculated using the difference between the groundwater and seawater level data observed based on the reference sea level. Each groundwater level station was compared with the sea level data of the nearest seawater level station. In I, data collection is possible and the continuously observed electrical conductivity is used; I is described in detail in Section 3.2.2. T was calculated through the drilling log of the groundwater level station. T was calculated by estimating the height of the aquifer bottom of the drilling log. Through the observed L, I, and T data, the data from the station were used as point source data and spatially interpolated to evaluate GALDIT. Figure 2 shows a flowchart for calculating monthly GALDIT parameters.

2.2. Modification of Original GALDIT Method

GALDIT is a model employed for assessing the vulnerability of underground aquifers to seawater intrusion using six parameters related to seawater intrusion. Data about these six parameters were collected and scored according to the criteria, and maps for the parameters were generated by applying predefined weights. Scores of 2.5, 5, 7.5, and 10 were determined according to the criteria. A weight of four was assigned to the index having the greatest effect on seawater intrusion, and the weight varied by the effect. The relationship between the index according to the criteria and weight of the index is expressed as follows:

$$\text{GALDIT} = \frac{\sum_{i=1}^{6}(W_i \times R_i)}{\sum_{i=1}^{6} W_i} \tag{1}$$

where W_i is parameter i's weights, and R_i is parameter i's importance rating.

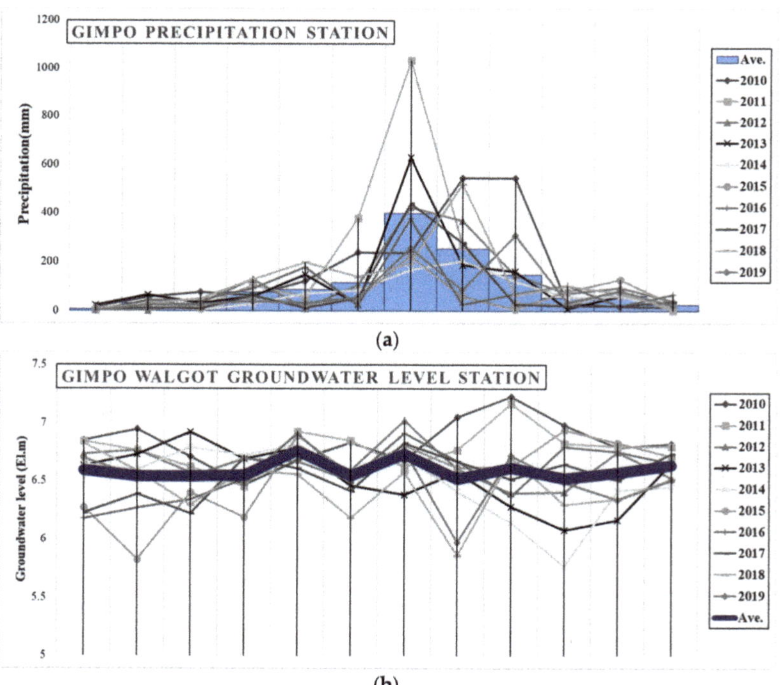

Figure 1. *Cont.*

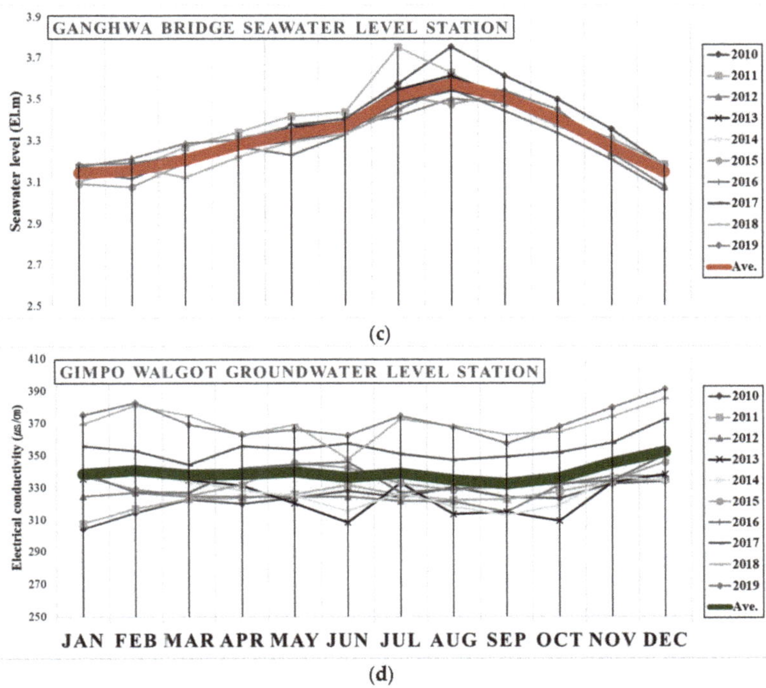

Figure 1. Example of 10-year-averaged monthly (**a**) precipitation, (**b**) groundwater level, (**c**) seawater level, and (**d**) electrical conductivity observation data.

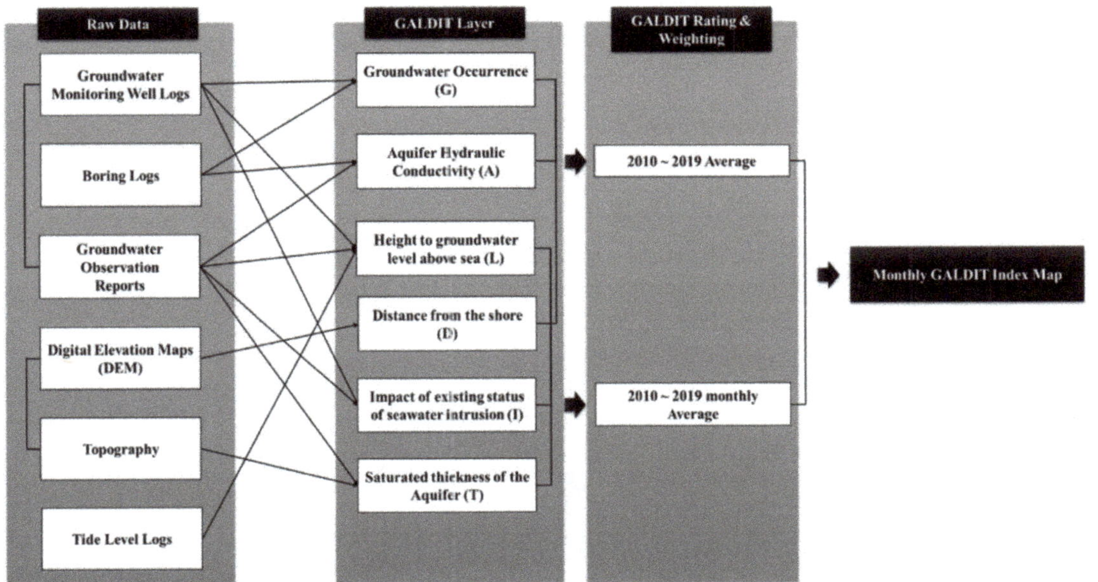

Figure 2. Procedure of monthly-based GALDIT index assessment.

Table 1 lists the ratings and weights according to the criteria for each parameter. According to the scoring method of the existing GALDIT, the highest score was 10, whereas the lowest was 2.5. The highest score was divided using the quartile method into 10, 7.5, 5, and 2.5, according to the parameter range. In this study, the GALDIT factor variable range and the importance rating in the two columns were modified in the right column of Table 1. Thus, the highest score of 10 was divided by using the decile method, and a score between 1 and 10 was assigned for the modified importance rating. The existing quartile method was used to classify aquifers according to their hydrogeological characteristics. The new distributions of scores for other parameters are listed in Table 2.

Table 1. Theoretical weights and rates for GALDIT.

Parameter	Weight	GALDIT Factor Variable Range	Importance Rating
Groundwater occurrence	1	Confined aquifer	10
		Unconfined aquifer	7.5
		Leaky confined aquifer	5
		Bounded aquifer	2.5
Aquifer hydraulic conductivity (m/day)	3	>40	10
		10–40	7.5
		5–10	5
		<5	2.5
Height of groundwater level above sea level (m)	4	<1.0	10
		1.0–1.5	7.5
		1.5–2.0	5
		>2.0	2.5
Distance from shore (m)	4	<500	10
		500–700	7.5
		750–1000	5
		>1000	2.5
Impact of existing status of seawater intrusion (μs/m)	1	>3000	10
		2000–3000	7.5
		1000–2000	5
		<1000	2.5
Saturated thickness of aquifer (m)	2	>10	10
		7.5–10	7.5
		5–7.5	5
		<5	2.5

Table 2. Modified rates for GALDIT.

Parameter	Modified Variable Range	Modified Importance Rating
Groundwater occurrence	Confined aquifer	10
	Unconfined aquifer	7.5
	Leaky confined aquifer	5
	Bounded aquifer	2.5
Aquifer hydraulic conductivity (m/day)	>40	10
	34–40	9
	28–34	8
	22–28	7
	16–22	6
	10–16	5
	8–10	4
	6–8	3
	4–6	2
	<4	1

Table 2. *Cont.*

Parameter	Modified Variable Range	Modified Importance Rating
Height of groundwater level above sea level (m)	<1.0	10
	1.0–1.2	9
	1.2–1.4	8
	1.4–1.6	7
	1.6–1.8	6
	1.8–2.0	5
	2.0–2.2	4
	2.2–2.4	3
	2.4–2.6	2
	>2.6	1
Distance from shore (m)	<500	10
	500–600	9
	600–700	8
	700–800	7
	800–900	6
	900–1000	5
	1000–1100	4
	1100–1200	3
	1200–1300	2
	>1300	1
Impact of existing status of seawater intrusion (μs/m)	>3000	10
	2600–3000	9
	2200–2600	8
	1800–2200	7
	1400–1800	6
	1000–1400	5
	600–1000	4
	200–600	3
	<200	2
	-	1
Saturated thickness of aquifer (m)	>10	10
	9–10	9
	8–9	8
	7–8	7
	6–7	6
	5–6	5
	4–5	4
	3–4	3
	2–3	2
	<2	1

2.3. Study Area

Korea has long coastlines and various coastal terrains, as it is surrounded by sea on three sides. According to recent data, the lengths of the western, southern, and eastern coastlines span approximately 4900, 3300, and 600 km, respectively [51]. The west coast has numerous bays, peninsulas, capes, and islands due to the crooked and broken coastline. In particular, a ria coast is developing in the coastal areas of the Dadohae and Taean Peninsula, where extremely crooked coastlines are present. Furthermore, this is also a place where numerous soils of terrestrial origin are transported and deposited owing to the gentle the terrain slope, severe tidal differences, and the flow through of Korea's great rivers. Consequently, large tidal flats develop along the west coast, and low hilly mountains or large and small coastal plains are distributed inland.

Among the sea areas under the influence of seawater intrusion in South Korea, the western coast with severe tidal differences was targeted, where the coastline is long and the terrain slope is gentle. The selected study area was a coastal area of the inland, excluding islands in the north, where urban areas are concentrated on the western coast of South Korea. The study area consists of nine administrative districts: Incheon, Asan, Ansan, Gimpo, Hwaseong, Siheung, Pyeongtaek, Dangjin, and Osan. In all of these nine areas, the manufacturing industry is developing, and urbanization is accelerating, leading to a continuous influx of population. Incheon is the third most-populated city in South Korea after Seoul and Busan and has a developed logistics industry, as it hosts the Incheon International Airport and Incheon Port. Ansan is a planned industrial city, where a population of a similar size to the residential one flows during the day due to numerous manufacturing plants. The manufacturing industry is also developed in Gimpo, Hwaseong, Siheung, Pyeongtaek, Dangjin, and Osan. There is Sihwa Lake Seawall in Siheung, Hwaseong, and Ansan, where industrial clusters and tourist attractions are developed. Daebu Island, the only island in the study area, was included in the study area because it is connected to a freshwater lake through the Sihwa Lake Seawall. The total area of the study is 3976.59 km^2, and the length of the coastline is 608.1 km. The automatic monitoring data and drill logs of the National Groundwater Information Center were used to assess the seawater intrusion vulnerability of the study area. There were 58 groundwater level stations in total. Further, seawater levels were observed at nine seawater level stations. Figure 3 shows the locations of the study area, rivers, coastlines, groundwater level stations, and sea water level stations.

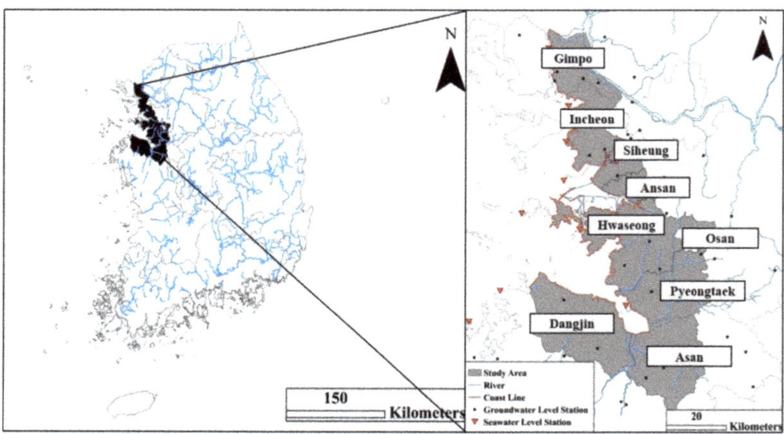

Figure 3. Study area in the western coastal region of South Korea.

3. Results

3.1. Static GALDIT Parameters

In this section, we examine the analysis results for parameters with slight changes over time, i.e., groundwater occurrence (G), aquifer hydraulic conductivity (A), and distance from the shore (D). The histogram colors in Figure 4 represent the ratings divided by the decile method, and the values were divided into 11 levels by adding 7.5 of the unconfined included in G. The y-axis represents the cumulative ratio of the area according to the rating, the secondary y-axis represents the index score, and the values indicated by the broken line represent the total average of the study area.

Groundwater occurrence (G) is caused by confined, unconfined, leaky, and bounded aquifers, out of which the confined aquifer is the most vulnerable. The scores of groundwater occurrence were divided into four levels, in the same way as the existing scoring method, using a theoretical weight of one. We assessed the alluvial layer in a free-surface aquifer of a shallow area. The aquifer type in all areas was an unconfined aquifer, and the G score was 7.5. Figure 5a shows the G index.

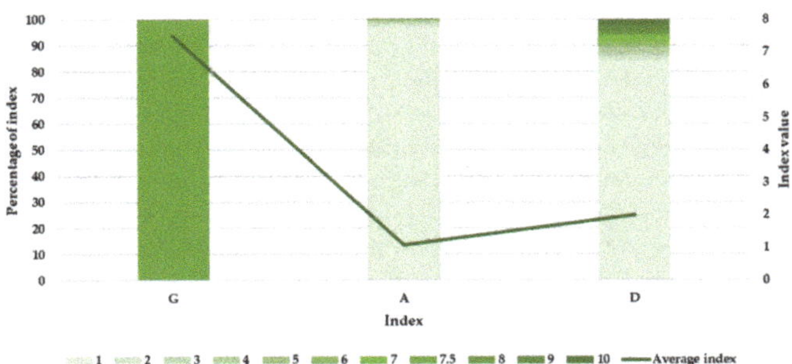

Figure 4. Distribution of static parameters' ratings for groundwater occurrence (G), aquifer hydraulic conductivity (A), and distance from the shore (D).

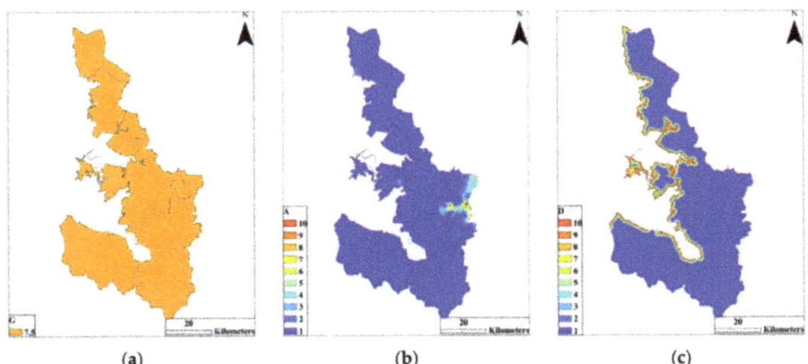

Figure 5. Thematic maps of static parameters' rating for (**a**) groundwater occurrence, G, (**b**) aquifer hydraulic conductivity, A, and (**c**) distance from shore, D.

A higher A induces smoother groundwater flow and exhibits larger vulnerability to the SWI. The theoretical weight was three in this case. When the aquifer hydraulic conductivity exceeds 40 m/day, it is the most vulnerable, and a score of 10 is to attributed it. When the aquifer hydraulic conductivity was less than 4 m/day, a score of one was given. Examination of the area ratios of each score yields A below 4 m/day at the largest ratio of 94.23%. Some areas have a high aquifer hydraulic conductivity near Pyeongtaek. The total average score of the study area was determined as 1.07. Figure 5b shows the distribution of A.

D was determined based on the observed coastline. Areas below 500 m away from the shore are most vulnerable to SWI, and those more than 1300 m from the coast are given a score of one. The theoretical weight, an index that can intuitively show vulnerability, was four. Most parts of the study area, accounting for 84.14%, were 1300 m from the shore. Areas 1000 m or less away from the shore with a score of less than five accounted for 10.88%, and those less than 500 m away accounted for 2.46%. The average score of the study area was 1.86. Figure 4 shows the percentages and average scores of the static GALDIT parameters. Figure 5c shows the distribution of the D index.

3.2. Dynamic GALDIT Parameters

3.2.1. Height of Groundwater Level above Sea Level (L)

In this section, we examine the analysis results of monthly averages for the data from 2010 to 2019 regarding the parameters that change significantly over time, that is, the height of the groundwater level above sea level (L), the impact of the existing status of seawater intrusion (I), and saturated thickness of the aquifer (T).

L was most vulnerable to SWI when it was less than 1 m, and the score was 10 in this case. When L exceeds 2.6 m, the score is the lowest at one, and the theoretical weight is four. This index changes monthly. The existing calculation method is used to observe L and compare the range values. In this study, the score was determined by comparing the measured groundwater level from the sea level, with the monthly sea level height measured at the sea water level station. The sea level observation data of the sea water level station were interpolated by setting the coastline as the domain. Figure 6 shows the dynamic parameter's rating for 10-year-averaged height of groundwater level above sea level for each month. L was determined by the minimum distance to the interpolated coastline. The histogram in Figure 7 shows the distribution of scores represented by the decile method. The y-axis on the left is the cumulative ratio of data L according to the rating, and the secondary y-axis represents the average of the data expressed in a straight line. As a consequence of calculating the monthly groundwater height relative to sea level, April exhibited the highest average of the study area, at 1.25, and August had the lowest average, at 1.21. Owing to the nature of Korea's climate, the groundwater level of the unconfined aquifer rises during the rainy season or when intensive rainfall from June to August occurs. The sea level also rose the most in August; however, the increase in the sea level was smaller than that of the groundwater level, and it was the lowest. The variations in groundwater level differed depending on the area, albeit the sea level was the lowest in January and February. Under low rainfall in April and concentrated use of groundwater, the groundwater level drops significantly, making it a period that is most vulnerable to SWI.

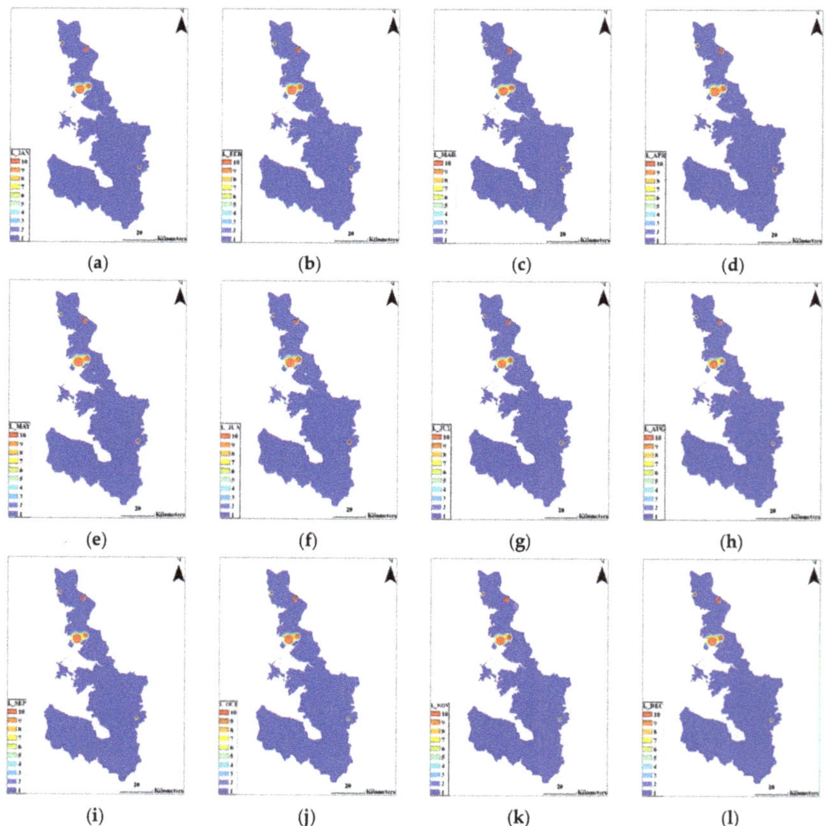

Figure 6. Thematic maps of the dynamic parameter's rating for 10-year-averaged monthly height of groundwater level above sea level (L); (**a**) JAN; (**b**) FEB; (**c**) MAR; (**d**) APR; (**e**) MAY; (**f**) JUN; (**g**) JUL; (**h**) AUG; (**i**) SEP; (**j**) OCT; (**k**) NOV; (**l**) DEC.

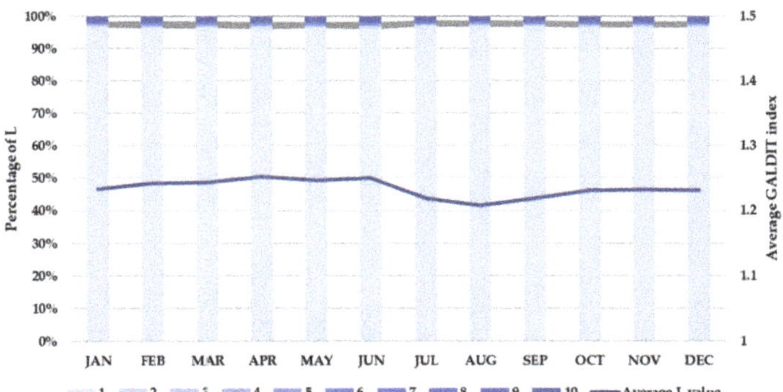

Figure 7. Distributior. of the dynamic parameter's rating for 10-year-averaged monthly height of groundwater level above sea level (L).

3.2.2. Impact of the Existing Status of Seawater Intrusion (I)

For the current SWI situation, we used electrical conductivity data, which is easily obtained from national groundwater monitoring network in Korea. Existing studies used the molar ratio of Cl^-; however, observations were irregular, and the requirement for the length of continuous data could not be satisfied. Chang et al. [38] used the electrical conductivity data obtained from the seawater intrusion monitoring network as input data for the I parameter in the GALDIT assessment. Here, high-quality electrical conductivity data obtained from the National Groundwater Monitoring Network were used. Based on the rating of Chang et al. [38], electrical conductivity above 3000 μs/m led to the highest vulnerability, and a score of two was attributed if it was below 200 μs/m. This index can indicate monthly changes, and the theoretical weight of the current SWI situation is one. When the thematic map in Figure 8 is examined with the naked eye, one can observe that the change in parameter values is not large in most areas. However, the area in the north of the study area shows higher values in March, and the values remained high until April and dropped from May. The values increased again in September, slightly decreased in October, and were maintained at 3 to 4 in November. The straight line in the graph in Figure 8 represents the average parameter value for each month. The colors in the histogram in Figure 9 express the rating divided by the decile method. The *y*-axis on the left is the cumulative ratio of data I according to the rating, and the secondary *y*-axis on the right represents the average of the data expressed by a straight line. Comparing the monthly average parameters, we see that the most vulnerable month is September, with the average of all areas at 3.65, whereas the least vulnerable month is February, with the average at 3.32.

3.2.3. Saturated Thickness of the Aquifer (T)

T was determined by using drill logs (www.gims.go.kr (accessed on 22 February 2021)). The bottom point of the unconfined aquifer was estimated by analyzing the sample and stratum composition from the drill logs. T was calculated from the height of the groundwater level observed in real time. High-quality data over a continuous period of 10 years were used among the groundwater level data observed in real-time. The saturated thickness of the aquifer is most vulnerable to SWI when it exceeds 10 m, and it is satisfactory when it is less than 2 m, and the score is one. The theoretical weight of T was two.

Figure 10 shows the dynamic parameter's rating for 10-year-averaged saturated thickness of aquifer for each month. The colors in the histogram in Figure 11 express the ratings divided by the decile method. The *y*-axis represents the cumulative ratio of the area according to the rating, the secondary *y*-axis represents the scores of the index, and the values indicated by the broken line indicate the total average of the study area. The analysis reveals the minimum score to be four, which indicates all T values were above 4 m. August was the most vulnerable month, at 9.12, and April had the lowest score, at 8.92.

T is an index related to the real-time groundwater level. It is believed that T increases in July and August when significant recharge occurs due to rainfall, and it decreases from February to April when there is fewer rainfall and increased pumping of groundwater.

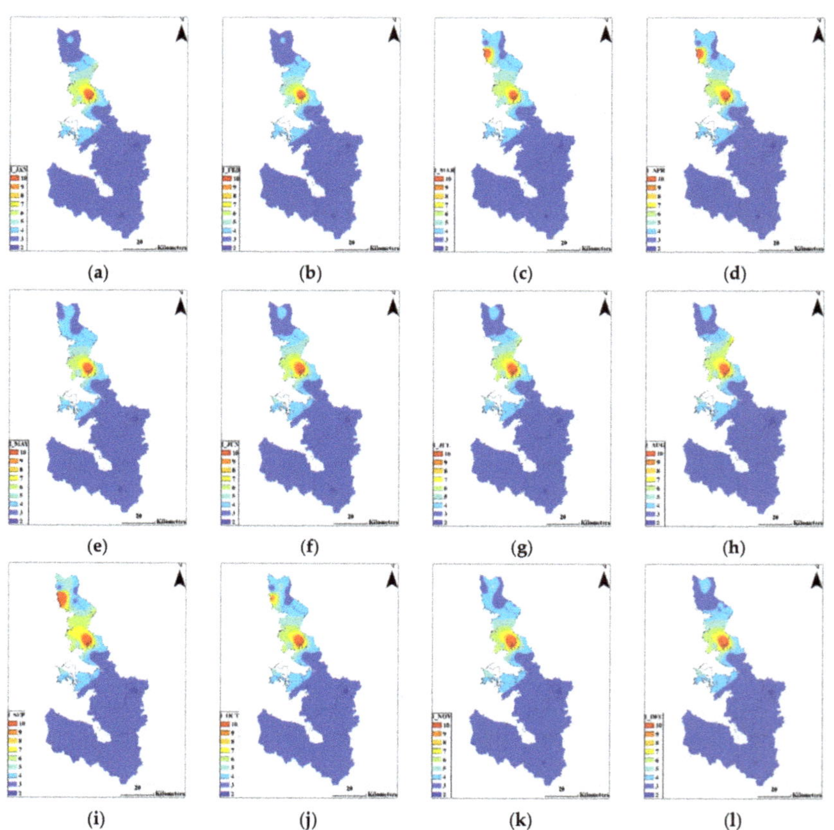

Figure 8. Thematic maps of the dynamic parameter's rating for 10-year-averaged monthly impact of existing status of seawater intrusion (I). (**a**) JAN; (**b**) FEB; (**c**) MAR; (**d**) APR;(**e**) MAY;(**f**) JUN; (**g**) JUL; (**h**) AUG; (**i**) SEP; (**j**) OCT; (**k**) NOV; (**l**) DEC.

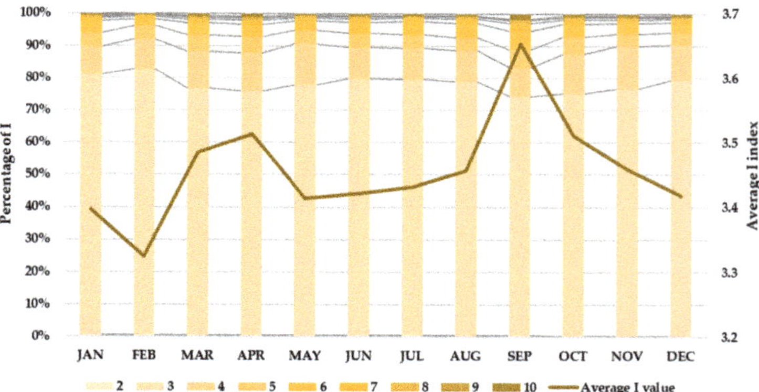

Figure 9. Distribution of the dynamic parameter's rating for 10-year-averaged monthly impact of existing status of seawater intrusion (I).

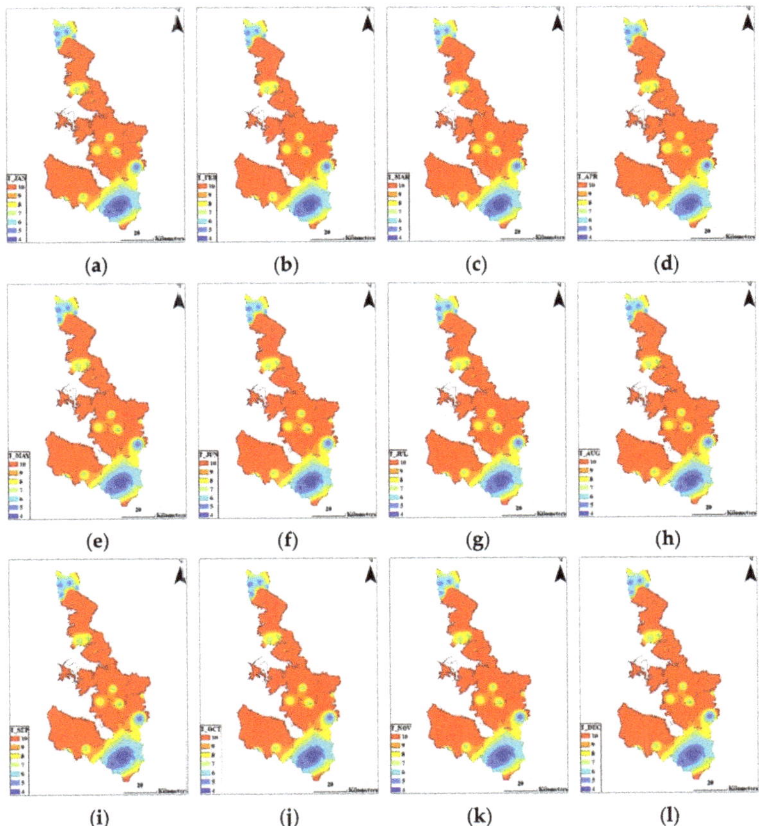

Figure 10. Thematic maps of the dynamic parameter rating for 10-year-averaged monthly saturated thickness of aquifer (T). (**a**) JAN; (**b**) FEB; (**c**) MAR; (**d**) APR;(**e**) MAY;(**f**) JUN; (**g**) JUL; (**h**) AUG; (**i**) SEP; (**j**) OCT; (**k**) NOV; (**l**) DEC.

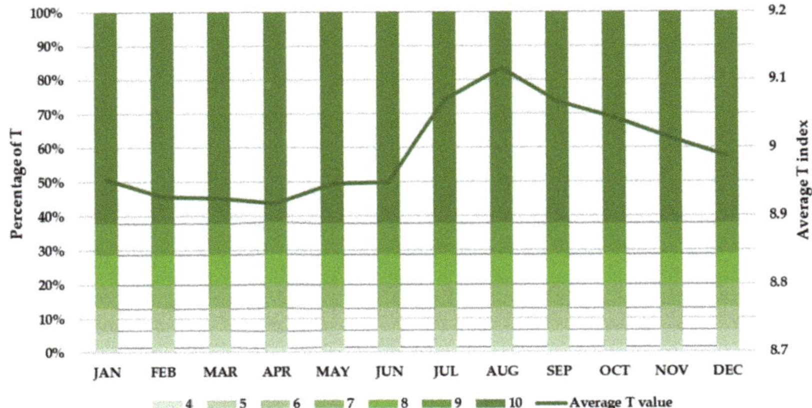

Figure 11. Distribution of the dynamic parameter's rating for saturated thickness of aquifer (T).

3.3. SWI Assessment of Study-Site Based on Monthly GALDIT Index

For the monthly GALDIT index to which the theoretical weight was applied, the average of observations for 10 years was applied for G, A, and D, whereas the parameters that changed monthly were applied for L, I, and T. The colors of the histogram in Figure 12 express the range of the calculated GALDIT index. The y-axis represents the cumulative ratio of the area according to the index range, the secondary y-axis represents the GALDIT index, and the value indicated by the broken line represents the average GALDIT index of the entire study area. The score range that occupied the highest proportion was 2 to 3, which accounted for 78% each in February, May, and July. The calculation results show that the most vulnerable month was September, when the average GALDIT index of the study area was 3.03. The ratio of areas estimated as moderate or high vulnerable with scores ≥ 5 was the highest in September (8.87%), followed by October (8.64%), and April (8.63%). In September, L was relatively robust at 1.19 as the third place from the bottom. However, it was the most vulnerable month in terms of I and the third from the top in terms of T. Thus, it was the most vulnerable month when the theoretical weight was applied. Figure 13a,b shows the original quartile results, and Figure 13c,d shows the decile results. Figure 13a,c shows the result of 10-year-averaged GALDIT maps, and Figure 13b,d shows that of the GALDIT map of September, which is the most vulnerable month as per monthly GALDIT results. In Figure 13a, the coastal area in the south of Incheon and areas near Incheon and Siheung are vulnerable, where the index value is approximately 8 to 9. Upon comparison of the GALDIT map for September in Figure 13d with the original, we see that the index of the Gimpo area indicated in yellow in the northeastern side of the study area is 5 to 6 in most periods, exhibiting a moderate vulnerability. The western coast area of Gimpo, not appearing in the original GALDIT, shows an index value of 7 to 8, indicating the boundary between moderate and high vulnerability. The SWI-vulnerable area on the southern coast of Incheon has an index value of 6 to 7 at both edges around the area protruding to the coast, showing a different pattern from the original GALDIT map. In the Pyeongtaek area, the eastern inland area exhibited low vulnerability. However, in the monthly GALDIT map, the yellow parts showed index values of 3–5, revealing differences in the seawater intrusion vulnerability. Most of the areas excluding the coast showed little differences in the degree; however, in the monthly GALDIT map, they showed partial differences, indicating a low vulnerability. In the entire study area, the inland areas up to 1 km away from the coastline are considered areas of moderate-to-high vulnerability.

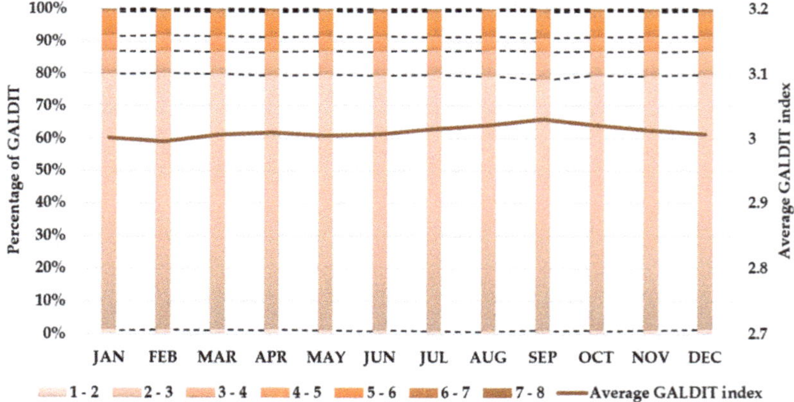

Figure 12. Distribution of 10-year-averaged monthly GALDIT index.

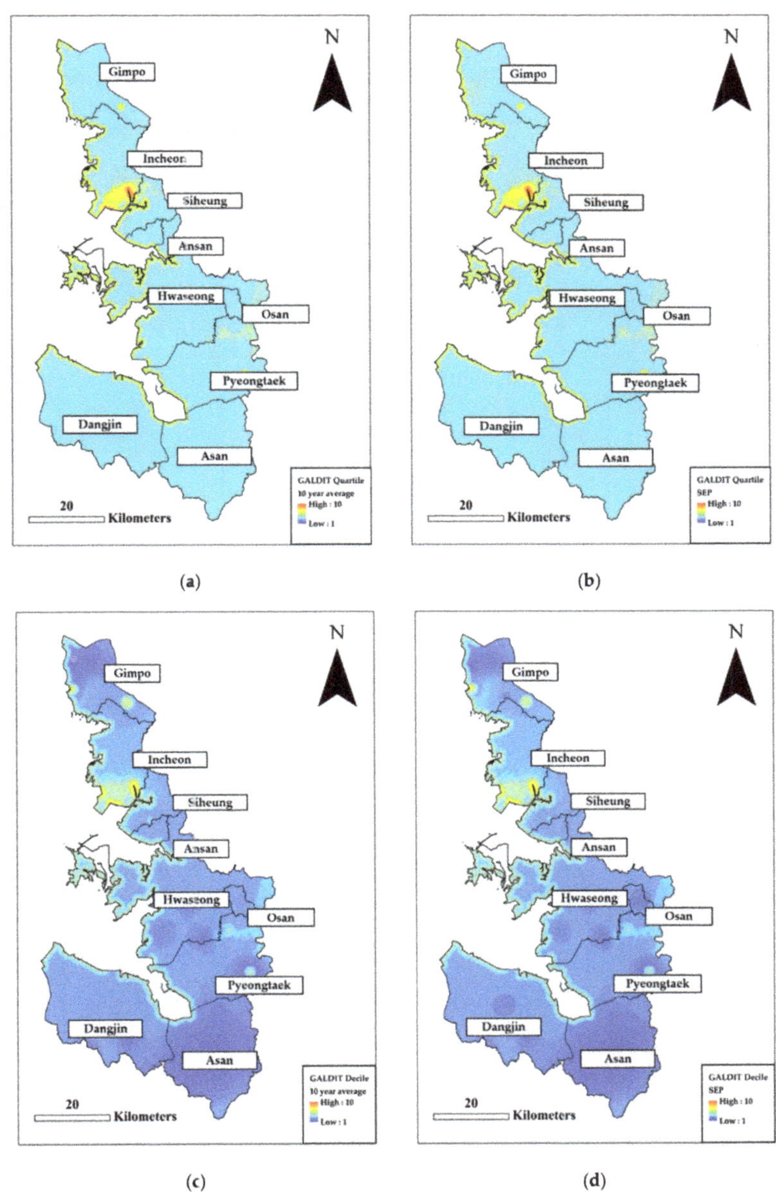

Figure 13. Comparison of (**a**) 10-year-averaged GALDIT in original quartile classification, (**b**) GALDIT in September in original quartile classification, (**c**) 10-year-averaged GALDIT in decile classification, and (**d**) GALDIT in September in decile classification.

4. Conclusions

SWI into aquifers is accelerating the depletion of coastal groundwater resources. An accurate diagnosis of SWI vulnerability is required for the sustainable utilization of groundwater resources in coastal areas. GALDIT is an SWI vulnerability assessment method that shows representative values, using a statistical test of observed data. The existing assessment method has a wide range of scores to express changing observation values.

To accurately represent regional characteristics or extreme climate patterns, there has been a continuing need for improved vulnerability assessments. Therefore, we developed a method to assess SWI vulnerability for averaged seasonality based on the original GLADIT. The analysis method was differentiated by classifying the six parameters of the existing GALDIT into static and dynamic parameters. For the static parameters—G (groundwater occurrence), A (aquifer hydraulic conductivity), and D (distance from the shore)—similar to the existing method, the annual average or short-term observed values was used. For the dynamic parameters—L (height to groundwater level above sea level), I (impact of existing status of seawater intrusion), and T (thickness of the aquifer)—10-year-averaged monthly data were used to reflect the observed values that change every month. In the existing score range, GALDIT values from the most vulnerable to least vulnerable values are divided by quartiles for the assessment of SWI vulnerability. The values assigned to each parameter were divided by using the decile method to sensitively reflect the degree of vulnerability that changes every month in areas having seasonal variation of the dynamic index, as in South Korea.

As a result of calculating the 10-year-averaged monthly GALDIT index by applying the existing weight, September was determined as the most vulnerable month, having a value of 3.03. In September, the ratio of areas with a score of five or higher was the highest at 8.87%, and it was the most vulnerable month for I and the third most vulnerable month for T. Based on this suggested method, areas that were particularly vulnerable in September, such as Gimpo and Pyeongtaek, were identified on the GALDIT map.

In the 10-year-averaged monthly seawater intrusion assessment, the saturated thickness of the aquifer is somewhat high, and most areas can be easily intruded by seawater. Because of T, which is mostly thick, it is difficult to indicate the differences in vulnerability even if time variability is considered. L was determined based on the sea level observed by sea water level stations near the groundwater level measuring site.

Our SWI vulnerability assessment method can prioritize the most vulnerable places and times according to month. If the countermeasures to the spatially vulnerable areas are prioritized and an operation management schedule is established according to the vulnerable period, the damage from seawater intrusion can be minimized. In addition, establishing a groundwater development and management plan for an area unaffected by SWI is possible.

Recently, various attempts have been made to mitigate and prevent seawater intrusion damage. Mitigation methods for the impact of the existing status of SWI after the occurrence of SWI include installing a freshwater injection well or a seawater pumping well in the aquifer, where SWI has progressed [52]. To mitigate SWI damage in advance, seawater pumping at the wedge part of SWI at the bottom [53,54], and conversely, injecting freshwater at the boundary of the SWI area are being researched [55,56]. If intensive response measures are applied to vulnerable areas using the methodology of this study and operational plans are established by considering the vulnerable period, the SWI damage could be effectively reduced, and sustainable utilization of groundwater in the coastal areas could be realized.

Note that this proposed monthly GALDIT method assessed the SWI vulnerability to the relatively large area with sufficient manpower and national monitoring network to detect the temporal variations of groundwater resources. In regions with scarce data or limited infrastructures, the observation period of the data can be selected flexible under efficient data-management plan. For example, long-term historical or future vulnerability can be estimated statistically if data obtained from areas with only recently installed monitoring wells show a constant change per year or show seasonally repeated fluctuation patterns by groundwater exploitation. The applicability of this study could be further expanded if site-specific postprocessing of dynamic parameters, such as moving average.

To employ GALDIT analysis as an assessment tool supporting water-resource management plans, considering vulnerable periods for sustainable operation of coastal groundwater, improving the theoretical weight, and expanding the variable range may be needed

Water **2021**, *13*, 1820

considering static and dynamic parameters representing extreme situations due to climate change in future. Furthermore, a follow-up study is required to improve the equations for calculating GALDIT parameters according to site characteristics.

Author Contributions: Conceptualization, S.W.C. and I.-M.C.; methodology, S.W.C.; validation, I.H.K., S.W.C., and I.-M.C.; investigation, I.H.K.; resources, I.H.K.; data curation, I.H.K.; writing-original draft preparation, I.H.K.; writing—review and editing, S.W.C. and I.-M.C.; visualization, I.H.K.; supervision, I.-M.C.; project administration, S.W.C.; funding acquisition, S.W.C. All authors have read and agreed to the published version of the manuscript.

Funding: This research was supported by a grant from the Development Program of Minimizing of Climate Change Impact Technology funded through the National Research Foundation of Korea (NRF) of the Korean government (Ministry of Science and ICT, Grant No. NRF-2020M3H5A1080735).

Institutional Review Board Statement: Not applicable.

Informed Consent Statement: Not applicable.

Data Availability Statement: Not applicable.

Conflicts of Interest: The authors declare no conflict of interest.

References

1. Nicholls, R.J.; Nicholls, P.P.W.; Burkett, V.; Colin, D.W.; Hay, J. Climate change and coastal vulnerability assessment: Scenarios for integrated assessment. *Sustain. Sci.* **2008**, *3*, 89–102. [CrossRef]
2. Howard, K. Urban Groundwater Issues—An Introduction. In *Current Problems of Hydrogeology in Urban Areas, Urban Agglomerates and Industrial Centres*; Springer: Dordrecht, The Netherlands, 2002; pp. 1–15. [CrossRef]
3. Chang, N.-B. *Effects of Urbanization on Groundwater: An Engineering Case-Based Approach for Sustainable Development*; American Society of Civil Engineers: Washington, DC, USA. 2010; ISBN 10:078441078X.
4. Chamine, H.I. Water resources meet sustainability: New trends in environmental hydrogeology and groundwater engineering. *Environ. Earth Sci.* **2015**, *73*, 2513–2520. [CrossRef]
5. Bernard-Jannin, L.; Sun, X.; Teissier, S.; Sauvage, S.; Sanchez-Perez, J.M. Spatiotemporal analysis of factors controlling nitrate dynamics and potential denitrification hot spots and hot moments in groundwater of an alluvial floodplain. *Ecol. Eng.* **2017**, *103*, 372–384. [CrossRef]
6. Azimi, S.; Moghaddam, M.A.; Monfared, S.A.H. Spatial assessment of the potential of groundwater quality using fuzzy AHP in GIS. *Arab. J. Geosci.* **2018**, *11*, 142. [CrossRef]
7. Mondal, I.; Bandyopadhyay, J.; Chowdhury, P. A GIS based DRASTIC model for assessing groundwater vulnerability in Jangalmahal area, West Bengal, India. *Sustain. Water Res. Manag.* **2019**, *5*, 557–573. [CrossRef]
8. Ray, S.S.; Ray, A. Major ground water development issues in South Asia: An overview. In *Ground Water Development-Issues and Sustainable Solutions*; Springer: Singapore, 2019; pp. 3–11.
9. Custodio, E. Prediction methods, in Hydrology. In *Groundwater Problems in Coastal Areas*; Bruggeman, A.G., Custodio, C., Eds.; Intergovernmental Hydrological Programme, UNESCO: Paris, France, 1987.
10. Herzberg, A. Die Wasserversorgung einiger Nordseebäder. *J. Gasbeleucht Wasserversorg* **1901**, *44*, 815–819.
11. Ghyben, B.W. Nota in Verband met de Voorgenomen Putboring Nabij Amsterdam. *Tijdschr. K. Inst. Ing.* **1988**, *9*, 8–22.
12. Todd, D.K.; Mays, L.W. *Groundwater Hydrology*, 3rd ed.; John Wiley & Sons: New York, NY, USA, 2005.
13. Sherif, M.M.; Singh, V.P. Effect of climate change on seawater intrusion in coastal aquifers. *Hydro. Proc.* **1999**, *13*, 1277–1287. [CrossRef]
14. Scanlon, B.; Faunt, C.; Longuevergne, L.; Reedy, R.; Alley, W.; McGuire, V.; Mcmahon, P. Groundwater depletion and sustainability of irrigation in the US high plains and central valley. *Proc. Nat. Acad. Sci. USA* **2012**, *109*, 9320–9325. [CrossRef]
15. Bobba, A.G. Numerical modelling of salt-water intrusion due to human activities and sea-level change in the Godavari Delta, India. *Hydrol. Sci. J.* **2002**, *47*, S67–S80. [CrossRef]
16. Loáiciga, H.A.; Pingel, T.J.; Garcia, E.S. Seawater intrusion by sea-level rise: Scenarios for the 21st century. *Groundwater* **2012**, *50*, 37–47. [CrossRef]
17. Carretero, S.; Rapaglia, J.; Bokuniewicz, H.; Kruse, E. Impact of sea-level rise on saltwater intrusion length into the coastal aquifer, Partido de La Costa, Argentina. *Cont. Shelf Res.* **2013**, *61–62*, 62–70. [CrossRef]
18. Langevin, C.D.; Zygnerski, M. Effect of sea-level rise on salt water intrusion near a coastal well field in Southeastern Florida. *Groundwater* **2013**, *51*, 781–803. [CrossRef]
19. Rasmussen, P.; Sonnenborg, T.O.; Goncear, G.; Hinsby, K. Assessing impacts of climate change, sea level rise, and drainage canals on saltwater intrusion to coastal aquifer. *Hydro. Earth Syst. Sci.* **2013**, *17*, 421–443. [CrossRef]
20. Sefelnasr, A.; Sherif, M. Impacts of seawater rise on seawater intrusion in the Nile Delta aquifer, Egypt. *Groundwater* **2014**, *52*, 264–276. [CrossRef]

21. El-Kadi, A.I.; Tillery, S.; Whittier, R.B.; Hagedorn, B.; Mair, A.; Ha, K.; Koh, G.-W. Assessing sustainability of groundwater resources on Jeju Island, South Korea, under climate change, drought, and increased usage. *Hydrogeol. J.* **2014**, *22*, 625–642. [CrossRef]
22. Polemio, M.; Walraevens, K. Recent research results on groundwater resources and saltwater intrusion in a changing environment water. *Water* **2019**, *11*, 1118. [CrossRef]
23. National Research Council. *Ground Water Vulnerability Assessment: Predicting Relative Contamination Potential under Conditions of Uncertainty*; National Academy Press: Washington, DC, USA, 1993; pp. 42–63.
24. Gogu, R.C.; Dassargues, A. Current trends and future challenges in groundwater vulnerability assessment using overlay and index methods. *Environ. Geol.* **2000**, *39*, 549–559. [CrossRef]
25. Uricchio, V.F.; Giordano, R.; Lopez, N. A fuzzy knowledge-based decision support system for groundwater pollution risk evaluation. *J. Environ. Manag.* **2004**, *73*, 189–197. [CrossRef] [PubMed]
26. Babiker, I.S.; Mohamed, M.A.; Hiyama, T.; Kato, K. A GIS-based DRASTIC model for assessing aquifer vulnerability in Kakamigahara Heights, Gifu Prefecture, Central Japan. *Sci. Total Env.* **2005**, *345*, 127–140. [CrossRef] [PubMed]
27. Aller, L.; Bennett, T.; Lehr, J.H.; Petty, R.J.; Hackett, G. *DRASTIC: A Standardized System for Evaluating Groundwater Pollution Potential Using Hydrogeologic Settings*; Environmental Protection Agency NWWA/EPA Series EPA-600/2-87-035; National Water Well Association: Dublin, Ireland, 1987.
28. Civita, M. *Le Carte Della Vulnerabilità Degli Acquiferi All'inquinamento: Teoria&Pratica [Aquifer Vulnerability to Pollution Maps: Theory and Practice]*; Pitagora: Bologna, Italy, 1994.
29. Chachadi, G.; Lobo-Ferreira, J.P. Sea water intrusion vulnerability mapping of aquifers using the GALDIT method. *COASTIN Newsl.* **2001**, *4*, 7–9.
30. Kardan Moghaddam, H.; Jafari, F.; Javadi, S. Vulnerability evaluation of a coastal aquifer via GALDIT model and comparison with DRASTIC index using quality parameters. *Hydrol. Sci. J.* **2017**, *62*, 137–146. [CrossRef]
31. Bordbar, M.; Neshat, A.; Javadi, S. Modification of the GALDIT framework using statistical and entropy models to assess coastal aquifer vulnerability. *Hydrol. Sci. J.* **2019**, *64*, 1117–1128. [CrossRef]
32. Bordbar, M.; Neshat, A.; Javadi, S.; Pradhan, B.; Aghamohammadi, H. Meta-heuristic algorithms in optimizing GALDIT framework: A comparative study for coastal aquifer vulnerability assessment. *J. Hydrol.* **2020**, *585*, 124768. [CrossRef]
33. Klassen, J.; Allen, D. Assessing the Risk of Saltwater Intrusion in Coastal Aquifers. *J. Hydrol.* **2017**, *551*, 730–745. [CrossRef]
34. Luoma, J.; Ruutu, S.; King, A.W.; Tikkanen, H. Time delays, competitive interdependence, and firm performance. *Strateg. Manag. J.* **2017**, *38*, 506–525. [CrossRef]
35. Hallal, D.; El Amine, M.K.; Zahouani, S.; Benamghar, A.; Haddad, O.; Ammari, A.; Lobo-Ferreira, J. Application of the GALDIT method combined with geostatistics at the Bouteldja aquifer (Algeria). *Environ. Earth Sci.* **2019**, *78*, 22. [CrossRef]
36. Lee, J.-Y.; Yi, M.-J.; Song, S.-H.; Lee, G.-S. Evaluation of seawater intrusion on the groundwater data obtained from the monitoring network in Korea. *Water Int.* **2008**, *33*, 127–146. [CrossRef]
37. Shin, J.; Hwang, S. A Borehole-Based Approach for Seawater Intrusion in Heterogeneous Coastal Aquifers, Eastern Part of Jeju Island, Korea. *Water* **2020**, *12*, 609. [CrossRef]
38. Chang, S.W.; Chung, I.-M.; Kim, M.-G.; Tolera, M.; Koh, G.-W. Application of GALDIT in Assessing the Seawater Intrusion Vulnerability of Jeju Island, South Korea. *Water* **2019**, *11*, 1824. [CrossRef]
39. Kim, I.-H.; Yang, J.-S. Prioritizing countermeasures for reducing seawater-intrusion area by considering regional characteristics using SEAWAT and a multi-criteria decision-making method. *Hydrol. Proc.* **2018**, *32*, 3741–3757. [CrossRef]
40. Chun, J.A.; Lim, C.; Kim, D.; Kim, J.S. Assessing impacts of climate change and sea-level rise on seawater intrusion in a coastal aquifer. *Water* **2018**, *10*, 357. [CrossRef]
41. Kim, K.-Y.; Seong, H.; Kim, T.; Park, K.H.; Woo, N.; Park, Y.S.; Koh, G.W.; Park, W.-B. Tidal effects on variations of fresh-saltwater interface and groundwater flow in a multilayered coastal aquifer on a volcanic island (Jeju Island, Korea). *J. Hydrol.* **2006**, *330*, 525–542. [CrossRef]
42. Kim, Y.; Yoon, H.; Kim, G.P. Development of a novel method to monitor the temporal change in the location of the freshwater-saltwater interface and time series models for the prediction of the interface. *Environ Earth Sci.* **2016**, *75*, 882. [CrossRef]
43. Kuan, W.K.; Jin, G.; Xin, P.; Robinson, C.; Gibbes, B.; Li, L. Tidal influence on seawater intrusion in unconfined coastal aquifers. *Water Res. Res.* **2012**, *48*, W02502. [CrossRef]
44. Recinos, N.; Kallioras, A.; Pliakas, F.-K.; Schüth, C. Application of GALDIT index to assess the intrinsic vulnerability to seawater intrusion of coastal granular aquifers. *Environ. Earth Sci.* **2014**, *73*, 1017–1032. [CrossRef]
45. Allouche, N.; Maanan, M.; Gontara, M.; Rollo, N.; Jmal, I. A global risk approach to assessing groundwater vulnerability. In *Environmental Modelling and Software*; Elsevier: Amsterdam, The Netherlands, 2017; Volume 88, pp. 168–182.
46. Gundogdu, K.S.; Guney, I. Spatial analyses of groundwater levels using universal kriging. *J. Earth Syst. Sci.* **2007**, *116*, 49–55. [CrossRef]
47. Sun, Y.; Kang, S.; Li, F.; Zhang, L. Comparison of interpolation methods for depth to groundwater and its temporal and spatial variations in the Minqin oasis of northwest China. *Environ. Model. Soft.* **2009**, *24*, 1163–1170. [CrossRef]
48. National Groundwater Information Center. Available online: http://www.gims.go.kr (accessed on 22 February 2021).
49. National Geographic Information Institute. Available online: http://www.ngii.go.kr (accessed on 22 February 2021).
50. Korea Hydrographic and Oceanographic Agency. Available online: http://www.khoa.go.kr (accessed on 22 February 2021).

51. Ministry of Oceans and Fisheries. *Coast Condition Survey Report-West Coast*; Ministry of Oceans and Fisheries: Busan, Korea, 2003.
52. Abarca, E.; Vázquez-Suñé, E.; Carrera, J.; Capino, B.; Gámez, D.; Batlle, F. Optimal design of measures to correct seawater intrusion. *Water Res. Res.* **2006**, *42*. [CrossRef]
53. Sriapai, T.; Walsri, C.; Phueakphum, D.; Fuenkajorn, K. Physical model simulations of seawater intrusion in unconfined aquifer. *Songklanakarin J. Sci. Technol.* **2012**, *34*, 679–687.
54. Pool, M.; Carrera, J. Dynamics of negative hydraulic barriers to prevent seawater intrusion. *Hydrogeol. J.* **2010**, *18*, 95–105. [CrossRef]
55. Botero-Acosta, A.; Donado, L.D. Laboratory scale simulation of hydraulic barriers to seawater intrusion in confined coastal aquifers considering the effects of stratification. *Proc. Env. Sci.* **2015**, *25*, 36–43. [CrossRef]
56. Luyun, R.; Momii, K.; Nakagawa, K. Effects of recharge wells and flow barriers on seawater intrusion. *Groundwater* **2011**, *49*, 239–249. [CrossRef] [PubMed]

water

Article

Groundwater of the Modder River Catchment of South Africa: A Sustainability Prediction

Saheed Adeyinka Oke [1,*] and Rebecca Alowo [2]

1 Department of Civil Engineering, Centre for Sustainable Smart Cities, Central University of Technology, Bloemfontein 9301, Free State, South Africa
2 Department of Civil Engineering, University of Johannesburg, Doorfontein 2094, Gauteng, South Africa; ralowo@uj.ac.za
* Correspondence: soke@cut.ac.za

Abstract: This paper presents a spatial interpolation of the hydrological and socioeconomic processes impacting groundwater systems to predict the sustainability of the Modder river catchment of South Africa. These processes are grouped as climatic (factor A), aquifer sustainability (factor D), social-economic and land use (factor B), and the human-induced parameters of rights and equity (factor C). The parameters evaluated for factors A and D included climatic zones, precipitation, sunshine, evapotranspiration, slope, topography, recharge, yields, storativity, aquifer types, and lithology/rock types. Factors B and C included population in the catchment, use per capita, water uses, tariffs and duration of the permits, pump rate per year, number of issued permits per year in the catchment, and number of boreholes in the sub-catchment. This paper, therefore, looks at the impact of the average values of the chosen set of parameters within the given factors A, B, C and D on groundwater in the C52 catchment of the Modder River, as modelled in a sustainability index. C52 is an Upper Orange catchment in South Africa. The results are presented in sustainability maps predicting areas in the catchment with differing groundwater dynamics. The Modder River groundwater sustainability ranged between low and moderate sustainability. The sustainability maps were validated with actual field groundwater recharge and surface water, a comparison between storativity and licensed volume, and a comparison of sustainability scores and storativity. The key finding in this paper will assist groundwater managers and users to adequately plan groundwater resources, especially on licensing and over pumping.

Keywords: groundwater recharge; groundwater sustainability; hydrology models; Modder River; sustainability index

Citation: Oke, S.A.; Alowo, R. Groundwater of the Modder River Catchment of South Africa: A Sustainability Prediction. *Water* 2021, 13, 936. https://doi.org/10.3390/w13070936

Academic Editors: Evangelos Tziritis and Andreas Panagopoulos

Received: 18 January 2021
Accepted: 22 March 2021
Published: 29 March 2021

1. Introduction

Groundwater typically forms through the concept of recharge. Surface water and rainfall form most of the groundwater recharge. Precipitation that infiltrates and percolates the earth's surface has three paths: (1) Capillary action forcing water into the vadose zone, (2) high temperature causing evapotranspiration, and (3) infiltration and percolation contributing to the water table [1]. In the arid and semi-arid areas of South Africa, farmers and communities have merely a limited number of water provision points [2]. This increases the pressure on groundwater and has increased the number of wells and boreholes being drilled to access the groundwater needed for multiple purposes, particularly for agriculture and drinking water [3]. The limited resources of water provision have put undue pressure on aquifers, such as those in the C52 Modder River catchment [4]. The Modder River catchment is part of the broader Orange River system termed the C5 secondary catchment. The sub-catchment (Modder River) is termed the C52 catchment [4], which has the following drainage regions: C52G, C52H, C52J, C52B, C52A, C52E and C52F. The Modder-Riet River catchment is a combined system. C51 is termed the Riet catchment while C52 is the

Modder River catchment. Both belong to the bigger Upper Orange catchment. The Orange catchment itself is further divided into the Upper and Lower Orange River system.

Groundwater sustainability is important because groundwater, as a global asset, is the world's most consumed natural resource. Globally, the withdrawal is estimated at 600–700 km^3/year [5]. This extraction affects the balance between space and time in its natural occurrence. It is further increasingly found that groundwater development (drilling, usage, quality, etc.) in most places takes place without understanding this balance. How groundwater is recharged and its impact on the environment are complex [1,4,6,7]. As a result, groundwater is excessively pumped, leading to depletion. Low water levels in aquifers are responsible for salinity intrusion in coastal aquifers, land subsidence, and the decline in the yield of water wells. This is a major global challenge [6–8].

The undue pressures in the catchment have led to excessive pumping and increased abstraction. The ability to supply water directly from groundwater aquifers to the farmers and other water users depends fundamentally on the rainfall, which is a major source of groundwater recharge [9]. With the low rainfall in South Africa, groundwater recharge is low [9]. High groundwater abstraction, also known as excessive pumping, has several negative effects. One of these effects is groundwater depletion, a result of fast-rate groundwater extraction from an aquifer. Fast-rate extraction does not allow for adequate recharge of the aquifer [1]. Other secondary effects of depletion are related to climate change, including surface albedo distortion, increased groundwater salinity, the high cost associated with pumping, poor operation and maintenance of the wells, and increased damage of built-up wells [1]. There have been changes, trends and threats that negatively affect water resources in the Modder River catchment [4]. These include increased population growth and increasing urbanisation. The population of the Modder River catchment increased from 618,566 in 2001 to an estimated 1,083,886 in 2016 [4]. This resulted in increasing water demands and excessive pumping. The Modder River catchment, particularly the Bloemfontein area, has the highest demand for water in the Upper Orange River (C5) catchment, at 351 million litres of the total local requirement. Other negative effects are the increased degradation of the environment, high levels of man-made climate change, and a high variability of the natural climate. The net effects have been the depletion of aquifers and prolonged periods of drought [4]. It is therefore necessary to develop groundwater sustainability models/indices to make informed decisions for improved groundwater management.

2. Materials and Methods

2.1. Study Area

The Modder River Basin (see Figure 1) is situated in the south western part of the Free State Province, South Africa, forming some portion of the Upper Orange Water Management Area (WMA) [10]. The Upper Orange WMA expands further into parts of the Eastern and Northern Cape areas [11]. The Modder River Basin is located from latitude 28°50″ to 29°40″ South and from longitude 24°40″ West to 27°00″ East, covering a total area of approximately 17,366 km^2. The altitude ranges from 1057 to 2106 m above sea level (m.a.s.l.) with the Riet and Modder River and its tributaries being the main drainage system [12]. The highest areas (Maluti Mountains), close to South Africa (Lesotho boundary around Dewetsdorp on the eastern end of the basin), are characterised by flat-topped hills. The lowest area lies to the southern side of Kimberley. The Modder River originates near Dewetsdorp and then flows to the North, thereafter heading west. After about 340 km, the river flows into the Riet River that joins the Oranje-Vaal River. The Modder River was generally, as most inland rivers in South Africa, a regular stream, yet because of the development of three critical dams, namely the Rustfontein, Mockes and Krugersdrift Dams, the waterway currently looks like a perennial river [13]. The dams' levels may drop to as low as 30% during the dry season. Water in the lower side stagnates in winter [14]. According to researchers [12], the savannah grassland is predominant in the eastern part of the catchment. The result is a Karoo shrubbery to the South and West of the catchment.

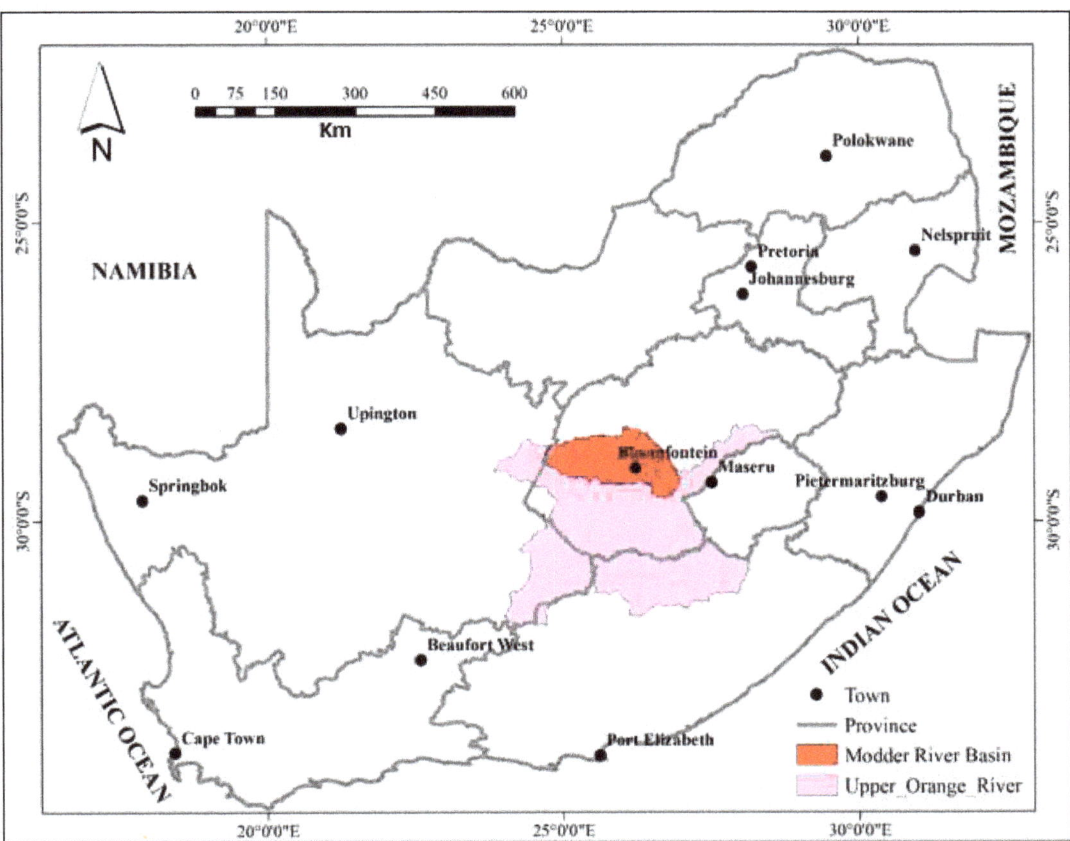

Figure 1. The location of the Modder River catchment showing boundaries within the Upper Orange River catchment and the Free State province of South Africa.

The Modder River catchment has both private boreholes and government monitoring boreholes. The study looked at the private boreholes whose owners gave consent. The private boreholes were of interest because the government is not in a position to monitor them. However, both private and government boreholes were used for the study. A monitoring system for all boreholes in the catchment is yet to be established [9]. The C52 catchment is further sub-divided into drainage regions: C52A to C52J. Figure 2 shows the boreholes available in the study area. The boreholes include both government monitoring boreholes (orange) and private boreholes (green).

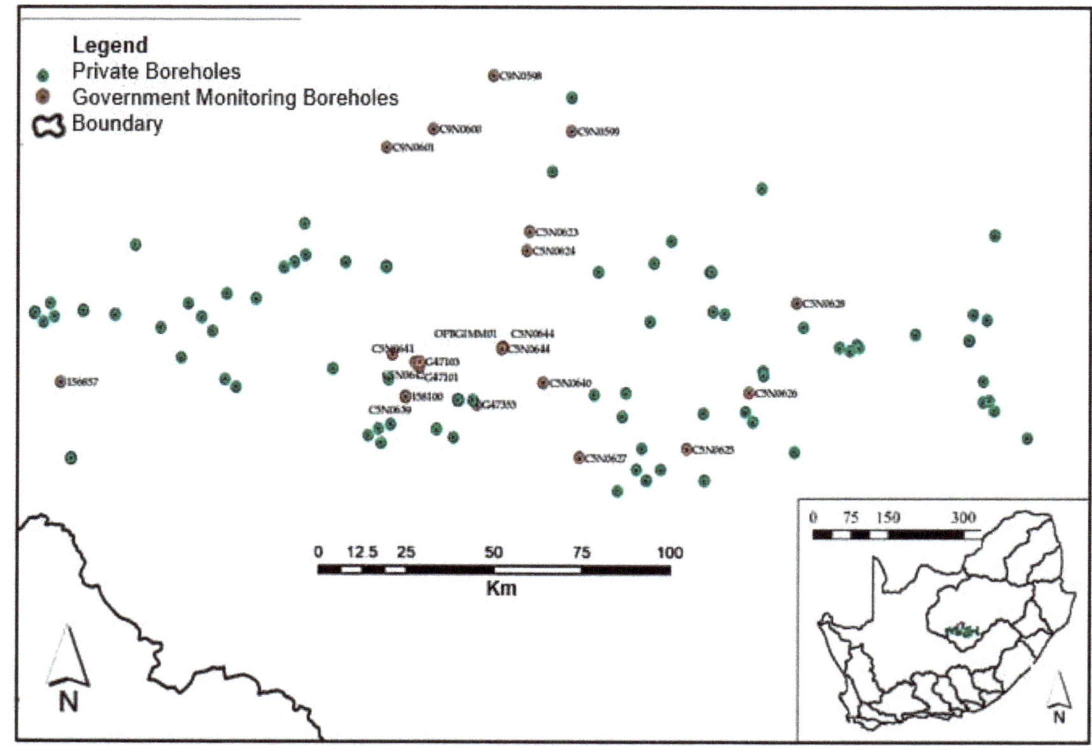

Figure 2. Map showing borehole distribution in the Modder River catchment.

2.2. Geological Formation Present in the Modder River Catchment

According to [15], the general geology of the Modder River catchment comprises mainly sedimentary rocks of the Karoo Super group, which were formed before the breakup of the Gondwana supercontinent. The Karoo Super group [16] covers around 66% of the present land surface of southern Africa. Its strata record a relatively endless glacio-marine to terrestrial succession. It started in the Permo-Carboniferous era (280 Ma) and ended 100 million years after the early Jurassic era. Its silt achieved an extreme combined thickness of up to 12 km in some places [17]. Overlying basaltic magmas, which denote the Lesotho area, are 1.4 km thick. These were collected in a retro-circular segmented basin [18], named "Karoo Basin." Along its southern fringe, the Karoo Basin is bordered by the Cape Fold Belt [19]. It was created amid a progression of compressional beats, beginning in the Late Carboniferous era and ending in the Late Triassic era [20]. The Late Palaeozoic advancement was started by plate assembly, subduction and growth along the palaeo-Pacific edge in the south western part of Gondwana [21]. The Cape Fold Belt comprises an E–W striking southern branch, with north-skirting folds and a N–S striking the western part of open folds, which converge into a 100 km-wide syntaxis zone. Its mountain ranges consist of siliciclastic silt, with an aggregate thickness of 8000 m, placing it within the Ordovician–Carboniferous Cape Supergroup.

As evidenced in Figure 3, the deposition of the Karoo Supergroup began after a rest at the Cape/Karoo Supergroup limit with the Dwyka Group. After glaciation, a broad sea remained facilitated by the melt water. Clays and muds of the Lower Ecca Group were aggregated. Deformation of the southern edge of the basin caused elevation and disintegration of the mountains towards the south. Quick down-warping of the basin was the result of thrust sheets in the nearby Cape Fold Belt [18]. The southwestern part of the

Karoo Basin was isolated by the Cape Fold Belt syntaxis into the Laingsburg and Tanqua sub-basins [22].

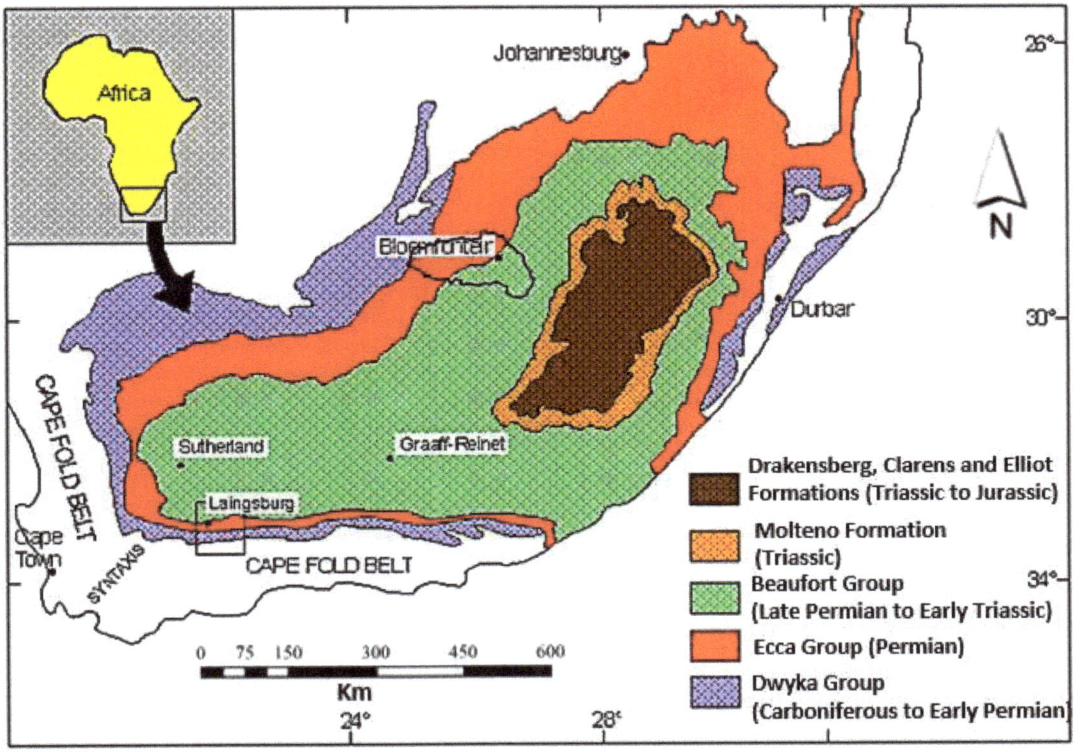

Figure 3. Regional geology of the Karoo Basin with inset of Modder River catchment location.

Deltaic progradation involved the filling of the sub-basins by a thick submarine fan and deltaic silt of the Upper Ecca Group [23]. Progressive shallowing of the fore took place within the Late Permian timeframe because of the rate of sedimentation surpassing the rate of subsidence [24]. The expansive scale backward succession protected the formation of the fluvial-lacustrine Beaufort Group. The Early Triassic lifted the Cape Fold Belt deposition in wide regions of the Karoo Basin [25]. In the focal piece of the Karoo Basin, the fluvial, alluvial, and aeolian residue of the Triassic Molteno, Elliot, and Clarens Formations were stored [26]. Regionally, the study area lies beneath the Beaufort Group on the east, the Ecca Group in the centre and the Dwyka Group on the western side [27]. The Main Karoo Basin becomes thinner from the south to the north [28].

Locally, according to GeoScience South Africa, the 1:1 million freely available geological information pieces, covering the Modder River catchment, five Karoo-aged rocks and Transvaal rocks are found in the catchment area. The Karoo-aged rocks are the Beaufort (Adelaide, Tarkastad), Dwyka, and Ecca Groups, as well as dolerite intrusive rocks [29]. The Karoo mafic intrusive rocks (dykes, sills) are scattered throughout [30]. The intrusive rocks are most pronounced on the northeastern side of Bloemfontein and areas surrounding Kimberley [31]. The Dwyka Group sediments, which are recorded to be stratified in a few places, are known to consist of diamictite (tillite) [32]. The Ecca Group (Tierberg) consists of undifferentiated shales, with interbedded siltstone [33]. Between Bloemfontein and Kimberley, Transvaal calcareous (limestone and calcarenite) rocks are exposed [34]. A small exposure of Transvaal Ventersdorp lava is found south of Kimberley [35].

2.3. Hydrogeology of the Modder River Catchment

The Modder River catchment is situated between the Ecca and Beaufort aquifer system, which will be discussed later. The underlying geology of the Modder River catchment is mainly sedimentary rocks intruded by the massive dolerite's dykes [36]. These numerous intrusive rocks reduce the pore spaces of the host rocks, thereby reducing the aquifer potential of the rocks. Therefore, the fractures are the only sources and target for abstracting large amounts of groundwater [37]. This suggests that the recharge rates and sustainable yields are relatively low in the catchment area in general even though some towns are using the small amount for rural water supply [38,39]. As a result of the rock type and minimal polluting surface activities, the quality of groundwater in the Modder River catchment is naturally satisfactory [40]. The eastern parts of the catchment have high rainfall with acceptable quality regarding taste, smell and colour [41]. The drier parts of the catchment, as well as areas with salt pan occurrence, are highly mineralised with brackish water [42].

The Ecca Group aquifers are mapped as combined fractured, as well as fractured and intergranular (Figure 4), with yields ranging between 0.5 and 2 L/s [43]. Burger [44] further concluded that the chances of obtaining an appreciable amount of water are where there are interlayered coal layers in the sediments than where there are solely sandstone and ordinary shale layers (Figure 3). It has been reported that the yield decreases as boreholes drilled closer to the dolerites dykes contact in the Ecca Group [45], which has been ascribed to small potential water-bearing fractures filled up with secondary materials, as well as baking of the contacts surrounding rocks with the dolerites intrusive rock [46]. This was observed in dry riverbeds where the various layers are exposed [47] and reported to be the best hydrogeological target for groundwater with good water quality [48,49].

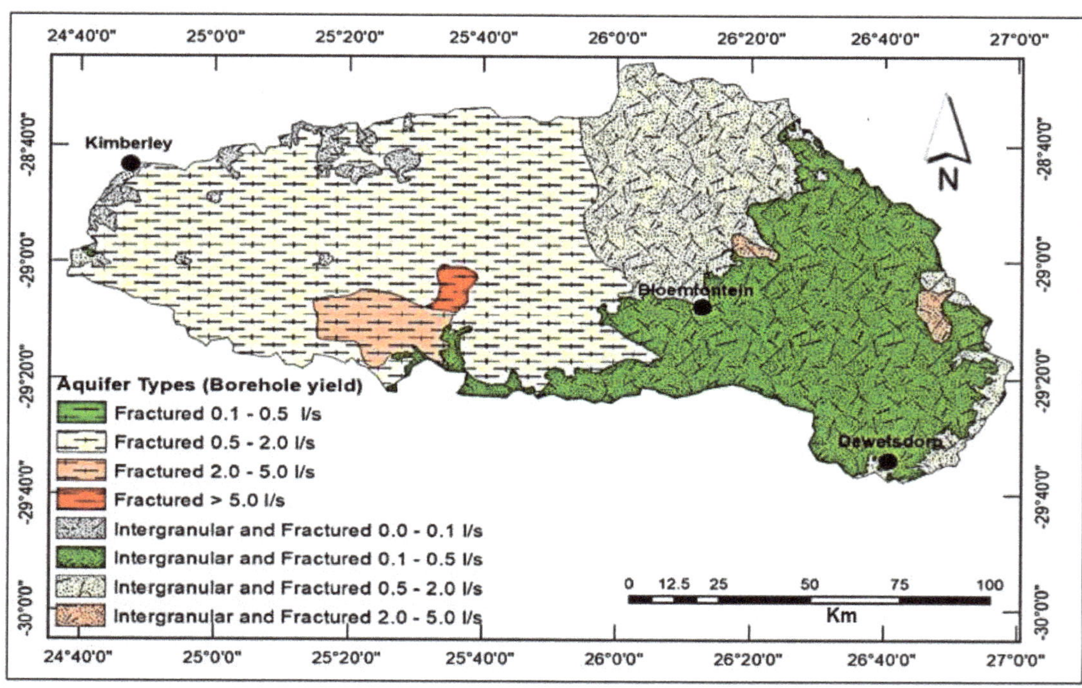

Figure 4. Aquifer types found within the C52 catchment area.

Beaufort Group aquifers are multi-layered, as well as multi-porous, with different thicknesses, owing to the complex geology of the formations of the Beaufort Group [50]. The contact plane between two different sedimentary layers will cause a discontinuity in the hydraulic properties of the composite aquifer [51]. This complex behaviour of aquifers in the Beaufort Group is further complicated by the fact that many of the coarser and thus more permeable sedimentary bodies are lens-shaped. The lifespan of a high-yielding borehole in the Beaufort Group may therefore be limited, if the aquifer is not recharged frequently [10,50].

Dolerite intrusions occur predominantly as sills and dykes throughout the Karoo sequence, with typical borehole yields of less than 2 L/s [12]. If the contact (side of the sill) is targeted, the factors to consider will be the same as when targeting a dyke (i.e., dip, recharge potential, static water level, the type of host rock and the distance to drill from the contact zone) [51]. When a source needs to be developed on a sill, three main targets should be investigated, namely, the upper weathered and fractured zone, the upper contact zone and the lower contact zone [52]. A number of factors has been stated to influence the success rate of boreholes on a sill [53].

2.4. Factors Affecting Groundwater Sustainability in Modder River Catchment

In reference to [54], three types of rainfall are mentioned: Orographic, cyclonic and conventional. The rainfall type determines the extent of the infiltration and percolation of rainwater to recharge the groundwater system [55]. Rainfall is an important groundwater sustainability factor because it is key to assessing any rainfall-runoff models [56]. The availability of precipitation data, its intensity and its duration are vital for hydrologic analysis during the design and management of water resources systems [57].

Topography defines the formation of the land surface. This includes its relief and the position of its natural and man-made features [56,58]. Topographic maps are usually used to show areas of different elevations. The elevations of mountains and valleys, steepness of slopes, and the direction of stream flow can be determined by studying topographic maps [56,59].

Topography is a key factor in groundwater sustainability because hydrologists use topographic and soil maps to understand an area [56,60,61]. In groundwater sustainability, topography dictates the direction of groundwater flow. Topography as reported by researchers [56,58] affects groundwater recharge and discharge. The impact of topography on rainfall distribution can be linked to different mechanisms, such as wind-driven effects and the small-scale topographic effects [56]. Topography is known to contribute to the base flow after the water table of groundwater in an aquifer has been satisfied.

Larger slopes generate more speed than smaller slopes. This can create faster runoff. Smaller slopes balance the rainfall input and the runoff rate that gets stored temporally over the area. With time, it can drain out gradually. This is an important consideration for groundwater sustainability. It is stated that a rise in surface slope showed a rise in surface runoff [56,62–64]. More runoff means less accumulation of groundwater in aquifers.

Land cover refers to natural vegetation cover and the human impact through several activities that are directly related to land occupation. Human activities make use of land resources and interferes in the ecological process that determines the functioning of land cover [56,65]. Land cover and therefore land use is one of the key parameters in the sustainable use of groundwater, particularly in the hydrologic cycle [56,66]. The effect of land use, land cover change and urbanisation on the hydrologic modalities and processes in catchments was studied in terms of vegetation conversion during the 1980s and 1990s [56,66]. In an investigation carried out by [67], it was found that urbanisation led to a 2.9% rise in the peak flow. In addition, a decrease of 14% on peak flows due to increased afforestation was also reported by [67]. A 5–12% increase in runoff was due to urbanisation [68]. Urbanisation causes an increase in storm flows in relation to the increased amount of surface runoff [69].

The assessment into the negative effects of farming on hydrological processes are very important for groundwater sustainability modelling. Most assessments conclude that high grazing pressure lowers infiltration rates, increases run-off from the ground surface as it lowers vegetation cover, and increases soil/ground compaction [56,70]. The changes in the land use and cover result in changes in the distribution of surface runoff within the catchment affecting groundwater infiltration rates [71]. Furthermore, the effects of land use on runoff generation by taking infiltration measurements on different land use categories were studied [72]. The results were that surface runoff was generated in varying magnitudes for different land-use types, with farmland being highest. High run-off coefficients were reported for different uses of farmland: 8.40% for cropland, 7.16% for pastureland, 2.61% for shrubland, 5.46% for woodland and 3.91% for grassland [73].

Soil is an important factor for ground water sustainability. The texture of a soil says a lot about its hydraulic conductivity and its grain-size distribution [56,74,75]. Soil texture and its structural content are two important properties for groundwater recharge and discharge and sustainability. This is because it affects water flow through the soil. In addition, it also sets out the amount of water retained in the soil, contributing to the water table of the aquifer in a catchment [56,76]. Different soil types affect runoff characteristics and generation. In an analysis of Trinidadian soils, [77] reported mean runoff values of 22.2, 22.9 and 40.9 mm for the loamy sandy, loamy clay and clay soils, respectively. Therefore, clay soil has the highest value compared to sandy soil.

The permit/licence system is a vital part of the whole groundwater sustainability system. Permits are set to guard the quality of groundwater resources and monitor the duration of groundwater extractions. The permits also ensure that the distribution rates and sizes/magnitudes work within limits which are politically acceptable, socially and environmentally viable and technically feasible. The importance of permits regarding groundwater sustainability includes economic instruments, demarcating groundwater rights and groundwater licensing. Groundwater restrictions and rights enable effective groundwater management [78]. Groundwater licensing guarantees groundwater abstraction with water management plans. Permits from an economic standpoint are important as they conserve groundwater and its quality, and control groundwater extraction. Various methods have been used to lessen abstraction. These include enabling water right trading, subsidies, and taxation. Caution is advised when taking water licensing measures. These measures must consider the intrinsic value of groundwater for all sectors of the economy [78,79].

The National Water Act of South Africa (1998) [80] gives the country and government ownership of water resources. The country links groundwater directly to land surface. As such, whoever owns the land has rights to the ground water below it. To date, arrangements and provision for trading groundwater rights do not exist. This is important to note regarding groundwater sustainability because the National Water Act (NWA) promotes efficiency, equity, and sustainability as paramount to water resources development management in South Africa. However, equity has not received the desired attention according to [79], resulting in inequitable water allocation. Equity has been deemed by the government as necessary for promoting sustainable economic growth and eradicating poverty.

2.5. Theory/Calculation

2.5.1. Sustainability Concepts and Index

Through the conceptual framework for sustainable groundwater in catchment management, there are several physical processes governing hydrological cycles in relation to groundwater sustainability in an aquifer. Some of these processes are land use-to-groundwater interactions, land use and climate interactions, and surface-to-groundwater interactions. A conceptual framework helps to represent these processes as factors. For this paper, the factors were grouped as: climatic, aquifer sustainability, right/equity of resources and socioeconomics (Figure 5). This is set within the context of the environment, economy and society, which are all at play in the catchment.

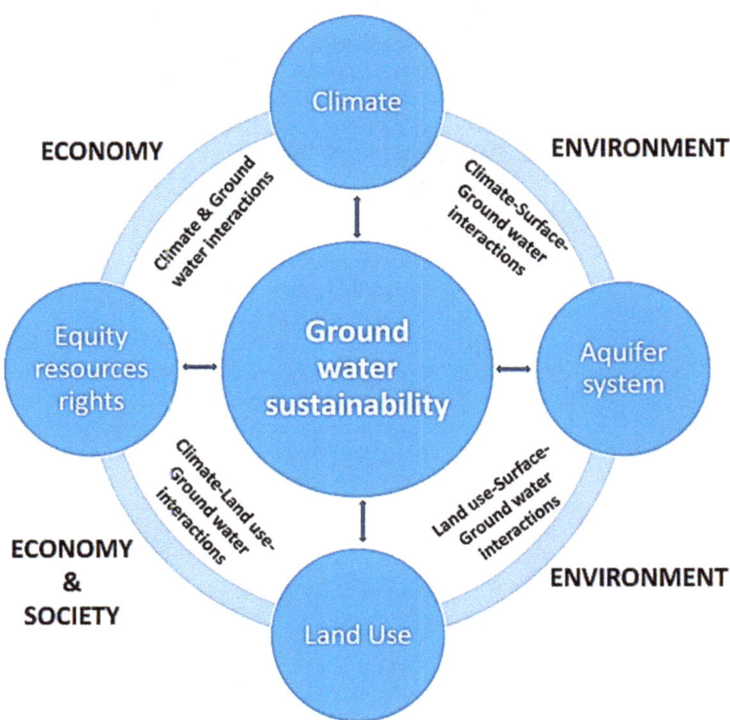

Figure 5. Supporting conceptual framework of predicting aquifer sustainability of Modder River catchment.

Sustainability in this context is a summative outcome of the link between the hydrological interactions. Water is extracted from an aquifer through well pumping, which is because of land use activity. Ownership and rights effectively shortcut the natural processes of recharge as the land is used. The amount of groundwater recharge and discharge in an aquifer is important to groundwater sustainability within this conceptual framework.

The conceptual framework was designed to trace the major relationships and interactions within the groundwater system in an aquifer that serves a catchment. This conceptual framework supports the making of inductions, deriving concepts from the data. It is also linked to the making of deductions directed at hypothesising the relationships between processes governing groundwater within the framework.

2.5.2. Rating of Hydrological Parameters

To achieve the sustainability prediction, important hydrological factors were rated. These ratings involve the assignment of values to the overall elements in factors A, B, C and D as follows:

2.5.3. Factor (A) Climate

Rainfall is assumed as critical in the catchment hydrological processes. Rainfall supports aquifer recharge, making it an important parameter in the overall sustainability prediction. A maximum score value of five is assigned to areas exceeding 3000 mm, and the lowest score of one is assigned to areas receiving less than 400 mm of rainfall. A similar approach is taken to assigning the maximum and minimum scores to evapotranspiration, sunshine, slope, vegetation and climate zones. Data used for these parameters were sourced from the South Africa Meteorological Centre office in Bloemfontein South Africa. In regions where the amount of rainfall is annually very low (< 400 mm/year), the lowest score value will be assigned. In general, when the rainfall rate is higher than the infiltration rate (intensity), the rainwater is likely to run off rather than infiltrate to recharge in aquifer, thus being stored as groundwater. This will also depend on the nature of the surface topography. The limitation of using rainfall in the sustainability design is that it does not consider a rating for rainfall intensity and the number of rainy days. This has all been captured by the annual volume an aquifer receives.

For factor A, the slope/topography is derived from the differences in contour values. Based on the slopes, scores were assigned as shown in Table 1. Lowest scores of 1 correspond to areas with the highest slopes greater than 50 m. These areas will encourage run-off and lower infiltration, while the lowest slopes of zero to 5 m correspond to areas with contours equal to those of water bodies, which encourage ponding. This means more infiltration and a high score of 5 assigned.

Slope/topography is derived from the difference between the highest topographic points to the lowest topographic point of an area. Run-off, infiltration, and recharge are influenced by the areal slope. Areas of low slope encourage ponding and retain water for a longer period, thereby increasing the possibility of percolation and infiltration and increase the potential for contaminated water migration. More runoffs occur in areas with steep slopes. This reduces the possibility of groundwater contamination. Flat slopes are prone to flooding and groundwater contamination because ponded surface water will readily infiltrate to groundwater. In summary, the equation that represents the evaluation of factor A is:

$$A = \sum_{k=0}^{n} k(R + E + S + T + V + C) \tag{1}$$

where:

A is the total score of all the parameters considered under Factor A;
K is the sum of the score of all the parameters considered under Factor A;
R is the rainfall;
E is the evapotranspiration;
S is the sunshine;
T is the topography;
V is the prevailing vegetation type;
C is the climatic zones assessed.

2.5.4. Factor (B) Rights and Equity versus Resources

Rights and equity to resources define the characteristics in an aquifer. The number of permits issued in the catchment per year is assumed to be important because it represents the abstraction activity in the catchment. Scores are within one and five. The maximum score value of five is assigned to areas receiving permits less than one, while the lowest score of one is assigned to areas receiving more than five permits. The same is considered in the assignment of scores for the duration of the permits, number of boreholes and pump rate. Based on global reports and databases, the values and score were assigned with the lowest scores corresponding to poor practices and the highest score of 5 to mean good practices. During analysis and the application of factors on the Modder catchment, the figures used were obtained from the Department of Water and Sanitation Affairs databases. In summary, the equation that represents the evaluation of factor B is:

$$B = \sum_{k=0}^{n} k(N + L + B + P) \tag{2}$$

where:

B is the total score of all the parameters considered under Factor B;
K is the score of all the parameters considered under Factor B;
N is the number of the permit;
L is the length or duration of permit years;

B is the number of boreholes in the sub-catchment;
P is the pumping rate.

2.5.5. Factor (C) Socioeconomics

Human activity affects an aquifer's sustainability through their use per capita. The use per capital of the catchment was assigned the maximum score of five to a use of less than 25 litres/capita/day, which is good practice for using little water. A per capita use of more than 100 is assigned a score of less than 1. This is the lowest score, and it indicates over-abstracting or using too much water. The scoring for the population in the catchment, water uses and tariffs takes the same trend as per capita use. These values/figures are from global reports and databases. However, in terms of the analysis of Modder catchment, the values assigned are derived from the Department of Water and Sanitation Affairs databases. Based on these data, scores were assigned with the lowest scores of 1, corresponding to poor practices, and the highest score of 5, denoting an acceptable practice. In summary, the equation that represents the evaluation of factor C is:

$$C = \sum_{k=0}^{n} k(U + P + W + T) \tag{3}$$

where:

C is the total score of all the parameters considered under Factor C;
k is the score of all the parameters considered under Factor C;
U is the use per capital of the catchment;
P is the population present in the catchment;
W is the water uses;
T is the tariffs.

2.5.6. Factor (D) Aquifer Sustainability

The pattern of groundwater recharges is found to be upstream supported by the discharge downstream. Most of the recharge areas correspond to mountain peaks where rainfall is higher; it is also at these points that higher recharge occurs as compared to low-laying plains. Aquifers are defined by their rock types. The extent to which rock type affects groundwater sustainability is dependent on hydraulic conductivity and permeability. Unfractured basement rock has little sustainability. The sustainability of fractured basement rock depends on the frequency variation, and the distribution and range of widths of the fractures. A score of five was assigned to intergranular rocks due to the expected longer time it will take for water to infiltrate into the groundwater. Water percolating through dense consolidated rocks is assumed to flow as surface run-off or subsurface horizontal flow, rather than as vertical infiltration flow, irrespective of the permeability of the topsoil. The dolerite dykes represent the lowest percolation in all geological rocks, and the low infiltration is due to the small pore spaces and lower permeability present in most of them. The dolerite formation is given a low score of 0.5 to a maximum score of 1. The water quality scoring will depend on the state of the water. If the water smells, tastes bad and is coloured, it is not fit for use and will frequently be wasted; therefore, a score of one was assigned; however, if the water is good in taste with no smell and colour, it is good and sustainable. A high-recharge aquifer (above 300 mm) is regarded as a sustainable aquifer and was assigned a high score, while a recharge of below 2 mm per year is unsustainable and was assigned a low score. During analysis of the Modder catchment, the values that are applied were obtained from databases from the Department of Water and Sanitation Affairs. In summary, the equation that represents the evaluation of factor D is:

$$D = \sum_{k=0}^{n} k(A + R + W + Y + R + S) \dots \tag{4}$$

where:

D is the total score of all the parameters considered under factor D;
k is the score of the parameter.
A is the aquifer system in place;
R is the rock type present;
W is the water quality;
Y is the aquifer yield;
R is the recharge condition;
S is the storage volume of the aquifer.

2.5.7. Sustainability Index

The index comprises climatic conditions, aquifer sustainability/system, rights/resources and socioeconomics (Figure 6). The overall objective of the index is to assess the sustainability of groundwater management in an aquifer in a catchment through analysis of a hydrological model, using predetermined parameters. It is therefore a major decision support system in the development of sustainability analysis methods. It considers the availability of input data for the hydrogeological system under consideration. The developed sustainability method targets the assessment of resources locally, regionally and globally.

The methodology requires an in depth understanding of the parameters and ranking of the physical processes affecting the groundwater system of the Modder catchment. These include the climatic factors (precipitation, evapotranspiration, sunshine, slope, topography and climatic zones) and aquifer system (recharge, yields, storativity, aquifer types and lithology/rock types). The methodology looks at how these factors work together and relate to give a picture of the status of sustainability in a catchment.

The formula includes human-induced parameters such as rights and equity. These human factors include the number of issued permits per year in the catchment, duration of the permits, number of boreholes in the sub-catchment, pump rate per year, socioeconomic and land use, use per capita, population in the catchment, water uses and tariffs.

$$\text{Sustainability } S = \sum A + B + C + D \tag{5}$$

where:

- A = Total score of the climatic condition factor (Factor A)
- B = Total score of the rights/equity factor (Factor B)
- C = Total score of the socioeconomic factor (Factor C)
- D = Total score of the aquifer sustainability factor (Factor D)

A and D have a scoring of 30 each, while B and C have a scoring of 20 each.

The final sustainability factors were added up because they all impact groundwater and therefore aquifer sustainability.

A and D factors (aquifer sustainability and climatic conditions) are complex in the natural context and, therefore, responsible for percolation and infiltration. As with the previously discussed principles of sustainability analysis, all factors have equal weighting. A carries an equal score to D because the sustainability methods assume rainfall evapotranspiration, sunshine and slope: As the principal climatic condition and initiator of the infiltration and subsequent percolation process which contributes to recharge and later becomes groundwater. The implication is that if rainfall is absent, there is no groundwater formation. The impact of B and C on groundwater sustainability may be higher on analysis as human activity will deplete whatever groundwater is available and not replenish it. For this reason, the scores of the two are the same.

The sustainability index was grouped into five classes. The classes and sustainability values are presented in Figure 6. A final groundwater sustainability index class score of 19–35 means a class of very low sustainability, 35–51 means a class of low sustainability, 51–67 means a class of moderate sustainability, 67–83 means a class of high sustainability and 83–100 means very high sustainability (Figure 7). The sustainability index acronym is derived from the initial letters of the factors used in ABC and D. The sustainability index method is designed to calculate the sustainability impact on groundwater.

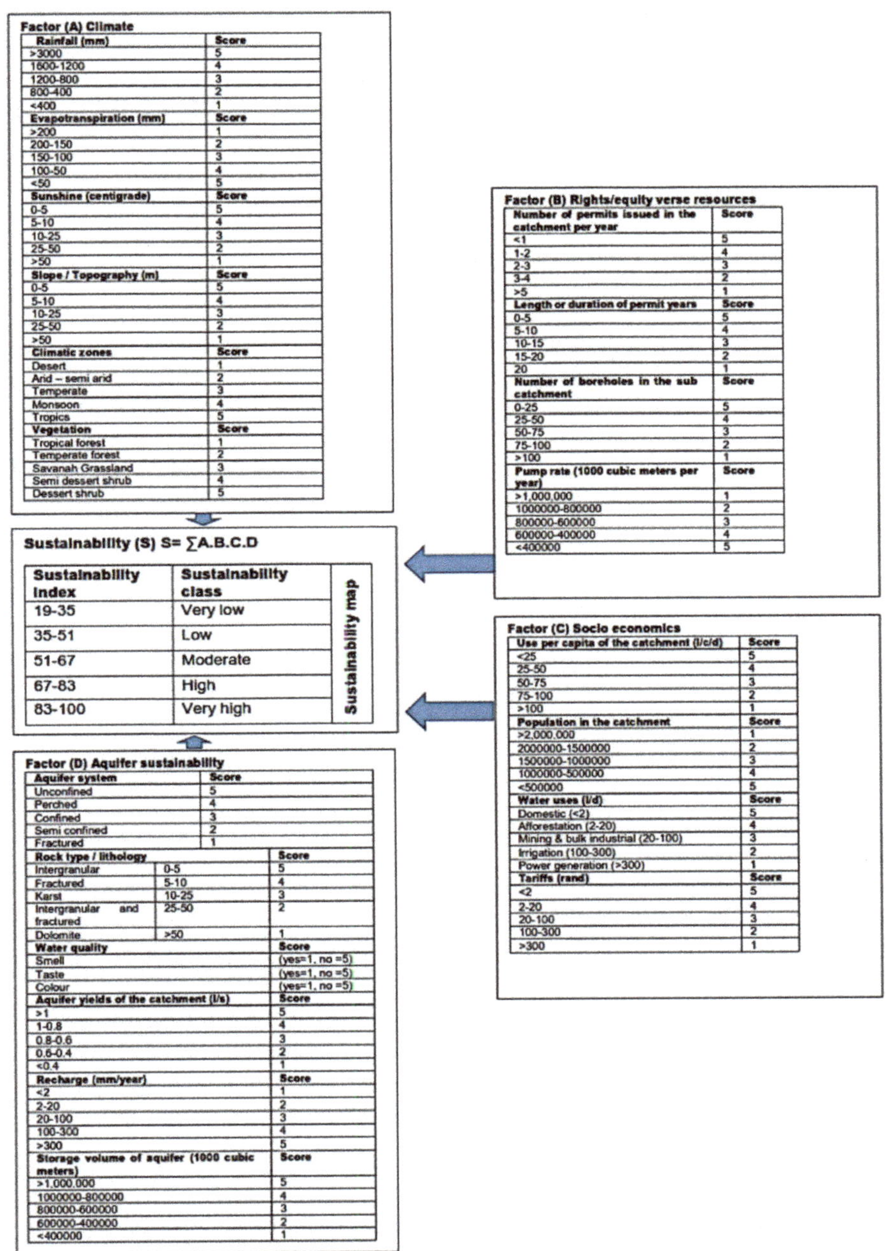

Factor (A) Climate

Rainfall (mm)	Score
>3000	5
1600-1200	4
1200-800	3
800-400	2
<400	1

Evapotranspiration (mm)	Score
>200	1
200-150	2
150-100	3
100-50	4
<50	5

Sunshine (centigrade)	Score
0-5	5
5-10	4
10-25	3
25-50	2
>50	1

Slope / Topography (m)	Score
0-5	5
5-10	4
10-25	3
25-50	2
>50	1

Climatic zones	Score
Desert	1
Arid – semi arid	2
Temperate	3
Monsoon	4
Tropics	5

Vegetation	Score
Tropical forest	1
Temperate forest	2
Savanah Grassland	3
Semi dessert shrub	4
Dessert shrub	5

Factor (B) Rights/equity verse resources

Number of permits issued in the catchment per year	Score
<1	5
1-2	4
2-3	3
3-4	2
>5	1

Length or duration of permit years	Score
0-5	5
5-10	4
10-15	3
15-20	2
20	1

Number of boreholes in the sub catchment	Score
0-25	5
25-50	4
50-75	3
75-100	2
>100	1

Pump rate (1000 cubic meters per year)	Score
>1,000,000	1
1000000-800000	2
800000-600000	3
600000-400000	4
<400000	5

Factor (C) Socio economics

Use per capita of the catchment (l/c/d)	Score
<25	5
25-50	4
50-75	3
75-100	2
>100	1

Population in the catchment	Score
>2,000,000	1
2000000-1500000	2
1500000-1000000	3
1000000-500000	4
<500000	5

Water uses (l/d)	Score
Domestic (<2)	5
Afforestation (2-20)	4
Mining & bulk industrial (20-100)	3
Irrigation (100-300)	2
Power generation (>300)	1

Tariffs (rand)	Score
<2	5
2-20	4
20-100	3
100-300	2
>300	1

Sustainability (S) S= ∑A.B.C.D

Sustainability Index	Sustainability class
19-35	Very low
35-51	Low
51-67	Moderate
67-83	High
83-100	Very high

Sustainability map

Factor (D) Aquifer sustainability

Aquifer system	Score
Unconfined	5
Perched	4
Confined	3
Semi confined	2
Fractured	1

Rock type / lithology		Score
Intergranular	0-5	5
Fractured	5-10	4
Karst	10-25	3
Intergranular and fractured	25-50	2
Dolomite	>50	1

Water quality	Score
Smell	(yes=1, no =5)
Taste	(yes=1, no =5)
Colour	(yes=1, no =5)

Aquifer yields of the catchment (l/s)	Score
>1	5
1-0.8	4
0.8-0.6	3
0.6-0.4	2
<0.4	1

Recharge (mm/year)	Score
<2	1
2-20	2
20-100	3
100-300	4
>300	5

Storage volume of aquifer (1000 cubic meters)	Score
>1,000,000	5
1000000-800000	4
800000-600000	3
600000-400000	2
<400000	1

Figure 6. Idealised illustration of the sustainability index/model.

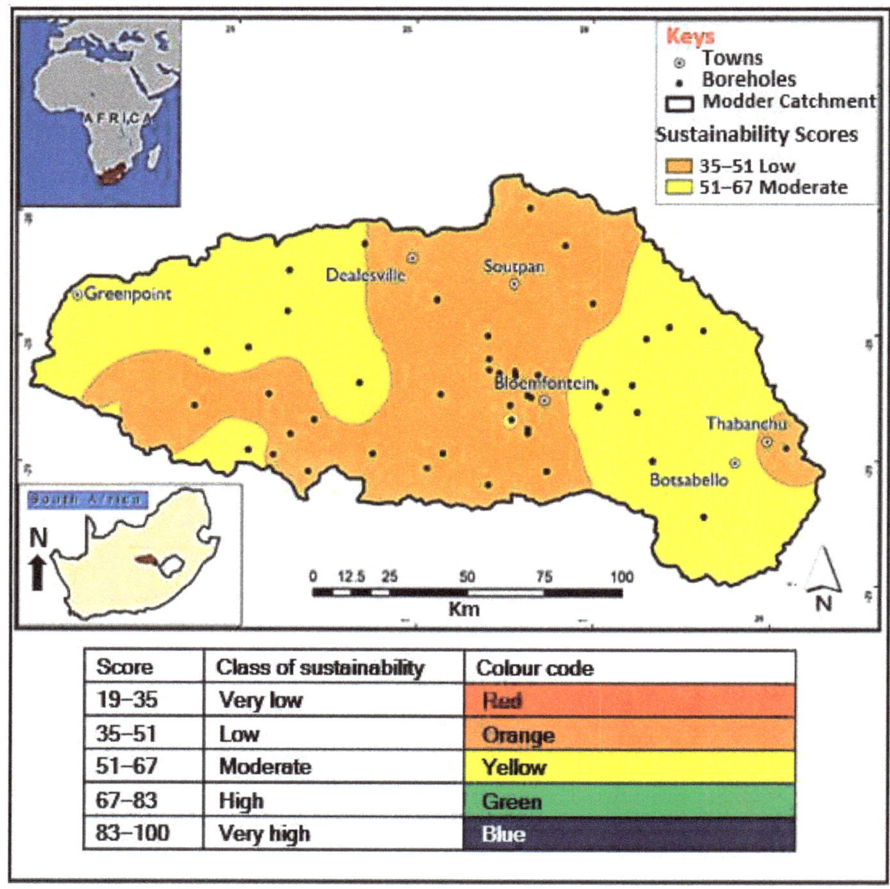

Score	Class of sustainability	Colour code
19–35	Very low	Red
35–51	Low	Orange
51–67	Moderate	Yellow
67–83	High	Green
83–100	Very high	Blue

Figure 7. Groundwater sustainability map of the Modder River catchment.

There are challenges when using the most established data to assess areas. As such, the developed sustainability method has been designed to use a few climate-based hydrogeological parameters and fuzzy logic parameters such as rights and social-economic data to the assess groundwater sustainability of a delineated catchment like the Modder River. The delineation exercise caters for inflows and outflows from the catchments and caters for other processes that affect the characteristics of a sub-catchment. This allows for the identification of a unit of analysis such as the Modder River that is representative of the rest of the larger catchment such as the Upper Orange.

2.5.8. Groundwater Recharge Calculation of Modder Catchment

The Department of Water and Sanitation conducted a study to generate a groundwater recharge map for South Africa, with a 1 km by 1 km grid cell size [81,82]. The method used to produce the recharge map is GIS-based, while Quaternary catchments were used as the unit of measure [83]. The recharge method essentially comprises four main components [84]:

- Chloride mass balance (CMB) approach
- Empirical rainfall/recharge relationships
- Layer model (GIS based) approach
- Cross-calibration of the results with field measurements and detailed catchment studies.

The results obtained from the recharge study by the Department of Water and Sanitation (DWS) agreed with the results obtained from earlier recharge studies [81]. Though the approach did not differentiate between the preferred path and matrix diffusion recharge, it is GIS-based, making it sufficiently flexible to include updates and new datasets. Part of Figure 8C shows the recharge map of the study area, as extracted from the countrywide recharge map of South Africa. In the study area, high recharge areas (over 19 mm) are found in the eastern and south-eastern parts of the study area. The central and western parts of the study area are marked by lower recharge values (below 19 mm).

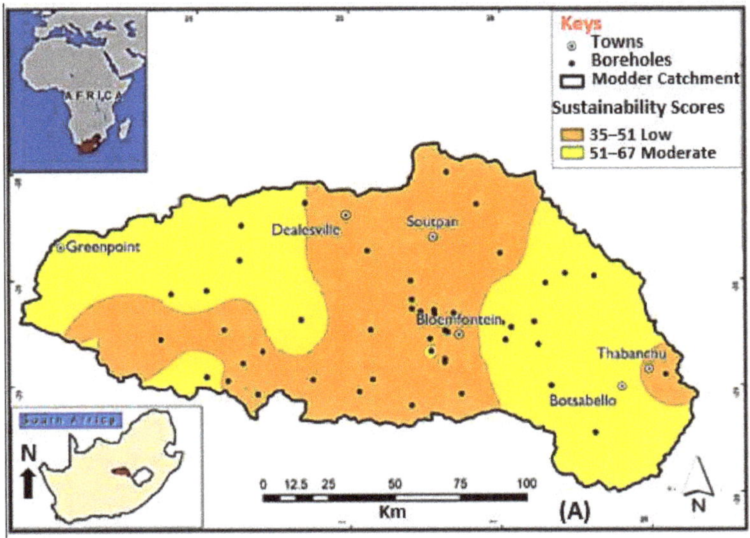

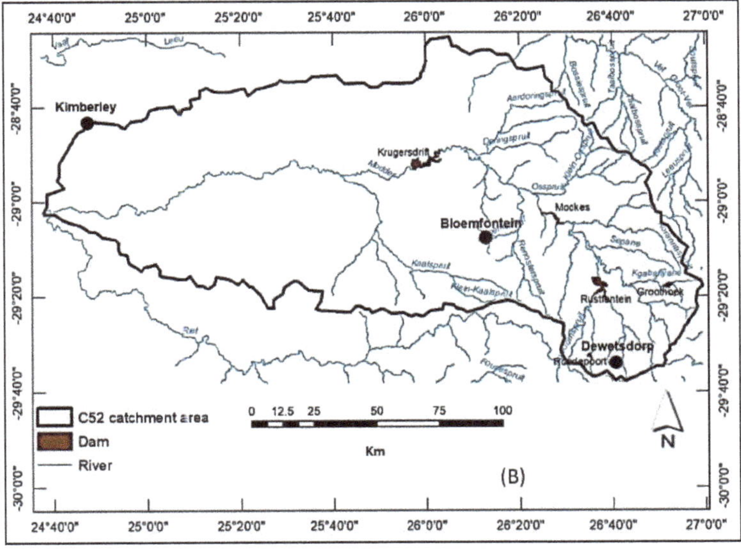

Figure 8. *Cont.*

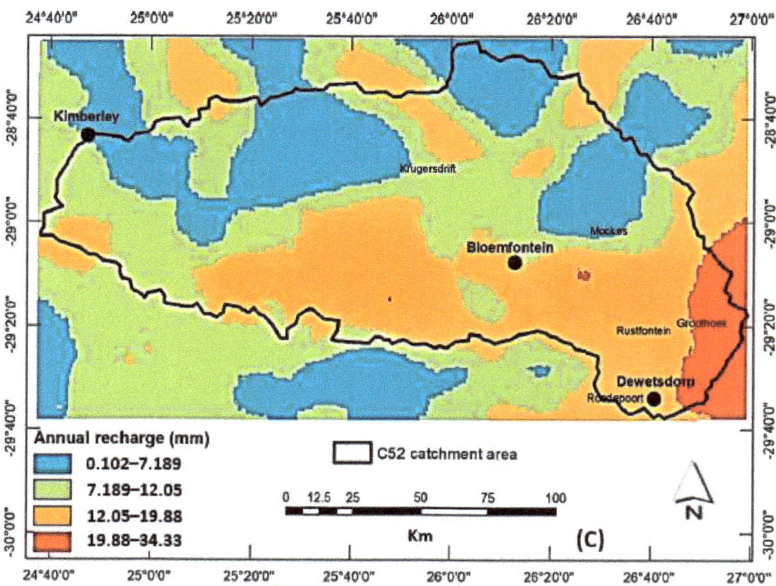

Figure 8. Final sustainability map (**A**), the river network map (**B**) and recharge map (**C**) of the Modder River catchment.

The chloride mass balance method was applied to the Modder River catchment. According to [81], it takes more than two years to measure the variability of recharge. This method uses chloride routing, as shown in Figure 5. This routing is the basis of the estimate of the net recharge of the catchment. For the Modder River, this is a collection from the slopes and channels without the losses through base flow discharged from the catchment. When applying the chloride model, Equation (6) is used. This equation calculates the net groundwater recharge volume denoted as R. It assumes a given number of years of rainfall in the catchment for which discharge takes place.

$$R = \frac{(P)(Cl_p) - (Q)(Cl_q)}{Cl_r} \tag{6}$$

In this equation:

- P is rainfall;
- Cl_p is the catchment's chloride concentration of rainfall;
- Cl_r is the measured groundwater chloride concentration of the catchment;
- Q is the total discharge from the catchment;
- Cl_q is the average chloride concentration in the stream discharge, i.e., the Modder River.

The application of this equation requires that the catchment's chloride concentrations for groundwater account for the evapotranspiration effects.

3. Results

3.1. Sustainability Map

Figure 7 represents the results from the application of the designed framework on the C52 catchment. The high-and very high-sustainability classes correspond to areas with favourable climatic conditions and favourable groundwater interaction and processes (fast recharge). It is also expected to have less abstraction and socioeconomic activity. The moderate to low classes suggest areas with the opposite of the previous scenario: Too much abstraction activity, unfavourable climatic conditions and slow or little groundwater recharge processes and interactions (high or steep slopes and low rainfall). Areas around Bainsvile west of Bloemfontein, South Africa (Figure 7) were declared as stressed; likewise, many boreholes in ThabaNchu, South Africa were declared stressed in 2015 during the last drought period.

3.2. Comparison of Recharge and Sustainability Map

The sustainability map was compared with the drainage map and groundwater recharge map for validation purposes. Figure 8 shows the final sustainability map (Figure 8A), the river network map (Figure 8B) and the chloride mass balance recharge map (Figure 8C) of the study area, as extracted from the countrywide recharge map of South Africa. The results obtained from the recharge study by the DWS correlated with the results obtained from the final sustainability map. In the study area, high recharge areas (over 20 mm) are found in the eastern and southeastern parts of the Modder River, which has high surface water and river tributaries. This suggests that the groundwater of the southeastern part of the Modder River is possibly recharged through base flow interactions. This assertion requires further investigations.

The central and western parts of the study area are marked by lower recharge values (below 7 mm). This is consistent with the final sustainability index maps and no river presence. There is a high correlation between the sustainability index (Figure 8A), the drainage basin (Figure 8B) and the recharge map (Figure 8C). The orange shaded areas in the sustainability map (Figure 8A) indicate low recharge. In relation to the river network maps (Figure 8B), surface water concentration suggests that the boreholes in the yellow zones are abstracting shallow river water that has percolated in the soil, especially along the streams and river levees, and the source of the deeply percolated water in boreholes. Therefore, there could be an interchange of groundwater–surface water exchanges. The groundwater recharge map (Figure 8C) shows red zones, which have slightly higher groundwater recharge. This is consistent with the yellow zones in the sustainability map (Figure 8A) that are consistent with moderate sustainability. The implications are that the Modder catchment has low ground water sustainability in the densely populated area of Bloemfontein. New developments relying on groundwater will not be viable in the low-sustainability zones.

4. Discussion

4.1. How Sustainable Will Groundwater Be in the Catchment?

It is important to explore the aquifer system sustainability in the catchment separately because of the possibility of some aquifers sustaining themselves, irrespective of the socioeconomic activities that the aquifer supports. Aquifers are recharged through several processes, including rainfall, infiltration and percolation and through recharge mechanisms and other hydrological processes (base flow, artificial recharge) that affect groundwater sustainability. The higher the population number in a catchment, the higher the socioeconomic activities (agriculture, mining, industrial and domestic); the higher their demands for groundwater, the more permits issued, which put the aquifers in vulnerable and less sustainable conditions.

The sustainability index scores of the study area (Table 1) consist of aspects such as the issuing of groundwater abstraction permits in the Modder River catchment, which therefore, serves as a measure of knowing the abstraction activity in the aquifer. This permit is issued by the regulatory agencies at Water Affairs in the Free State. The sustainability index scores in one of its parameters consider the rights and equity of the water users relating to the groundwater resources of the catchment. These comprise the number of borehole permits issued per year, duration or length of the borehole permit issued, number of actual boreholes drilled and pump rate per year currently in the Upper Orange River. These indicators are monitored and available at the Department of Water and Sanitation (DWS). The sustainability index, therefore, represents the groundwater abstraction rate in the aquifer. In addition to the DWS regulating the right of the water users to exploit the groundwater resources, it further issues permits and keeps records of the groundwater yields. Table 1 shows that most of the boreholes have low sustainability index scores and a corresponding low storativity. A few boreholes with higher storativity record high moderate sustainability index scores (boreholes 26, 27, 31, 32, 33, 37, 42, 44, 50). It should be noted that a borehole might have a higher storativity value but low sustainability scores because of over-abstraction taking place in the borehole.

Table 1. Calculated sustainability scores and classes linked to boreholes and storativity.

Drainage Region	Storativity (in 1000 cm^3/year)	Borehole	Sustainability Index	Sustainability Class
C52G	74,320.5	1	44.5	Low
C52G	74,320.5	2	43.5	Low
C52G	74,320.5	3	44.5	Low
C52H	160.51	4	44.5	Low
C52H	160.51	5	42.5	Low
C52H	160.51	6	42.5	Low
C52J	22,668.67	7	43.5	Low
C52H	160.51	8	44.5	Low
C52H	160.51	9	44.5	Low
C52H	160.51	10	46.5	Low
C52H	160.51	11	47.5	Low
C52H	160.51	12	44.5	Low
C52J	22,668.67	13	43.5	Low
C52J	22,668.67	14	42.5	Low
C52H	160.51	15	42.5	Low
C52H	160.51	16	42.5	Low
C52H	160.51	17	44.5	Low
C52H	160.51	18	44.5	Low
C52J	22,668.67	19	41.5	Low
C52J	22,668.67	20	44.5	Low
C52J	22,668.67	21	42.5	Low
C52J	22,668.67	22	42.5	Low
C52H	160.51	23	43.5	Low
C52B	62.58	24	45.5	Low
C52A	73,4547.6	25	49.5	Low
C52A	73,4547.6	26	52.5	Moderate
C52A	73,4547.6	27	51.5	Moderate
C52E	52,7202.8	28	46.5	Low
C52E	52,7202.8	29	48.5	Low
C52E	52,7202.8	30	48.5	Low
C52E	52,7202.8	31	52.5	Moderate
C52E	52,7202.8	32	52.5	Moderate
C52F	36,8407.7	33	52.5	Moderate
C52F	36,8407.7	34	46.5	Low
C52F	36,8407.7	35	46.5	Low
C52F	36,8407.7	36	51.5	Moderate
C52F	36,8407.7	37	51.5	Moderate
C52F	36,8407.7	38	47.5	Low
C52F	36,8407.7	39	47.5	Low
C52F	36,8407.7	40	46.5	Low
C52F	36,8407.7	41	47.5	Low
C52F	36,8407.7	42	51.5	Moderate
C52F	36,8407.7	43	46.5	Low
C52F	36,8407.7	44	51.5	Moderate
C52F	36,8407.7	45	46.5	Low
C52F	36,8407.7	46	48.5	Low
C52F	36,8407.7	47	46.5	Low
C52F	36,8407.7	48	47.5	Low
C52F	36,8407.7	49	47.5	Low
C52F	36,8407.7	50	51.5	Moderate
C52F	36,8407.7	51	49.5	Low

In addition, the sustainability index, as explained earlier, also reflects tariffs and the commercial use of water in the Modder River catchment, which therefore, represents the abstraction activity in the aquifer that supports economic growth. The sustainability index scores sum up the socioeconomic activities of the catchment, as related to their impact on the groundwater resources. The indicators considered in the sustainability index consist of the per capita use of the catchment, where boreholes are located, the population of the Modder River catchment, tariffs paid based on the use, the economic activities of the users, and the purpose of the groundwater use (mining, agriculture, and domestic and energy development). The sustainability index represents the potential use of the water, the economic growth that relies on the groundwater abstracted and the activities that impact the groundwater abstraction. Generally, these indicators represent the main activities that influence the groundwater sustainability of the Modder River catchment. These indicators are monitored and available at the DWS, as stated in methodology section. This makes the plot of sustainability index scores versus pump rate of great significance.

A further step in this plot involves a comparison of the calculated indices with the measured physical values such as storativity. An analysis of the actual end users' validation was carried out to assess the ability of the sustainability index, to calculate values that are a close reflection of the physical system (sustainability index scores vs. storativity). Figure 9 shows a plot of the final sustainability index value against a measurable parameter (storativity). There is a high correlation between the storativity and the sustainability index.

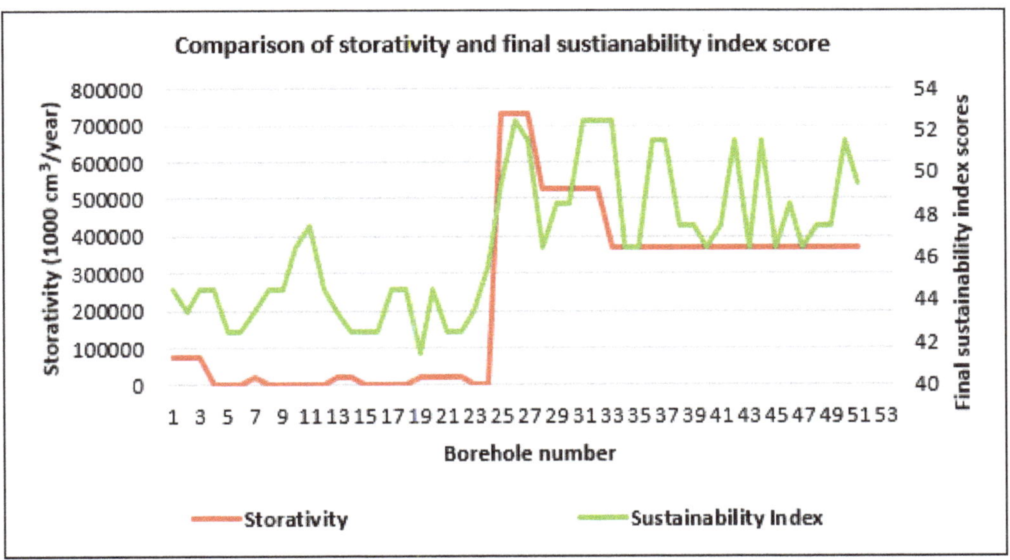

Figure 9. Plot of final sustainability scores against the storativity value of the catchment.

4.2. Groundwater Use versus Availability

A further validation tool is a critical look at the groundwater sustainability of the catchment as viewed through the socioeconomic activities of the catchment. This includes aspects such as the tariffs and commercial use of water in the Modder River catchment, which therefore, represents the abstraction activity in the aquifer that supports economic growth. This is of significance and considered part of validating the sustainability prediction. Figure 10 details the socioeconomic activities of the catchment as related to their impact on the groundwater resources. As explained earlier, these socioeconomics includes the per capita use of the catchment where boreholes are located, the population of the C52 catchment that relies on groundwater, the tariffs paid based on the use, the economic activities of the users and the purpose of the groundwater/land use (mining, agriculture, domestic use and energy development). Furthermore, this plot represents the potential use of the water, the economic growth that relies on the groundwater abstracted, and the activities that impact the groundwater abstraction. Generally, these are the main activities that influence the groundwater sustainability of the catchment.

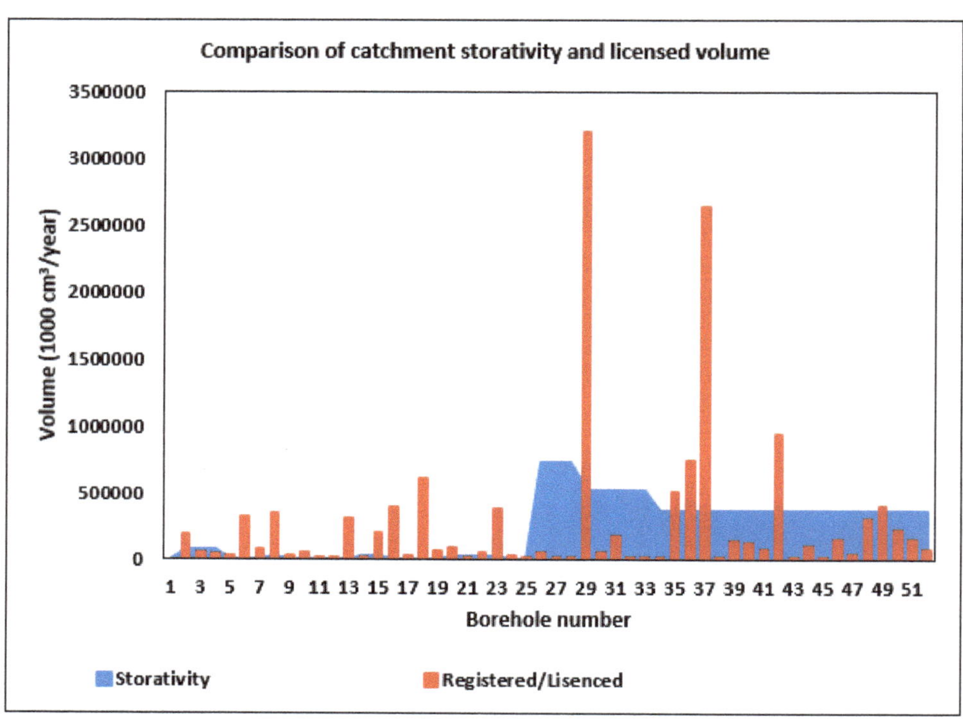

Figure 10. Relationship between storativity and licensed volume.

The graph of storativity versus licensed volume (Figure 10) shows that higher volumes are licensed as compared to the storage volume of the aquifer. The findings on the final sustainability map further correlate with this. There is no correlation between the pump rate/licensed volume and the storativity of Modder River catchment.

5. Conclusions

The sustainability method discussed in this paper was used to assess the sustainability of groundwater in the Modder River catchment. The sustainability class in the Modder River ranges from low sustainability to moderate sustainability. The moderate-to-low sustainability class is typical of areas with extensive dolerite, low slopes and low recharge.

The created sustainability index was applied to 52 boreholes in the Modder River catchment of South Africa. ArcView GIS (version 8, Esri, Redlands, CA, USA) was used in demonstrating the results for each of the 52 boreholes towards the groundwater sustainability concept. The results show that 9 boreholes are sustainable, implying that they are reasonably maintainable, while 43 boreholes have low sustainability scores, which indicates that they are not viable.

The sustainability method employed in this paper is designed from general techniques of indexing and rating, and the sustainability concept is simplified based on groundwater interactions with land and climate. The sustainability index method makes use of both the subjective and physically based techniques. The sustainability method and its evaluation are easy to collate, calculate and apply. Its successful application to the Modder River catchment shows that it can be further extended to groundwater in the Upper Orange River Basin and other similar catchments, particularly those in South Africa and also globally.

Author Contributions: S.A.O. and R.A. designed the conceptual framework and modelling framework. R.A. designed the sustainability index with contribution from S.A.O., R.A. and S.A.O. prepared the manuscript. Both authors have read and agreed to the published version of the manuscript.

Funding: The second author was funded by the Central University of Technology, Free State, for her doctoral studies. The APC was funded by the University of Johannesburg.

Conflicts of Interest: No conflict of interest.

References

1. Ponce, V.M. Groundwater Utilization and Sustainability: 78. Available online: http://groundwater.sdsu.edu/ (accessed on 12 July 2017).
2. Rao, N.; Singh, C.; Solomon, D.; Camfield, L.; Sidiki, R.; Angula, M.; Poonacha, P.; Sidibé, A.; Lawson, E.T. Managing risk, changing aspirations and household dynamics: Implications for wellbeing and adaptation in semi-arid Africa and India. *World Dev.* **2020**, *125*, 104667. [CrossRef]
3. Döring, S. Come rain, or come wells: How access to groundwater affects communal violence. *Political Geogr.* **2020**, *76*, 102073. [CrossRef]
4. DWA (Department of Water Affairs). *National Water Resource Strategy*, 2nd ed.; Department of Water Affairs: Pretoria, South Africa, 2013.
5. Wani, S.P.; Garg, K.K.; Singh, A.K.; Rockström, J. Sustainable Management of Scarce Water Resource in Tropical Rainfed Agriculture. In *Soil Water and Agronomic Productivity. Advances in Soil Science*; CRC Press: Boca Raton, FL, USA, 2012; pp. 347–408.
6. Akurugu, B.A.; Chegbeleh, L.P.; Yidana, S.M. Characterisation of groundwater flow and recharge in crystalline basement rocks in the Talensi District, Northern Ghana. *J. Afr. Earth Sci.* **2020**, *161*, 103665. [CrossRef]
7. Sophocleous, M. Groundwater recharge and sustainability in the High Plains aquifer in Kansas, USA. *Hydrogeol. J.* **2005**, *13*, 351–365. [CrossRef]
8. Reddy, P.J.R. *A Textbook of Hydrology*; University Science Press: New Delhi, India, 2006; pp. 341–342.
9. South Africa Department of Water and Sanitation (DWS). Groundwater Resource Assessment II. Available online: https://www.dws.gov.za/groundwater/GRAII.aspx (accessed on 18 January 2021).
10. Molaba, G.L. Investigating the Possibility of Targeting Major Dolerite Intrusives to Supplement Municipal Water Supply in Bloemfontein: A Geophysical Approach. Master's Thesis, University of the Free State, Bloemfontein, South Africa, 2017.
11. Du Plessis, A. Current Water Quality Risk Areas for Vaal, Pongola-Mtamvuna and Orange WMAs. In *Water as an Inescapable Risk*; Springer: Berlin/Heidelberg, Germany, 2019; pp. 213–246.
12. Du Toit, W.H.; Sonneku, C.J. *An Explanation of the 1:500 000 General Hydrogeological Map, Nelspruit 2530*; Department of Water and Sanitation: Pretoria, South Africa, 2014.
13. Oldknow, C.; Hooke, J. Alluvial terrace development and changing landscape connectivity in the Great Karoo, South Africa. Insights from the Wilgerbosch River catchment, Sneeuberg. *Geomorphology* **2017**, *288*, 12–38. [CrossRef]
14. Bezuidenhout, C. Macrophytes as indicators of physico-chemical factors in South African Estuaries. Ph.D. Thesis, Nelson Mandela Metropolitan University, Port Elizabeth, South Africa, 2011.
15. Van der Zaag, P.; Carmo Vaz, Á. Sharing the Incomati waters: Cooperation and competition in the balance. *Water Policy* **2003**, *5*, 349–368. [CrossRef]
16. Bicca, M.M.; Philipp, R.P.; Jelinek, A.R.; Ketzer, J M.M.; Scherer, C.M.D.S.; Jamal, D.L.; dos Reis, A.D. Permian-Early Triassic tectonics and stratigraphy of the Karoo Supergroup in northwestern Mozambique. *J. Afr. Earth Sci.* **2017**, *130*, 8–27. [CrossRef]
17. South African Committee for Stratigraphy (SACS). *Lithostra-tigraphy of the Republic of South Africa, South West Africa/Namibia, and the Republics of Bophutatswana, Transkei and Venda*; Republic of South Africa, Departament of Mineral and Energy Affairs, Geological Survey: Pretoria, South Africa, 1980.
18. Vieira, H.M.; Weschenfelder, J.; Fernandes, E.H.; Oliveira, H.A.; Möller, O.O.; García-Rodríguez, F. Links between surface sediment composition, morphometry and hydrodynamics in a large shallow coastal lagoon. *Sediment. Geol.* **2020**, *398*, 105591. [CrossRef]
19. Fuchs, S.; Cole, S. Making Science: Between Nature and Society. *Soc. Forces* **1994**, *72*, 929. [CrossRef]
20. Milani, E.J.; De Wit, M.J. Correlations between the classic Paraná and Cape–Karoo sequences of South America and southern Africa and their basin infills flanking the Gondwanides: Du Toit revisited. *Geol. Soc. London, Spéc. Publ.* **2008**, *294*, 319–342. [CrossRef]
21. Hälbich, I.W.; Fitch, F.J.; Miller, J.A. *Dating the Cape Orogeny*; The Geological Society of South Africa: Johannesburg, South Africa, 1983.

22. Foden, J.; Elburg, M.; Turner, S.; Clark, C.; Blades, M.L.; Cox, G.; Collins, A.S.; Wolff, K.; George, C. Cambro-Ordovician magmatism in the Delamerian orogeny: Implications for tectonic development of the southern Gondwanan margin. *Gondwana Res.* **2020**, *81*, 490–521. [CrossRef]

23. Wickens, H.V. Submarine fans of the Permian Ecca Group in the SW Karoo Basin: Their origin and reflection on the tectonic evolution of the basin and its source areas. In *Inversion Tectonics of the Cape Fold Belt, Karoo and Cretaceous Basins of Southern Africa*; De Wit, M.J., Ransome, I.G.D., Eds.; Balkema: Rotterdam, The Netherlands, 1992; pp. 117–126.

24. Kingsley, C.S. A composite submarine fan-delta-fluvial model for the Ecca and lower Beaufort Groups of Permian age in the Eastern Cape Province, South Africa. *South Afr. J. Geol.* **1981**, *84*, 27–40.

25. Osleger, D.; Hallam, A. Phanerozoic Sea-Level Changes. *PALAIOS* **1993**, *8*, 399. [CrossRef]

26. Veevers, J.J.; Cole, D.I.; Cowan, E.J. Southern Africa: Karoo Basin and Cape Fold Belt. *Geol. Soc. Am. Mem.* **1994**, *184*, 223–280. [CrossRef]

27. Madi, K. Neotectonics and its Applications for the Exploration of Groundwater in the Fractured Karoo Aquifers in the Eastern Cape, South Africa. Ph.D. Thesis, University of Fort Hare, Alice, South Africa, 2010.

28. De Nora, O. Oronzio de Nora Impianti Elettrochimici SpA, Novel Electrolysis Cell. U.S. Patent 4,343,690, 8 November 1982.

29. Catuneanu, O.; Hancox, P.J.; Rubidge, B.S. Reciprocal flexural behaviour and contrasting stratigraphies: A new basin development model for the Karoo retroarc foreland system, South Africa. *Basin Res.* **1998**, *10*, 417–439. [CrossRef]

30. Andersen, T.; Kristoffersen, M.; Elburg, M.A. How far can we trust provenance and crustal evolution information from detrital zircons? A South African case study. *Gondwana Res.* **2016**, *34*, 129–148. [CrossRef]

31. Duncan, R.A.; Hooper, P.R.; Rehacek, J.; Marsh, J.S.; Duncan, A.R. The timing and duration of the Karoo igneous event, southern Gondwana. *J. Geophys. Res. Space Phys.* **1997**, *102*, 18127–18138. [CrossRef]

32. Swanevelder, C.J. Utilising South Africa's largest river: The physiographic background to the Orange River scheme. *GeoJournal* **1981**, *2*, 29–40. [CrossRef]

33. Woodford, A.C.; Chevallier, L.P. *Regional Characterization and Mapping of Karoo Fractured Aquifer Systems: An Integrated Approach Using a Geographical Information System and Digital Image Processing*; Water Research Commission: Pretoria, South Africa, 2002.

34. Chere, N.; Linol, B.; De Wit, M.; Schulz, H.-M. Lateral and temporal variations of black shales across the southern Karoo Basin-Implications for shale gas exploration. *South Afr. J. Geol.* **2017**, *120*, 541–564. [CrossRef]

35. Cooper, M.R. *Cretaceous Fossils of South-Central Africa: An Illustrated Guide*; CRC Press: Boca Raton, FL, USA, 2018.

36. Hawthorne, J.B. Model of a kimberlite pipe. In *Physics and Chemistry of the Earth*, 1st ed.; Pergamon: Oxford, UK, 1975; ISBN 978-0-08-018017-5. [CrossRef]

37. Dalton, T.J.S.; Paton, D.A.; Needham, D.T. Influence of mechanical stratigraphy on multi-layer gravity collapse structures: Insights from the Orange Basin, South Africa. *Geol. Soc. Lond. Spec. Publ.* **2017**, *438*, 211–228. [CrossRef]

38. Hazell, J.R.T.; Cratchley, C.R.; Jones, C.R.C. The hydrogeology of crystalline aquifers in northern Nigeria and geophysical techniques used in their exploration. *Geol. Soc. Lond. Spec. Publ.* **1992**, *66*, 155–182. [CrossRef]

39. Usher, B.; Pretorius, J.; Van Tonder, G. Management of a Karoo fractured-rock aquifer system—Kalkveld Water User Association (WUA). *Water SA* **2007**, *32*, 9–19. [CrossRef]

40. Deal, J. *Timeless Karoo*; Penguin Random House South Africa: Pretoria, South Africa, 2016.

41. Woyessa, Y.E.; Pretorius, E.; Van Heerden, P.S.; Hensley, M.; Van Rensburg, L.D. *Impact of Land Use on River Basin Water Balance: A Case Study of the Modder River Basin, South Africa*; IMWI: Colombo, Sri Lanka, 2006.

42. Van Ginkel, C.E.; O'Keeffe, J.H.; Hughes, D.A.; Herald, J.R.; Ashton, P.J. *A Situation Analysis of Water Quality in the Catchment of the Buffalo River, Eastern Cape with Special Emphasis on the Impacts of Low Cost, High-density Urban Development on Water Quality*; Water Research Commission: Grahamstown, South Africa, 1996.

43. Salama, R.B.; Otto, C.J.; Fitzpatrick, R.W. Contributions of groundwater conditions to soil and water salinization. *Hydrogeol. J.* **1999**, *7*, 46–64. [CrossRef]

44. Sibley, D.F.; Blatt, H. Intergranular pressure solution and cementation of the Tuscarora orthoquartzite. *J. Sediment. Res.* **1976**, *46*, 881–896.

45. Burger, M. Geohydrological Report for the Proposed Usutu Opencast Colliery on the Farms Jan Hendriksfontein 263 IT, Ermelo, Mpumalanga. Available online: https://sahris.sahra.org.za/sites/default/files/additionaldocs/Geohydrological_Report_-Usutu_Opencast_Colliery_May_2012_Draft_version_1.1.1-40.pdf (accessed on 18 January 2021).

46. Aarnes, I.; Svensen, H.; Polteau, S.; Planke, S. Contact metamorphic devolatilization of shales in the Karoo Basin, South Africa, and the effects of multiple sill intrusions. *Chem. Geol.* **2011**, *281*, 181–194. [CrossRef]

47. Swart, C.J.U.; James, A.R.; Kleywegt, R.J.; Stoch, E.J. The future of the dolomitic springs after mine closure on the Far West Rand, Gauteng, RSA. *Environ. Earth Sci.* **2003**, *44*, 751–770. [CrossRef]

48. Schieferstein, B.; Loris, K. Ecological investigations on lichen fields of the Central Namib. *Vegetatio* **1992**, *98*, 113–128. [CrossRef]

49. Vermeulen, P.D. A South African perspective on shale gas hydraulic fracturing. Proceedings of the International Mine Water Association Annual Conference 2012. pp. 146–149. Available online: http://www.mwen.info/docs/imwa_2012/IMWA2012_Vermeulen_149.pdf (accessed on 26 March 2021).

50. Adams, S.; Titus, R.; Pietersen, K.; Tredoux, G.; Harris, C. Hydrochemical characteristics of aquifers near Sutherland in the Western Karoo, South Africa. *J. Hydrol.* **2001**, *241*, 91–103. [CrossRef]

51. Steyl, G.; De Lange, F.; Mbinze, A. *Hydrocensus and Groundwater Potential Assessment Bloemfontein*; Mangaung Local Municipality, Institute for Groundwater Studies: Bloemfontein, South Africa, 2011.
52. Nel, G.P. Geohydrological Characteristics of the Msikaba, Dwyka and Ecca Groups in the Lisikisiki area. Ph.D. Thesis, University of the Free State, Bloemfontein, South Africa, 2007.
53. Singhal, B.B.S.; Gupta, R.P. *Applied Hydrogeology of Fractured Rocks*; Springer Science & Business Media: Berlin/Heidelberg, Germany, 2010.
54. Botha, F.; Van Rooy, J. Affordable water resource development in the northern province, South Africa. *J. Afr. Earth Sci.* **2001**, *33*, 687–692. [CrossRef]
55. Lyon, J.G.; Trimble, S.W.; Ward, A.D.; Burckhard, S.R. *Environmental Hydrology*; CRC Press: Boca Raton, FL, USA, 2015.
56. Sanfo, S.; Barbier, B.; Dabiré, I.W.; Vlek, P.L.; Fonta, W.M.; Ibrahim, B.; Barry, B. Rainfall variability adaptation strategies: An ex-ante assessment of supplemental irrigation from farm ponds in southern Burkina Faso. *Agric. Syst.* **2017**, *152*, 80–89. [CrossRef]
57. Zhao, J.; Ma, Y.; Luo, X.; Yue, D.; Shao, T.; Dong, Z. The discovery of surface runoff in the megadunes of Badain Jaran Desert, China, and its significance. *Sci. China Earth Sci.* **2017**, *60*, 707–719. [CrossRef]
58. Tetsoane, S.T. Evaluation of the Swat Model in Simulating Catchment Hydrology: Case Study of the Modder River Basin. Ph.D. Thesis, Central University of Technology, Bloemfontein, South Africa, 2013.
59. Zulkafli, Z.; Perez, K.; Vitolo, C.; Buytaert, W.; Karpouzoglou, T.; Dewulf, A.; De Bièvre, B.; Clark, J.; Hannah, D.M.; Shaheed, S. User-driven design of decision support systems for polycentric environmental resources management. *Environ. Model. Softw.* **2017**, *88*, 58–73. [CrossRef]
60. Krause, T. Topography and its Effect on Groundwater. Available online: http://www.co.portage.wi.us/Groundwater/undrstnd/topo.html (accessed on 12 May 2017).
61. Gabler, R.E.; Peterson, J.F.; Trapasso, L.M. *Essentials of Physical Geography*, 7th ed.; Brooks/Cole: Pacific Grove, CA, USA, 2004.
62. Hasanuzzaman, M.; Song, X.; Han, D.; Zhang, Y.; Hussain, S. Prediction of Groundwater Dynamics for Sustainable Water Resource Management in Bogra District, Northwest Bangladesh. *Water* **2017**, *9*, 238. [CrossRef]
63. Ward, R.C.; Robinson, M. *Principles of Hydrology*; McGraw-Hill: London, UK, 1990.
64. Xanke, J.; Liesch, T.; Goeppert, N.; Klinger, J.; Gassen, N.; Goldscheider, N. Contamination risk and drinking water protection for a large-scale managed aquifer recharge site in a semi-arid karst region, Jordan. *Hydrogeol. J.* **2017**, *25*, 1795–1809. [CrossRef]
65. Haggard, B.E.; Moore, P.A.; Brye, K.R. Effect of slope on runoff from a small variable slope boxplot. *J. Environ. Hydrol.* **2005**, *13*, 25.
66. Khan, F.; Waliullah, M.N.; Bhatti, A.U. Maize Cultivar Response to Population Density and Planting Date for Grain and Biomass Yield. *Sarhad J. Agric.* **2007**, *23*, 25–30.
67. Niehoff, D.; Fritsch, U.; Bronstert, A. Land-use impacts on storm-runoff generation: Scenarios of land-use change and simulation of hydrological response in a meso-scale catchment in SW-Germany. *J. Hydrol.* **2002**, *267*, 80–93. [CrossRef]
68. Giertz, S.; Junge, B.; Diekkrüger, B. Assessing the effects of land use change on soil physical properties and hydrological processes in the sub-humid tropical environment of West Africa. *Phys. Chem. Earth Parts A/B/C* **2005**, *30*, 485–496. [CrossRef]
69. Hundecha, Y.; Bárdossy, A. Modeling of the effect of land use changes on the runoff generation of a river basin through parameter regionalization of a watershed model. *J. Hydrol.* **2004**, *292*, 281–295. [CrossRef]
70. Tang, Z.; Engel, B.; Pijanowski, B.; Lim, K. Forecasting land use change and its environmental impact at a watershed scale. *J. Environ. Manag.* **2005**, *76*, 35–45. [CrossRef]
71. Klöcking, B.; Haberlandt, U. Impact of land use changes on water dynamics—a case study in temperate meso and macroscale river basins. *Phys. Chem. Earth, Parts A/B/C* **2002**, *27*, 619–629. [CrossRef]
72. Girmay, G.; Singh, B.; Nyssen, J.; Borrosen, T. Runoff and sediment-associated nutrient losses under different land uses in Tigray, Northern Ethiopia. *J. Hydrol.* **2009**, *376*, 70–80. [CrossRef]
73. Kashaigili, J. Impacts of land-use and land-cover changes on flow regimes of the Usangu wetland and the Great Ruaha River, Tanzania. *Phys. Chem. Earth Parts A/B/C* **2008**, *33*, 640–647. [CrossRef]
74. Giertz, S.; Diekkrüger, B. Analysis of the hydrological processes in a small headwater catchment in Benin (West Africa). *Phys. Chem. Earth Parts A/B/C* **2003**, *28*, 1333–1341. [CrossRef]
75. Wei, W.; Chen, L.; Fu, B.; Huang, Z.; Wu, D.; Gui, L. The effect of land uses and rainfall regimes on runoff and soil erosion in the semi-arid loess hilly area, China. *J. Hydrol.* **2007**, *335*, 247–258. [CrossRef]
76. Jensen, J.L.; Schjønning, P.; Watts, C.W.; Christensen, B.T.; Munkholm, L.J. Soil texture analysis revisited: Removal of organic matter matters more than ever. *PLoS ONE* **2017**, *12*, e0178039. [CrossRef] [PubMed]
77. Pidwirny, M. The Hydrologic. CycleFundamentals of Physical Geography, 2nd ed. Available online: http://www.physicalgeography.net/fundamentals/8b.htm (accessed on 6 February 2017).
78. Schaetzl, R.; Anderson, S. *Soils: Genesis and Geomorphology*, 1st ed.; Cambridge University Press: Cambridge, UK, 2005.
79. Ekwue, E.; Harrilal, A. Effect of soil type, peat, slope, compaction effort and their interactions on infiltration, runoff and raindrop erosion of some Trinidadian soils. *Biosyst. Eng.* **2010**, *105*, 112–118. [CrossRef]
80. National Water Act of South Africa, Pretoria. 2014. Available online: https://www.gov.za/documents/national-water-act# (accessed on 18 January 2021).
81. South Africa Department of Water and Sanitation (DWS). *National Groundwater Resource: Assessment Phase II*; Department of Water and Sanitation: Pretoria, South Africa, 2016.

82. Woodford, A.; Girman, J. *How Much Groundwater Does South Africa Have?* Department of Water Affairs & Forestry: Pretoria, South Africa, January 2006; pp. 1–6. Available online: https://dxi97tvbmhbca.cloudfront.net/upload/user/image/1_A_Woodford20200304180745931.pdf (accessed on 18 January 2021).
83. Pandey, V.P.; Shrestha, S.; Chapagain, S.K.; Kazama, F. A framework for measuring groundwater sustainability. *Environ. Sci. Policy* **2011**, *14*, 396–407. [CrossRef]
84. Sami, K.; Hughes, D. A comparison of recharge estimates to a fractured sedimentary aquifer in South Africa from a chloride mass balance and an integrated surface-subsurface model. *J. Hydrol.* **1996**, *179*, 111–136. [CrossRef]

Article

Hydrogeochemical and Hydrodynamic Assessment of Tirnavos Basin, Central Greece

Ioannis Vrouhakis [1,2,*] , Evangelos Tziritis [2] , Andreas Panagopoulos [2] and Georgios Stamatis [1]

1. Sector of Geological Sciences, Mineralogy and Geology Laboratory, Department of Natural Resources & Agricultural Engineering, Agricultural University of Athens, Iera Odos 75, 11855 Athens, Greece; stamatis@aua.gr
2. Hellenic Agricultural Organisation "Demeter", Soil & Water Resources Institute, Sindos, 57400 Thessaloniki, Greece; e.tziritis@swri.gr (E.T.); a.panagopoulos@swri.gr (A.P.)
* Correspondence: i.vrouhakis@swri.gr; Tel.: +30-2310-798-790

Abstract: A combined hydrogeochemical and hydrodynamic characterization for the assessment of key aspects related to groundwater resources management was performed in a highly productive agricultural basin of the Thessaly region in central Greece. A complementary suite of tools and methods—including graphical processing, hydrogeochemical modeling, multivariate statistics and environmental isotopes—have been applied to a comprehensive dataset of physicochemical analyses and water level measurements. Results revealed that the initial hydrogeochemistry of groundwater was progressively impacted by secondary phenomena (e.g., ion exchange and redox reactions) which were clearly delineated into distinct zones according to data processing. The progressive evolution of groundwater was further verified by the variation of the saturation indices of critical minerals. In addition, the combined use of water level measurements delineated the major pathways of groundwater flow. Interestingly, the additional joint assessment of environmental isotopes revealed a new pathway from E–NE (which had never before been validated), thus highlighting the importance of the joint tools/methods application in complex scientific tasks. The application of multivariate statistics identified the dominant processes that control hydrogeochemistry and fit well with identified hydrodynamic mechanisms. These included (as dominant factor) the salinization impact due to the combined use of irrigation water return and evaporitic mineral leaching, as well as the impact of the geogenic calcareous substrate (mainly karstic calcareous formations and dolostones). Secondary factors, acting as processes (e.g., redox and ion exchange), were identified and found to be in line with initial assessment, thus validating the overall characterization. Finally, the outcomes may prove to be valuable in the progression toward sustainable groundwater resources management. The results have provided spatial and temporal information for significant parameters, sources, and processes—which, as a methodological approach, could be adopted in similar cases of other catchments.

Keywords: hydrochemistry; hydrodynamics; groundwater; environmental isotopes; Tirnavos basin

Citation: Vrouhakis, I.; Tziritis, E.; Panagopoulos, A.; Stamatis, G. Hydrogeochemical and Hydrodynamic Assessment of Tirnavos Basin, Central Greece. *Water* **2021**, *13*, 759. https://doi.org/10.3390/w13060759

Academic Editor: Dongmei Han

Received: 31 December 2020
Accepted: 8 March 2021
Published: 11 March 2021

Publisher's Note: MDPI stays neutral with regard to jurisdictional claims in published maps and institutional affiliations.

1. Introduction

Groundwater is a critical natural resource that needs to be properly managed in order to sustain its paramount aspects of quantity and quality. In service of this goal, rational groundwater management requires accurate information and adequate knowledge about the processes affecting groundwater evolution in time and space. Therefore, scientists (and eventually stakeholders) need robust tools to efficiently evaluate the status of groundwater and decipher the governing factors that regulate its hydrogeochemical and hydrodynamic regimes. The acquisition and interpretation of such information is a demanding and complex task, which requires multidisciplinary approaches, different perspectives and varied methodologies [1]. Essentially, hydrogeochemical and hydrodynamic information needs to be dealt with holistically, with special emphasis put on their interactions. The methodological approach of the present work dictated the synergetic and combinational consideration of various tools and methods, e.g., classic hydrogeochemical approaches, the use of bivariate plots and/or molar ratios [2–6], hydrodynamic characterizations [7–9], hydrogeochemical modelling [10–12], multivariate statistics [13–18] and environmental isotopes [19–23].

As a case study, we focused on the Tirnavos basin, which is part of the greater Thessaly plain in central Greece and is among the most productive regions of the country. The study area was favored for its relative abundance of water resources, most of which are used for irrigation purposes. Systematic exploitation of the system has been subject to the water authorities for more than half a century. This paper attempted to comprehensively elucidate the dominant attributes of the aquifer system that control its hydrodynamic evolution and shape its groundwater quality characteristics, based on historic and contemporary data.

The main goals of this research may be summarized as follows:

- Combine and test variable methodologies and tools through a specific proposed workflow, which may act a basic methodological array for assessing the hydrogeochemical and hydrodynamic conditions in similar cases.
- Develop and optimize a conceptual model for the groundwater resources of the study area, based on previous literature and the newly applied combined methods/tools.

Performed analyses served as a basis for developing a deep understanding of the system's characteristics, in order to identify and adopt optimal solutions for rational groundwater management. The latter is important, especially in environmentally vulnerable arid or semiarid Mediterranean areas—like the Tirnavos basin—which are expected to face adverse impacts (e.g., increased temperatures, decreased precipitation, and reduced recharge of groundwater resources) in the future due to climate change [24]. Therefore, comprehensive planning in the framework of climate change is regarded as highly important for safeguarding food safety and overall sustainability of the water-ecosystem-food nexus on which the socioeconomic stability and welfare of the region depend.

It is apparent that, in complex aquifer systems that are strongly controlled by active tectonics, co-evaluation of numerous analysis methodologies of hydrological, hydrolithological, hydrogeochemical and isotopic data need to be conjunctively considered. That is the optimal way to get safe results and decipher controlling mechanisms that may prove to be critical in intensively stressed systems. The performed analysis confirmed known key hydrodynamic evolution mechanisms; it also revealed significant elements that were not apparent—having been masked by the controlling hydraulic and hydrogeochemical mechanisms.

2. Study Area

Thessaly plain is the largest alluvial basin in Greece, with a total area of 13,142 km^2 and an average altitude of 427 m. It is in central Greece and—through the Mid-Thessalic hills—it is divided into two sub-basins: the western Thessaly basin and the eastern Thessaly basin, each of which developed in a NW-SE direction as part of a wider tectonic trough. The Tirnavos subbasin, from a hydrological perspective, is essentially the northwest part of the eastern Thessaly basin. It includes sections of the Titarisios River basin to the north and the Pinios River basin to the south. Its area is estimated to 251 km^2, or about 2.35% of the total area of the Pinios basin (Figure 1). The perimeter of the study area was 106 km and the average altitude 76 m. The smallest and largest morphological gradients recorded are 0 and 45.93%, respectively, while the mean slope is 5.3%. The relief of the wider region is depicted in Figure 1.

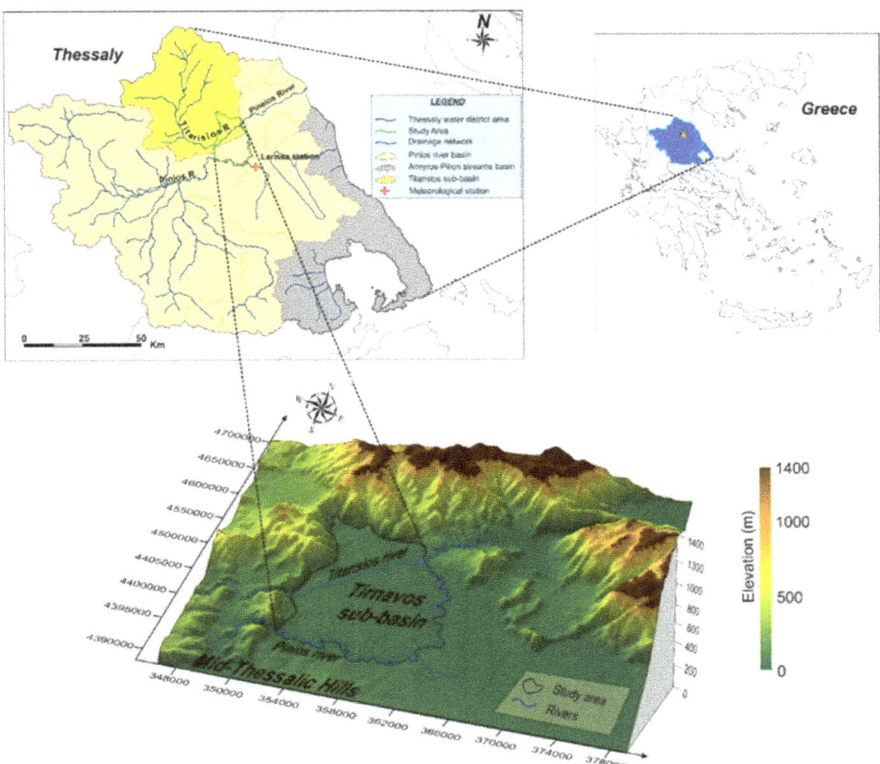

Figure 1. Location map of the study area and geomorphological illustration of the wider region.

The climate is typical Mediterranean, with annual rainfall from 400 mm to 500 mm, distributed almost entirely during the wet hydrological period, without any significant precipitation during summer. Thus, several irrigation systems have been developed to support the cultivation of highly productive summer crops [25]. With regards to the meteorological data, Larissa station is the closest to the study area, and has gathered continuous and reliable data over several decades.

Agriculture is the dominant land use, covering approximately 165 km^2 or 65.73% of the Tirnavos sub-basin. Intensified agricultural activities, including both cultivation and livestock, are a major source of groundwater contamination by nitrogen compounds. Manure waste and the often excessive, improper use of nitrogen fertilizers aiming to improve agricultural production have led to the occurrence of elevated concentrations of nitrates in groundwater [26].

3. Geology—Hydrogeology

The Tirnavos subbasin forms the northwestern part of the eastern Thessaly plain of Central Greece. It is filled by Quaternary alluvial formations which are bounded along the southwest part of the basin by Neogene marls and sandy-clay deposits. The western margins consist of karstified marbles of middle-upper Cretaceous origin. The crystalline bedrock is composed of mica-schists and gneisses of upper Paleozoic and Paleozoic age, respectively, forming the northern boundary of the subbasin (Figure 2). Two major springs (Mati Tirnavou and Agia Anna) emerge at the contact of the karstified system with the alluvial deposits. The Pinios and Titarisios rivers flow across the subbasin which, as already mentioned, hydrologically part of the wider Pinios River basin.

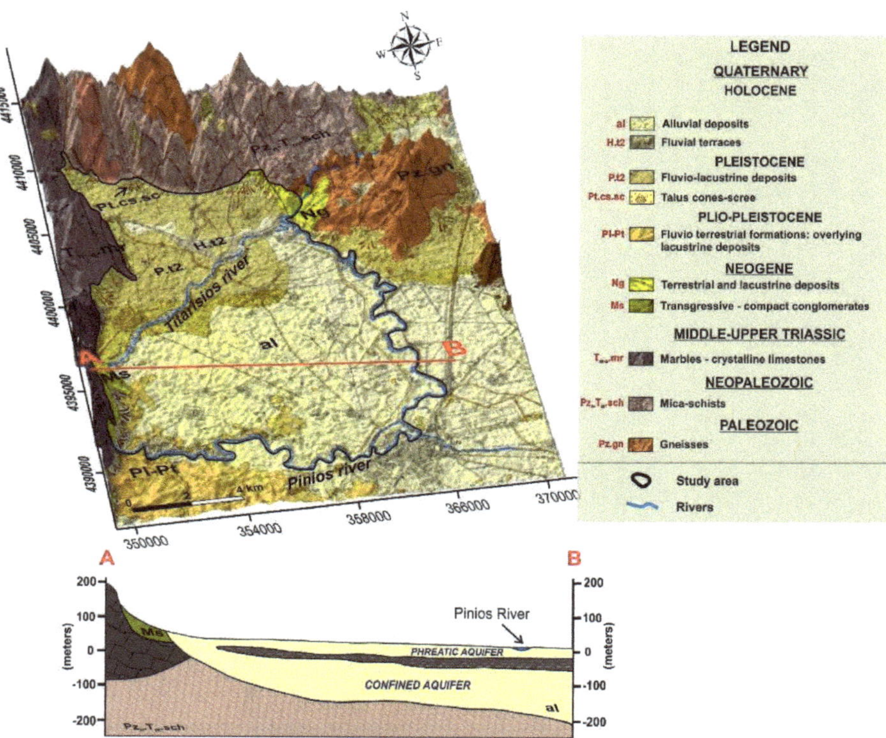

Figure 2. Geological map and conceptualized cross section of the study area, based on [27,28] and [29], respectively.

The Quaternary deposits host an unconfined aquifer near the talus cones of Titarisios River in the NW, which converts to a phreatic and deeper confined aquifer toward the center of the basin. These are separated by a sequence of clay layers, which form an aquitard [29,30] and have a maximum thickness of over 550 m at their central parts [25,31]. Despite the limited potential of the phreatic aquifer nowadays, its importance is paramount as it forms an effective buffer zone that protects the confined units from potential surface-released contaminants of anthropogenic origin (e.g., due to agricultural activities). In addition, the marbles of the western margins host a karstic aquifer of great potential, which recharges the alluvial system by lateral crossflows.

Due to the occurrence of marls along the southern part of the karst, the inflow from this part is reduced compared to that of the northern parts. Crossflow from the crystalline bedrock at the northern margins of the basin also occurs but is of minor importance. In addition to the above, the southern extent of the karst system—and to a minor extent the Mid-Thessalic hills—recharges the central plain parts by crossflow from the southwest and southern parts of the area.

In the western part of the basin, there is an extensive marginal cone through which the alluvial system receives significant amounts of recharge, as crossflow through the Titarisios River gorge sediments. A smaller volume recharges the aquifer system as crossflow from the Pinios River gorge sediments to the south [29].

4. Methodology

4.1. Methodological Array

A suite of different methodological tools was efficiently combined to provide a robust array of methodological workflows (Figure 3), having as their goal the construction of a comprehensive conceptual model for the evolution of groundwater resources at the study area. The bases of this workflow were the physicochemical analyses of groundwater samples, which were subsequently processed with the aid of (i) graphical methods (expanded Durov diagram), (ii) multivariate statistics (R-mode factor analysis and cluster analysis) and (iii) hydrochemical sections and bivariate plots. This combined approach was essential for grouping samples of similar hydrogeochemical identity (physicochemical fingerprint and processes). It identified the dominant previous and/or ongoing geochemical processes which strongly influenced the hydrogeochemical status of the study area. It also revealed the main sources of enrichment and/or contamination to/from specific parameters and allowed for the an assessment of their spatial evolution—progressively with the groundwater flow of the system.

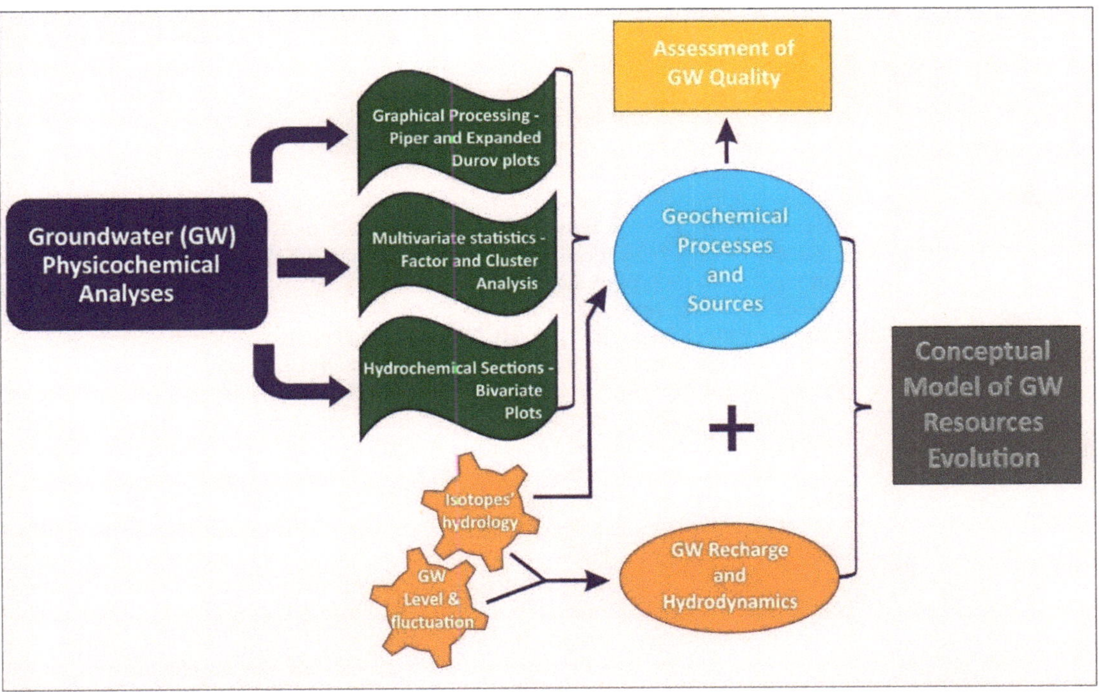

Figure 3. Methodological workflow for the development of groundwater resources conceptual model.

As a first result of this assessment, an evaluation of the groundwater quality was achieved, based on a comparison of the parametric values with relative standards and legislation. An additional set of raw data and field measurements—including groundwater level registrations—was then used for deciphering the basic hydrogeological characteristics (e.g., groundwater flow). This, combined with the results from stable isotope analysis (oxygen and deuterium) can provide insight into the recharge and hydrodynamic conditions of the system. Eventually, the combination and co-assessment of the hydrodynamic conditions with the dominant geochemical processes will constitute the fundamental aspect of the conceptual model for the groundwater resources of the area. Based on distinct steps and specified tools, this methodological array may be used as a model in similar cases where the groundwater conceptual model is the main scientific quest.

4.2. Groundwater Sampling, Analyses and Measurements

In total, 174 water samples were collected during 4 sampling periods (2 dry and 2 wet periods) between September 2016 and April 2018. Specifically, 153 samples were collected from boreholes within the study area and 21 surface water samples were taken from the rivers. Water samples were collected in double-capped polyethylene bottles of 1000 mL and stored in cool conditions during transport to a laboratory at The Soil and Water Resources Institute (SWRI) for analyses. Electrical conductivity (EC), pH, dissolved oxygen (DO) and oxidation-reduction potential (ORP) were determined in the field using the ProDSS (YSI inc., Yellow Springs, OH, USA) multiparametric probe. Laboratory analyses determined 27 parameters, including major and minor ions and trace elements; additional parameters were also calculated subsequently (Appendix A, Table A1). Calcium, magnesium and trace elements (except boron) were determined using an atomic absorption spectrophotometer by flame method. Sodium and potassium were analyzed through flame photometry while sulphate, nitrate, nitrite and ammonium were analyzed using spectrophotometer UV-VIS measurement. Bicarbonates, carbonate anions (neutralization by H_2SO_4) and chloride (neutralization by $AgNO_3$) were determined volumetrically. The reliability of the results was determined by ionic balance error, which was found to be less than 10% for all samples, with a median value of -3%.

In addition, twenty-six (26) samples were collected for isotopic analyses from selected sites of the monitoring network in April (n = 15) and September (n = 11) 2018, respectively, according to the International Atomic Energy Agency's sampling specifications [32]. The first isotope sampling period took place in April 2018; 15 samples were collected. The second sampling period took place in September 2018; only 11 out of 15 samples were collected because some of the wells were not operating. Additionally, the Titarisios River was dry during that period. In detail, the following samples were collected: (a) 21 from groundwater sites (12 in the first period, 9 in the second); (b) 2 from the Mati Tirnavou spring (1 for each period); (c) 2 from the Pinios River (1 for each period); and (d) 1 from the Titarisios River (in the first period). Fully filled 100 mL double cap polyethylene bottles were used for sampling, and samples were analyzed for oxygen and hydrogen isotopic compositions (^{18}O and ^{2}H) at the Hydrology Laboratory of Lubeck Technical University in Germany. Analyses were conducted via off-axis integrated cavity output spectroscopy (OA-ICOS, DLT-100 Liquid Water Isotope Analyzer, Los Gatos Research Inc., Mountain View, CA, USA) and reported in per mil (‰). The analytical precision for $\delta^{18}O$ and $\delta^{2}H$ was 0.2‰ and 0.6‰, respectively.

In addition to water sampling, a groundwater level monitoring network comprising 46 wells was compiled (Figure 4) and operated to facilitate the objectives of the research. At all groundwater level measuring stations (boreholes), the altitude was accurately recorded using the Real Time Kinematic (RTK)—GPS, Epoch 50 by Spectra Geospatial. Water level monitoring was performed in April and September 2017, to reflect representative conditions of the wet and dry hydrological periods. Based on the compiled piezometric maps (Figure 4), a relative drawdown is evident, ranging from a few to ten meters locally. This indicated a seasonal effect on water level and consequently volume of water stored in the aquifer system.

4.3. Data Processing

Groundwater chemistry data was processed with graphical approaches and multivariate statistics. An expanded Durov diagram [34] was used to identify the ongoing hydrogeochemical processes in the study area. In this diagram, the cation and anion triangles were recognized and distinguished along the 25% axes so that the main field was conveniently divided. The expanded Durov diagram has a distinct advantage over the Piper diagram [35] in that it provides a better display of hydrochemical water types [36] and has the potential to reveal geochemical processes that could affect groundwater evolution [37].

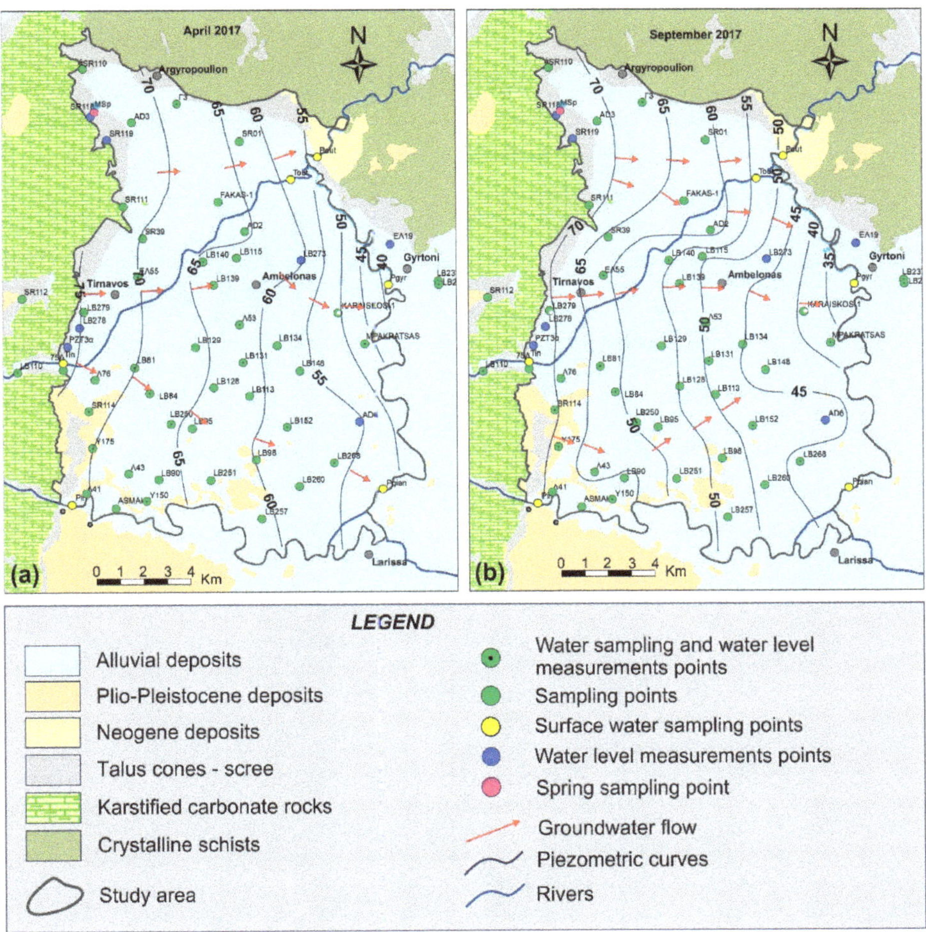

Figure 4. Monitoring network, piezometric curves and dominant flow lines for: (**a**) April 2017; and (**b**) September 2017 (Hydrolithology map based on [33]).

To get a better insight on the potential correlation between the examined parameters, the Pearson correlation coefficient (r) was calculated. If the correlation coefficient (r) is greater than 0.7, parameters are considered strongly correlated; if it ranges between 0.5 and 0.7, it indicates a moderate correlation at a significance level of $p < 0.05$. [38]. In addition, a classic robust multivariate statistical technique (R-mode factor analysis) was applied. Factor analysis (FA) has been widely used in environmental sciences and hydrogeochemical research [1,14,39,40]. It is a multivariate statistical technique that involves linear combinations of variables through a correlation-focused approach. FA seeks to reproduce intercorrelations among variables, in which the factors represent their common variance. The relationship among several observed quantitative variables is represented in terms of a few underlying, independent variables, called factors, which may not be directly measured or even measurable [41]. The exact number of factors was chosen by Kaiser criterion [42] in which factors with eigenvalues smaller than 1 were eliminated. Prior to processing, initial data was standardized through log transformation and z-scores to eliminate the influence of different units between variables and acquire a normal or log normal distribution [43]. Variables that failed to meet that criterion were omitted from further data processing. Finally, 19 parameters compiled the correlation matrix that accounted for the degree of mutually shared variability between individual pairs of groundwater quality parameters. The higher the factor loading of a parameter, the greater its participation to the examined factor. Factor loadings above 0.750 were considered high, those between 0.500 and 0.750 were considered moderate, and those between 0.400 and 0.500 were considered weak [40]. To further assist the performed assessments, the saturation indices of critical minerals were calculated with the aid of PHREEQC software [44].

5. Results and Discussion

5.1. Physicochemcial Analyses

Electric conductivity (EC) varied from 253 to 1821 μS/cm (Figure 5a). The higher EC values occurred in the southeastern part of the study area, where most of the region's small-to-medium sized industrial units are located [45]. The lowest values of EC are noted in the eastern and central part of the study area, where no significant industrial activity occurs.

Groundwater temperature is known to be an important driver for water quality [46,47], and therefore a crucial parameter for groundwater quality management. In addition, groundwater temperature is one of the best environmental tracers for detecting water flux, because heat in aquifers is transported both by conduction and advection caused by groundwater flow water temperature [48]; hence it may be directly related to the hydrodynamic evolution of a system. Low groundwater temperatures normally indicate recharge waters while higher temperatures indicate greater distance from the recharge source. As shown in Figure 5b, the lower groundwater temperatures were observed in the western margins of the basin, at the contact of the alluvial with the carbonate formations, while more elevated temperatures occurred toward the central and southern parts of the basin. This spatial distribution of temperature—combined with the piezometric data— indicated that the karstic system to the west is the main recharge source. On the other hand, low temperatures were also observed in the eastern parts of the basin, where alluvial deposits exist between metamorphic formations of crystalline bedrock, probably denoting a potential recharge effect from the east. In the absence of monitoring points at that part of the basin, however, that hypothesis could not be verified by piezometric data.

The main recharge areas of the basin were also indicated by the spatial patterns of DO and pH. For DO, higher values suggested enrichment in oxygen; thus, recharge conditions were present in the western and northern parts, while DO values were progressively reduced toward the center of the basin (Figure 5c). Deviations from that pattern (e.g., NW part of the basin) may be attributed to potential secondary geochemical processes (e.g., redox reactions) which create locally reducing conditions. Regarding the pH values, it was evident that in the recharge areas (west and north), pH was circumneutral, but progressively more alkaline moving toward the central and eastern parts of the basin (Figure 5d).

Based on the analytical results (Appendix A, Table A1), groundwater samples were slightly alkaline (median value: med = 7.61) with relatively low electrical conductivity (med = 486 μS/cm). The order of abundance for cations is $Ca^{2+} > Mg^{2+} > Na^+ > K^+$, and for anions $HCO_3^- > NO_3^- > SO_4^{2-} > Cl^-$. With regard to calcium and bicarbonates (which constitute the most abundant ions), it was evident that the main origin of their concentrations was the karstic system, cropping out at the western parts of the basin. That, as indicated in the previous section, is considered a major pathway for recharge of the system. Interestingly, nitrates are the second most abundant anion, clearly reflecting the anthropogenic impact due to widespread agricultural activities. In total, 72% of samples exhibited concentrations above 10 mg/L, which is an indicative threshold for nitrate contamination in natural systems [49]. Over 9% exceeded the Maximum Admissible Concentration (MAC) for drinking water (50 mg/L), according to [50,51], reaching up to 145 mg/L. However, their spatial distribution was scattered, without any profound pattern, elucidating that, apart from diffuse agricultural impact, contamination is point-source. This could be explained by agricultural malpractice and/or septic tanks, reflecting the potential influence of locally developed favorable conditions for percolation of pollutants through the vadose zone—which is in line with the above-described characteristics of the sediments that fill the basin. Other outlying values and cooccurrences of elevated ion concentrations (e.g., maximum values of Na^+ (287 mg/L) and SO_4^{2-} (604 mg/L)) were also related to local factors (e.g., soil amendments rich in Na^+ and SO_4^{2-} or dissolution of evaporitic minerals such as thenardite (Na_2SO_4), gypsum ($CaSO_4$), and halite ($NaCl$)) which could have occurred due to previous paleoenvironmental evaporitic conditions [1]. Nevertheless, their distribution (and thus their impact) was very limited. Finally, concentrations of heavy metals and metalloids were generally low; the only exception was an outlier for B (1.25 mg/L), most likely attributed to local impact of boron-rich soil amendments.

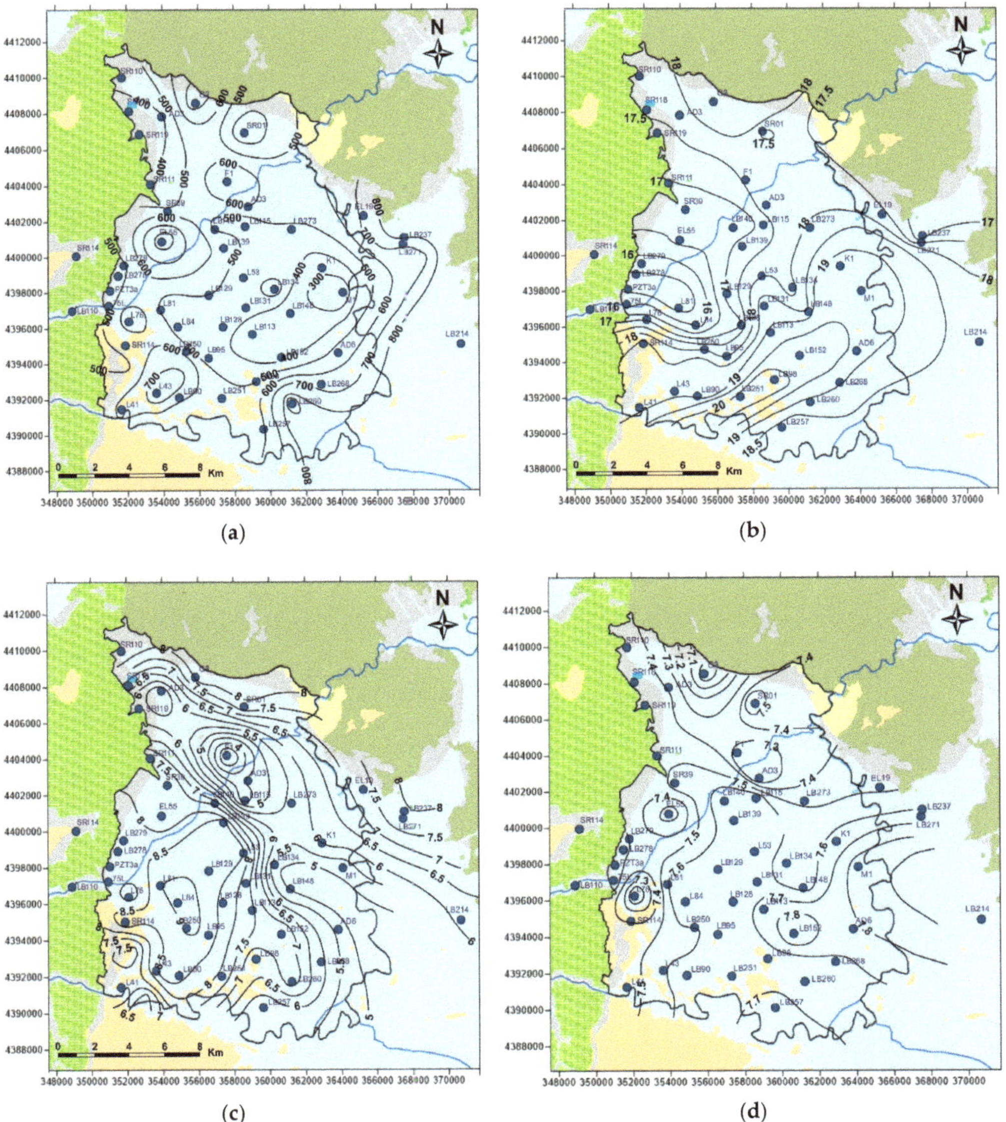

Figure 5. Spatial distribution of physicochemical parameters at the study area: (**a**) electrical conductivity (µS/cm); (**b**) water temperature (°C); (**c**) dissolved oxygen (mg/L); (**d**) pH.

5.2. Identification of Governing Processes and Spalital Evolution of Hydrogeochemistry

The results of chemical analyses were plotted in the expanded Durov diagram with the help of the DurovPwin application [52]. To ensure the representativeness of the results, the median values of the four periods were considered in the process. According to the expanded Durov diagram (Figure 6), two basic hydrochemical characters are identified, which can be further separated into subgroups: 1a, 1b, 2, 3, 5, 6. These characters are described below, while in Figure 6 the spatial distribution of the identified groups is illustrated.

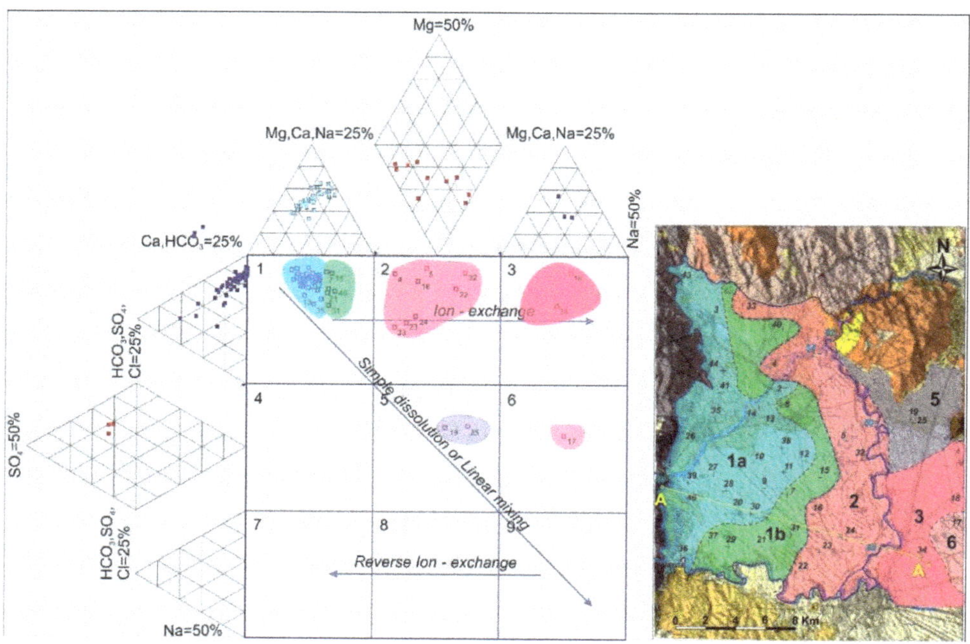

Figure 6. Expanded Durov plot illustrating the hydrogeochemical processes [36] and their spatial distribution in the study area.

- Hydrochemical Type 1 (Ca-Mg-HCO$_3$): In Subgroup 1a, Ca^{2+} and HCO$_3{}^-$ were dominant, indicating that this was a recharge zone, which was verified by high dissolved oxygen and low water temperature values in this area (according to field measurements analyzed in previous paragraphs). Based on the dominant cations in the formation of the hydrochemical character, group 1b did not differ substantially from 1a. However, there was a slight increase in the concentrations of Na$^+$ and Mg^{2+} and, with respect to the anions, higher SO$_4{}^{2-}$ values. Spatially, this hydrochemical type occupied two zones on either side of the Titarisios River. The first began north of it and extended to the northwestern part of the basin, while the second began from the central part of the basin (south of Titarisios) and extended to the south and southwestern part of the basin. The dominance of Ca^{2+} and HCO^{3-} in this subgroup also indicated a recharge zone, possibly lower than that of group 1a. The involvement of Mg^{2+} in the formation of the hydrochemical type was consistent with the mild dolomitization process which characterized the carbonate system [27]. The reason for the observed differentiation between the two subgroups may have been due to the limited system recharge rate from the karst (which allows mixing of recharge water with the aquifer system and/or the development of a small degree of ion exchange processes). In particular, the development of this subgroup on its southwestern boundary may have been justified by the emergence of Neogene deposits in this area, which limited the rate of groundwater replenishment in this part of the aquifer. At the same time, in this area, according to earlier studies [53], lateral recharge from the Mid-Thessalic hills was reported—of limited extent and intensity and qualitatively inferior to that of the karst system.

- Hydrochemical Type 2 (Ca-Mg-Na-HCO$_3$-SO$_4$): The second hydrochemical type was characterized by higher Na$^+$ and SO$_4^{2-}$ concentrations. The involvement of Ca^{2+} and HCO$_3^-$ in the formation of the hydrochemical type of subgroup 2 (Mg-Ca-HCO$_3$-SO$_4$) was still significant; however, it was less than that of subgroups 1a and 1b. The projection position of subgroup 2 in the expanded Durov diagram indicated progressive involvement of ion exchange and perhaps also mixing mechanisms. At least locally and seasonally [27], the mixing of Pinios River water with aquifer water through filtration along its bed in the formation of the observed hydrochemical type cannot be excluded. It was also characterized by low analogues with subgroups 1a and 1b at Ca^{2+} and HCO$_3^-$ concentrations. Subgroup 2 (Ca-Mg-Na-HCO$_3$) can therefore be considered as representative of a transition zone from intense recharge zones (subgroups 1a and 1b) to a restricted recharge zone and longer groundwater residence time in the aquifer (subgroups 3, 5, 6). In subgroup 3 (Na-HCO$_3$), HCO$_3^-$ and Na$^+$ dominated, and the ion exchange phenomenon was fully evolved, as indicated by the projection of its representative samples (wells 18 and 34) on the expanded Durov diagram. Subgroup 5 (Na-Mg-Ca-SO$_4$-HCO$_3$) represented mixing or dissolution waters, where Mg^{2+} (but mainly Na$^+$ from cations) and Cl$^-$ (but mainly SO$_4^{2-}$ and HCO$_3^-$, to a lesser extent from anions) dominated, thus suggesting a contributing recharge mechanism in the form of lateral influx from surrounding formations that contributed to the aquifer balance and obviously affected its hydrochemical identity. Last, subgroup 6 (Na-SO$_4$-HCO$_3$), which was only represented by a single but distinct sample, plotted on an uncommon part of the graph for groundwaters, which is often a product of mixing.

The above assessments may be further supported by the physicochemical evolution of groundwater along its flow path (Figures 7 and 8), from the identified recharge area to the end of the transition zone, as defined by the six (6) selected (representative) wells (A-A' axis) depicted in Figure 6. With regard to the physicochemical parameters (Figure 7), an increasing trend was identified for the EC (like Total Dissolved Solids -TDS-, as expected), following the typical ion enrichment of groundwater from the recharge areas to the central parts of the hydrological basin. Bicarbonates (HCO$_3^-$) progressively decreased and then subsequently increased, reaching a higher value than the recharge area; this evolution probably denoted an external impact from CO$_2$ content. Specifically, the reaction of calcite from the karstic substrate in the recharge area (west) with the CO$_2$ derived from the respiration of organic matter, produced carbonic acid as an intermediate step and then bicarbonates—with simultaneous release of calcium, as shown in Equation (1).

$$CO_{2(g)} + H_2O + CaCO_3 \rightarrow Ca^{2+} + 2HCO_3^-, \tag{1}$$

Progressively, along the groundwater flow path, the CO$_2$ content increased (presumably using as its source the occurrence of organic matter in the sediments of the aquifer matrix due to the sedimentation process of the fluviolacustrine deposits under C rich paleogeographic conditions), leading to an increase of CaCO$_3$ solubility [54] and shifting the Saturation Index of calcite (SI$_{CaCO3}$) values from supersaturated to slightly saturated. The latter was in accordance with Figure 8, indicating sharp SI$_{CaCO3}$ reduction.

With respect to Cl$^-$: a constant increase toward groundwater flow was identified. Bearing in mind that Cl$^-$ is a conservative ion, the increase of its concentration should be attributed to external factors which—based on the dominant land use activities and site characteristics—may be related to manure leaching and/or septic tanks. However, the dissolution of evaporitic minerals such as halite (NaCl) could not be excluded. Interestingly, the concentrations of SO$_4^{2-}$ changed without a significant pattern, with notable variations.

The primary source of sulfates was geogenic, stemming from the weathering of S-bearing minerals (e.g., gypsum, pyrite, etc). Changes in their concentration were likely due to further impact from the S-bearing minerals, as well as from external factors, e.g., soil amendments. However, their concentration along the dominant groundwater flow path may also change due to variations in redox conditions. The latter was also evidenced by the similar trend in nitrate concentrations—which may act as electron donors. Nevertheless, in the case of nitrates, the changes in their concentrations were less sharp than sulfates. The redox impact was also evidenced in some wells through the elevated values of NH_4^+ and the relatively low values of NO_3^-. Potential effects due to leaching of N-fertilizers could not be excluded.

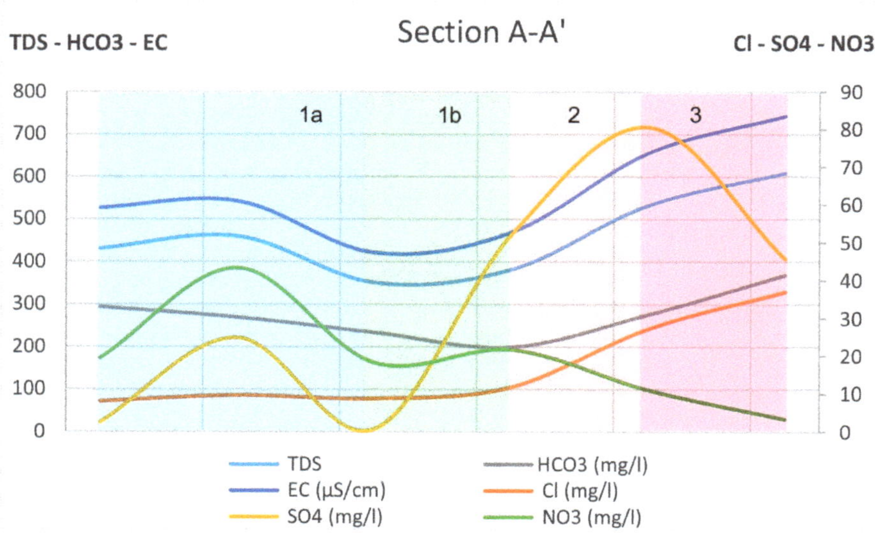

Figure 7. Variations in the concentration/values of major anions and physicochemical parameters along the groundwater flow.

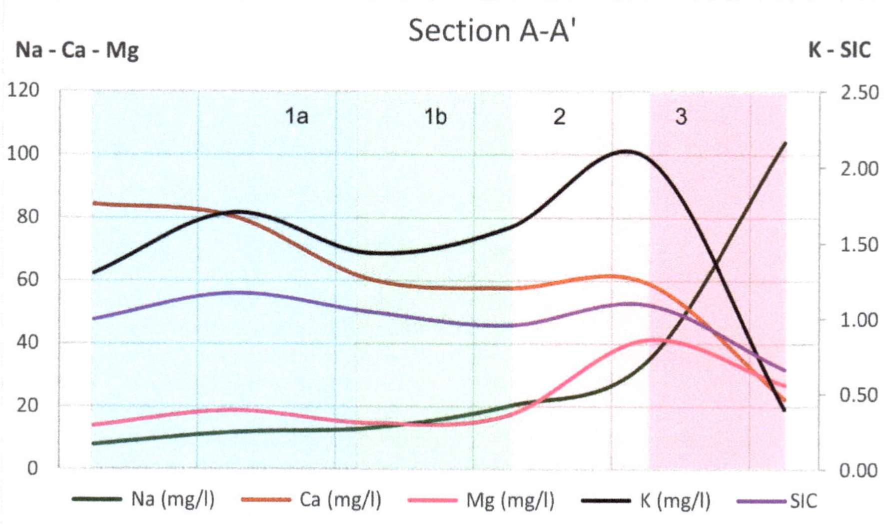

Figure 8. Variations in the concentration/values of major cations and SI (Saturation Index) of calcite along the groundwater flow.

Regarding the major cations (Figure 8), there was an antithetic change in the concentrations of Ca^{2+} and Na^+, explained by the process of ion exchange. The fresh (recharge) groundwater (which was enriched in Ca^{2+} due to the karstic substrate) is subject to ion exchange toward its flow with the Na^+ content of silicate minerals, leading progressively to a decrease in calcium and increase in sodium. This was also reflected in the shift of water types, from Ca-HCO_3 to Na-HCO_3. The magnesium (Mg^{2+}) also showed a slight increase toward flow, probably enriched by the weathering of dolomite or other Mg-bearing aluminosilicate minerals. Potassium (K^+) was nearly constant, exhibiting only minor changes in concentration. Overall, the selected parameters' evolution along the regional groundwater flow direction (Figures 7 and 8) agreed with the findings and observations discussed above, supporting the concept of strong recharge from the karstic domain in the form of lateral crossflow which then mixes with resident groundwater and is impacted locally by leachates from anthropogenic activities. Moreover, it clearly denoted the progressive shift from mixing to ion exchange process, which was also affected by oxygen depletion and the relative dominance of reducing conditions.

The dominance of the ion exchange process was further verified by the major ion relation analysis conducted and illustrated in Figure 9. Specifically, the relation between $(Ca+Mg-SO_4-HCO_3)$ and $(Na+K-Cl)$ in meq/L was examined, as suggested by [55,56]. The product of $(Na+K-Cl)$ represented the excessive sodium originating from sources other than halite dissolution (assuming that all chloride is derived from halite). In addition, the product of $(Ca+Mg-SO_4-HCO_3)$ represented the calcium and/or magnesium which originated from sources other than gypsum and carbonate dissolution (calcite and/or dolomite). If these processes were significant in defining the hydrogeochemistry of groundwater, the relation between the two products should have been linear with a slope of -1. As seen in Figure 9, many data points lie close to a straight line (r = +0.93) with a slope of -0.80, which clearly points to the existence of cation exchange [57,58].

According to [58] the plot of $(Ca + Mg)$ vs. $(SO_4 + HCO_3)$ could be a reliable indicator—helping to distinguish between ion exchange and reverse ion exchange processes, if these are active in a given study area. The points plotted below the 1:1 line toward the $Mg + Ca$ axis suggest prevalence of the ion exchange process, while those plotted above the 1:1 line towards the $SO_4 + HCO_3$ axis denote prevalence of reverse ion exchange. The results of the plotted samples (Figure 10) are in accordance with the outcomes of the expanded Durov diagram (Figure 6). The relative positions of the points in the plot indicate that excessive calcium and magnesium in groundwater were exchanged with sodium from the aquifer matrix. Figure 10 shows the amount of $Ca + Mg$ gained or lost relative to that provided by calcite, dolomite, and gypsum. When $HCO_3 + SO_4$ is low (< 5 meq/L) and the samples plot on 1:1 line, dissolution of calcite and dolomite is the major process influencing water chemistry; on the contrary, when $HCO_3 + SO_4$ is greater than 5 meq/L, in addition to calcite and dolomite, dissolution of gypsum is likely to occur [59].

According to the calculated Pearson coefficient (r), there were strong positive correlations between Na-Cl (+0.847), Na-SO_4 (+0.889), Cl-SO_4 (+0.855), B-Na (+0.899), and B-SO_4 (+0.904); there were moderate positive correlations between Ca-Mg (+0.532), Mg-Cl (+0.683), Mg-SO_4 (+0.612), B-Cl (+0.737) and Mg-NO_3 (+0.648). Most of the correlated parameters were indicative of a salinization impact, which, according to the dominant land use and site specs, can probably be attributed to leachates rich in high salinity fertilizers washed off through irrigation water return flow. It is interesting, though, that boron (B) was strongly correlated with Na, Cl and SO_4, denoting a similar origin or enrichment process. Based on the hypothesis addressed above (occurrence of evaporitic minerals), B could potentially derive from the dissolution of buried evaporitic minerals such as borax ($Na_2B_4O_7 \cdot 10H_2O$), considering the paleoclimatic conditions of the area. Nevertheless, this cannot solely explain the correlation with Cl and SO_4. Most likely, B is also related to fertilizers/soil amendments, whose leached products may contain boron, apart from other saline-related parameters.

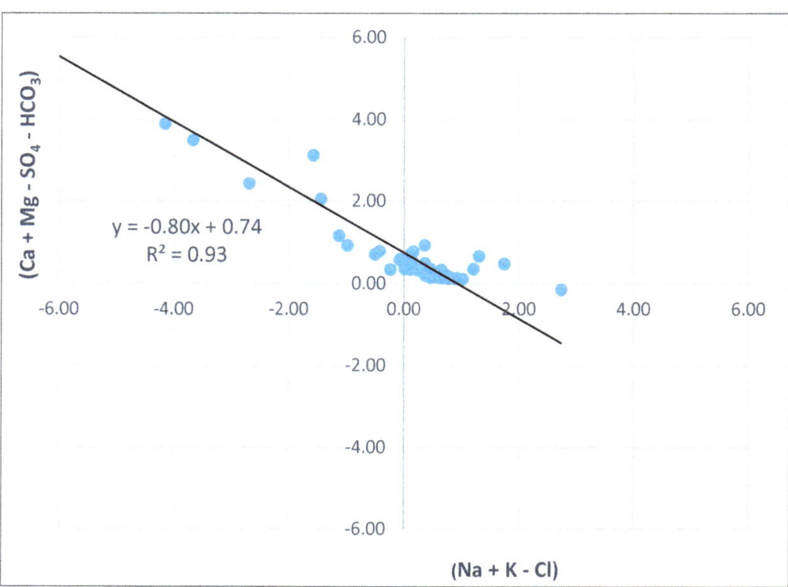

Figure 9. Plot of (Ca+Mg-SO$_4$-HCO$_3$) and (Na+K-Cl) in meq/L.

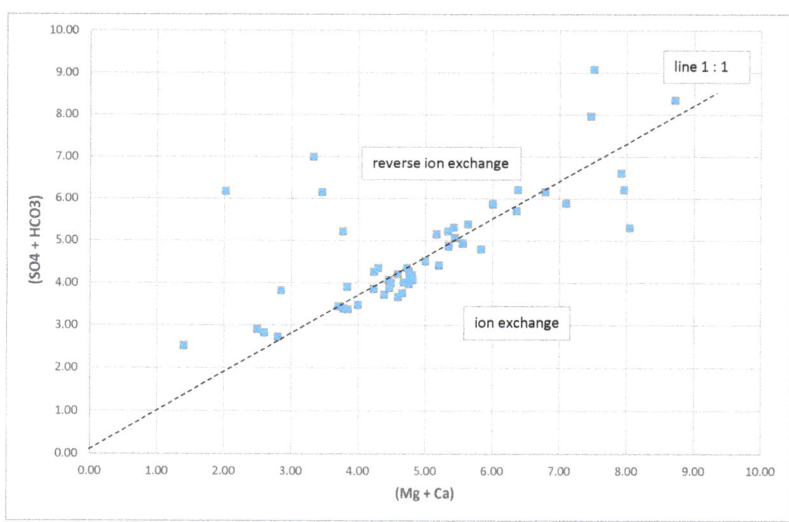

Figure 10. Plot of (Ca + Mg) and (SO$_4$ + HCO$_3$) in meq/L.

5.3. Multivariate Statistics

The results of the FA (Table 1) identified five (5) factors that define the hydrogeochemical regime and explained 83.3% of the total variance. The first factor (F1) accounted for the 32.8% of total variance and denoted the cumulative effect of groundwater salinization by variable sources. It included (with strong positive loadings) the parameters of EC, Na, Cl, B, and SI$_{halite}$; it also included (with medium loadings) the parameters of Mg and SI$_{gypsum}$, reflecting the impact from irrigation water return flow and the dissolution of evaporitic minerals. It should be noted that higher values of SIs correlate with a greater state of saturation.

Table 1. Results of Factor Analysis (FA) with rotated (varimax) factor loadings and communalities. Medium-to-high positive correlation is shown in bold, while medium-to-high negative correlation is shown in bold italics.

Variable	Factor 1	Factor 2	Factor 3	Factor 4	Factor 5	Communality
pH	0.149	*−0.867*	0.333	0.131	0.046	0.903
EC	**0.921**	0.29	0.165	−0.130	−0.006	0.976
K	0.084	**0.693**	0.009	0.055	0.256	0.556
Na	**0.918**	−0.252	0.094	−0.157	0.062	0.944
Ca	0.017	**0.883**	0.232	0.133	0.085	0.859
Mg	**0.673**	**0.421**	0.147	−0.232	−0.312	0.803
Cl	**0.941**	0.17	0.019	0.02	−0.015	0.915
HCO$_3$	0.075	**0.508**	**0.547**	−0.52	−0.09	0.841
SO$_4$	**0.965**	−0.073	0.017	0.001	0.074	0.942
B	**0.892**	−0.251	−0.034	−0.114	0.068	0.877
Cu	−0.189	0.181	0.082	0.091	**0.772**	0.680
Fe	0.375	−0.115	0.068	*−0.541*	**0.551**	0.754
Mn	0.191	−0.225	−0.18	*−0.804*	0.045	0.767
NO$_3$	0.017	**0.815**	−0.052	0.189	−0.06	0.707
NH$_4$	−0.013	−0.036	0.083	*−0.897*	−0.104	0.823
SI$_{Calcite}$	−0.042	0.036	**0.945**	0.128	0.175	0.942
SI$_{Dolomite}$	0.291	−0.142	**0.927**	−0.081	−0.055	0.973
SI$_{Gypsum}$	**0.595**	**0.492**	0.143	0.171	−0.238	0.702
SI$_{Halite}$	**0.893**	0.016	0.066	−0.098	−0.224	0.862
Variance	6.241	3.719	2.352	2.264	1.251	15.827
% Var	32.8	19.6	12.4	11.9	6.6	83.3

The second factor (F2) accounted for the 19.6% of total variance and included: Ca and NO$_3$ with strong positive loadings; K and HCO$_3$ with medium ones; Mg and SI$_{gypsum}$ with weak ones; and pH with a strong negative loading. This factor may possibly interpret the hydrogeochemistry in the recharge areas, where Ca and HCO$_3$ contents are elevated (higher values), and pH (as shown previously) is circumneutral (lower values). This sufficiently explained the antithetic loading. The strong positive loading of nitrates possibly denoted an additional dominant process of nitrate enrichment in these areas. This was possibly due to existence of two additional criteria: the relative oxidizing conditions and the external source of NO$_3$ (e.g., agricultural activities and/or septic tanks); the latter being profoundly supported by the geometry of the aquifer system at that part of the basin, characterized by a phreatic unit of high hydraulic parameters.

The third factor (F3) explained 12.4% of total variance and included (with strong positive loadings) the SIs of calcite and dolomite. It probably denoted the areas which were supersaturated in the above minerals because of the calcareous substrate (limestones and dolostones). The weak correlation with the dominant cations of these formations (r = +0.35 for Ca-SI$_{calcite}$, r = +0.45 for Mg-SI$_{dolomite}$) probably reflected the existence of additional sources for Ca and Mg (e.g., F1 and F2) which masked the direct covariance of these parameters due to impact from limestones and dolostones.

The fourth (F4) factor explained 11.9% of total variance, with strong negative loadings for Mn and NH$_4$ and a medium one for Fe. This factor clearly reflected the local reducing conditions, in which dominance of Fe^{2+} and Mn^{2+} prevailed along with ammonium (NH$_4$). These areas were probably affected by the increased organic content that creates reducing conditions and/or the anoxic environments caused by limited groundwater recharge (e.g., semiconfined and/or confined aquifers).

Finally, the fifth (F5) factor explained a minor percentage (6.6%) of total variance and included (with strong and medium positive loadings) the parameters of Cu and Fe, respectively. It probably denoted the occurrence of a weak local sulfide mineralization (e.g., chalcopyrite—$CuFeS_2$ or other) without excluding the potential impact from the use of Cu as a soil conditioner (mainly as a major constituent in several plant protection products, especially for orchards and vineyards).

5.4. Stable Isotopes

With regard to the stable isotopes, the Local Groundwater Isotope regression Line (LGIL) was compiled and plotted, along with the Global Meteoric Water Line (GMWL), Greek MWL and Thessalian MWL—as depicted in Figure 11. Isotope values of δ^2H and $\delta^{18}O$ were expressed as the difference between the measured ratios of the sample and reference divided by the measured ratio of the reference, which was in turn expressed as VSMOW values (Vienna-standard mean ocean water). The isotopic ratio of $\delta^{18}O$ in the study area ranged between −9.8‰ and −6.9‰ with an average value of −8‰. This value almost coincided with the average value recorded from the spring waters of Thessaly (−8.29‰) [60]. Similarly, the isotopic ratio of δ^2H ranged between −66.2‰ and −47.5‰ with an average value of −51.3‰ which was almost the same as the average value recorded in Thessaly (−51.1‰), based on the reference. Based on these observations, the isotope composition of groundwater in the Tirnavos subbasin seemed to fit perfectly with the obtained levels from a previous study on the Thessaly region. The function that expressed the LGIL ($\delta^2H = 6.71 \times \delta^{18}O + 2.09$) compared to Thessaly's MWL ($\delta^2H = 6.48 \times \delta^{18}O + 1.7$) presented a similar slope (6.71 and 6.48) and slightly higher value of the d-excess (2.09 and 1.7). Generally, deuterium excess is primarily controlled by kinetic effects associated with evaporation of water at the surface of the oceans or inland. It increases with an increase in the moisture deficit of oceanic air masses [61]. Differences in d-excess arise because of varying temperature, relative humidity, and wind speed at the sea surface, whereas global atmospheric moisture mainly originates from admixture of recycled continental vapor [19,60]. Hence, the value of d-excess (close to 2) in the study area, compared to the GMWL value (10) could probably be attributed to environmental conditions.

Based on the results of the ^{18}O isotope analyses, a spatial distribution map of groundwater samples was constructed (Figure 12), aiming to facilitate the identification and evolution of the active recharge mechanisms in the Tirnavos alluvial basin in support of hydrochemical and piezometric data analyses. For the construction of this map, the monitoring period of April 2018 was used because of the larger number of available samples.

Due to the influence of altitude and continentality on the isotopic composition of precipitation [64,65]—and based on the recharge mechanisms already identified and substantiated—it would be expected that the most negative values of $\delta^{18}O$ were observed along the western part of the basin, where the alluvial aquifer system receives most of its natural recharge in the form of lateral crossflows from the karstic system of Tirnavos. This is the most distant boundary of the alluvial system from the sea, and receives water from direct infiltration of precipitation at altitudes that reach as high as 900 m. Likewise, moving to the downstream parts of the basin, towards the sea, the $\delta^{18}O$ values would have been expected to become less negative. From the spatial distribution map (Figure 12), however, this only happened in the NW part of the basin while, as we move to its central and southeastern parts, the values were increasingly negative—with the minimum figure (the most negative value) found at monitoring point LB214, the easternmost of the compiled network.

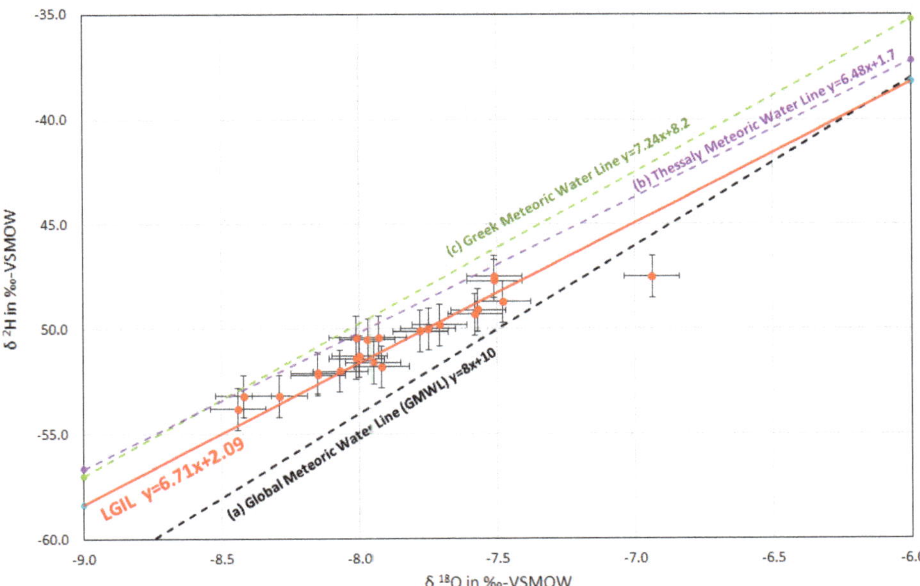

Figure 11. Local groundwater isotope regression line (Tirnavos subbasin) along with (a) GMWL [62], (b) Thessaly Meteoric Water Line [60] (c) Greek Meteoric Water Line [63].

This created a need for further study of the wider area to the east of the basin, in order to obtain a more satisfactory interpretation of this isotopic spatial distribution. About 18 km NE of monitoring point LB214 (where the $\delta^{18}O$ value had the lowest negative value (−9.82‰)) is the Ossa mountain range. The Ossa range rises to 1978 m elevation. Ossa is mainly structured by pre alpine and alpine formations which, geotectonically, are integrated in three units (Figure 12, [66]).

Regarding the tectonics of the area: according to [66], the most important transverse structures resulting from the statistical analysis of the Ossa tectonic elements are illustrated in Figure 12. These are: A, B, C, D, and E (the main fault zones); as well as F, and G (the secondary fault zones). The intense disruption observed along these zones (especially A, B and C, being the main fault zones), in combination with the anticlinal structure of Ossa, was a key factor in the formation of preferential groundwater flow paths to the eastern Thessaly basin (i.e., towards the study area). However, the rate of this recharge (and thus the exact contribution of this mechanism to the balance of the system) is yet to be determined.

The high altitude of Ossa carbonate formations (maximum altitude 1978 m) justified the existence of the lightest ^{18}O isotopes in the wells of the Pinios area in the eastern part of the basin, giving the spatial distribution of $\delta^{18}O$ as shown in Figure 12. High negative values of $\delta^{18}O$, close to (−10), indicated water of meteoric origin. Considering the literature [67], for regions of Thessaly with similar characteristics, the composition of the precipitation indicated a recharge altitude greater than 1200 m. Hence, originating from higher elevations of the Ossa mountain, this recharge mound—in the form of lateral crossflows—does explain the observed $\delta^{18}O$ spatial distribution anomalies and indeed the apparent anomaly in the spatial distribution of the hydrochemical water types that were earlier discussed.

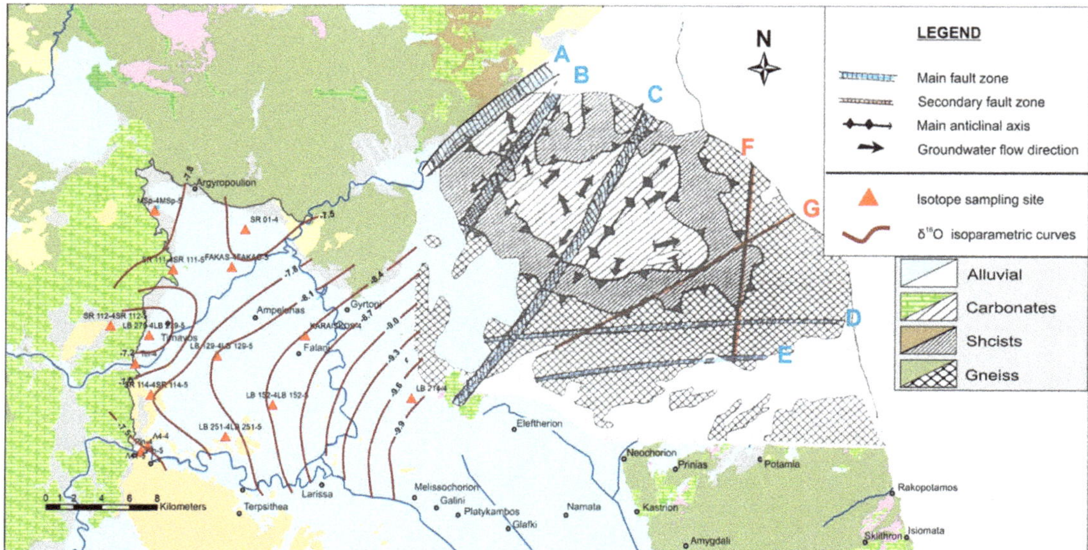

Figure 12. Spatial distribution of $\delta^{18}O$ for groundwater (April 2018) and conceptualization of geological structure tectonic features and flow paths in the Ossa mountain, after [66].

5.5. Conceptual Model of Groundwater Resources

Geometry: The Tirnavos subbasin is the northwesternmost part of the eastern Thessaly basin and is filled by alluvial sediments of varying thickness that reach a maximum of over 550 m towards its central parts [25,31]. The deposition environment progressively turns from fluvial/terrestrial to lacustrine toward the central parts of the basin. A sequence of confining units (which, on a regional scale, could be assumed to form a uniform layer of high clay content,) separated the alluvial sediments in a phreatic aquifer overlain by a thick and high-potential confined aquifer system (Figure 13).

Recharge: Talus cones and scree along the margins of the basin act as a favorable medium for the aquifer system's recharge, especially the extensive fluvial deposits at the exit of the Titarisios River to the basin. The aquifer system exhibits limited hydraulic interactions with the Titarisios River and the Pinios River. These var spatiotemporally in direction and magnitude [53]. The main recharge mechanism to the aquifer system is the lateral inflow across the western boundary with the karstic system of Tirnavos. Lateral recharge from the karstic massif of Mt Ossa, at the northeastern extent of the eastern Thessaly basin, through preferential flow paths developed along the heavily tectonized and disrupted zones [66], is also a significant recharge source to the system. Deep percolation of precipitation forms a source of recharge to the system, and so do irrigation returns. The latter is of progressively reducing importance as more efficient and less water-consuming irrigation systems are being employed.

Discharge: Numerous production wells operate to cover irrigated agriculture and, to a lesser extent, domestic and industrial demands. These account for the majority of discharge from the aquifer system, which has led to the establishment of negative water budgets and the exhaustion of the phreatic aquifer. Natural discharge occurs in the form of lateral crossflows to the southeastern extension of the eastern Thessaly alluvial basin. Seasonal discharge occurs at temporally variable rates to the Pinios River upstream and the town of Larissa and the confluence of the Titarisios and Pinios Rivers downstream [53].

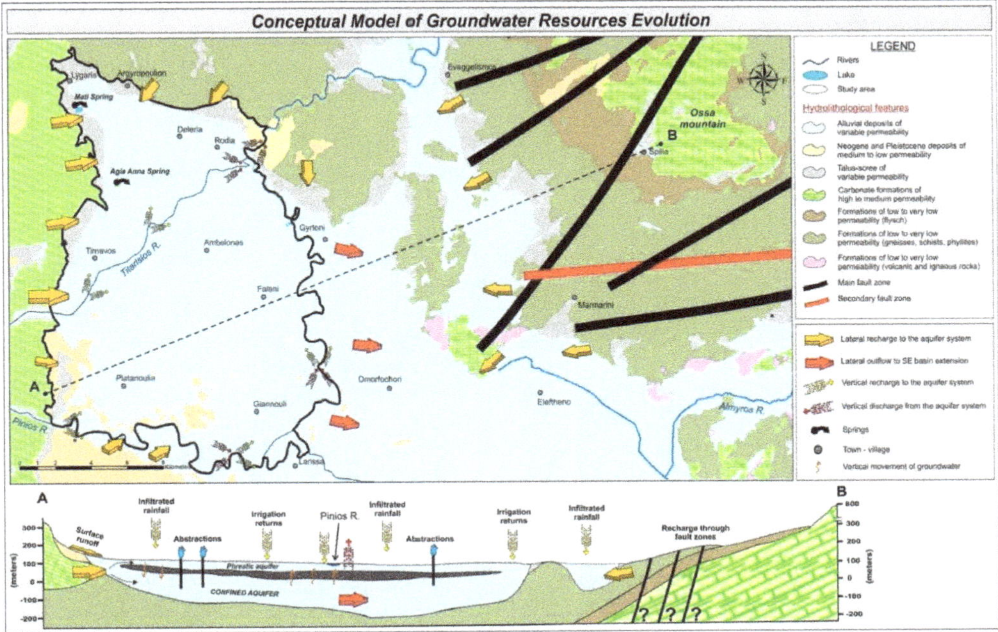

Figure 13. Groundwater resources evolution conceptualization.

Flow domain: Dominant groundwater flow direction is from NW to SE, towards the extension of the eastern Thessaly basin. A key hydrodynamic feature in the basin is the development of a depression cone during the irrigation period, at the central parts of the subbasin, as a result of extensive groundwater abstractions. During hydrologically dry years, the depression cone expands and becomes sharper, characterized by higher hydraulic gradients [33]. Over prolonged droughts characterized by increased water demands, outflow to the SE extension of the basin reduces considerably and is even reported to be completely interrupted [25].

Quality characteristics & Land use impact: Excessive abstractions cause considerable groundwater heads decline and distort the spatial distribution and values of hydraulic gradients. Calcium carbonate is the dominant hydrochemical character attributed to the origin of recharge to the aquifer system. Regionally, residence times increase toward the central parts of the basin where ion exchange mechanisms become important, along with reducing redox conditions. Mixing of resident water with leachates of agricultural activity— and possibly leaks of septic tanks and/or domestic effluent treatment plants—does bring in groundwater contamination issues. These predominantly take the form of elevated nutrients and sporadic findings of high Cu concentrations attributable mainly to orchards and cultivars. Small industrial unit activity is also reflected on groundwater quality in the form of elevated EC values.

6. Conclusions

Rational and sustainable management of groundwater resources requires knowledge of the geometric and hydraulic characteristics of the aquifer system that hosts the reserves, as well as a deep understanding of the hydrogeochemical and hydrodynamic mechanisms that control their evolution. This is even more true in cases of complex geotectonic environments, under intensive and prolonged exploitative conditions—especially within the climate change framework.

The alluvial system of the Tirnavos subbasin has a long record of exploitation, dating to the early 1970s, in support of the development of the region through irrigated agriculture. Anthropogenic activities have had a profound impact in shaping up groundwater reserves' availability and groundwater quality. Even though the Tirnavos subbasin is among the oldest and best-studied basins in Greece, this paper brings up new insights into its hydrodynamic evolution and controlling mechanisms. This enlightens and explains hydrogeochemical signatures and hydrodynamic behaviors which have been noticed in the past but disregarded.

The exercise performed in this research—characterizing and assessing the groundwater resources of the Tirnavos alluvial subbasin—consisted of a comprehensive overview of a suite of methodologies and approaches that acted complementarily to each other, leading to a clear and well-justified overview of flow and hydrochemical evolution mechanisms. Evidently, no single method is able to provide comprehensive, reliable answers or fully explain the observed spatiotemporal distribution of data. In this work, we have successfully made conjunctive use of hydrogeochemical, hydrodynamic, geological, structural, geotectonic and isotopic data—all analyzed in their spatiotemporal distribution, assessed in statistical indices, and evaluated in hydrogeochemical indices and tri-linear diagrams.

The key outcome of the present paper is the development of an integrated conceptual model for groundwater resource evolution, through the application of a specific methodological framework. This framework may be used globally as a generic model in other cases addressing similar scientific quests. The main findings of the research which compose this conceptual model, are briefly described below:

- The dominant ions (Ca^{2+} and HCO_3^-) of groundwater were indicative for the main recharge mechanisms—which were related to the karstic substrate.
- The recharge areas were delineated with the joint use of variable tools and concluded in two major axes in an E–W direction. The western direction was related to the recharge from the karstic system of Titarisios and the eastern one was related to the recharge through preferential flow paths from the karstic massif of Mt. Ossa.
- The prevalence of carbonate formations was also reflected in the dominant hydrochemical type (Ca-Mg-HCO_3) which also encompassed the Mg content from relevant Mg-rich carbonate formations (dolostones).
- Groundwater quality in the Tirnavos subbasin was in a generally good state, with few exceptions (elevated values of NO_3 and Cu) which indicate local impact due to anthropogenic activities.
- Nitrates can be considered the main adverse environmental aspect, occurring locally in hot spots at the area, due to irrational agricultural activities.
- The groundwater quality was also impacted by salinization due to the combined use of irrigation water returns (agricultural leachates) and evaporitic mineral leaching.
- The governing hydrogeochemical process identified was ion exchange, which progressively alters the chemical composition of groundwater from west to east.
- A secondary process which seems to affect hydrogeochemistry was redox, which locally controls speciation of groundwater solute parameters.
- The physicochemical evolution mechanisms were jointly assessed and verified by the hydrochemical sections of major ions and saturation indices of critical mineralogical phases.

Even though the presented work sheds light on—and further proves the existence of—several major controlling evolution mechanisms, there is still room for further elaboration and in-depth analysis. This is important to capture potential alterations in the significance each identified mechanism might have on the system's evolution because of ongoing continuous changes in land and water use—and in the framework of climate change. Moreover, overexploitation of the phreatic aquifer—to the point of depletion and destruction or abandonment of shallow wells and boreholes—makes it imperative to seek out replacements for functional and aquifer unit-specific monitoring points, to ensure accurate, representative, and meaningful measurements are made. Moreover, the authors of the presented work propose further detailed analyses of redox processes and status. This will bring added value and knowledge and help to further decipher details of the controlling factors that shape the established hydrogeochemical signatures captured in the analyzed water samples.

Author Contributions: I.V. conceived the methodological approach, performed the data processing and supervised the drafting and revision of the final text. E.T., A.P. and G.S. participated in data processing, verified and optimized the methodology, contributed to the discussion parts, and revised the final text. All authors have read and agreed to the published version of the manuscript.

Funding: This research received no external funding.

Institutional Review Board Statement: Not applicable.

Informed Consent Statement: Not applicable.

Data Availability Statement: Data is contained within the article.

Acknowledgments: The authors wish to acknowledge the valuable contribution of Hydrology Laboratory of Lubeck Technical University in analyzing the collected samples for isotopes—and especially C. Kuells for offering valuable guidance in the assessment of the results. This work was supported by the SWRI infrastructure, with the aid of which sampling and analysis of the collected samples was made possible. Special thanks are due to the laboratory staff of the Institute that performed the presented determinations. Last but not least, the technical and administrative staff of the Local Irrigation Organizations of Tirnavos, Ampelona and Agia Sofia are acknowledged for their support and guidance in the conducted sampling campaigns.

Conflicts of Interest: The authors declare no conflict of interest.

Appendix A

Table A1. Basic descriptive statistics of analyzed samples.

	pH	EC (μS/cm)	K (mg/L)	Na (mg/L)	Ca (mg/L)	Mg (mg/L)	Tot. Hardness (mg CaCO$_3$/L)	Cl (mg/L)	HCO$_3$ (mg/L)	CO$_3$ (mg/L)	SO$_4$ (mg/L)
MIN	6.92	253.00	0.40	6.86	14.66	8.10	69.96	4.77	118.81	0.00	0.11
MAX	7.96	1821.00	7.21	286.69	120.15	52.98	436.05	88.62	453.87	0.00	604.34
MEDIAN	7.59	473.04	1.68	13.60	67.69	16.67	234.53	10.82	238.37	0.00	12.21
STDEV	0.21	246.39	1.11	44.50	23.27	10.84	83.51	14.06	64.37	0.00	99.21

	NO$_3$ (mg/L)	NH$_4$ (mg/L)	B (mg/L)	Cu (μg/L)	Fe (μg/L)	Mn (μg/L)	Pb (μg/L)	Cd (μg/L)	As (μg/L)	SAR	TDS
MIN	0.22	0.01	0.00	0.00	0.81	0.07	0.00	0.00	0.00	0.20	216.00
MAX	145.37	5.76	1.25	24.22	100.85	150.83	3.12	0.13	4.22	6.76	1374.87
MEDIAN	19.08	0.12	0.02	1.65	22.26	1.27	0.08	0.04	1.62	0.39	390.39
STDEV	27.89	0.98	0.23	5.14	17.70	28.32	0.72	0.03	1.06	1.19	187.85

References

1. Tziritis, E.P.; Datta, P.S.; Barzegar, R. Characterization and assessment of groundwater resources in a complex hydrological basin of central Greece (Kopaida basin) with the joint use of hydrogeochemical analysis, multivariate statistics and stable isotopes. *Aquat. Geochem.* **2017**, *23*, 271–298. [CrossRef]
2. Kim, Y.; Lee, K.S.; Koh, D.C.; Lee, D.H.; Lee, S.G.; Park, W.B.; Woo, N.C. Hydrogeochemical and isotopic evidence of groundwater salinization in a coastal aquifer: A case study in Jeju volcanic island, Korea. *J. Hydrol.* **2003**, *270*, 282–294. [CrossRef]
3. Barzegar, R.; Moghaddam, A.A.; Tziritis, E.; Fakhri, M.S.; Soltani, S. Identification of hydrogeochemical processes and pollution sources of groundwater resources in the Marand plain, northwest of Iran. *Environ. Earth Sci.* **2017**, *76*, 297. [CrossRef]
4. Iqbal, J.; Nazzal, Y.; Howari, F.; Xavier, C.; Yousef, A. Hydrochemical processes determining the groundwater quality for irrigation use in an arid environment: The case of Liwa Aquifer, Abu Dhabi, United Arab Emirates. *Groundw. Sustain. Dev.* **2018**, *7*, 212–219. [CrossRef]
5. Wang, Z.; Torres, M.; Paudel, P.; Hu, L.; Yang, G.; Chu, X. Assessing the Karst Groundwater Quality and Hydrogeochemical Characteristics of a Prominent Dolomite Aquifer in Guizhou, China. *Water* **2020**, *12*, 2584. [CrossRef]

6. Zancanaro, E.; Teatini, P.; Scudiero, E.; Morari, F. Identification of the Origins of Vadose-Zone Salinity on an Agricultural Site in the Venice Coastland by Ionic Molar Ratio Analysis. *Water* **2020**, *12*, 3363. [CrossRef]

7. Domingo-Pinillos, J.C.; Senent-Aparicio, J.; García-Aróstegui, J.L.; Baudron, P. Long term hydrodynamic effects in a semi-arid Mediterranean multilayer aquifer: Campo de Cartagena in south-eastern Spain. *Water* **2018**, *10*, 1320. [CrossRef]

8. Li, M.; Liang, X.; Xiao, C.; Cao, Y.; Hu, S. Hydrochemical Evolution of Groundwater in a Typical Semi-Arid Groundwater Storage Basin Using a Zoning Model. *Water* **2019**, *11*, 1334. [CrossRef]

9. Cao, F.; Jaunat, J.; Vergnaud-Ayraud, V.; Devau, N.; Labasque, T.; Guillou, A.; Ollivier, P. Heterogeneous behaviour of unconfined Chalk aquifers infer from combination of groundwater residence time, hydrochemistry and hydrodynamic tools. *J. Hydrol.* **2020**, *581*, 124433. [CrossRef]

10. Murgulet, D.; Murgulet, V.; Spalt, N.; Douglas, A.; Hay, R.G. Impact of hydrological alterations on river-groundwater exchange and water quality in a semi-arid area: Nueces River, Texas. *Sci. Total Environ.* **2016**, *572*, 595–607. [CrossRef] [PubMed]

11. Barbieri, M.; Nigro, A.; Petitta, M. Groundwater mixing in the discharge area of San Vittorino Plain (Central Italy): Geochemical characterization and implication for drinking uses. *Environ. Earth Sci.* **2017**, *76*, 393. [CrossRef]

12. Maest, A.; Prucha, R.; Wobus, C. Hydrologic and Water Quality Modeling of the Pebble Mine Project Pit Lake and Downstream Environment after Mine Closure. *Minerals* **2020**, *10*, 727. [CrossRef]

13. Güler, C.; Kurt, M.A.; Alpaslan, M.; Akbulut, C. Assessment of the impact of anthropogenic activities on the groundwater hydrology and chemistry in Tarsus coastal plain (Mersin, SE Turkey) using fuzzy clustering, multivariate statistics and GIS techniques. *J. Hydrol.* **2012**, *414*, 435–451. [CrossRef]

14. Voutsis, N.; Kelepertzis, E.; Tziritis, E.; Kelepertsis, A. Assessing the hydrogeochemistry of groundwaters in ophiolite areas of Euboea Island, Greece, using multivariate statistical methods. *J. Geochem. Explor.* **2015**, *159*, 79–92. [CrossRef]

15. Barzegar, R.; Moghaddam, A.A.; Tziritis, E. Assessing the hydrogeochemistry and water quality of the Aji-Chay River, northwest of Iran. *Environ. Earth Sci.* **2016**, *75*, 1486. [CrossRef]

16. Celestino, A.E.M.; Ramos-Leal, J.; Cruz, D.A.M.; Tuxpan, J.; Bashulto, J.D.L.; Ramírez, J.M. Identification of the hydrogeochemical processes and assessment of groundwater quality, using multivariate statistical approaches and water quality index in a wastewater irrigated region. *Water* **2019**, *11*, 1702. [CrossRef]

17. Walter, J.; Chesnaux, R.; Gaboury, D.; Cloutier, V. Subsampling of Regional-Scale Database for improving Multivariate Analysis Interpretation of Groundwater Chemical Evolution and Ion Sources. *Geosciences* **2019**, *9*, 139. [CrossRef]

18. Chai, Y.; Xiao, C.; Li, M.; Liang, X. Hydrogeochemical Characteristics and Groundwater Quality Evaluation Based on Multivariate Statistical Analysis. *Water* **2020**, *12*, 2792. [CrossRef]

19. Tziritis, E. Stable isotope study of a karstic aquifer in Central Greece. Composition, variations and controlling factors. In *Advances in the Research of Aquatic Environment*; Springer: Berlin/Heidelberg, Germany, 2011; pp. 193–200. [CrossRef]

20. Joshi, S.K.; Rai, S.P.; Sinha, R.; Gupta, S.; Densmore, A.L.; Rawat, Y.S.; Shekhar, S. Tracing groundwater recharge sources in the northwestern Indian alluvial aquifer using water isotopes ($\delta^{18}O$, δ^2H and ^3H). *J. Hydrol.* **2018**, *559*, 835–847. [CrossRef]

21. Liu, F.; Wang, S.; Wang, L.; Shi, L.; Song, X.; Yeh, T.C.J.; Zhen, P. Coupling hydrochemistry and stable isotopes to identify the major factors affecting groundwater geochemical evolution in the Heilongdong Spring Basin, North China. *J. Geochem. Explor.* **2019**, *205*, 106352. [CrossRef]

22. Zamora, H.A.; Eastoe, C.J.; Wilder, B.T.; McIntosh, J.C.; Meixner, T.; Flessa, K.W. Groundwater Isotopes in the Sonoyta River Watershed, USA-Mexico: Implications for Recharge Sources and Management of the Quitobaquito Springs. *Water* **2020**, *12*, 3307. [CrossRef]

23. Wu, C.; Wu, X.; Mu, W.; Zhu, G. Using Isotopes (H, O, and Sr) and Major Ions to Identify Hydrogeochemical Characteristics of Groundwater in the Hongjiannao Lake Basin, Northwest China. *Water* **2020**, *12*, 1467. [CrossRef]

24. Panagopoulos, A.; Arampatzis, G.; Tziritis, E.; Pisinaras, V.; Herrmann, F.; Kunkel, R.; Wendland, F. Assessment of climate change impact in the hydrological regime of River Pinios Basin, central Greece. *Desalination Water Treat.* **2016**, *57*, 2256–2267. [CrossRef]

25. Alexandridis, T.; Panagopoulos, A.; Galanis, G.; Alexiou, I.; Cherif, I.; Chemin, Y.; Stavrinos, E.; Bilas, G.; Zalidis, G. Combining remotely sensed surface energy fluxes and GIS analysis of groundwater parameters for irrigation assessment. *Irrig. Sci.* **2014**, *32*, 127–140. [CrossRef]

26. Vrouhakis, I.; Panagopoulos, A.; Stamatis, G. Current quality and quantity status of Tirnavos sub-basin water system—Central Greece. In Proceedings of the 11th International Hydrogeological Congress of the Greece, Athens, Greece, 4–6 October 2017.

27. Plastiras, V. *Geological Map of Greece, Larissa Sheet*; Institute of Geology and Mineral Exploitation: Athens, Greece, 1982.

28. Miggiros, G. *Geological Map of Greece, Gonnoi Sheet*; Institute of Geology and Mineral Exploitation: Athens, Greece, 1980.

29. Panagopoulos, A. A methodology for groundwater resources management of a typical alluvial aquifer system in Greece. Ph.D. Thesis, School of Earth Sciences, Faculty of Science, University of Birmingham, Birmingham, UK, 1995.

30. Vrouhakis, I.; Tziritis, E.; Panagopoulos, A.; Kulls, C.; Stamatis, G. The use of environmental stable isotopes at the Tirnavos alluvial basin (Central Greece). In Proceedings of the 15th International Congress of the Geological Society of Greece, Athens, Greece, 22–24 May 2019.

31. Demitrack, A. The Late Quaternary Geologic History of the Larissa Plain, Thessaly, Greece: Tectonic, Climatic, and Human Impact on the Landscape. Ph.D. Thesis, Stanford University, Stanford, UK, 1986. (Unpublished).

32. Aggarwal, P.K.; Araguas, L.; Garner, W.A.; Groeninig, M.; Kulkarni, K. Introduction to Water Sampling Analysis for Isotope Hydrology. Water Resources Programme-IAEA. Available online: www-naweb.iaea.org/napc/ih/documents/other/Sampling%20booklet%20web.pdf (accessed on 1 December 2009).

33. Ministry of the Environment Energy and Climate Change. *Development of river Basin Management Plans for the Water District of Thessalia, Epirus, Western Sterea Ellada, in Accordance with the Directive 2000/60/EC, the Law 3199/2003 and the P.D. 51/2007*; Special Secretariat for Water: Athens, Greece, 2014.

34. Burdon, D.; Mazloum, S. *Some Chemical Types of Ground-Water from Syria*; David, J.B., Soubhi, M., Eds.; UNESCO: Paris, France, 1958.

35. Piper, A.M. A graphic procedure in the geochemical interpretation of water-analyses. *Trans. Am. Geophys. Union* **1944**, *25*, 914–928. [CrossRef]

36. Lloyd, J.; Heathcote, J. *Natural Inorganic Hydrochemistry in Relation to Ground Water*; Oxford Science Publications: Oxford, UK, 1985.

37. Ravikumar, P.; Somashekar, R.K.; Prakash, K.L. A comparative study on usage of Durov and Piper diagrams to interpret hydrochemical processes in groundwater from SRLIS river basin, Karnataka, India. *Elixir Int. J.* **2015**, *80*, 31073–31077.

38. Shyu, G.S.; Cheng, B.Y.; Chiang, C.T.; Yao, P.H.; Chang, T.K. Applying factor analysis combined with kriging and information entropy theory for mapping and evaluating the stability of groundwater quality variation in Taiwan. *Int. J. Environ. Res. Public Health* **2011**, *8*, 1084–1109. [CrossRef] [PubMed]

39. Liu, C.W.; Lin, K.H.; Kuo, Y.M. Application of factor analysis in the assessment of groundwater quality in a blackfoot disease area in Taiwan. *Sci. Total Environ.* **2003**, *313*, 77–89. [CrossRef]

40. Panda, U.C.; Sundaray, S.K.; Rath, P.; Nayak, B.B.; Bhatta, D. Application of factor and cluster analysis for characterization of river and estuarine water systems–a case study: Mahanadi River (India). *J. Hydrol.* **2006**, *331*, 434–445. [CrossRef]

41. Tziritis, E.P. Environmental monitoring of Micro Prespa Lake basin (Western Macedonia, Greece): Hydrogeochemical characteristics of water resources and quality trends. *Environ. Monit. Assess.* **2014**, *186*, 4553–4568. [CrossRef] [PubMed]

42. Kaiser, H.F. The application of electronic computers to factor analysis. *Educ. Psychol. Meas.* **1960**, *20*, 141–151. [CrossRef]

43. Reimann, C.; Filzmoser, P.; Garrett, R.; Dutter, R. *Statistical Data Analysis Explained: Applied Environmental Statistics with R.*; John Wiley & Sons: Hoboken, NJ, USA, 2011.

44. Parkhurst, D.L.; Appelo, C.A.J. *PHREEQC2 User's Manual and Program*; Water-Resources Investigations Report; US Geological Survey: Denver, CO, USA, 2004.

45. Ministry of Environment; Energy and Climate Change; Special Secretariat for Water. River Basin Management Plan of Thessaly Water District (GR08). In *Analysis of Anthropogenic Pressures and Their Impact on Surface Water Systems and Aquifer Systems*; Special Secretariat for Water: Athens, Greece, 2014.

46. Sharma, L.; Greskowiak, J.; Ray, C.; Eckert, P.; Prommer, H. Elucidating temperature effects on seasonal variations of biogeochemical turnover rates during riverbank filtration. *J. Hydrol.* **2012**, *428*, 104–115. [CrossRef]

47. Menberg, K.; Blum, P.; Kurylyk, B.L.; Bayer, P. Observed groundwater temperature response to recent climate change. *Hydrol. Earth Syst. Sci.* **2014**. [CrossRef]

48. Taniguchi, M. Analysing the long term reduction in groundwater temperature due to pun pumping. *Hydrol. Sci. J.* **1995**, *40*, 407–421. [CrossRef]

49. Panagopoulos, A.; Kassapi, K.A.; Arampatzis, G.; Perleros, B.; Drakopoulou, S.; Tziritis, E.; Chrysafi, A.; Vrouhakis, I. Assessment of chemical and quantitative status of groundwater systems in Pinios hydrological basin-Greece. In Proceedings of the XI Int. Conference Protection and Restoration of the Environment, Thessaloniki, Greece, 3–6 July 2012.

50. Directive 98/83/EC of 3 November 1998 on the Quality of Water Intended for Human Consumption. Available online: https://eur-lex.europa.eu/legal-content/EN/TXT/?uri=CELEX%3A31998L0083 (accessed on 19 March 2019).

51. World Health Organization (WHO). *Guidelines for Drinking Water Quality*, 2nd ed.; Health Criteria and Other Supporting Information; World Health Organization (WHO): Vienna, Austria, 1996; Volume 2.

52. Al-Bassam, A.M.; Khalil, A.R. DurovPwin: A new version to plot the expanded Durov diagram for hydro-chemical data analysis. *Comput. Geosci.* **2012**, *42*, 1–6. [CrossRef]

53. Panagopoulos, A.; Lloyd, J.; Fitzsimons, V. Groundwater evolution of the Tirnavos alluvial basin, central Greece, as indicated by hydrochemistry. In Proceedings of the 3rd Hydrogeological Conference of the Hellenic Chapter of IAH, Heraklion, Greece, 3–5 November 1995.

54. Appelo, C.A.J.; Postma, D. *Geochemistry, Groundwater and Pollution*; CRC Press: Boca Raton, FL, USA, 2004.

55. Boghici, R.; Van Broekhoven, N.G. Hydrogeology of the Rustler Aquifer, Trans-Pecos, Texas; in Aquifers of West Texas. *Tex. Water Dev. Board Rep.* **2001**, *356*, 207–225.

56. Jalali, M. Hydrochemical characteristics and sodification of groundwater in the Shirin Sou, Hamedan, Western Iran. *Nat. Resour. Res.* **2012**, *21*, 61–73. [CrossRef]

57. Jalali, M. Major ion chemistry of groundwaters in the Bahar area, Hamadan, western Iran. *Environ. Geol.* **2005**, *47*, 763–772. [CrossRef]

58. Esmaeili-Vardanjani, M.; Rasa, I.; Amiri, V.; Yazdi, M.; Pazand, K. Evaluation of groundwater quality and assessment of scaling potential and corrosiveness of water samples in Kadkan aquifer, Khorasan-e-Razavi Province, Iran. *Environ. Monit. Assess.* **2015**, *187*, 53. [CrossRef] [PubMed]

59. Kalantary, N.; Rahimi, M.; Charchi, A. Use of composite diagram, factor analyses and saturation index for quantification of Zaviercherry and Kheran plain groundwaters. *J. Eng. Geol.* **2007**, *2*, 339–356.

60. Dotsika, E.; Lykoudis, S.; Poutoukis, D. Spatial distribution of the isotopic composition of precipitation and spring water in Greece. *Glob. Planet. Chang.* **2010**, *71*, 141–149. [CrossRef]
61. Sharp, Z. *Stable Isotope Geochemistry*; Pearson Education; Prentice Hall: New York, NY, USA, 2007; p. 344.
62. Craig, H. Isotopic variations in meteoric waters. *Science* **1961**, *133*, 1702–1703. [CrossRef] [PubMed]
63. Argiriou, A.A.; Lykoudis, S. Isotopic composition of precipitation in Greece. *J. Hydrol.* **2006**, *327*, 486–495. [CrossRef]
64. Clark, I.D.; Fritz, P. *Environmental Isotopes in Hydro-Geology*; CRC Press: New York, NY, USA, 1997.
65. Matiatos, I. Hydrogeological and Isotopic Investigations at Regions of the Argolis Peninsula. Ph.D. Thesis, National and Kapodistrian University of Athens, Athens, Greece, 2010.
66. Stamatis, G.; Miggiros, G. The relationship between fractured tectonic and groundwater reservoir of massive formations of Ossa Mountain (E. Thessaly, Greece). *Bull. Geol. Soc. Greece* **2004**, *36*, 2077–2086. [CrossRef]
67. Payne, B.; Dimitroula, C.; Leondiadis, I.; Kallergis, G. *Environmental Isotope Data in the Western Thessaly Valley, Greece: Use of Mathematical Model for Quantitative Evaluations with Tritium*; Bulletin of the Geological Society of Greece: Athens, Greece, 1976.

Article

An Integrated Modeling System for the Evaluation of Water Resources in Coastal Agricultural Watersheds: Application in Almyros Basin, Thessaly, Greece

Aikaterini Lyra [1,*], Athanasios Loukas [2], Pantelis Sidiropoulos [1], Georgios Tziatzios [1] and Nikitas Mylopoulos [1]

1 Laboratory of Hydrology and Aquatic Systems Analysis, Department of Civil Engineering, School of Engineering, University of Thessaly, 38334 Volos, Greece; psidirop@uth.gr (P.S.); getziatz@uth.gr (G.T.); nikitas@civ.uth.gr (N.M.)
2 Department of Rural and Surveying Engineering, Aristotle University of Thessaloniki, 54124 Thessaloniki, Greece; agloukas@topo.auth.gr
* Correspondence: klyra@uth.gr; Tel.: +30-242-107-4153

Abstract: This study presents an integrated modeling system for the evaluation of the quantity and quality of water resources of coastal agricultural watersheds. The modeling system consists of coupled and interrelated models, including (i) a surface hydrology model (UTHBAL), (ii) a groundwater hydrology model (MODFLOW), (iii) a crop growth/nitrate leaching model (REPIC, an R-ArcGIS-based EPIC model), (iv) a groundwater contaminant transport model (MT3DMS), and (v) a groundwater seawater intrusion model (SEAWAT). The efficacy of the modeling system to simulate the quantity and quality of water resources has been applied to the Almyros basin in Thessaly, Greece. It is a coastal agricultural basin with irrigated and intensified agriculture facing serious groundwater problems, such as groundwater depletion, nitrate pollution, and seawater intrusion. Irrigation demands were estimated for the main crops cultivated in the area, based on precipitation and temperature from regional weather stations. The models have been calibrated and validated against time-series of observed crop yields, groundwater table observations, and observed concentrations of nitrates and chlorides. The results indicate that the modeling system simulates the water resources quantity and quality with increased accuracy. The proposed modeling system could be used as a tool for the simulation of water resources management and climate change scenarios.

Keywords: integrated water resources management; coastal agricultural basin; groundwater nitrate pollution; seawater intrusion

Citation: Lyra, A.; Loukas, A.; Sidiropoulos, P.; Tziatzios, G.; Mylopoulos, N. An Integrated Modeling System for the Evaluation of Water Resources in Coastal Agricultural Watersheds: Application in Almyros Basin, Thessaly, Greece. *Water* **2021**, *13*, 268. https://doi.org/10.3390/w13030268

Received: 5 January 2021
Accepted: 18 January 2021
Published: 22 January 2021

Publisher's Note: MDPI stays neutral with regard to jurisdictional claims in published maps and institutional affiliations.

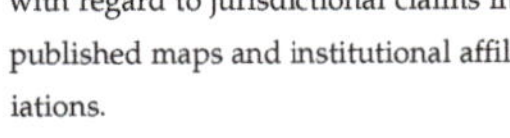

1. Introduction

Around the semi-arid Mediterranean basin, complex cases of water resources degradation, also characterized by poor quantity and quality status, are currently met in coastal agricultural basins. The complexity of the problems of such water systems arises mainly from: (i) the limited use of surface water, (ii) the excessive groundwater abstractions for irrigation, and (iii) the over-fertilization practices for crop yield magnification [1]. These actions cause the lowering of the water table of aquifers, increase nitrate groundwater pollution, and invoke seawater intrusion to groundwater systems [2]. Groundwater pollution has an important role among water resources management strategies due to the majority of the semi-arid regions it is encountered. Some known examples of coastal water systems where intensive agricultural practices, for irrigation and fertilization purposes, have caused serious degradation of the groundwater resources and documented in the international literature, including Italy (Nurra region of Sardinia [3] and central-southern Italy [4]), south-eastern France (Lower Var Valley) [5], Spain (e.g., Mancha Oriental System in Jucar River Basin, Oropesa Plain, Vinaroz Plain) [6,7], Portugal (Tagus catchment) [8], Egypt (e.g.,

Bagoush plain, North Sinai area, East Nile Delta aquifer), Tunisia (Jerba Island), Algeria (Nador plain), Morocco (Bou-Areg aquifer) [9], Lebanon (Akkar and Damour aquifers), Syria (e.g., Latakia and Tartous groundwaters), Palestine (Gaza aquifer), Jordan (northern Jordan, Yarmouk Basin) [10], Cyprus (Magosa aquifer) [11], Turkey (e.g., Silifke-Goksu Deltai, Serik, and Tarsus Plains) [12], and in the Greek coastal areas.

According to the Greek Ministry of the Environment, many coastal aquifer systems experience groundwater depletion, nitrate pollution, and seawater intrusion in Greece [13–15]. Examples of water-stressed and polluted groundwater systems are located in Macedonia (Axios River basin, Galikos River basin), in Chalkidiki peninsula (Moudania watershed and Havrias River basin), in central Greece (Sperheios River basin, Voiotikos Kifisos River basin (the Kopaida Plain)), in Peloponnese (Plain of Argos and Pinios River basin), in Greek Islands (Crete), and in Thessaly (Pinios River basin and Almyros basin) [13–15].

In nitrate contaminated waters, the "threshold of concern" is 25 NO_3 mg/L, and mostly refers to the suitability of the drinking water. The upper safe concentration is set at 50 NO_3 mg/L while higher nitrates concentrations endanger human health, surrounding ecosystems, and local biodiversity [15,16]. Thresholds of chloride concentrations have been defined as well, for the groundwater systems of more than 10 Member States of the EU. According to the Directive 2006/118/EC, the maximum allowable concentrations of chlorides range between 24 Cl mg/L and 12,300 Cl mg/L, in reliance on the diversity of characteristics of groundwater systems. The maximum allowable concentration of chlorides for urban water supply is set at 250 Cl mg/L, while in higher concentrations its consumption is traceable in taste and is also linked to health problems [17].

Hence, research must be done in the direction of the development of efficient software and tools that take into account the impacts of water abstractions, the water balance deficit, the nitrate leaching, the nitrate pollution, and the seawater intrusion [18–21]. Until recently, most of the hydrological processes and human interactions either for surface water resources [22] or groundwater [23], or under climate change [6] or land uses [24], or landscape and population changes [25] are often studied separately. The connection and communication of surface water and groundwater [26], the unsustainable water use [1,27], the irrational fertilization of crops [2], the groundwater contamination, the head dropdown of coastal aquifers [28], their salinization [29,30], and the reduced yields of crops [31] are rarely modeled in an integrated modeling system. Consequently, the simulation of the spatiotemporal responses of the water resources systems, in view of an integrated modeling system, can advance the reliability of fully understanding and modeling the complex interactions of water systems. A holistic approach of the natural and socio-economic drivers that may cause quantitative and qualitative problems of water resources, can further promote their effective management and the configuration of improvement strategies [32]. In this direction, very recent examples of integrated modeling of water resources quantity and quality are documented in the literature; such as the Community Water Model for the quantitative simulation of water resources [33], the remote sensing and GIS-based conceptual model of land use change impacts on groundwater quantity and salinity [34], the coupled models of groundwater flow and particle tracking of contaminants in karstic groundwater bodies [35], the Bow River Integrated Model (BRIM) for the simulation and management of water resources at basin scale [36], the FREEWAT, a QGIS based simulation and management system of coupled models of surface and groundwater quality and quantity, and agricultural water uses [37], the national-scale conceptual model for nitrate transport and dilution in aquifers [38], the modeling system of nonconservative contaminants in variable-density flow [4], and the Integrated Hydrological Modeling System (IHMS) for the simulation and management of surface and groundwater resources and saltwater intrusion [39]. Under the concept of the development of an integrated modeling system, adequate resolution in space and in time is of critical importance, in order to achieve faithful results, reduced inherent uncertainty, effective modeling of heterogeneity and of parameter variability, and reduced calibration/validation times [32].

The major research question that the paper tries to answer is how to efficiently simulate the integrated water balance and water quality of groundwater resources of coastal agricultural watersheds at a basin/watershed scale. This study aims to address this question with the development of an integrated modeling system, which includes coupled and interrelated models of surface and groundwater hydrology, a crop growth/nitrate leaching model, and groundwater models for the simulation of groundwater, nitrate contamination, and seawater intrusion, set up for use in agricultural coastal watersheds. The modeling system has been applied in the Almyros basin, a significant coastal region with intense agricultural activity, located in Thessaly, central Greece. The modeling system has been calibrated and validated against observations and applied to reproduce the water resources, mainly groundwater, quantity, and quality for the period of October 1991 to September 2018.

The simulation results indicated that the proposed modeling system simulates the water resources quantity and quality, with increased accuracy, and could be used as a tool for the simulation of alternative water resources management scenarios and water resources management. Especially, for the Almyros basin, the causes of the deficit water balance are determined to be (i) the increased water well pumping during the crop growth summer periods, (ii) the absent use of surface water for irrigation, and (iii) the possible unsuitability of crop types to the regional semi-arid climate. This, in turn, led to the inland salinization of the coastline areas. Furthermore, the nitrates assimilation into the aquifer originates from the nitrates leached during crop irrigation and rainfall as a result of the excess fertilization applications. The integrated modeling system and its application in the Almyros Basin is described in the following sections.

2. Models and Methods

2.1. Modeling System

The scope of this study is to present the development of a modeling system for the evaluation of the surface water balance on a sub-basin scale, the crop water demands, the nitrates leached from the unsaturated zone, the groundwater quantity, and nitrate pollution and salinization in an aquifer system, on a grid-scale. The developed integrated modeling system consists of coupled simulation models of the natural processes of the hydrologic cycle, and the contamination events and their impacts. The components of the system are the models of surface hydrology (UTHBAL) [1], groundwater hydrology (MODFLOW) [40], crop growth/nitrate leaching (REPIC), which is an innovative R-ArcGIS user interface with the EPIC model [41], contaminant transport/ nitrate pollution (MT3DMS) [42], and seawater intrusion/aquifer salinization (SEAWAT) [43].

The mean monthly areal precipitation, temperature, and evapotranspiration are used as inputs in the surface hydrology model (UTHBAL). The model estimates the monthly surface runoff and the natural groundwater recharge per sub-basin. The weighted average irrigation return flow per sub-basin and main crop category are, then, added to the recharge rate, as calculated in the previous step. The sum of natural recharge and irrigation return flow is the input flows of the groundwater model (MODFLOW). To address the crop water demands, abstraction water well flows are determined as the outflows of the aquifer. The input of the sea level head completes the water balance of the model. In the context of this research, the distributed crop growth/nitrate leaching simulation model REPIC was developed, along with tools for efficient and easy interaction with the models of UTHBAL and MT3DMS. REPIC model is a spatially distributed model, similar to GEPIC [44], as it simulates every grid cell area as a field. It is based on the R-programming language [45] with the R-ArcGIS Bridge [46] in ArcGIS 10+/Pro, and the public domain Environmental Policy Integrated Climate (EPIC) model. EPIC model is used as is, compiled in Fortran and as provided individually by [47]. It is an agro-hydrological model that simulates and estimates, among other parameters, the nitrates leached into the groundwater. Subsequently, the nitrates leached are the input contaminant flows of the MT3DMS model. The modeling system is completed with the simulation of the seawater intrusion of the aquifer. The chloride concentration of the sea is the input chloride inflow of the SEAWAT model. Finally, the quantity and quality status of the coastal water resources can be defined. More details about the models are described in the next paragraphs. The flow chart of the integrated modeling system is illustrated in Figure 1.

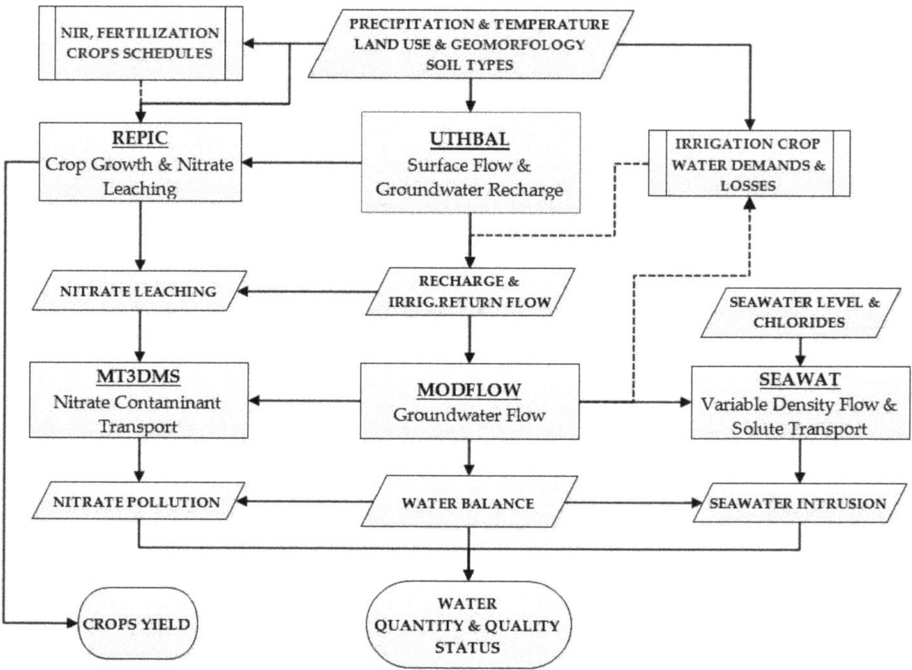

Figure 1. Flow Chart of the Integrated Modeling System.

2.1.1. Surface Hydrology Simulation Model Description

The UTHBAL model is a surface hydrology model that can simulate the surface runoff and groundwater recharge, developed by Loukas et al. in 2007 [1]. UTHBAL uses as inputs monthly time series of precipitation, mean temperature, and potential evapotranspiration. The water balance model separates the total precipitation into rainfall and snowfall and calculates the snowpack and snowmelt. The model divides the total watershed runoff into three components: the surface runoff, the interflow, and the baseflow using a soil moisture mechanism. The first priority of the model is to fulfill the actual evapotranspiration. The output of the model is watershed runoff, actual evapotranspiration, groundwater recharge, and soil moisture.

The model can be applied as a lumped, semidistributed, and fully-distributed model depending on the available data. The model contains six parameters that are estimated during the calibration process, based on surface runoff monthly observations. These are, the parameter of monthly melt rate factor, Cm, the coefficient of actual evapotranspiration, α, the coefficient of interflow β, the coefficient of baseflow, γ, the coefficient of groundwater recharge, K, and the Curve Number of the US Soil Conservation Service [48].

In this study, the surface hydrology model (UTHBAL) has been used for the estimation of the monthly surface runoff and the monthly groundwater recharge. The UTHBAL model has been applied as a semidistributed model simulating the surface hydrological processes of six sub-basins of the Almyros basin from October 1961 to September 2018. For the estimation of areal precipitation, the gradient method modified with the Thiessen polygon method were combined [49–51]. The steps followed in the procedure are:

- Multiplication of the precipitation time-series of each station with the respective Thiessen polygon ratio of a sub-basin. The Thiessen areal precipitation, P_{th}, is considered at the mean elevation of the sub-basin.

- The correction of the estimation of the mean areal precipitation is performed with the monthly precipitation gradient of the whole basin. The reduction to the mean elevation of the sub-basin, Y_b, from the elevation of each station, Y_{st}, is equal to their difference, dh:

$$dh = \sum (Y_b - Y_{st}) \quad [\text{m}] \tag{1}$$

- The corrected areal precipitation, P_b, attributed to the mean elevation of each sub-basin is given by the equation:

$$P_b = P_{th} + \beta \cdot (Y_b - Y_{st}) \quad [\text{mm}] \tag{2}$$

The estimation of the mean monthly temperature was estimated with the gradient method and the potential evapotranspiration was calculated with the Thorthwaite method [52]. The UTHBAL model has been successfully applied, calibrated, and validated in various Mediterranean regions, like Crete, Cyprus, Nestos/Mesta Basin, Thessaly [1], and in the neighboring areas of Pinios River Basin and Karla Basin [16,53].

2.1.2. Groundwater Flow Model Description

MODFLOW is a software application that mathematically resolves the three-dimensional groundwater flow equation in a porous medium [40]. The partial differential equation that describes groundwater flow, Equation (3), results from the application of the equation of conservation of mass and Darcy's law. McDonald and Harbaugh [54] have originally reported that MODFLOW simulates steady and transient water flow by using the finite difference method with a block-centered approach. The model was updated by Harbaugh and McDonald in 2000 and this version became the most used by scientists since.

$$\frac{\partial}{\partial x}\left(K_{xx}\frac{\partial h}{\partial x}\right) + \frac{\partial}{\partial y}\left(K_{yy}\frac{\partial h}{\partial y}\right) + \frac{\partial}{\partial z}\left(K_{zz}\frac{\partial h}{\partial z}\right) + W = S_s\frac{\partial h}{\partial t} \tag{3}$$

where, (i) xx, yy, zz are the axes of the Cartesian system, in which the components of the hydraulic conductivity of the aquifer, are attributed parallel to their positive directions, (ii) h is the hydraulic head, (iii) W is the water flux into or out of the system per time-step of each stress period, (iv) S_s is the specific storage of the hydrogeological formation, and (v) t is the time-step of the stress period simulated.

In a variably structured model grid, the aquifer is divided in layer(s), which can be confined, unconfined, or a mixture of confined and unconfined representing a homogeneous-heterogeneous isotropic, or anisotropic aquifer system. Furthermore, the model layer(s) can be inclined, or not, and have different cell sizes and layer thickness. MODFLOW is capable of simulating water flow originating from sources out of the aquifer model, and from physical causes of groundwater movement; such are the water abstraction flow rates of wells, constant hydraulic head bodies, and groundwater recharge. These are represented mathematically with boundary conditions and incorporate the solution of Equation (1). The main water flow boundary conditions are the hydraulic head along an aquifer's margins (Dirichlet Condition), the hydraulic head gradient across an aquifer's boundaries (Neumann Condition), and the combination of the two (Cauchy Condition). The hydraulic head variations are, then, calculated for each time-step of the simulated stress period [40].

The groundwater hydrology model MODFLOW has been used for the simulation of groundwater flow and the estimation of the water balance of the Almyros aquifer. MODFLOW has been successfully applied in aquifer systems in Greece such as in the Lake Karla aquifer [16,53,55], in Eidomeni-Evzones region in Axios basin [56], in Moudania aquifer [57], on Thira aquifer in Santorini island [58,59], in Glafkos aquifer in Patras Gulf [60], in Cyclades [61], and in Messara aquifer system in Crete [62].

2.1.3. Nitrate Leaching Simulation Model Description

The REPIC model is a new model developed in the context of this research and firstly introduced in this study. The REPIC model is a spatially distributed nitrate leaching and crop growth model, based on the Environmental Policy Integrated Climate (EPIC) model and the R-programming language with the R-ArcGIS Bridge in ArcGIS 10+/Pro. REPIC also integrates the groundwater recharge from the surface hydrology model, UTHBAL, in the estimation of the nitrate leaching in mg/L.

The EPIC model was initially introduced for the simulation of the impacts of soil erosion on soil productivity in small watersheds, for up to 100 ha, by Williams et al. in 1984 [63]. The model was introduced to be the Environmental Policy Integrated Climate model, when it was enriched with functions for many environmental problems. Some of the modules that have been incorporated in the EPIC model, by William et al. in 1995 [41], are the crop growth and cultivation management practices, and the nitrogen and pesticide transport functions [64]. The operation results of crop growth and nitrate leaching, which are closely related to the quantity and quality of water resources, are the ones used in the REPIC model. Particularly, in the EPIC model, the estimation of the nitrate leaching quantity from the soil layers of the unsaturated zone is calculated by the Equation (4) [65]:

$$C_{NO_3} = \frac{V_{NO_3}}{QT} \tag{4}$$

where, C_{NO3} is the average daily concentration in a quantity of water height, QT, and V_{NO3} is the amount of NO_3-N lost from the unsaturated zone towards groundwater.

Similarly, the crop yield of the simulated crop types is calculated by the Equation (5) [65]:

$$YLD_j = HI_j \cdot B_{AG} \tag{5}$$

where, YLD_j is the crop yield of crop j, HI_j is the harvest index of the crop j, and B_{AG} is the above-ground biomass extracted in harvest.

Considering the aforementioned, the spatially distributed modeling of the EPIC model is performed with programming languages that form a Graphical User Interface with the EPIC model. Such programming languages are the Visual Basic and Python, and the R. Previous examples of spatially distributed modeling of the EPIC model are the GEPIC model, a Visual Basic-based EPIC model, and the PEPIC model, a Python-based EPIC model [66].

The REPIC model is an R-ArcGIS based model. The REPIC model simplifies the procedure of the creation of the input data files, because it uses as input data a point shapefile with all the required characteristics of the grid-cell areas. The points are the coordinates of the centroids of the grid cells, their elevation, area, crop type, maximum NIR, nitrate loading, weather file code, and soil file code of the EPIC model. The simulation is performed considering different crop types in every grid cell, as opposed to GEPIC that simulates all grid cells as only one crop type per simulation run. REPIC also solves the restrictions of the GEPIC model for regional simulations, which are: (i) the stepping into the GEPIC's VBA code to change the geographical extent of the simulated area and also to increase the grid's resolution, and (ii) the dependence from the UTIL executable that inputs the data to the EPIC files. Moreover, relevant tools were designated for the manipulation of the input data. These are (i) the EPIC Parameter tool, which changes the values of hydrological parameters, (ii) the NIR tool, which assigns the NIR to each crop type, (iii) the NLD tool, which assigns the nitrate loading per crop type, (iv) the Nitrate Leaching tool, that reads the results, integrates the groundwater recharge from the UTHBAL model, and calculates the nitrate leaching in mg/L, (v) the Crop Yield tool, which calculates the spatial distribution of crop yields, and the (vi) DataForMT3DMS tool, which produces the estimated nitrate leaching in mg/L of all grid cells, on a monthly step, in .txt format ready for direct input into the MT3DMS model. Especially, the flexibility of the R programming language and the type of input file, which is a shapefile, creates the opportunity to downscale to finer resolutions if available data exist, and even to field-scale simulations of hydrological basins.

Spatially distributed modeling of the EPIC model, especially of the GEPIC model, has been successfully applied, calibrated, and validated in global gridded scale for wheat yield [44], maize [67], rice [68], crop water productivity, and drought risk assessment [44] in Sub-Saharan Africa [69], in country scale, in China [70], and in regional scale in Jordan River Basin [71], and in Karla Basin, Greece [53]. The PEPIC model has been also successfully applied in global scale for nitrogen losses [66].

The nitrate leaching/crop growth model REPIC has been used for the simulation of nitrate inflows in the groundwater system of Almyros, and the simulation and validation of crop yields in the Almyros basin. The nitrates leached into the groundwater, calculated with the REPIC model, are then added as input contaminant fluxes into the MT3DMS model.

2.1.4. Nitrate Transport and Dispersion Model Description

MT3DMS is a structural, three-dimensional, multispecies contaminant transport model, which can simulate advection, dispersion, and chemical responses of a groundwater system with dissolved compounds [42]. MT3DMS code is able to simulate pollutant and solute concentrations, while based on a solved problem of groundwater flow, most often provided by MODFLOW. MT3DMS is designed for interaction with any finite difference model, similar to MODFLOW. This linkage is feasible under the premise that concentration variations in space and time have negligible impact on the regional water flow pattern and that both codes share the same structure of the aquifer model [40,54].

MT3DMS solves the three-dimensional transport and dispersion of pollutants in the groundwater with the partial differential Equation (6) [42,72]:

$$\frac{\vartheta\left(\vartheta C^k\right)}{\vartheta t} = \frac{\vartheta}{\vartheta x_i}\left(\theta D_{ij}\frac{\vartheta C^k}{\vartheta x_j}\right) - \frac{\vartheta}{\vartheta x_i}\left(\theta v_i C^k\right) + q_s C_s^k + \sum R_n \tag{6}$$

where, (i) D_{ij} is the hydrodynamic dispersion coefficient tensor, (ii) C^k is the pollutant concentration in the aquifer system, (iii) C_s is the recharge, or outflow, concentration of the pollutant k, (iv) θ is the porosity of the hydrogeological formation, (v) $x_{i,j}$ is the distance paved by the pollutant parallel to a Cartesian axis (here is xx axis), (vi) q_s is the volume of the pollutant's flow rate attributed to each volumetric water flux, of an aquifer's discrete grid-cell, (vii) v_i is the seepage or water velocity, (iix) $\sum R_n$ is the component for the contaminant chemical production for n reactions, and (ix) t is the time-step of the stress period simulated.

The MT3DMS code simulates the flow of pollutants in the groundwater taking into consideration the advection, dispersion, and diffusion, and even chemical reactions. The differential term of Equation (6), $\partial(\theta v_i C^k)/\partial x_i$, expresses the advection of the pollutant. The pollutant concentrations flow along with the transport medium, meaning at the same velocity as the groundwater flow. The hydrodynamic dispersion, D_{ij}, describes the characteristic of the pollutant to spread along the area of its location and cannot be estimated with the groundwater flow [42,73]. Hydrodynamic dispersion is equal to the sum of the mechanical dispersion and molecular diffusion. The mechanical dispersion stretches the pollutant concentrations along the vectors of the velocity deviations of groundwater velocity on the microscale. The molecular diffusion drives the pollutant molecules from areas with higher concentrations to areas with lower concentrations but is considered negligible, unless the groundwater velocity is very small. The parameter of longitudinal dispersivity (α_L) was calculated with the Neuman formula (1990) for water flow distance smaller than 3500 m [74]:

$$\alpha_L = 0.0175 \cdot L^{1.46} \tag{7}$$

where L is the water flow length, in this case, the grid size on the x-axis, and then it was calibrated to the hydraulic conductivity zones. The ratios of the horizontal transverse dispersivity to longitudinal dispersivity and the vertical transverse dispersivity to longitudinal dispersivity are kept in the default values of 1 and 0.1, respectively. The parameter of porosity is uniformly set to 0.3 according to literature for Neogene and Quaternary formations.

MT3DMS also simulates the sources of pollution as pollutant loadings in volume per unit volume of water flux. Moreover, sources of pollution are represented with concentration boundary conditions. Similarly to MODFLOW, these are the concentration along the aquifer's margins (Dirichlet Condition), the concentration gradient across the aquifer's boundaries (Neumann Condition), and the combination of the two (Cauchy Condition).

The groundwater contaminant transport model MT3DMS has been used for the simulation of nitrates transport and dispersion in the Almyros aquifer. MT3DMS has been successfully applied for the simulation of nitrate pollution in the Lake Karla aquifer system in Thessaly in Greece [16,53,55], in Vocha plain in Korinthos [75].

2.1.5. Chloride Solute Transport and Dispersion Model Description

SEAWAT is a finite difference, three-dimensional, modular transport model that simulates the variable density flow of water and solutes in porous aquifers. The concept of solving the variable density flow with the combination of MODFLOW and MT3DMS was first introduced by Guo and Bennett in 1998 [76]. The source code underwent several updates and its final version was developed by Guo and Langevin [43]. The code of SEAWAT merges the codes of MODFLOW and MT3DMS, while maintaining the consistency of the models' structural characteristics and of the assumptions for groundwater flow and contaminant transport, regarding the advection and the hydrodynamic dispersion. The boundary conditions considered in a simulation with SEAWAT are exactly similar to the MT3DMS boundary conditions.

The presence of solutes in the groundwater in low concentrations does not have any effect on the fluid's density, because the mass of the contaminant molecules is negligible. However, when the solute concentrations rise excessively, then their mass and density are increased accordingly and cause the water to move slower than the freshwater. This differentiation of the flow velocity of contaminated with solutes water, and of the freshwater, is studied as variable density flow [43]. Variable density flow is based on the concept of equivalent freshwater head. Equivalent freshwater head of a salinized hydraulic head, is the hydraulic head it would have, if there was not any solute contamination, if the fluid pressure is considered stable in the two states. The Equation (8) represents the dependance of salinized hydraulic head on the different fluid densities and freshwater head [43]:

$$ h = \frac{\rho_f}{\rho} h_f + \frac{\rho - \rho_f}{\rho} Z \tag{8} $$

Initially, the groundwater flow with freshwater head, h_f, density ρ_f, and elevation, Z, is calculated by MODFLOW. The MT3DMS performs an update of the estimation of the fluid density, ρ, based on the solute concentrations. Then, the difference of the updated fluid density and the freshwater density is integrated in MODFLOW, which calculates the final water flow field due to variable fluid density.

Especially, seawater intrusion is a representative problem of variable density flow often simulated by SEAWAT. Due to the extravagant concentration of solutes in seawater, as related to fresh groundwater the fluid densities differ substantially. The fluid density of freshwater is 1000 kg/m^3 and the fluid density of seawater is 1025 kg/m^3.

SEAWAT has been successfully applied in coastal aquifer systems in Greece such as in Santorini island in Thira aquifer [58,59], in Cyclades [61], in Nea Moudania [57,77], in Glafkos aquifer in Gulf of Patras [60].

2.2. Statistical and Graphical Evaluation of the Models

Criteria of the goodness of fit of the simulated values against the observed measurements were incorporated, to evaluate the performance of the models REPIC, MODFLOW,

MT3DMS, and SEAWAT in simulating the annual crop yields and the monthly groundwater flow, the monthly nitrate transport, and dispersion and the monthly chloride solute transport. The normal errors' estimation of the Nash–Sutcliffe [78] Model Efficiency (*Eff*) (Equation (9)), the Coefficient of Determination (R^2) [79] (Equation (10)), and the Index of Agreement (*IA*) [80] (Equation (11)) indicate the fit of the modelled to the observed values of variables.

$$Eff = 1 - \frac{\sum_{i=1}^{n}(y_i - f_i)^2}{\sum_{i=1}^{n}(y_i - \overline{y})^2} \tag{9}$$

$$R^2 = 1 - \frac{SS_{res}}{SS_{tot}} \tag{10}$$

$$IA = 1 - \frac{\sum_{i=1}^{n}(y_i - f_i)^2}{\sum_{i=1}^{n}(|f_i - \overline{y}| + |y_i - \overline{y}|)^2} \tag{11}$$

where (Y, F) represent the simulated and the observed values, SSres the sum of squares of residuals, and SStot the total sum of squares of the data.

The perfect agreement between the simulated and observed variables, all statistical criteria/measures (i.e., *Eff*, R^2, *IA*) take the value of 1. Values of the statistical criteria bellow 0,5 indicate a problematic/bad model and, as the values of the statistical criteria get values closer to 1, the model accurately simulates the observed variables.

Apart from the statistical evaluation of the modelled variables, the modelled and observed values of variables were drawn on maps and visually compared. Moreover, scatterplots were used to compare the modelled and observed values of the variables. The slope and intercept of the regression lines were statistically tested against the slope and intercept of the line of the perfect agreement (1:1 line) were tested using the *t*-test at the 5% significance level ($\alpha = 0.05$) [81].

3. Study Area and Database

3.1. Study Area

The Almyros basin is located in the central region of Greece, Thessaly. The total area of the basin is approximately 856 km^2. The geomorphology of the basin is characterized by the presence of ephemeral streams and the absence of surface water storage bodies. It is the only plain of the coastal Thessaly and consists of six sub-basins, namely, Kazani, Lahanorema, Holorema, Xiria, Platanorema, and Xirorema (Figure 2). The basin consists of about 30% plain areas (elevation lower than 150 m), 57% semi-mountainous areas (elevation 150–800 m), and 13% mountainous areas (elevation higher than 800 m). The aquifer covers the lowest and coastal part of the basin and has an area of 293 km^2 with about 71% of its extent area to be in planar and 29% in hilly terrain. The Almyros basin area is delineated by the Chalkodonion or Mavrovouni mountains in the north and the Othrys Mountain of the Pindus Mountain Range in the west, while the coastline forms the eastern boundary of the Pagasitikos Gulf [82].

The climate of the basin is semi-arid Mediterranean climate with hot and dry summers and cold and wet winters. Mean annual precipitation, for the meteorological station in N.Aghialos, the only station located in the basin (Figure 2), is about 491 mm with a standard deviation of 111 mm, and mean annual temperature is 16.5 °C, with a standard deviation of 0.6 °C. The mean annual precipitation is distributed in time by 12.3% in October, 11.8% in November, 13.4% in December, 9.5% in January, 9.7% in February, 10.3% in March, 6.7% in April, 7.7% in May, 4.5% in June, 3.9% in July, 3.3% in August, and 6.9% in September. The monthly mean temperature varies from the mean annual temperature by 3.3% in October, −26.4% in November, −50.5% in December, −58.9% in January, −51.9% in February, −36.6% in March, −11.2% in April, 20.0% in May, 51.4% in June, 65.0% in July, 60.4% in August, and 35.4% in September.

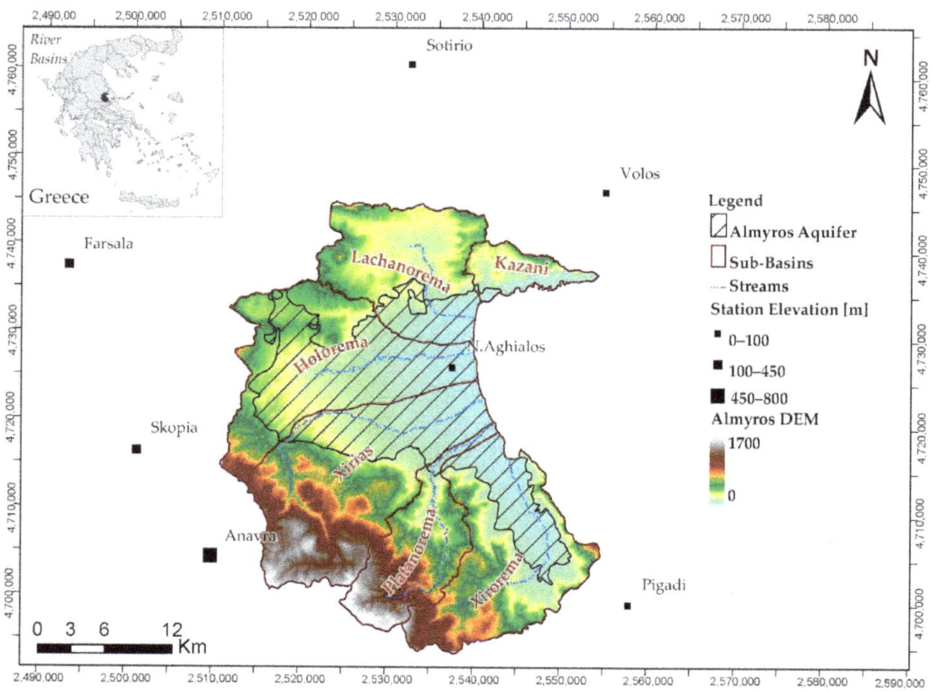

Figure 2. Almyros aquifer system, elevation and streams of sub-basins, and locations of weather stations.

The basin includes an intensive irrigated agricultural area of about 205 km^2. The agricultural area mostly covers the area of the aquifer. The main crops cultivated are alfalfa, cereals, cotton, maize, trees, olive groves, vegetables, vineyards, and wheat. The main land uses and the cultivated corps are presented in Table 1. In the Almyros basin, there is no surface storage project for the storage and use of surface water and all irrigation water demands and urban water supply (and other water uses) are covered by groundwater pumping.

Table 1. Percent land use area irrigated by the Almyros aquifer.

Main Land Use/Crop	2010 (% Area)	2018 (% Area)	Irrigation Return Flow Coefficient
Alfalfa	7.74	16.83	0.15
Cereals	10.33	25.16	0.15
Cotton	8.55	8.40	0.20
Maize	2.55	1.62	0.35
Olives	10.86	12.91	0.13
Trees	1.34	2.36	0.13
Vegetables	1.62	6.56	0.24
Vineyards	2.02	2.46	0.13
Wheat	32.75	9.92	0.19

The main soils types in the study basin and, especially, the area of the aquifer are clay loam, clay, and silt loam (Table 2). Most of the materials are alluvials deposited in the middle and low elevation areas of the basin by the streams and torrents of the area. The hydrological soil group of type B (medium low runoff potential when saturated), based on data from the previous study of Thessaly [1], covers the 254 km^2 of the overlaying aquifer area, while soil groups A (low runoff potential when saturated), C (medium high runoff potential when saturated), and D (high runoff potential when saturated) occupy 12.3 km^2, 7.3 km^2, and 19.1 km^2, respectively.

Table 2. Areal percentage coverage of soil types of the Almyros basin according to the U.S. Department of Agriculture (U.S.D.A.) classification.

Soil Texture Class	% Area
Sandy Loam	1.1
Loam	10.4
Silt Loam	21.4
Sandy Clay Loam	2.6
Clay Loam	35.2
Silty Clay Loam	6.5
Sandy Clay	0.2
Silty Clay	1.3
Clay	21.3
Sandy Loam	1.1

The coastal low-land area of the basin mostly comprises of sandy permeable materials with clay lenses. Clay layers and the intercalations of clay, sand, gravel with volcanic rocks and conglomerates, form low permeability structures in the western area and higher elevation areas of the basin. In the southern part of the basin, small areas of limestone form kasts, which are direct communication with the sea [82]. The geological and hydrogeological setting of the study area and the relative data are presented in the Section 3.2.4 of the paper and in Figure 3.

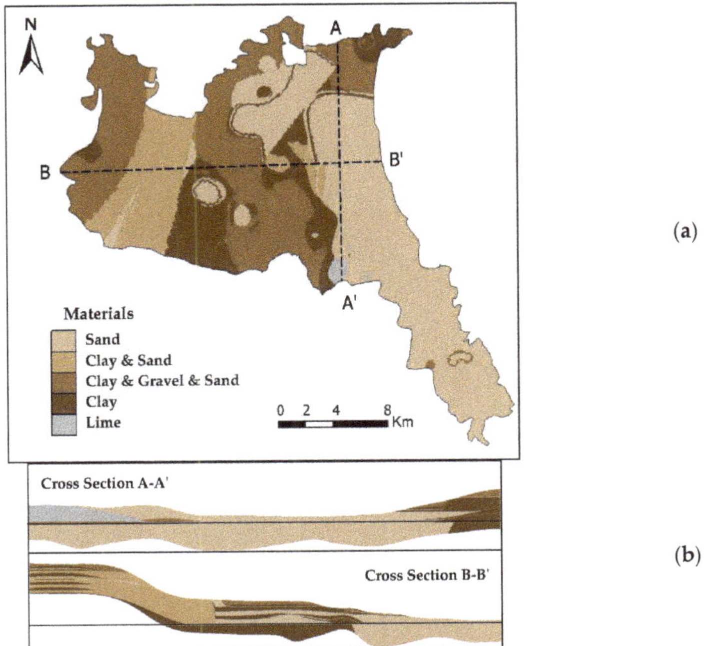

Figure 3. (**a**) Geological Top View and (**b**) Section A-A' (vertical) and Section B-B' (horizontal), of the permeable formations of the Almyros aquifer system.

3.2. Database

The simulation of water resources requires databases of a wide range of measured variables that span over many years of monitoring for calibration and validation purposes of the models. Climatic variables (e.g., precipitation and temperature), water table measurements of wells, and groundwater nitrates and chloride concentrations were also required and used in the analysis. Land use and crop type data were used to estimate the crop irrigation demands, number, and groundwater well abstractions.

3.2.1. Meteorological Data

Monthly precipitation and temperature collected in six (6) meteorological stations distributed in and around the basin were used in this study. The locations of the stations are depicted in Figure 2. The range of the elevation of the stations is among 3 m and 850 m. Daily and monthly precipitation data, daily and monthly minimum, maximum and mean temperature data were available for the station of N.Aghialos, the only station located in the basin, from 1961 to 2018. The meteorological data are collected by various governmental organizations and agencies. They have been pre-processed and validated.

3.2.2. Land Use

The agricultural land use occupies almost 70% of the aquifer area for the historical period. The spatial distribution of the cultivated fields across the aquifer counties was provided for the years 2010 and 2018, by the Greek Payment Authority of Common Agricultural Policy (C.A.P.) Aid Schemes (OPEKEPE). The crop data were grouped into nine main crop categories. Thus, the main crops cultivated in the aquifer area are alfalfa, cereals, cotton, maize, trees, olive groves, vegetables, vineyards, and wheat. The irrigation return flow was estimated with irrigation return coefficients as measured in fields with similar soil characteristics and climatic conditions, for every main crop category [83–85]. The distribution of the main crop categories for the years 2010 and 2018, and the respective irrigation return flow coefficients are shown in Table 1.

Moreover, crop yield data are publicly available by the Greek Payment Authority of Common Agricultural Policy (C.A.P.) Aid Schemes (OPEKEPE). The crop yield data span from 2000 to 2018 and refer to annual measured crop yields for various crop types.

3.2.3. Soil Characteristics of Unsaturated Zone

Soil physical and hydrological parameters of saturated conductivity, bulk density, soil water content at the wilting point, and field capacity, organic carbon concentration, exchangeable K concentration, electrical conductivity, and initial water storage were provided from European Soil Data Centre (ESDAC) [86,87]. The sand, clay, and silt content, and coarse fragments were estimated from point observation data provided by (OPEKEPE) and National Agricultural Research Foundation (NAGREF). According to the U.S. Department of Agriculture (U.S.D.A.) soil texture classification [88] the soils overlaying the Almyros aquifer are classified and presented in Table 2.

3.2.4. Geology and Hydrogeological Setting and Data

The coastal low-land area consists mostly of sandy permeable materials with clay lenses towards the western part of the basin, following the topographical elevation change. In the western and high elevation areas of the basin and the aquifer, the presence of a low permeability clay layers and the intercalations of clay, sand, gravel with volcanic rocks and conglomerates, form low permeability structures within the granular aquifer. Limestone is present at small areas of the southern part of the basin and the aquifer area, forms karsts, which are in direct communication with the sea and do not interact with the aquifer [82]. No significant hydraulic communication has been observed and documented for the northern, western, and southern part of the aquifer with the neighbouring aquifer systems.

Borehole data of 55 wells in the area of Almyros drilled between 1968 and 1990, from the Greek Ministry of Agriculture, were used to produce the stratigraphy of the aquifer. The main geological materials of the Almyros aquifer are classified into five categories, namely clay (Neogene), clay-gravel-sand (Neogene), sand (Quaternary), clay-sand (Neogene), and limestone and form the aquifer of the study basin. The top view and the two most representative cross-sections of the geological materials of the aquifer are presented in Figure 3.

3.2.5. Observation Data of Water Table, Nitrate Concentrations, and Chloride Concentrations

Well observation data regarding the water table, the nitrates concentrations, and the chloride concentrations were mostly performed and provided by the Institute of Geology and Mineral Exploration (IGME), the Regional Government of Thessaly, and the Magnesia Prefecture. The measurements span from 1991 to 2015. Additional nitrates concentrations measurements and chloride observation measurements for the period of 2013 to 2015 were performed in the Almyros aquifer, in a previous research study [82]. The locations of water table observation wells are illustrated in Figure 4a. Known locations of pumping wells were extracted by regional well maps and the National Register of Water Abstraction Points from Surface and Underground Water Bodies [89]. The distribution of pumping wells is presented in Figure 4b.

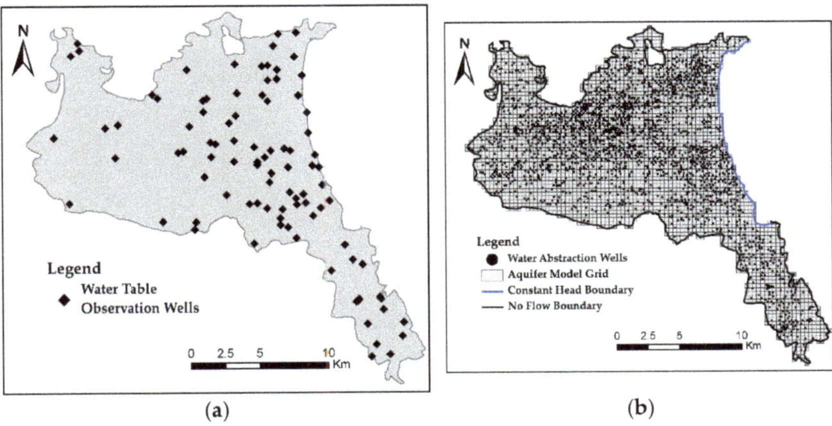

Figure 4. (**a**) Locations of water table observation wells and (**b**) locations of groundwater abstraction wells.

The distribution of wells of observed nitrates concentrations and chloride concentrations are presented in Figure 5a,b, respectively.

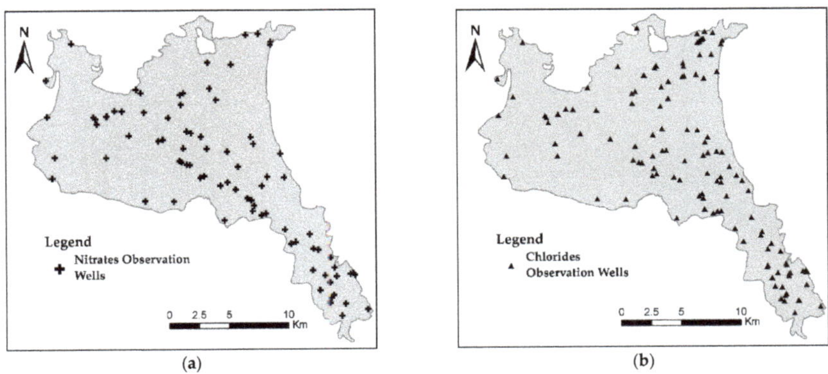

Figure 5. (**a**) Locations of nitrates observation wells and (**b**) locations of chlorides observation wells.

4. Results

4.1. Mean Areal Precipitation and Temperature

The daily meteorological data of the station in N.Aghialos were aggregated to produce monthly time-series of the variables of precipitation and temperature. The precipitation and temperature missing data of the surrounding stations were infilled using linear regression based on the N. Aghialos station, the only station in the region with no missing values. The coefficients and correlation of the stations used for the estimation of precipitation and temperature are shown in Tables 3 and 4.

Table 3. Linear regression coefficients of the stations for the precipitation variable.

Precipitation Station	Slope	Intercept	R^2	R
N. Aghialos	1.00	1.00	1.00	1.00
Anavra	0.93	15.49	0.74	0.86
Skopia	0.96	6.71	0.69	0.83
Volos	0.99	0.00	0.90	0.95
Pigadi	0.90	10.43	0.65	0.80

Table 4. Linear regression coefficients of the stations for the temperature variable.

Temperature Station	Slope	Intercept	R^2	R
N. Aghialos	1.00	1.00	1.00	1.00
Pigadi	0.84	2.94	0.16	0.40
Volos	0.93	2.12	0.29	0.54
Skopia	0.97	−1.10	0.37	0.61
Sotirio	0.11	1.00	0.40	0.63
Farsala	0.83	−0.86	0.42	0.64

For the estimation of areal precipitation, the gradient method modified with the Thiessen polygon method was used. The gradient for the annual precipitation to 100 m of elevation change in the Almyros basin is 15.9 mm (R^2 = 0.70), while the gradient for the annual temperature is −0.43 °C per 100 m increase in elevation for the whole basin (R^2 = 0.71). The spatial distribution of mean annual precipitation and temperature is presented in Figure 6.

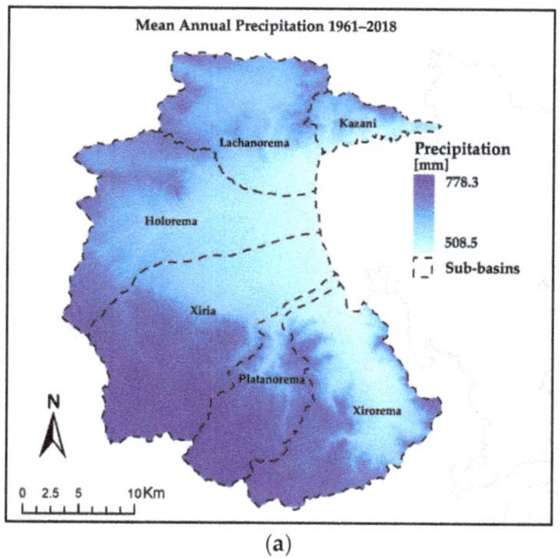

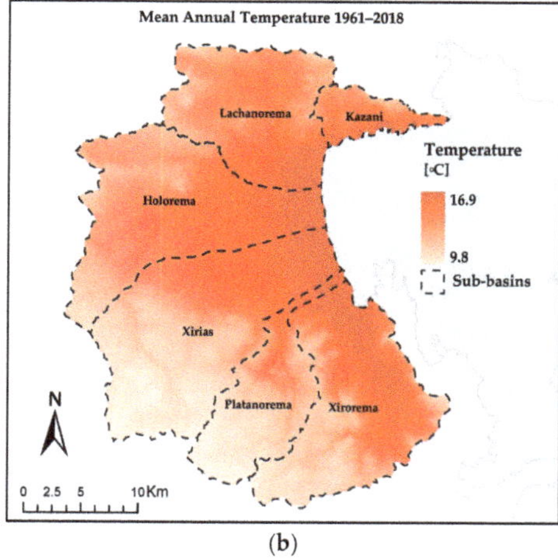

Figure 6. Spatial distribution of (**a**) mean annual precipitation and (**b**) mean annual temperature for the years of 1961–2018.

The monthly mean areal precipitation was estimated with the use of the gradient method and the Thiessen polygon method per-sub-basin. The mean annual precipitation of the Almyros basin has an average of 566 mm, a median value of 540 mm, and a standard deviation of 107.16 mm. The driest year with the least annual precipitation that exceeds more than 40% of the interannual average is the hydrological year of October 2006 to September 2007 with 355.9 mm annual precipitation. The wettest years with more than 40% exceedance from the average value are the hydrological years of 1968–1969, 1981–1982, 2002–2003, and 2017–2018. The monthly mean areal temperature was estimated with the gradient method. The mean annual temperature of the Almyros basin has an average of 15.0 °C, and a standard deviation of 0.64 °C. The hottest monthly areal temperatures were noted in July 1988 with 28.1 °C, in August 2010 with 28.6 °C, and in July 2012 with 28.8 °C at the mean elevation of the Almyros basin.

Mean annual areal precipitation for the simulation period of October 1991 to September 2018, is estimated at 561 mm and mean annual areal temperature is estimated at 15.41 °C. The monthly averages of areal precipitation and temperature are presented in Figures 7 and 8.

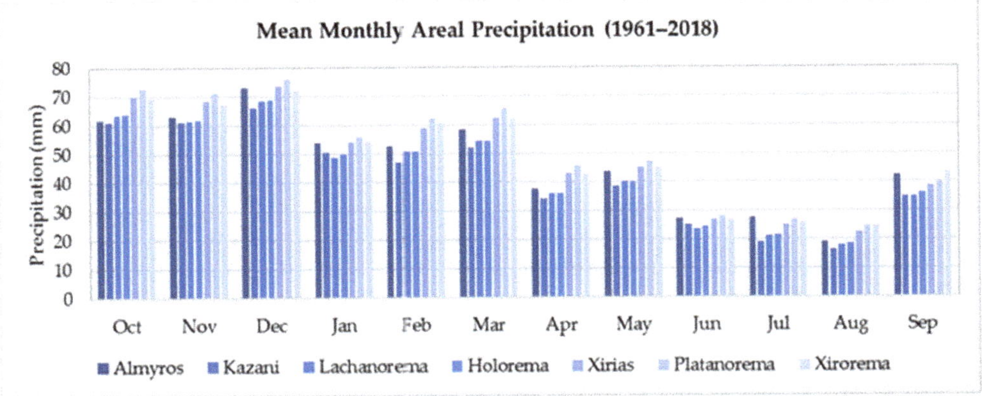

Figure 7. Mean monthly areal precipitation at the mean elevation of the Almyros basin and the sub-basins.

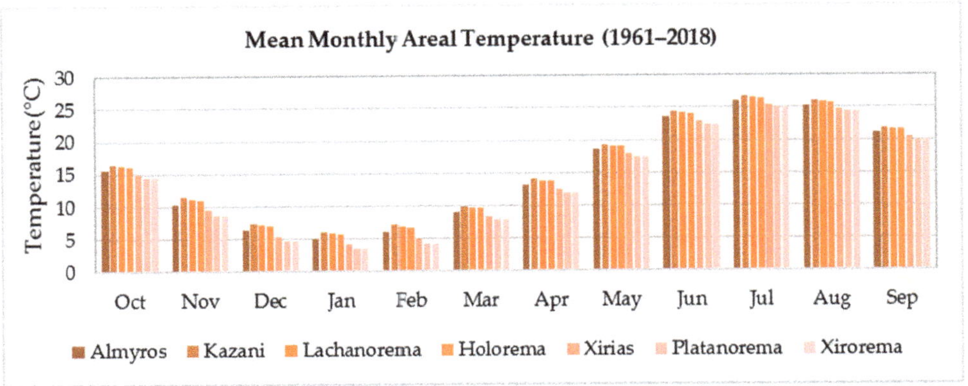

Figure 8. Mean monthly areal temperature at the mean elevation of the Almyros basin and the sub-basins.

4.2. Surface Hydrology-Groundwater Recharge

The UTHBAL model simulated the monthly surface hydrologic balance of the Almyros basin from October 1960 to September 2018. It was not possible to calibrate the model in the sub-basins of the Almyros basin because there are no streamflow measurements of the Almyros ephemeral stream discharge. For these reasons, most of the parameters of the model were taken the values found in a regional analysis of the model in Thessaly [28]. The values of the model used in the study were: the parameter of monthly melt rate factor, $Cm= 6$ mm/°C, the coefficient of actual evapotranspiration, $\alpha= 0.48$, the coefficient of interflow $\beta = 0.033$, the coefficient of baseflow, $\gamma= 0.203$, and the coefficient of groundwater recharge, $K = 0.68$. The parameter of the Curve Number, of the US Soil Conservation Service [48] was estimated for each sub-basin with the HEC-GeoHMS tool in ArcGIS [90], based on soil type (A, B, C, D) maps, Corine Land Cover uses, and the Digital Elevation Model of the area. The weighted average Curve Number per sub-basin is shown in Table 5.

Table 5. Curve Number of the Almyros basin and its sub-basins.

Basin	CN
Almyros	61.43
Kazani	67.93
Lahanorema	68.11
Holorema	68.47
Xirias	60.69
Platanorema	51.07
Xirorema	53.84

The mean annual surface runoff is 113.49 mm and the median annual surface runoff is 106.11 mm, found for the year 1999–2000. The wettest year with the largest cumulative annual runoff is 1962–1963, and the driest year with the least cumulative annual runoff is 2004–2005. The mean annual groundwater recharge is 54.1 mm for the whole basin. The wettest year is 1962–1963 with 214.4 mm. The median of the simulated years is noted in 2003–2004 with 43.2 mm of recharge, while the driest year is encountered in 2004–2005, with 0 mm of groundwater recharge. Interannual statistics of the simulated runoff, recharge, and precipitation to the input water of the region are depicted in Table 6.

Table 6. Mean annual statistics of precipitation (P_b), surface runoff (Q_c), and groundwater recharge (R_g) per sub-basin for the period of October 1961 to September 2018.

Sub-Basin	P_b [mm]	Q_c [mm]	R_g [mm]	Q_c/P_b	R_g/Q_c	R_g/P_b
Kazani	507.9	97.5	56.3	19.2%	57.7%	11.09%
Lachanorema	522.8	103.6	62.5	19.8%	60.4%	11.96%
Holorema	527.9	105.5	65.1	20.0%	61.7%	12.33%
Xirias	590.4	125.4	63.5	21.2%	50.7%	10.76%
Platanorema	617.8	127.46	34.2	20.6%	26.8%	5.54%
Xirorema	596.6	112.4	31.9	18.8%	28.4%	5.35%

4.3. Ground Water Flow

The MODFLOW model simulated the groundwater recharge, the well abstractions, the change of the storage of the aquifer, and the sea fluxes to the coastline, in a monthly transient mode, for the period of October 1991 to September 2018. The water table observations of 75 wells were used for the definition of the starting heads of the aquifer in October 1991, as well as, for the calibration and the validation of the model.

The model discretization forms a one-layer rectangular grid of 200 rows and 200 columns, with a cell size of approximately 150 m $\times$ 150 m and consists of 40,000 cells with 12,464 of them being active. Unconfined conditions were considered for the simulation of the aquifer with the Layer Property Flow package. The coastline at the eastern part of the aquifer was considered as a Constant-Head-Boundary, at zero m above sea level. In the western part, the boundary condition was set a No-Flow Boundary, according to the imperviousness of the adjacent geological formations at the margins of the aquifer. The hydraulic conductivity was simulated in zones, according to the hydrogeological characteristics of geological formations. In particular, the number of simulated water abstraction wells was estimated at 2072 wells, 28 of which are urban water supply wells and 2044 are irrigation wells. The estimated pumping rates were based on the water demands of crops and the water losses of the local irrigation private-owned systems. The crop water demands were estimated with the Near Irrigation Requirement (NIR) method [91] and are distributed in the sub-basins, on average, by 9% in Kazani, 20% in Lahanorema, 21% in Holorema, 19% in Xirias, 17% in Platanorema, and 14% in Xirorema. The averaged water losses of the irrigation systems are equal to 41% of the crop water demands, according to [28]. Measured pumping rates per county were also compared against the estimated pumping rates.

The model was calibrated for the period October 1991 to September 2009 and validated for the period October 2013 to September 2015. Calibration was performed using PEST for the coastal and central part of the aquifer, based on available hydraulic conductivity measurements of several boreholes, mostly located in the coastal region. The values of horizontal anisotropy, specific yield, specific storage, and the upland hydraulic conductivity were kept at the values set by the previous simulation of the Almyros groundwater flow [28]. Sensitivity analysis indicated that the most sensitive hydraulic conductivities of the hydrogeological formations are located in the Xirias and Xirorema sub-basin. Specifically, the sensitive regions are (i) in a very small area of hydraulic conductivity of 0.05 m/day, consisting of marl and lignite compounds, (ii) along the southern semi-mountainous boundary of Neogene formations with 1.0 m/d, (iii) the upper part of Xirias sub-basin consisting of calcareous conglomerates with 0.8 m/d, and (iv) the eastern along-side Neogene formations of Xirorema sub-basin with 1.2 m/d. Groundwater flows from the western higher part of the aquifer towards the low elevation eastern coastal region and higher velocities are encountered in the central Holorema and Xirias sub-basin following the distribution of hydraulic conductivity. The values of hydraulic conductivity range between 0.1–18.7 m/day, with a spatial average value of 2.3 m/d with the highest value encountered in the north-eastern part of the aquifer near the coastline. The calibration results for the groundwater model indicate that the Nash–Sutcliffe model efficiency, the Pearson correlation, and the Index of Agreement are, on average, for the simulation period 0.986, 0.989, and 0.996, respectively. Summary statistics for the calibration and the validation period are depicted in Table 7.

Table 7. Summary efficiency statistics of the MODFLOW model.

MODFLOW	Calibration Average 1991–2009	Validation Average 2013–2015
Eff	0.975	0.997
R^2	0.981	0.997
IA	0.993	0.999

Simulated contours of the water table of equal potential against the respectively observed contours are presented for July 1992, September 2006, and June 2015 in Figures 9–11. The simulated water table at the end of the simulation period in September 2018 is presented in Figure 12. The simulated water table contours are almost identical to the observed water table contours. Additionally, scatterplots of the simulated water head of wells against their observed water head have been plotted and the slope and intercept of the regression lines against the slope of the line of perfect agreement (1:1 line has been tested using the *t*-test).

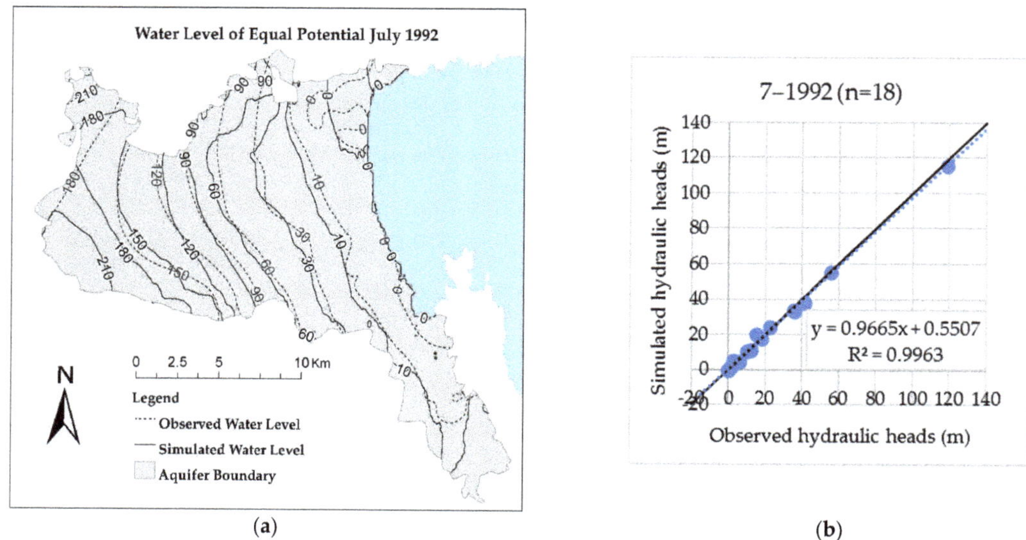

Figure 9. (**a**) The water level of equal potential for July 1992 and (**b**) scatterplot of simulated against observed water table values.

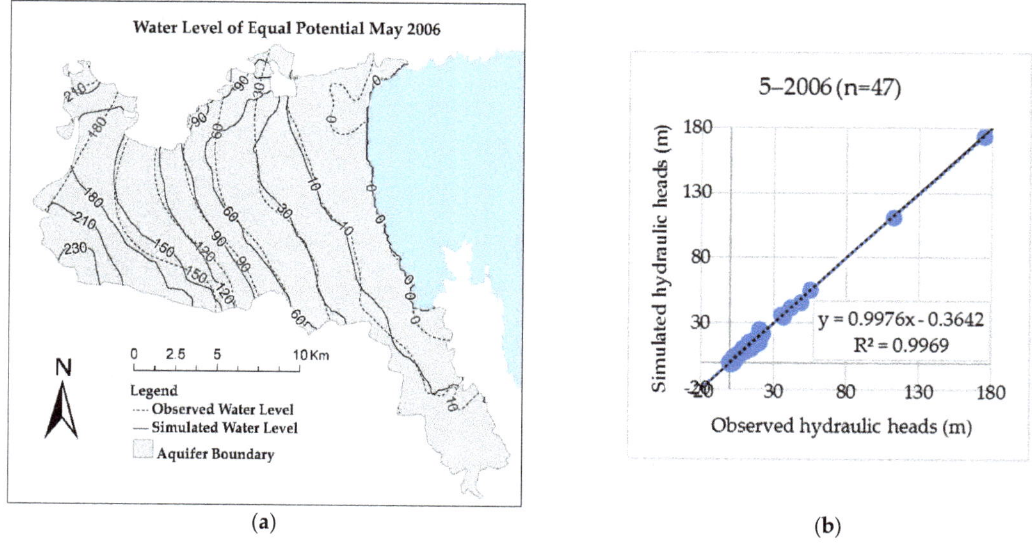

Figure 10. (**a**) The water level of equal potential for May 2006 and (**b**) scatterplot of simulated against observed water table values.

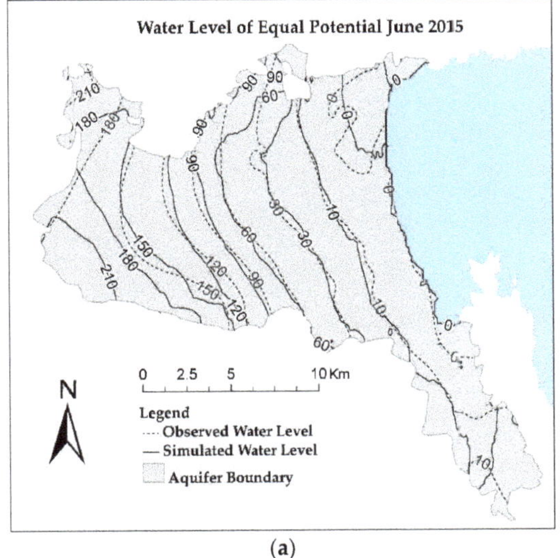

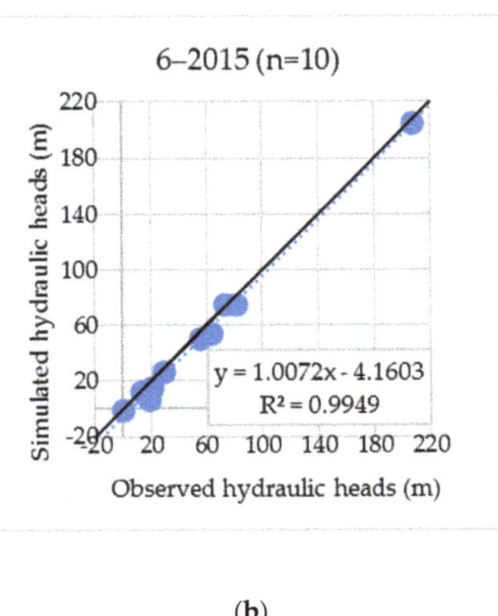

(a)   (b)

Figure 11. (a) The water level of equal potential for June 2015 and (b) scatterplot of simulated against observed water table values.

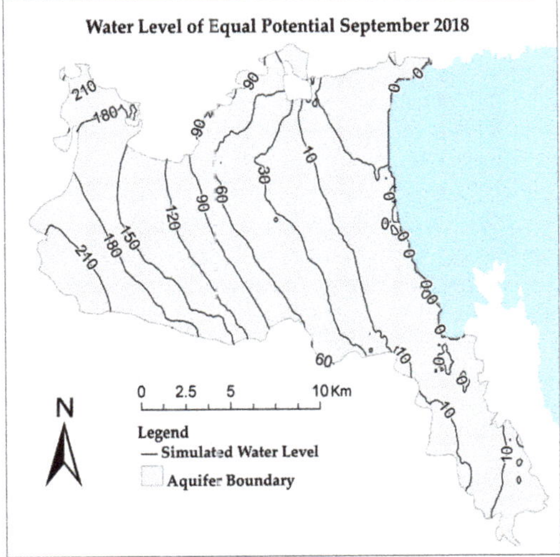

Figure 12. The water level of equal potential at the end of groundwater simulation in September 2018.

For July 1992, two-sided *t*-test was performed for 18 water table wells. The slope and the intercept do not differ significantly from the line of perfect agreement (1:1 line) at $\alpha = 0.05$ significance level.

For May 2006, two-sided *t*-test was performed for 47 water table wells. The slope and the intercept do not differ significantly from the line of perfect agreement (1:1 line) at $\alpha = 0.05$ significance level. For June 2015, two-sided *t*-test was performed for 10 water table wells. The slope and the intercept do not differ significantly from the line of perfect agreement (1:1 line) at $\alpha = 0.05$ significance level.

4.4. Nitrate Leaching Simulation

The discretization of the REPIC forms a grid of rectangular cells of approximately 300 m × 300 m and it consists of 3215 cells with one REPIC cell equal to four MT3DMS cells. The nitrates leaching into the Almyros aquifer from the fertilization practices [92] for crop growth were simulated for the stress period of October 1991 until September 2018 with the REPIC model. The simulation took place for four land use periods 1990–2000, 2001–2006, 2007–2012, and 2013–2018 for the main crop types of Almyros. The crop water requirements and fertilizer application are presented in Table 8.

Table 8. Maximum annual Near Irrigation Requirement (NIR) and typical nitrogen fertilizer loading.

Crop	NIR [mm]	N_{Fer} [Kg/ha]
Alfalfa	893	30
Cereals	336	100
Cotton	409	140
Maize	389	325
Olives	515	125
Trees	515	175
Vegetables	271	150
Vineyards	297	125
Wheat	336	160

Since there are no nitrate leaching observations of the unsaturated zone of the Almyros basin, the crop growth parameters were calibrated and validated against crop yield data for the periods of 2007–2012 and 2013–2018, respectively. Moreover, the model's recharge is calculated with a stochastic estimation of the Curve Number and the water balance parameters of the model were calibrated against the sum of recharge, as calculated by UTHBAL, and irrigation return flow. The calibration and validation procedures were performed with R-script in R-studio for each sub-basin for groundwater recharge, and, similarly, for the Almyros basin as a whole, for crop yields. Firstly, vectors of parameters were defined, secondly, a matrix of all their possible combinations was constructed, and then the model was run iteratively, while the code estimated the Nash–Sutcliffe efficiency and R-squared between the simulated and observed crop yields and groundwater recharge values. The parameters that resulted in the best statistical efficiency were considered the appropriate values of the REPIC model for the Almyros basin. Column diagrams of crop yields per crop type combined with the respective scatterplot of simulated crop yields against observed crop yields for the calibration period 2007–2012 are shown in Figure 13 and for the validation period in Figure 14. The slopes and the intercepts of the regression lines of the crop yield scatterplots do not differ significantly from the line of perfect agreement (1:1 line) at $\alpha = 0.05$ significance level using the two-sided t-test. The average values of statistical measures of efficiency, Nash–Sutcliffe, and R-squared, for the calibration and validation of the simulated crop yields, are shown in Table 9. The distributed nitrates leaching maps for the years 2010 and 2018 are depicted in Figures 15 and 16, respectively.

Table 9. Summary efficiency values of simulated against observed crop yields.

Crop	Calibration Average 2007–2012	Validation Average 2013–2018
Eff	0.98	0.92
R^2	0.99	0.96
IA	0.99	0.99

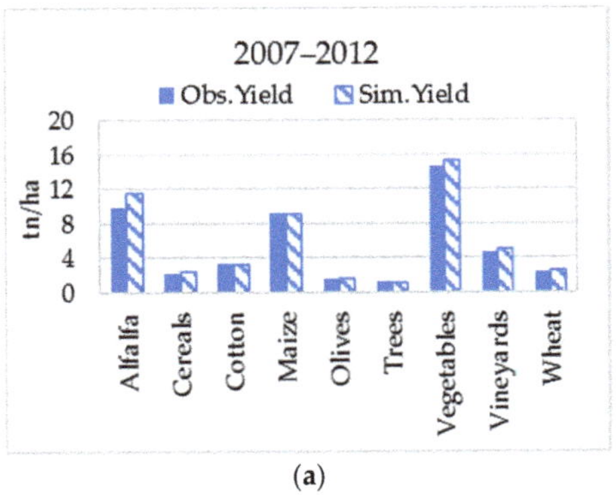

(**a**)

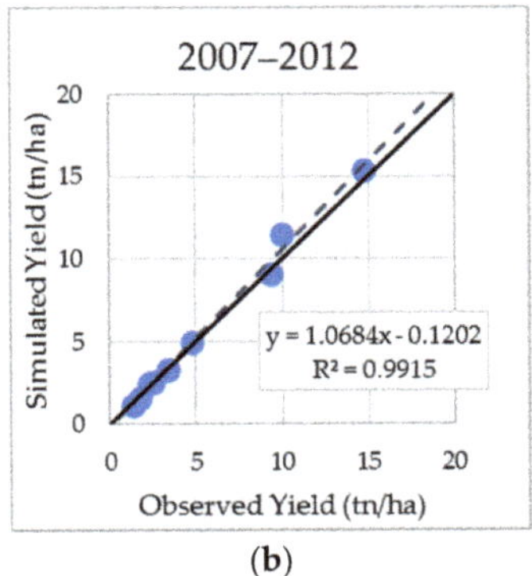

(**b**)

Figure 13. (**a**) Average simulated crop yield per crop type, and (**b**) spatially simulated crop yields against spatially observed crop yields for the calibration period 2007–2012.

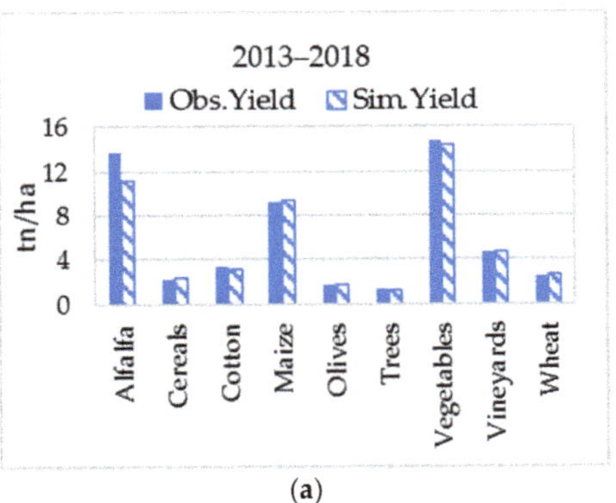

(**a**)

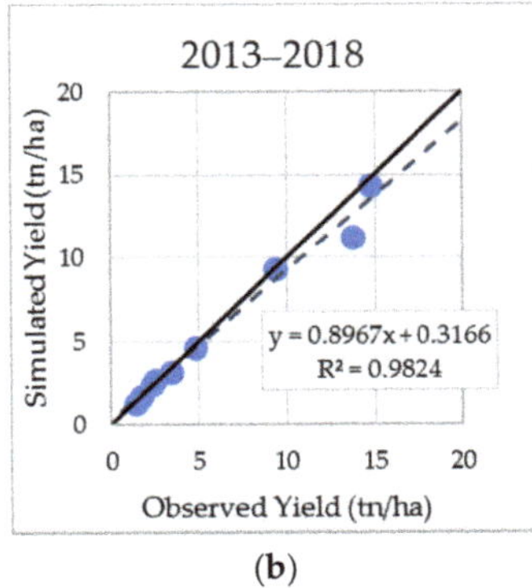

(**b**)

Figure 14. (**a**) Average simulated crop yield per crop type, and (**b**) spatially simulated crop yields against spatially observed crop yields for the validation period 2013–2018.

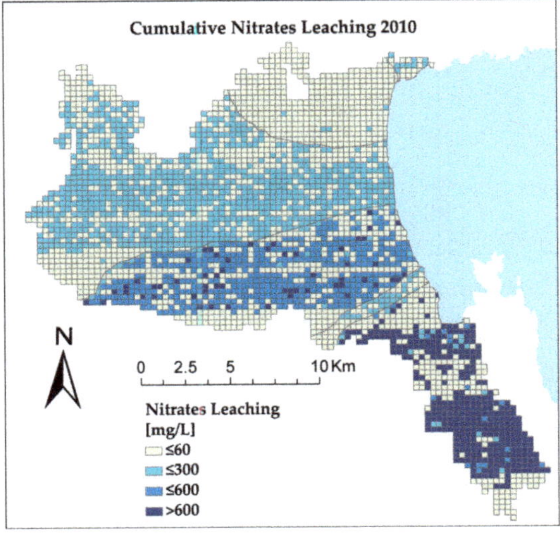

Figure 15. Simulated cumulative nitrates leached into the aquifer for the year of 2010.

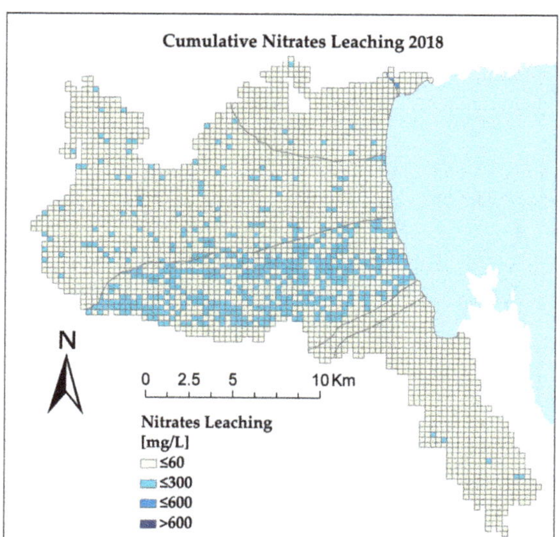

Figure 16. Simulated cumulative nitrates leached into the aquifer for the year of 2018.

4.5. Nitrate Transport and Dispersion

The fluxes of nitrates concentration in the Almyros aquifer were simulated in a monthly transient mode for the stress period of October 1991 until September 2018. The transient advection and dispersion of nitrates that flow with the groundwater were simulated with the MT3DMS model. The model was run in the Almyros aquifer for the stress periods from October 1991 until September 2018.

The MT3DMS model was calibrated from October 1992 to September 2004 and validated from October 2013 to September 2015. The calibration procedure was performed with trial-and-error for the specification of the starting concentrations and several hydrogeological zones of longitudinal dispersivity. The range of the longitudinal dispersivity values is 0.07–30 m with a spatial average of 3.5 m. The calibration results for the MT3DMS model indicate that the Nash–Sutcliffe model efficiency, the Pearson correlation, and the Index of Agreement are, on average, 0.81, 0.91, and 0.95, respectively. Summary statistics for the calibration and the validation period of the MT3DMS model are depicted in Table 10.

Table 10. Summary efficiency statistics of the MT3DMS model.

MT3DMS	Calibration Average 1992–2004	Validation Average 2013–1015
Eff	0.80	0.82
R^2	0.87	0.96
IA	0.95	0.95

Because of the intense hydraulic gradient of the western part of the aquifer, and even though the hydraulic conductivity is smaller than the coastal area, the nitrates are washed away with the groundwater towards the sea. The nitrate contamination that is observed in the aquifer is attributed to the agricultural fertilizers applied on the crops, and more specifically the spatial persistence of the pollution of the lower altitudes and hydraulic gradients follows inversely the magnitudes of the hydraulic conductivity zones.

The nitrates concentrations show a narrowing trend in the northern Lachanorema and Holorema sub-basin boundary in the winter of 2003, which is less discrete yet also evident in the rest of the aquifer. The central and central-coastal parts of the Almyros aquifer retain high nitrate concentrations for the simulation period 1991–2018. Nonetheless, the simulation results also indicate that the central Almyros aquifer has the potential to wash away the nitrate pollution, as it slowly does through the flow pathways, even in the hydrogeological clayish formations, but in the closed Xirorema sub-basin, the flow pattern of the region and the fertilizer applications impede the nitrates to fall more than 4 mg/L during the years 1991–2018.

Simulated isonitrate contours against the observed nitrate concentrations are presented for October 1992, September 2004, and March 2013 in Figures 17–19. Additionally, scatterplots of the simulated nitrates concentrations of wells against their observed values indicate the validity of the results. The slopes and the intercepts of the regression lines of the nitrate concentrations scatterplots do not differ significantly from the line of perfect agreement (1:1 line) at $\alpha = 0.05$ significance level using the two-sided *t*-test. The simulated nitrate pollution at the end of the simulation period in September 2018 is presented in Figure 20.

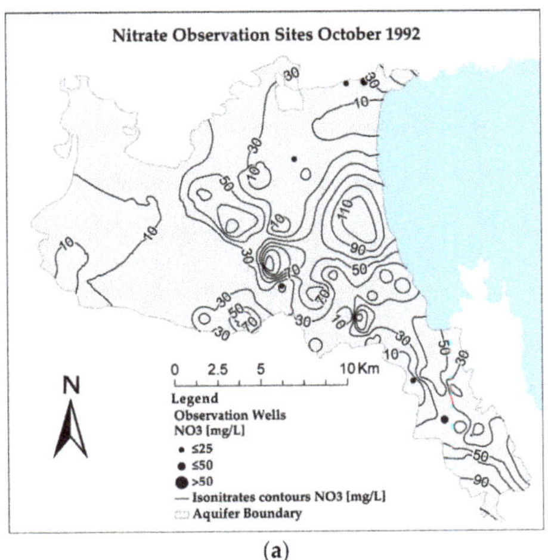

(a)

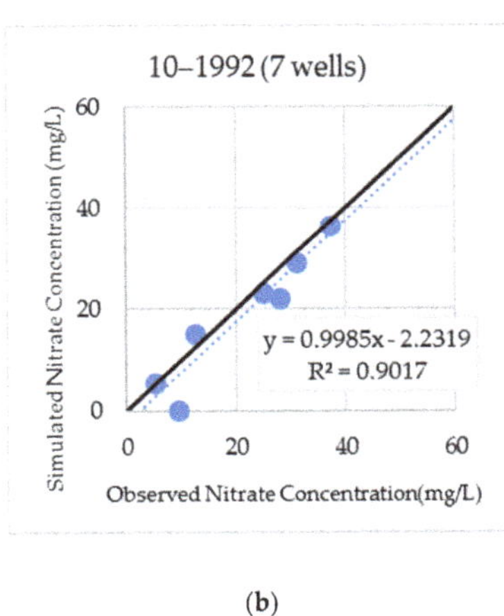

(b)

Figure 17. (a) Isonitrates contours for October 1992 and (b) scatterplot of simulated against observed nitrates concentration values.

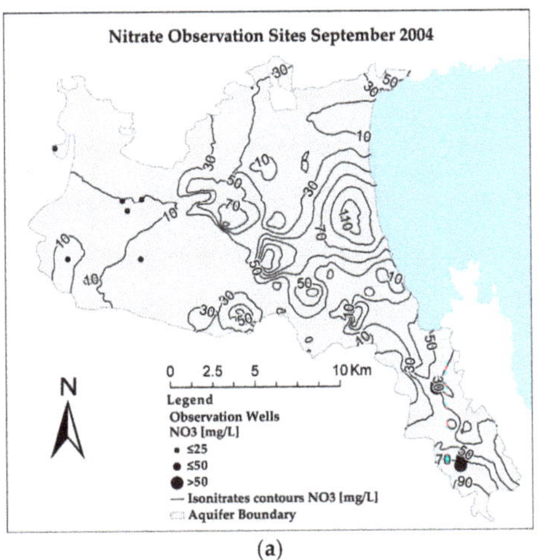

(a)

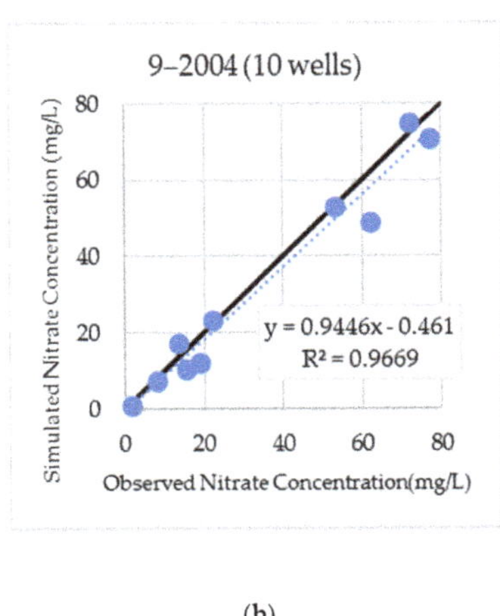

(b)

Figure 18. (a) Isonitrates contours for September 2004 and (b) scatterplot of simulated against observed nitrates concentration values.

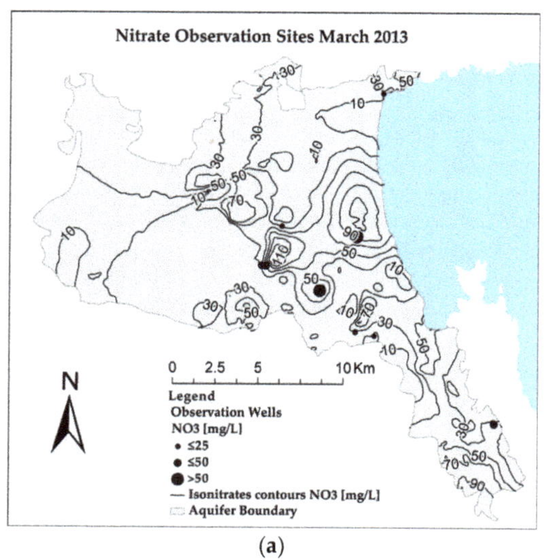

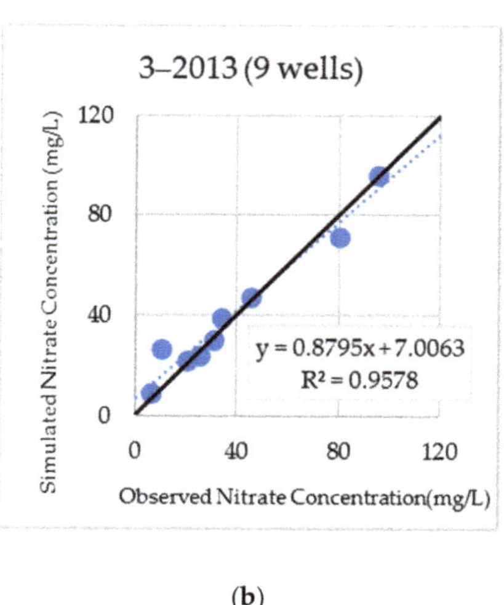

Figure 19. (**a**) Isonitrates contours for March 2013 and (**b**) scatterplot of simulated against observed nitrates concentration values.

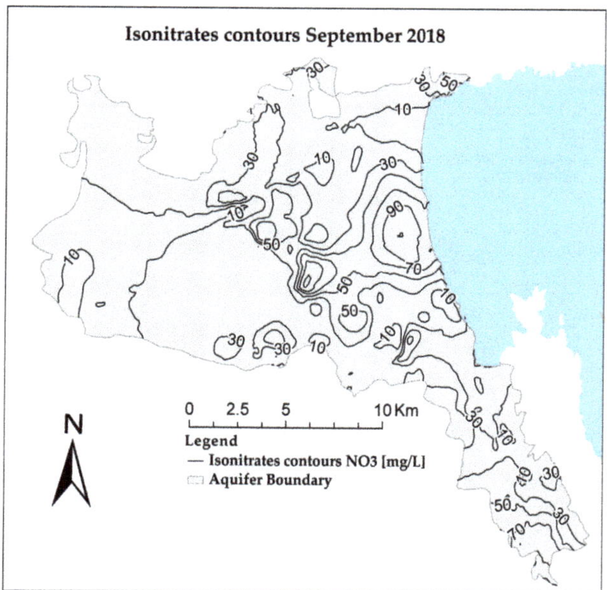

Figure 20. Isonitrates contours at the end of the simulation of nitrate pollution in September 2018.

4.6. Chloride Solute Transport and Dispersion

The solute transport of chlorides in the Almyros aquifer was simulated the SEAWAT model in a monthly transient mode for the stress periods of October 1991 until September 2018. The model was run in the Almyros aquifer for the stress periods from October 1991 until September 2018, in the variable density mode, without taking under consideration viscosity and thermal effects. The parameter of longitudinal dispersivity (aL) was set previously in the MT3DMS model. The parameter of the effective molecular diffusion coefficient, which expresses the reactivity of the pollutant, for chlorides, as them being very conservative anions, was set at the value 10^{-10} [93]. The reference fluid density for the freshwater is 1000 kg/m^3. The slope of density with the chloride concentration was estimated during the calibration/validation process of the model. The chloride concentration of the seawater for the Almyros coast, at the Constant-Head-Boundary that represents the sea, was set at 20,000 mg/L based on measurements of the study for the salinity of the Pagasitikos Gulf by [94].

The model was calibrated from October 1991 to September 2004 and validated for the period October 2005 to September 2007. The calibration procedure was performed with trial-and-error for the estimation of starting concentrations and of the slope of fluid density with the chloride concentrations. The slope of density with the chloride concentration was calibrated and validated at the value of 0.7143×10^{-6}, for the simulation period considering the seawater intrusion with the variable density flow package of SEAWAT. The calibration results for the SEAWAT model indicate that the Nash–Sutcliffe model efficiency, the R-squared, and the Index of Agreement are, on average, 0.905, 0.945, and 0.980, respectively. Summary statistics for the calibration and the validation period of the SEAWAT model are shown in Table 11.

Table 11. Statistical measures of the model's efficiency for the calibration and validation periods.

SEAWAT	Calibration Average 1991–2004	Validation Average 2005–2007
Eff	0.92	0.89
R^2	0.94	0.95
IA	0.98	0.98

The highest concentrations are generally observed in the northern part of the aquifer for the whole simulation period. Moreover, salinization occurs in the Platanorema and Xirorema basins in the south, especially because of the low altitude and the presence of lime formations close to the boundaries of the sub-basins. The seawater intrusion that is observed in the aquifer is attributed to the groundwater abstractions to address the crop water needs, the low altitude, the variability of the hydraulic conductivity of the coastal area, and the proximity to the sea. The seawater intrusion is exacerbated during the simulation years 1991–2018 on the northern coastline. Moreover, the groundwater in the northern sub-basins degraded from freshwater to brackish water, with an accelerating pace evident in 2001–2002, 2005–2006, and 2007–2008, while the rest of the simulation showed an accelerated trend but relatively stable variations of the chlorides' concentrations trend. The coastal zone, at the proximities of 150 m and 300 m from the shore, in Kazani and Lachanorema sub-basins, is characterized by seawater of more than 12,300 mg/L and 4000 mg/L, respectively, at the end of 2018. Simulated isochlorides contours against observed chloride concentrations are presented for July 1992, and April 2007 in Figures 21 and 22. Additionally, scatterplots of the simulated chloride concentrations against observed chloride concentrations indicate the validity of the simulation. The slopes and the intercepts of the regression lines of the chloride concentrations scatterplots do not differ significantly from the line of perfect agreement (1:1 line) at $\alpha = 0.05$ significance level using the two-sided t-test.

The simulated chloride concentrations at the end of the simulation period in September 2018 in the north coastline surpass the upper pan-European limit of 12,300 mg/L. The simulated seawater intrusion in September 2018 is presented in Figure 23.

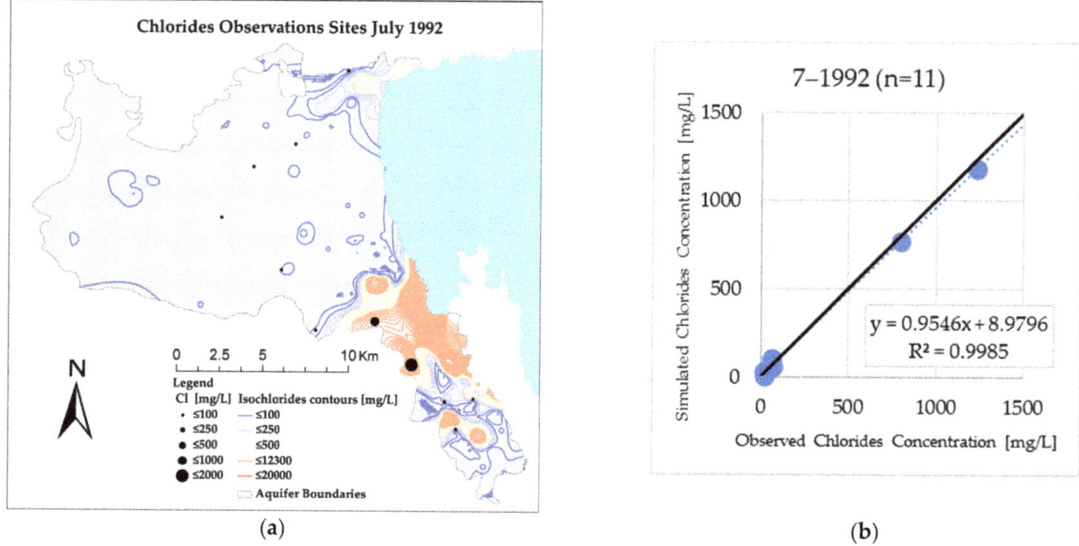

(**a**)

(**b**)

Figure 21. (**a**) Isochlorides contours for July 1992 and (**b**) scatterplot of simulated against observed chloride concentration values.

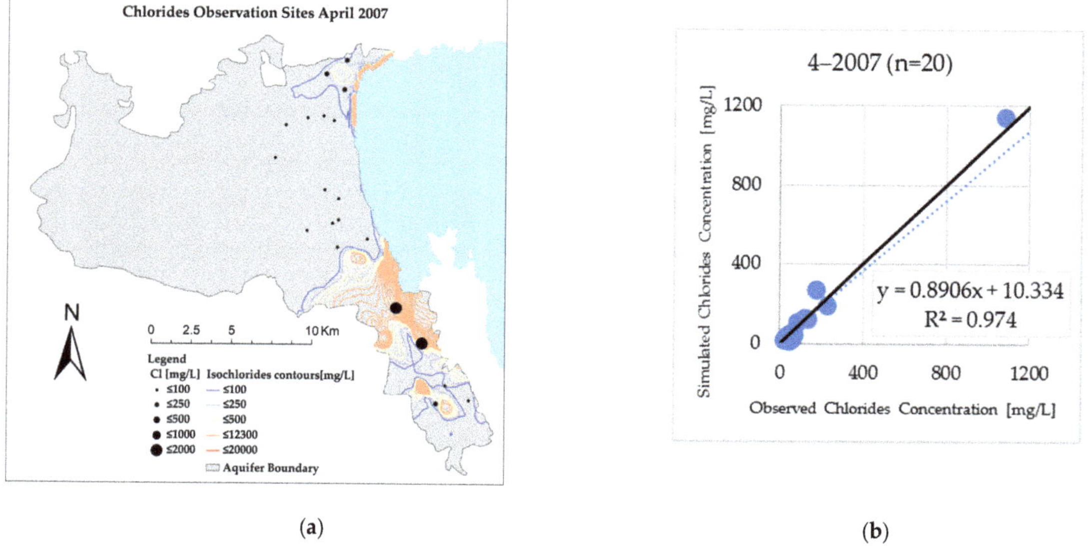

(**a**)

(**b**)

Figure 22. (**a**) Isochlorides contours for April 2007 and (**b**) scatterplot of simulated against observed chloride concentration values.

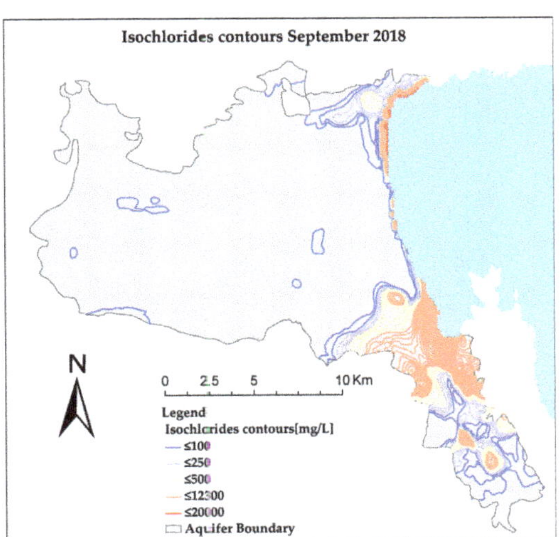

Figure 23. Isochlorides contours at the end of the simulation of seawater intrusion in September 2018.

5. Discussion

Complex problems of the quantity and quality of water resources of coastal agricultural watersheds, and especially of the groundwater regarding the water table depletion, the nitrate contamination, and the seawater intrusion, are encountered in the semi-arid Mediterranean coastal region [1–18,23,28,29,32,39]. The developed integrated modeling system consisting of coupled and interrelated models of surface and groundwater hydrology, crop growth/nitrate leaching, and groundwater contaminant transport and its application in the Almyros basin, emphasize the significance of the integrated modeling for the study and the effective and accurate analysis of the spatiotemporal patterns of groundwater flow, nitrate pollution originated by fertilizer practices, and seawater intrusion [32]. The calibrated results of the Integrated Modeling System validate the effective implementation of the developed crop growth/nitrate leaching model (REPIC) for the simulation of crop yields and nitrate leaching in grid and basin/watershed scale. The R-ArcGIS based EPIC model (REPIC), along with the data handling tools and the optimization procedures, establish an advanced intrinsic connection with the surface hydrology model (UTHBAL) and the contaminant transport/aquifer pollution model (MT3DMS).

The application of the integrated modeling system in the Almyros basin proves that the modeling system is able to reproduce, in a holistic way [18–20,32], the observed water quantity and quality variables of the groundwater resources in the study area. Notably, in all t-tests [81], the slope and the intercept do not differ significantly from the hypothetical line of absolute agreement (at $\alpha = 0.05$ significance level). The scores of all statistical measures of modeling efficiency are characterized by an excellent fit of the simulation parameters against the observed measurements [78–80] and validate the calibrated parameter values of the hydrological and the crop variables.

Hence, it is safe to estimate the water fluxes through the years of the simulation, taking into account the impacts of the variable density flow due to seawater intrusion on the aquifer balance [43]. The calculation of the water balance accounts for the recharge due to precipitation and irrigation return flows, the water abstractions due to agricultural and urban demands, and the seawater fluxes along the coastline boundary. The Almyros aquifer is recharged with 18.8 hm^3/yr on average, while the water abstractions reach up to 29.7 hm^3/yr resulting in a drop-down of the groundwater table and seawater intrusion of 0.3 hm^3/yr but with adverse effects on the groundwater quality. The average water balance of the aquifer is presented in Figure 24. The water deficit is, on average, 12.02 hm^3/yr. The aquifer's annual water balance and the cumulative water deficit for the simulation period are presented in Figure 25. The water quantity pumped out of the aquifer is much larger than the quantity that is naturally replenished into the aquifer through the recharge and the irrigation return flows. Even though there are three years of positive water balance, the time-series of the aquifer's water balance show an increasing trend of water deficit.

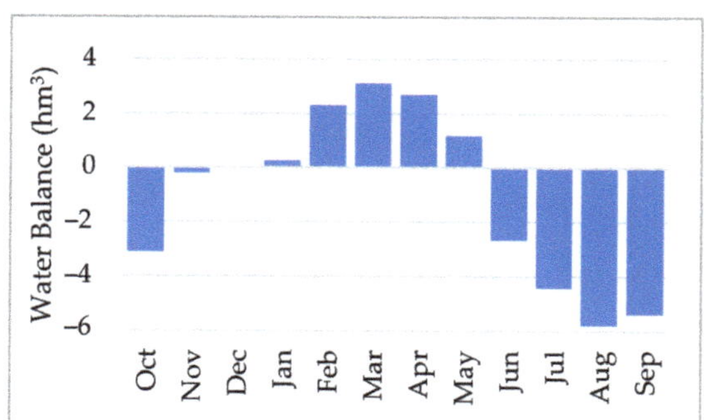

Figure 24. Average monthly water balance of the Almyros basin aquifer for the simulation years 1991–2018.

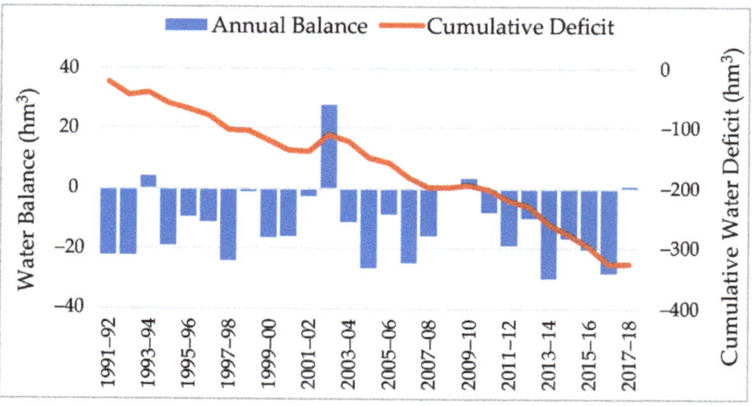

Figure 25. Annual water balance and cumulative water deficit for the simulation years 1991–2018.

6. Conclusions

An Integrated Modeling System has been developed and presented for the evaluation of the quantity and quality of water resources, mainly of the groundwater resources, in coastal agricultural watersheds to address the need for integrated approaches in the simulation of water resources at appropriate spatiotemporal scales. The modeling system consists of coupled models of surface hydrology (UTHBAL), groundwater hydrology (MODFLOW), crop growth/nitrates leaching (REPIC), contaminant transport/nitrate pollution (MT3DMS), and seawater intrusion/aquifer salinization (SEAWAT). The modeling system holistically estimates and reproduces the monthly water balance and the groundwater pollution by nitrates and chlorides in coastal watersheds and aquifers in grid and watershed/basin scale. The accuracy of the simulated groundwater flow, nitrate pollution, and seawater intrusion are evaluated and validated with statistical measures of modeling efficiency in the coastal agricultural Almyros basin.

The results indicate that the causes of the negative water balance and drop-down of the water table are the limited use of surface water, the groundwater abstractions for agricultural use, and the cultivation of water demanding crops. Groundwater nitrate pollution is attributed to nitrate leaching due to the fertilizer applications for maximizing crop production, the geological formations with low hydraulic conductivity that withhold the washing the nitrates out of the Almyros basin aquifer, and the morphology and water flow regime of the aquifer. The seawater intrusion observed in the Almyros aquifer is caused by the increased crop water demands during the dry season and the intensive irrigation, and the granular and sandy geological character of the Almyros aquifer, especially near the shoreline.

Overall, the results of the application of the Integrated Modeling System prove that the modeling system is capable of simulating effectively complex water resources with increased accuracy. The modeling system may be used to simulate and evaluate various water resources management scenarios and strategies to overcome the water quantity and quality problems of coastal agricultural watersheds and the impacts of climate change on surface water and groundwater resources. The developed Integrated Modeling System may help to develop sustainable water resources management and agricultural practices.

Author Contributions: Conceptualization, methodology, supervision, writing-review and editing, A.L. (Athanasios Loukas); methodology, software, writing—original draft preparation, investigation, data curation, formal analysis, A.L. (Aikaterini Lyra); data curation, writing—original draft preparation, G.T.; data curation, writing—original draft preparation, P.S.; writing-review and editing, N.M. All authors have read and agreed to the published version of the manuscript.

Funding: This research is co-financed by Greece and the European Union (European Social Fund—ESF) through the Operational Programme "Human Resources Development, Education and Lifelong Learning" in the context of the project "Strengthening Human Resources Research Potential via Doctorate Research" (MIS-5000432), implemented by the State Scholarships Foundation (IKY).

Institutional Review Board Statement: Not applicable.

Informed Consent Statement: Not applicable.

Data Availability Statement: The data (observations/measurements) used in this paper were taken after application from various European and Greek National organizations. These organizations have been cited in the paper and are responsible for the handling and providing the data. The results of this study are freely available.

Acknowledgments: The Hellenic National Meteorological Service for the provision of meteorological data of the stations N. Aghialos and Volos, K. Kagkadis for the provision of the meteorological data of his station in Pigadi, M. Spiliotopoulos for helping with weather station file format.

Conflicts of Interest: The authors declare no conflict of interest.

References

1. Loukas, A.; Mylopoulos, N.; Vasiliades, L. A Modeling System for the Evaluation of Water Resources Management Strategies in Thessaly, Greece. *Water Resour. Manag.* **2007**, *21*, 1673–1702. [CrossRef]
2. Daskalaki, P.; Voudouris, K. Groundwater quality of porous aquifers in Greece: A synoptic review. *Environ. Geol.* **2008**, *54*, 505–513. [CrossRef]
3. Ghiglieri, G.; Barbieri, G.; Vernier, A.; Carletti, A.; Demurtas, N.; Pinna, R.; Pittalis, D. Potential risks of nitrate pollution in aquifers from agricultural practices in the Nurra region, northwestern Sardinia, Italy. *J. Hydrol.* **2009**, *379*, 339–350. [CrossRef]
4. Colombani, N.; Mastrocicco, M.; Prommer, H.; Sbarbati, C.; Petitta, M. Fate of arsenic, phosphate and ammonium plumes in a coastal aquifer affected by saltwater intrusion. *J. Contam. Hydrol.* **2015**, *179*, 116–131. [CrossRef] [PubMed]
5. Potot, C.; Féraud, G.; Schärer, U.; Barats, A.; Durrieu, G.; Le Poupon, C.; Travi, Y.; Simler, R. Groundwater and river baseline quality using major, trace elements, organic carbon and Sr–Pb–O isotopes in a Mediterranean catchment: The case of the Lower Var Valley (south-eastern France). *J. Hydrol.* **2012**, *472–473*, 126–147. [CrossRef]
6. Pulido-Velazquez, M.; Peña-Haro, S.; García-Prats, A.; Mocholi-Almudever, A.F.; Henriquez-Dole, L.; Macian-Sorribes, H.; Lopez-Nicolas, A. Integrated assessment of the impact of climate and land use changes on groundwater quantity and quality in the Mancha Oriental system (Spain). *Hydrol. Earth Syst. Sci.* **2015**, *19*, 1677–1693. [CrossRef]
7. Giménez-Forcada, E. Use of the Hydrochemical Facies Diagram (HFE-D) for the evaluation of salinization by seawater intrusion in the coastal Oropesa Plain: Comparative analysis with the coastal Vinaroz Plain, Spain. *HydroResearch* **2019**, 76–84. [CrossRef]
8. Rosário Cameira, M.D.; Rolim, J.; Valente, F.; Mesquita, M.; Dragosits, U.; Cordovil, C.M. Translating the agricultural N surplus hazard into groundwater pollution risk: Implications for effectiveness of mitigation measures in nitrate vulnerable zones. *Agric. Ecosyst. Environ.* **2021**. [CrossRef]

9. Telahigue, F.; Mejri, H.; Mansouri, B.; Souid, F.; Agoubi, B.; Chahlaoui, A.; Kharroubi, A. Assessing seawater intrusion in arid and semi-arid Mediterranean coastal aquifers using geochemical approaches. *Phys. Chem. Earthparts A/B/C* **2020**, *115*, 102811. [CrossRef]

10. Mohammad, A.H.; Jung, H.C.; Odeh, T.; Bhuiyan, C.; Hussein, H. Understanding the impact of droughts in the Yarmouk Basin, Jordan: Monitoring droughts through meteorological and hydrological drought indices. *Arab. J. Geosci.* **2018**, *11*, 103. [CrossRef]

11. Rachid, G.; Alameddine, I.; El-Fadel, M. SWOT risk analysis towards sustainable aquifer management along the Eastern Mediterranean. *J. Environ. Manag.* **2021**, *279*, 111760. [CrossRef] [PubMed]

12. Cobaner, M.; Yurtal, R.; Dogan, A.; Motz, L.H. Three dimensional simulation of seawater intrusion in coastal aquifers: A case study in the Goksu Deltaic Plain. *J. Hydrol.* **2012**, *464–465*, 262–280. [CrossRef]

13. EASAC. *Groundwater in the Southern Member States of the European Union: An Assessment of Current Knowledge and Future Prospects*; Country Report for Greece; European Academies Science Advisory Council: Harley, Germany, 2010; pp. 18–19. Available online: https://easac.eu/ (accessed on 21 January 2021).

14. European Commission. *Verification of Vulnerable Zones Identified Under the Nitrate Directive, Greece*; European Commission: Luxembourg, 2003; Available online: https://ec.europa.eu/ (accessed on 21 January 2021).

15. NCESD. *Greece, State of the Environment Report, Summary*; National Center of Environment and Sustainable Development: Athens, Greece, 2018; pp. 70–71. ISBN 978-960-99033-3-2.

16. Sidiropoulos, P.; Tziatzios, G.; Vasiliades, L.; Papaioannou, G.; Mylopoulos, N.; Loukas, A. Modelling flow and nitrate transport in an over-exploited aquifer of rural basin using an integrated system: The case of Lake Karla watershed. *Proceedings* **2018**, *2*, 667. [CrossRef]

17. EU. *Report from the Commission: In Accordance with Article 3.7 of the Groundwater Directive 2006/118/EC on the Establishment of Groundwater Threshold Values*; European Commission: Brussels, Belgium, 2010.

18. Madani, K.; Mariño, M.A. System dynamics analysis for managing Iran's Zayandeh-Rud river basin. *Water Resour. Manag.* **2009**, *23*, 2163–2187. [CrossRef]

19. Mirchi, A. System Dynamics Modeling as a Quantitative-Qualitative Framework for Sustainable Water Resources Management: Insights for Water Quality Policy in the Great Lakes Region. Master's Thesis, Michigan Technological University, Horton, MI, USA, 2013. [CrossRef]

20. Zomorodian, M.; Lai, S.H.; Homayounfar, M.; Ibrahim, S.; Fatemi, S.E.; El-Shafie, A.J.J.o.e.m. The state-of-the-art system dynamics application in integrated water resources modeling. *J. Environ. Manag.* **2018**, *227*, 294–304. [CrossRef]

21. Medici, G.; Baják, P.; West, L.J.; Chapman, P.J.; Banwart, S.A. DOC and nitrate fluxes from farmland; impact on a dolostone aquifer KCZ. *J. Hydrol.* **2020**, 125658. [CrossRef]

22. Barthel, R.; Banzhaf, S. Groundwater and Surface Water Interaction at the Regional-scale—A Review with Focus on Regional Integrated Models. *Water Resour. Manag.* **2016**, *30*, 1–32. [CrossRef]

23. Milano, M.; Ruelland, D.; Fernandez, S.; Dezetter, A.; Fabre, J.; Servat, E.; Fritsch, J.-M.; Ardoin-Bardin, S.; Thivet, G. Current state of Mediterranean water resources and future trends under climatic and anthropogenic changes. *Hydrol. Sci. J.* **2013**, *58*, 498–518. [CrossRef]

24. Feng, D.; Zheng, Y.; Mao, Y.; Zhang, A.; Wu, B.; Li, J.; Tian, Y.; Wu, X. An integrated hydrological modeling approach for detection and attribution of climatic and human impacts on coastal water resources. *J. Hydrol.* **2018**, *557*, 305–320. [CrossRef]

25. Riad, P.; Graefe, S.; Hussein, H.; Buerkert, A. Landscape transformation processes in two large and two small cities in Egypt and Jordan over the last five decades using remote sensing data. *Landsc. Urban Plan.* **2020**, *197*, 103766. [CrossRef]

26. Bobba, A.G. Ground Water-Surface Water Interface (GWSWI) Modeling: Recent Advances and Future Challenges. *Water Resour. Manag.* **2012**, *26*, 4105–4131. [CrossRef]

27. Mylopoulos, N.; Kolokytha, E.; Loukas, A.; Mylopoulos, Y. Agricultural and water resources development in Thessaly, Greece in the framework of new European Union policies. *Int. J. River Basin Manag.* **2009**, *7*, 73–89. [CrossRef]

28. Sidiropoulos, P.; Loukas, A.; Georgiadou, I. Response of a degraded coastal aquifer to water resources management. *Eur. Water* **2016**, *55*, 67–77.

29. Daliakopoulos, I.N.; Tsanis, I.K.; Koutroulis, A.; Kourgialas, N.N.; Varouchakis, A.E.; Karatzas, G.P.; Ritsema, C.J. The threat of soil salinity: A European scale review. *Sci. Total Environ.* **2016**, *573*, 727–739. [CrossRef] [PubMed]

30. Ketabchi, H.; Mahmoodzadeh, D.; Ataie-Ashtiani, B.; Simmons, C.T. Sea-level rise impacts on seawater intrusion in coastal aquifers: Review and integration. *J. Hydrol.* **2016**, *535*, 235–255. [CrossRef]

31. Barthel, R.; Reichenau, T.G.; Krimly, T.; Dabbert, S.; Schneider, K.; Mauser, W. Integrated Modeling of Global Change Impacts on Agriculture and Groundwater Resources. *Water Resour. Manag.* **2012**, *26*, 1929–1951. [CrossRef]

32. Le Page, M.; Fakir, Y.; Aouissi, J. Chapter 7—Modeling for integrated water resources management in the Mediterranean region. In *Water Resources in the Mediterranean Region*; Zribi, M., Brocca, L., Tramblay, Y., Molle, F., Eds.; Elsevier: Amsterdam, The Netherlands, 2020; pp. 157–190. [CrossRef]

33. Burek, P.; Satoh, Y.; Kahil, T.; Tang, T.; Greve, P.; Smilovic, M.; Guillaumot, L.; Zhao, F.; Wada, Y. Development of the Community Water Model (CWatM v1.04)—A high-resolution hydrological model for global and regional assessment of integrated water resources management. *Geosci. Model Dev.* **2020**, *13*, 3267–3298. [CrossRef]

34. Odeh, T.; Mohammad, A.H.; Hussein, H.; Ismail, M.; Almomani, T. Over-pumping of groundwater in Irbid governorate, northern Jordan: A conceptual model to analyze the effects of urbanization and agricultural activities on groundwater levels and salinity. *Environ. Earth Sci.* **2019**, *78*, 40. [CrossRef]

35. Medici, G.; West, L.J.; Chapman, P.J.; Banwart, S.A. Prediction of contaminant transport in fractured carbonate aquifer types: A case study of the Permian Magnesian Limestone Group (NE England, UK). *Environ. Sci. Pollut. Res.* **2019**, *26*, 24863–24884. [CrossRef]

36. Wang, K.; Davies, E.G.R.; Liu, J. Integrated water resources management and modeling: A case study of Bow river basin, Canada. *J. Clean. Prod.* **2019**, *240*, 118242. [CrossRef]

37. Rossetto, R.; De Filippis, G.; Borsi, I.; Foglia, L.; Cannata, M.; Criollo, R.; Vázquez-Suñé, E. Integrating free and open source tools and distributed modelling codes in GIS environment for data-based groundwater management. *Environ. Model. Softw.* **2018**, *107*, 210–230. [CrossRef]

38. Wang, L.; Stuart, M.E.; Lewis, M.A.; Ward, R.S.; Skirvin, D.; Naden, P.S.; Collins, A.L.; Ascott, M.J. The changing trend in nitrate concentrations in major aquifers due to historical nitrate loading from agricultural land across England and Wales from 1925 to 2150. *Sci. Total Environ.* **2016**, *542*, 694–705. [CrossRef] [PubMed]

39. Ragab, R.; Bromley, J.; D'Agostino, D.R.; Lamaddalena, N.; Luizzi, G.T.; Dörflinger, G.; Katsikides, S.; Montenegro, S.; Montenegro, A. Water Resources Management Under Possible Future Climate and Land Use Changes: The Application of the Integrated Hydrological Modelling System, IHMS. In *Integrated Water Resources Management in the Mediterranean Region: Dialogue towards New Strategy*; Choukr-Allah, R., Ragab, R., Rodriguez-Clemente, R., Eds.; Springer: Dordrecht, The Netherlands, 2012; pp. 69–90. [CrossRef]

40. Harbaugh, A.W.; McDonald, M.G. *User's Documentation for MODFLOW-2000, an Update to the U.S. Geological Survey Modular Finite-Difference Ground-Water Flow Model*; United States Government Printing Office: Washington, DC, USA, 2000.

41. Williams, J.R. The EPIC Model. In *Computer Models of Watershed Hydrology*; Singh, V.P., Ed.; Water Resources Publisher: Highlands Ranch, CO, USA, 1995; pp. 909–1000.

42. Zheng, C.; Wang, P.P. *MT3DMS: A Modular Three-Dimensional Multi-Species Transport Model for Simulation of Advection, Dispersion and Chemical Reactions of Contaminants in Groundwater Systems, Documentation and User's Guide*; Contract Report SERDP-99-1; U.S. Army Engineer Research and Development Center: Vicksburg, MS, USA, 1999.

43. Guo, W.; Langevin, C.D. *User's Guide to SEAWAT: A Computer Program for Simulation of Three-Dimensional Variable-Density Ground-Water Flow*; Techniques of Water-Resources Investigation; U.S. Geological Survey: Reston, VA, USA, 2002.

44. Liu, J.; Williams, J.R.; Zehnder, A.J.B.; Yang, H. GEPIC—Modelling wheat yield and crop water productivity with high resolution on a global scale. *Agric. Syst.* **2007**, *94*, 478–493. [CrossRef]

45. Gardener, M. *Beginning R: The Statistical Programming Language*; John Wiley & Sons: Hoboken, NJ, USA, 2012.

46. Pobuda, M. Using the R-ArcGIS Bridge: The Arcgisbinding Package. Available online: https://r.esri.com/assets/arcgisbinding-vignette.html (accessed on 21 January 2021).

47. TexasA & Magriliferesearch. EPIC & APEX Models. Available online: https://epicapex.tamu.edu/ (accessed on 21 January 2021).

48. U.S. Soil Conservation Service. *National Engineering Handbook, Section 4—Hydrology*; United States Department of Agriculture: Washington, DC, USA, 1972.

49. Thiessen, A.H. Precipitation averages for large areas. *Mon. Weather Rev.* **1911**, *39*, 1082–1089. [CrossRef]

50. Fiedler, F.R. Simple, Practical Method for Determining Station Weights Using Thiessen Polygons and Isohyetal Maps. *J. Hydrol. Eng.* **2003**, *8*, 219–221. [CrossRef]

51. Şen, Z. Average areal precipitation by percentage weighted polygon method. *J. Hydrol. Eng.* **1998**, *3*, 69–72. [CrossRef]

52. Thornthwaite, C.W. An Approach toward a Rational Classification of Climate. *Geogr. Rev.* **1948**, *38*, 55–94. [CrossRef]

53. Sidiropoulos, P.; Tziatzios, G.; Vasiliades, L.; Mylopoulos, N.; Loukas, A. Groundwater Nitrate Contamination Integrated Modeling for Climate and Water Resources Scenarios: The Case of Lake Karla Over-Exploited Aquifer. *Water* **2019**, *11*, 1201. [CrossRef]

54. McDonald, M.G.; Harbaugh, A.W.; original authors of MODFLOW. The History of MODFLOW. *Groundwater* **2003**, *41*, 280–283. [CrossRef]

55. Tziatzios, G.; Sidiropoulos, P.; Vasiliades, L.; Mylopoulos, N.; Loukas, A. Simulation of Nitrate Contamination in Lake Karla Aquifer. In Proceedings of the 14th International Conference on Environmental Science and Technology (CEST2015), Rhodes, Greece, 3–5 September 2015; Available online: https://cest2015.gnest.org/papers/cest2015_00121_oral_paper.pdf (accessed on 21 January 2021).

56. Psilovikos, A.A. Optimization models in groundwater management, based on linear and mixed integer programming. An application to a Greek hydrogeological basin. *Phys. Chem. Earth* **1999**, *24*, 139–144. [CrossRef]

57. Siarkos, I.; Latinopoulos, P. Modeling seawater intrusion in overexploited aquifers in the absence of sufficient data: Application to the aquifer of Nea Moudania, northern Greece. *Hydrogeol. J.* **2016**, *24*, 2123–2141. [CrossRef]

58. Kopsiaftis, G.; Mantoglou, A.; Giannoulopoulos, P. Variable density coastal aquifer models with application to an aquifer on Thira Island. *Desalination* **2009**, *237*, 65–80. [CrossRef]

59. Kourakos, G.; Mantoglou, A. Pumping optimization of coastal aquifers based on evolutionary algorithms and surrogate modular neural network models. *Adv. Water Resour.* **2009**, *32*, 507–521. [CrossRef]

60. Kaleris, V.K.; Ziogas, A.I. Using electrical resistivity logs and short duration pumping tests to estimate hydraulic conductivity profiles. *J. Hydrol.* **2020**, *590*, 125–277. [CrossRef]

61. Kopsiaftis, G.; Tigkas, D.; Christelis, V.; Vangelis, H. Assessment of drought impacts on semi-arid coastal aquifers of the Mediterranean. *J. Arid Environ.* **2017**, *137*, 7–15. [CrossRef]
62. Kritsotakis, M.; Tsanis, I.K. An integrated approach for sustainable water resources management of Messara basin, Crete, Greece. *Eur. Water* **2009**, *27*, 15–30.
63. Williams, J.; Jones, C.; Dyke, P. The EPIC Model and its Application. In Proceedings of the ICRISAT-IBSNAT-SYSS S International Symposium on Minimum Data Sets for Agrotechnology Transfer, Patancheru, India, 21–26 March 1983; pp. 111–121.
64. Wang, X.; Williams, J.; Gassman, P.; Baffaut, C.; Izaurralde, R.; Jeong, J.; Kiniry, J.R. EPIC and APEX: Model use, calibration, and validation. *Trans. ASABE* **2012**, *55*, 1447–1462. [CrossRef]
65. Sharpley, A.N.; Williams, J.R. *EPIC-Erosion/Productivity Impact Calculator. I: Model Documentation. II: User Manual*; Agricultural Research Service: Washington, DC, USA, 1990.
66. Liu, W.; Yang, H.; Liu, J.; Azevedo, L.B.; Wang, X.; Xu, Z.; Abbaspour, K.C.; Schulin, R. Global assessment of nitrogen losses and trade-offs with yields from major crop cultivations. *Sci. Total Environ.* **2016**, *572*, 526–537. [CrossRef]
67. Yin, Y.; Zhang, X.; Yu, H.; Lin, D.; Wu, Y.; Wang, J.a. Mapping Drought Risk (Maize) of the World. In *World Atlas of Natural Disaster Risk*; Shi, P., Kasperson, R., Eds.; Springer: Berlin/Heidelberg, Germany, 2015; pp. 211–226. [CrossRef]
68. Zhang, X.; Lin, D.; Guo, H.; Wu, Y.; Wang, J.a. Mapping Drought Risk (Rice) of the World. In *World Atlas of Natural Disaster Risk*; Shi, P., Kasperson, R., Eds.; Springer: Berlin/Heidelberg, Germany, 2015; pp. 243–258. [CrossRef]
69. Liu, J.; Fritz, S.; van Wesenbeeck, C.F.A.; Fuchs, M.; You, L.; Obersteiner, M.; Yang, H. A spatially explicit assessment of current and future hotspots of hunger in Sub-Saharan Africa in the context of global change. *Glob. Planet. Chang.* **2008**, *64*, 222–235. [CrossRef]
70. Liu, J.; Wiberg, D.; Zehnder, A.J.; Yang, H.J.I.S. Modeling the role of irrigation in winter wheat yield, crop water productivity, and production in China. *Irrig. Sci.* **2007**, *26*, 21–33. [CrossRef]
71. Koch, J.; Wimmer, F.; Schaldach, R.; Onigkeit, J. An integrated land-use system model for the Jordan River region. In *Environmental Land Use Planning*; Seth Appiah-Opoku InTechOpen: Tuscaloosa, AL, USA, 2012; Available online: https://www.intechopen.com/books/environmental-land-use-planning/an-integrated-land-use-system-model-for-the-jordan-river-region (accessed on 21 January 2021).
72. Zheng, C.; Bennett, G. *Applied Contaminant Transport Modeling*, 2nd ed.; Wiley-Interscience: New York, NY, USA, 2002; Volume 34.
73. Anderson, M.P.; Cherry, J.A. Using models to simulate the movement of contaminants through groundwater flow systems. *Crit. Rev. Environ. Control* **1979**, *9*, 97–156. [CrossRef]
74. Neuman, S.P. Universal scaling of hydraulic conductivities and dispersivities in geologic media. *Water Resour. Res.* **1990**, *26*, 1749–1758. [CrossRef]
75. Psaropoulou, E.T.; Karatzas, G.P. Pollution of nitrates—contaminant transport in heterogeneous porous media: A case study of the coastal aquifer of Corinth, Greece. *Glob. Nest J.* **2014**, *16*, 9–23. [CrossRef]
76. Guo, W.; Bennett, G.D. Simulation of Saline/Fresh Water Flows Using MODFLOW. In Proceedings of the MODFLOW 98 Conference, Golden, CO, USA, 4–8 October 1998; pp. 267–274.
77. Siarkos, I.; Latinopoulos, D.; Mallios, Z.; Latinopoulos, P. A methodological framework to assess the environmental and economic effects of injection barriers against seawater intrusion. *J. Environ. Manag.* **2017**, *193*, 532–540. [CrossRef]
78. Nash, J.E.; Sutcliffe, J.V. River flow forecasting through conceptual models part I—A discussion of principles. *J. Hydrol.* **1970**, *10*, 282–290. [CrossRef]
79. Colin Cameron, A.; Windmeijer, F.A.G. An R-squared measure of goodness of fit for some common nonlinear regression models. *J. Econom.* **1997**, *77*, 329–342. [CrossRef]
80. Matthews, J.; Bendig, A.W. The index of agreement: A possible criterion for measuring the outcome of group discussion. *Speech Monogr.* **1955**, *22*, 39–42. [CrossRef]
81. Smalheiser, N.R. Chapter 9—Null Hypothesis Statistical Testing and the t-test. In *Data Literacy*; Smalheiser, N.R., Ed.; Academic Press: New York, NY, USA, 2017; pp. 127–136. [CrossRef]
82. Lyra, A.; Pliakas, F.; Skias, S.; Gkiougkis, I. Implementation of DPSIR framework in the management of the Almyros basin, Magnesia Prefecture. *Bull. Geol. Soc. Greece* **2016**, *50*, 825–834. [CrossRef]
83. Stevenson, D.S. Irrigation Efficiency in Orchards. *Can. Water Resour. J.* **1980**, *5*, 102–110. [CrossRef]
84. Dewandel, B.; Gandolfi, J.-M.; de Condappa, D.; Ahmed, S. An efficient methodology for estimating irrigation return flow coefficients of irrigated crops at watershed and seasonal scale. *Hydrol. Process.* **2008**, *22*, 1700–1712. [CrossRef]
85. Willis, T.M.; Black, A.S.; Meyer, W.S. Estimates of deep percolation beneath cotton in the Macquarie Valley. *Irrig. Sci.* **1997**, *17*, 141–150. [CrossRef]
86. Panagos, P.; Van Liedekerke, M.; Jones, A.; Montanarella, L. European Soil Data Centre: Response to European policy support and public data requirements. *Land Use Policy* **2012**, *29*, 329–338. [CrossRef]
87. ESDAC. European Soil Data Centre, Joint Research Centre, European Commission. Available online: esdac.jrc.ec.europa.eu (accessed on 21 January 2021).
88. Soil Science Division Staff. *Soil Survey Manual*; Ditzler, C., Scheffe, K., Monger, H.C., Eds.; USDA Handbook, 18; Government Printing Office: Washington, DC, USA, 2017. Available online: https://www.nrcs.usda.gov/wps/portal/nrcs/detail/soils/scientists/?cid=nrcs142p2_054262 (accessed on 21 January 2021).
89. MEE. Ministry of Environment and Energy, Secretariat of National Environment and Water. Available online: http://lmt.ypeka.gr/public_view.html (accessed on 21 January 2021).

90. Merwade, V. Creating SCS Curve Number Grid Using HEC-GeoHMS. 2012. Available online: http://web.ics.purdue.edu/~{}vmerwade/tutorial.html (accessed on 21 January 2021).
91. Dastane, N. *Effective Rainfall, FAO Irrigation and Drainage Paper No. 25*; Food and Agriculture Organization: Rome, Italy, 1974.
92. Wichmann, W. *World Fertilizer Use Manual*; International Fertilizer Industry Association (IFA): Paris, France, 1992.
93. Tang, A.; Sandall, O.C. Diffusion coefficient of chlorine in water at 25–60 °C. *J. Chem. Eng. Data* **1985**, *30*, 189–191. [CrossRef]
94. Petihakis, G.; Triantafyllou, G.; Pollani, A.; Koliou, A.; Theodorou, A. Field data analysis and application of a complex water column biogeochemical model in different areas of a semi-enclosed basin: Towards the development of an ecosystem management tool. *Mar. Environ. Res.* **2005**, *59*, 493–518. [CrossRef] [PubMed]

Article

Checking the Plausibility of Modelled Nitrate Concentrations in the Leachate on Federal State Scale in Germany

Tim Wolters [1], Nils Cremer [2], Michael Eisele [3], Frank Herrmann [1], Peter Kreins [4], Ralf Kunkel [1] and Frank Wendland [1,*]

1 Forschungszentrum Juelich, IBG-3, 52425 Juelich, Germany; t.wolters@fz-juelich.de (T.W.); f.herrmann@fz-juelich.de (F.H.); r.Kunkel@fz-juelich.de (R.K.)
2 Erftverband, Bereich: Gewässer, Abteilung, Grundwasser, 50126 Bergheim, Germany; Nils.Cremer@erftverband.de
3 Landesamt für Natur, Umwelt und Verbraucherschutz NRW, FB 52 Grundwasser, Wasserversorgung, Trinkwasser, Lagerstättenabbau, 47051 Duisburg, Germany; michael.eisele@lanuv.nrw.de
4 Thünen-Institut für Ländliche Räume, 38116 Braunschweig, Germany; peter.kreins@thuenen.de
* Correspondence: f.wendland@fz-juelich.de

Abstract: In Germany, modelled nitrate concentrations in the leachate are of great importance for the development of scenarios for the long-term achievement of the groundwater quality target according to the specific requirements of the EU Water Framework Directive as well as within the context of the recently adopted general administrative regulation for the designation of nitrate-polluted areas in Germany. For the German federal states of North Rhine-Westphalia (NRW) and Rhineland-Palatinate (RLP), an area-covering modelling of mean long-term nitrate concentrations in leachate with high spatial resolution was carried out using the model system RAUMIS-mGROWA-DENUZ. Hotspot regions with nitrate concentrations in the leachate of 50 mg NO_3/L and more were identified for intensively farmed areas in the Münsterland, Lower Rhine, and Vorderpfalz. The validity of modelled values was checked using measured values from 1119 preselected monitoring stations from shallow springs and aquifers filtered near to the surface with oxidizing properties. For the land use categories of urban areas, arable land, grassland, and forest, an at least good agreement of modelled nitrate concentrations in the leachate and measured nitrate concentrations in groundwater was obtained at numerous sites. An equally good agreement was obtained for 1461 measuring stations from the area of responsibility of the Erftverband, which is a major water supplier in the Lower Rhine region. Here, discrepancies have been analyzed in detail due to profound regional knowledge on observation sites. It turned out that in most cases, accuracy limitations of input data (e.g., N balance surpluses of agriculture at the municipal level, 1:50,000 soil map) have been the reason for larger deviations between observed and modelled values. In a broader sense, the case study has shown on the one hand that the model system RAUMIS-mGROWA-DENUZ is able to reliably represent interrelationships and influencing factors that determine simulated nitrate concentrations in the leachate. On the other hand, it has been proven that observed nitrate concentrations in groundwater may provide a solid data source for checking the plausibility of modelled nitrate concentrations in leachate in cases where certain preselection criteria are applied.

Keywords: nitrate; groundwater; leachate; modelling; validation; state scale

Citation: Wolters, T.; Cremer, N.; Eisele, M.; Herrmann, F.; Kreins, P.; Kunkel, R.; Wendland, F. Checking the Plausibility of Modelled Nitrate Concentrations in the Leachate on Federal State Scale in Germany. *Water* **2021**, *13*, 226. https://doi.org/10.3390/w13020226

Received: 10 December 2020
Accepted: 14 January 2021
Published: 18 January 2021

1. Introduction and Objective

Due to the fact that the nitrate threshold limit value for groundwater of 50 mg NO_3/L is continuously exceeded in some regions of Germany, the EU Commission already determined in 2016 that the Federal Republic of Germany had failed to take stricter and harmonized measures against water pollution by nitrates at the federal level, although Germany would have been obliged to do so according to the EU Nitrates Directive [1] by 2012 at the latest. Due to these omissions, Germany was sued by the EU Commission in 2016 for excessively high nitrate levels in groundwater [2], which was ruled by the European Court of Justice (CJEU) in 2018 [3].

As the latest joint nitrate report of the Federal Ministries for Environment, Nature Conservation and Nuclear Safety and for Food and Agriculture shows, still too much nitrate leaches into groundwater, although N-reduction measures have been implemented in all federal German states [4]. In June 2020, the Federal Republic of Germany reacted to this situation in accordance to § 13a paragraph 1 of the German Fertilization Ordinance [5] by drafting a general administrative regulation for standardized designation of nitrate-polluted and eutrophicated areas in Germany [6]. The Federal Council of Germany passed this administrative regulation on 18 September 2020 [7]. By doing so, a nationwide uniform methodology for the designation of nitrate polluted areas ("red areas"), where further requirements for groundwater protection are necessary, was introduced.

For designating areas polluted with nitrate, the administrative regulation [6] provides a three-stage procedure: starting points are the groundwater bodies that are at a poor status due to nitrate according to the Water Framework Directive [8] or in which groundwater monitoring stations polluted with nitrate occur. In the first step, sub-areas are defined for observed nitrate concentrations in the groundwater above 50 mg NO_3/L or above 37.5 mg NO_3/L with an increasing trend.

In the second step, a site-specific assessment of the nitrate pollution risk has to be carried out for these sub-areas. This may be completed for predefined areal units, for instance, of one hectare in size or field blocks. At this, the decisive indicator is the nitrate concentration in the leachate, which indicates the nitrate concentration occurring when the leachate leaves the soil zone. For agricultural reference parcels, this nitrate concentration in the leachate is the starting point for deriving the "maximum permissible N balance surplus", which is intended to ensure that the nitrate concentration in the leachate below the root zone does not exceed a value of 50 mg NO_3/L or a value of 37.5 mg NO_3/L with an increasing trend.

In the third step, the difference between the maximum permissible N-balance surplus and the current N-balance surplus is used to quantify the "agricultural N reduction requirement", which must be met to ensure that the nitrate concentration in the leachate below the root zone does not exceed 50 mg NO_3/L or a value of 37.5 mg NO_3/L with an increasing trend.

The three-stage procedure described in the administrative regulation [6] is not fundamentally new. In Germany, this procedure has already been used since 2008 to forecast and develop scenarios for the long-term achievement of the targets for groundwater and water bodies according to the specific requirements of the EU Water framework directive [7] and the EU Groundwater Directive [9], see [10–14]. The German Working Group on Water Issues of the Federal States and the Federal Government (LAWA) has taken up this procedure and suggested to use a nitrate concentration in the leachate of 50 mg NO_3/L as a reference value that should not be exceeded [15]. Similar aspects are considered in the groundwater protection strategies of other countries [16,17].

In two recently finalized R&D projects funded independently of each other on behalf of the Ministry for Climate Protection, Environment, Agriculture, Nature and Consumer Protection of North Rhine-Westphalia (NRW) [14,18] and the Ministry of Environment, Energy, Food and Forestry Rhineland-Palatinate (RLP), respectively, the mean long-term nitrate concentration in leachate was modelled as a basis for defining the maximum permissible N-balance surplus and deriving agricultural N-reduction requirements.

Against the background of these R&D projects and the recently adopted General Administrative Regulation [6], which is to be implemented nationwide, this article summarizes on the one hand a procedure for simulating nitrate concentration in the leachate on a state scale. On the other hand, questions making it possible to classify the relevance and the regional plausibility of modelled nitrate concentrations in the leachate are addressed using the federal German states of North Rhine-Westphalia and Rhineland-Palatinate as example regions:

1. What are the values of the modelled nitrate concentration in leachate? Can "hotspot" areas where the nitrate concentrations on a long-term average are expected to exceed 50 mg NO_3/L be identified?
2. How valid are state-wide modelled nitrate concentrations in the leachate? Are nitrate concentrations observed in groundwater monitoring stations operated by federal German states or water management associations suitable for such plausibility checks? If so, under what premises?

The working hypothesis of this paper is based on the assumption that the influencing factors that determine the nitrate concentration in leachate are adequately represented in the model if the magnitude of the modelled values corresponds spatially to the observed values in both federal states. Accordingly, it can be concluded that the modelled nitrate concentration in the leachate is a reliable indicator for the nitrate pollution of groundwater in a certain region. This again would confirm that the modelled nitrate concentrations in the leachate provide a reliable basis for both the designation of the N reduction requirements to achieve the groundwater protection objective according to LAWA [15] and for the uniform designation of nitrate-polluted areas [6].

It will be demonstrated that all these premises are mostly fulfilled for this special case. More generally, this paper addresses the question of whether the generally widely available data on nitrate concentrations in groundwater from official monitoring networks, which are not exploited for checking the validity of modelled nitrate concentrations in leachate so far, can be recommended for this purpose.

2. Methodology

For calculating nitrate concentrations in the leachate, simulation models have been applied worldwide since more than 30 years ago [19]. Each of these models has been developed against the background of both a specific research question and a certain scale range of application. Physically based models like HYDRUS-1D [20] or the Daisy model [21] may be suitable for the simulation of site-specific pore water fluxes of nitrate at field scale [22]; their applicability on the scale of Federal States or entire countries is, however, limited due to the fact that numerous input datasets are not available on this scale [11].

Models such as SWAT [23], HYPE [24], or MONERIS [25] are suitable for an application at the level of states. The spatial resolution of these models is limited to the level of sub-catchment areas however, which impedes the identification of site-specific hotspot areas of nitrate leaching below the sub-catchment level. The latter one is a pre-requisite, however, in order to fulfill the requirements of the German-wide administrative regulations [6,8].

The model system RAUMIS-mGROWA-DENUZ [10,12,26,27] (see also Figure 1) is not only suitable for applications on the state-scale; it is also able to determine the nitrate concentration in the leachate in the required high spatial resolution (here, 100×100 m grid). For detailed descriptions of individual sub-models, please refer to the relevant literature, e.g., for RAUMIS [26], for mGROWA [27], and for DENUZ [10,12].

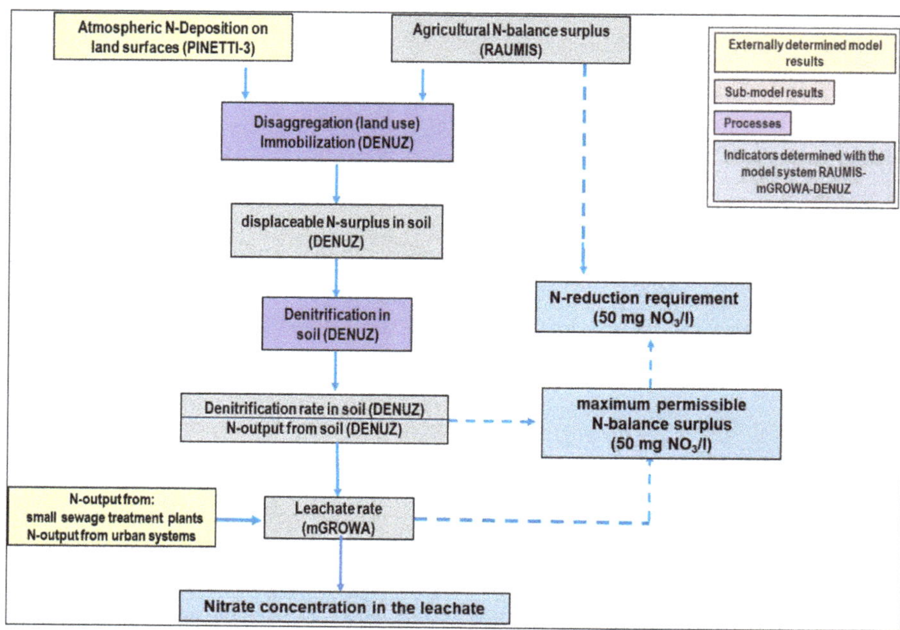

Figure 1. Schematic representation of the model system for deriving nitrate concentration in leachate and related parameters.

Relevant starting points for simulating the nitrate concentration in the leachate are the agricultural N-balance surplus, the N-deposition from atmosphere, N-emissions from urban systems and small sewage treatment plants, and the leachate rate in Equation (1).

$$c_{NO_3} = \frac{443}{Q_L} \cdot [N_{NBS} + N_{NH_x} + N_{NO_x} + N_{SSP} + N_{US} - I_{soil} - D_{soil}] \tag{1}$$

with:

c_{NO_3}: Nitrate concentration in the leachate (mg NO_3/L);
N_{NBS}: N-inputs into the soil from agricultural N-balance surplus (kg N/ha·a);
N_{NH_x}: N-inputs into the soil from atmospheric NH_x deposition (kg N/ha·a);
N_{NO_x}: N-inputs into the soil from atmospheric NO_x-deposition (kg N/ha·a);
N_{SSP}: N-inputs into the unsaturated zone from small sewage treatment plants (kg N/ha·a);
N_{US}: N-inputs into the unsaturated zone from urban systems (kg N/ha·a);
I_{soil}: N-immobilisation in the soil (kg N/ha·a);
D_{soil}: Denitrification in the soil (kg N/ha·a);
Q_L: Mean long-term leachate rate (mm/a).

The leachate rate is derived from the water balance model mGROWA [27]. Against the background that the model intends to depict a mean long-term-and thus regionally typical hydrological situation, the leachate rates simulated with mGROWA in daily resolution are aggregated to long-term mean values for the 30-year hydrological reference period 1981–2010.

For calculating nitrate concentrations in the leachate, the mGROWA leachate rates are coupled to N-emissions from soil and N-emissions from urban systems and small sewage treatment plants. The most important N source for assessing N emissions from soil are agricultural N balance surpluses. In both projects, these N balance surpluses were determined at the community level by the Thünen Institute in Braunschweig for the reference period 2014–2016 using the model RAUMIS [26]. Atmospheric NH_x deposition determined in the PINETI3 project [28] was taken into account as a further diffuse N source, which can be attributed to agriculture. As a non-agricultural share of atmospheric N deposition, the NOx emissions from transport, industry, and households assessed in the PINETI3 project [28] were considered.

Initially, the displaceable N surplus in soil is calculated under consideration of N immobilization in soils under grassland and forest areas. The displaceable N surplus, however, may be partly denitrified during transport through the soil. This process is accounted for in the DENUZ model [10,12,29] using Michaelis–Menten kinetics, which considers the amount of denitrified N in soils as a function of different influencing factors. Favorable factors for denitrification in soil are, for example, high soil moisture, high organic carbon content, and high soil temperatures. In contrast, inhibited denitrification in soil is likely a result of a lack of microbially readily available carbon sources (low humus content) in well aerated soils with a tendency to acidification [30,31].

In addition to the diffuse N emissions from soils, N emissions from urban systems and small sewage treatment plants contributing to the N load of the leachate below the soil zone have been considered. N emissions to groundwater from urban systems under settlement areas were determined in both Federal States on the assumption of a nitrogen release of 4 kg N per capita and year. For NRW, it was assumed that the N release to groundwater from urban systems amounts to 15%, i.e., 0.6 kg N per capita and year [32]. In Rhineland-Palatinate, losses from urban systems were included as lump sums according to the period the sewer systems were installed following suggestions from LAWA-AH [33]: 3% exfiltration for sewers installed before 1970, 1% exfiltration for sewers installed between 1971 and 2000, and 0.5% exfiltration for sewers installed since 2000.

In NRW, nitrogen emissions into groundwater from small sewage treatment plants were considered as an additional nitrogen source. Related nitrogen loads were derived from the state discharge register [34]. In Rhineland-Palatinate, in contrast, small sewage treatment plants were not explicitly considered as nitrogen emission sources discharging into groundwater.

3. Model Results for Nitrate Concentration in the Leachate

In Table 1, relevant input data sources for determining nitrate concentrations in the leachate are summarized. Figure 2 shows its spatial distribution.

As expected, N emissions from urban systems and small sewage treatment plants loading the leachate with N below the soil prevail in the metropolitan areas along the river Rhine and in the Ruhr area (see Figure 2A). In total, these N emissions sum up to approx. 11,500 t N/a.

Table 1. Input parameters for determining nitrate concentrations in the leachate.

	Parameter	Data source
Diffuse N-emission from soil caused by:	Agricultural N-balance surplus	RAUMIS: model result
	Atmospheric NH_x-deposition	PINETI-3: externally modelled by [28]
	Atmospheric NO_x-deposition	
N-emissions below the soil zone from:	Urban systems	Externally modelled according to [32,33]
	Small sewage treatment plants	Externally determined using [34]
Hydrology	Leachate rate	mGROWA: model result
	Denitrification conditions in soil	DENUZ: model result
	Residence time in soil	DENUZ: model result

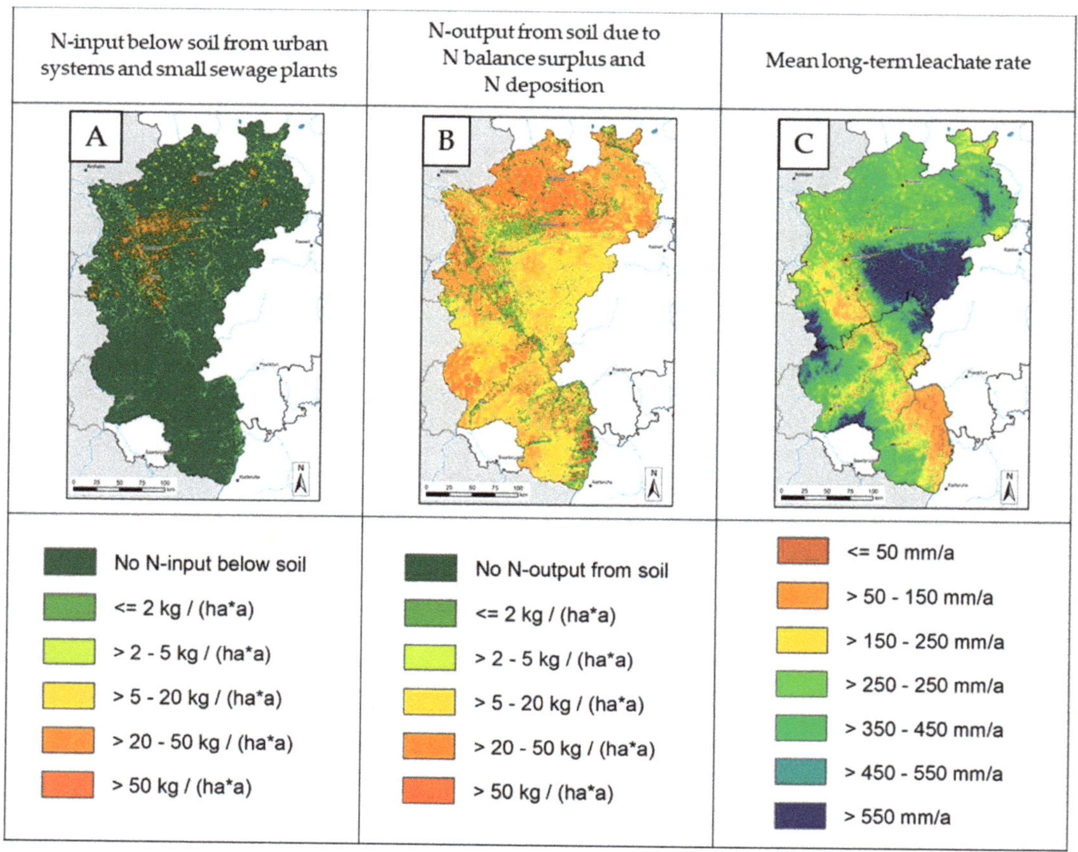

Figure 2. Relevant input parameters for calculating nitrate concentration in the leachate: Nitrogen input below soil from urban systems and small sewage plants (**A**), Nitrogen output from soil due to agricultural N balance surplus and atmospheric N deposition (**B**), and mean long-term leachate rate (**C**).

Nitrogen output from soil (Figure 2B) was determined with the DENUZ model on the basis of the N balance surpluses from agriculture as well as the atmospheric NHx and NOx deposition, taking into account N immobilization and nitrate degradation processes in soil. The resulting diffuse N emission from soil adds up to approx. 79,000 t N/a in NRW and 28,400 t N/a in Rhineland-Palatinate.

For the central part of the Münsterland and some regions in the Western Eifel, diffuse N emissions from soil > 50 kg N/(ha·a) occur due to intensive animal husbandry. N-emissions from soil > 50 kg N/(ha·a) can also be found in regions where the cultivation of special crops (e.g., Vorderpfalz, parts of the Upper Rhine Valley and the Lower Rhine Embayment) prevails. In the fertile Börde landscapes (e.g., Cologne-Aachener Bucht, Soester Börde), N emissions from soil of 20 up to 50 kg N/(ha·a) are observed. In low mountain ranges of NRW, N emissions from soil range between 10 and 25 kg N/(ha·a), while in corresponding areas of Rhineland-Palatinate, N emissions from soils of less than 10 kg N/(ha·a) are common. In urban areas, N emissions from soil are generally low and rarely exceed 5 kg N/(ha a).

Figure 2C shows the mean long-term leachate rate for the reference period 1981–2010. In low land regions consisting of unconsolidated fluviatile sand and gravel deposits, leachate rates between less than 100 and 200 mm/a dominate. While leachate rates up to 200 mm/a occur in the entire Lower Rhine Embayment, leachate rates rarely exceed

100 m/a in the Upper Rhine Valley and the Vorderpfalz. In low mountain ranges consisting of consolidated rocks (sandstones and slated shale stones), leachate rates up to 600 mm/a and more predominate. Consequently, the dilution of N emissions in the entire Rhenish Slate Mountain region is three to six times higher than in lowland regions.

In reality, N emissions from the various (agricultural and non-agricultural) N sources will mix in the leachate. To represent this in the model, N emissions from these individual N sources were added and combined with the mean long-term leachate rates for each individual grid. Figure 3A shows the corresponding nitrate concentration in the leachate.

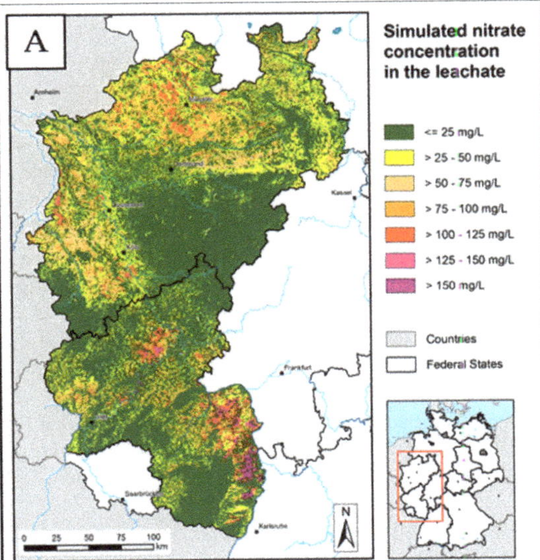

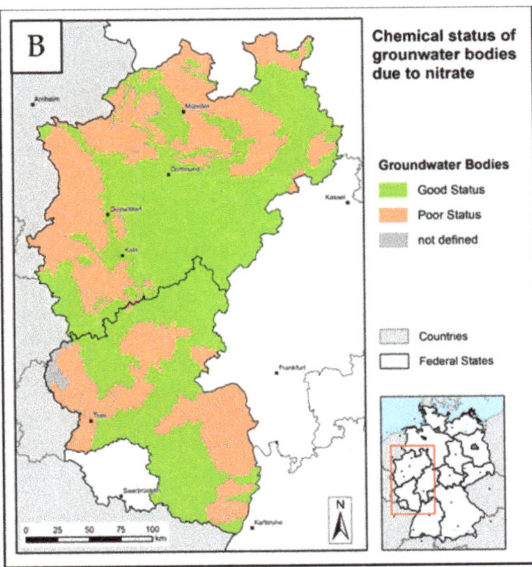

Figure 3. Mean long-term nitrate concentration in the leachate (**A**) and chemical status of groundwater bodies due to nitrate (status according to the EU Water Framework Directive (**B**): good status: <50 mg NO_3/L, poor status >50 NO_3 mg/L.

All areas in Figure 3A showing nitrate concentrations in the leachate of 50 mg NO_3/L and more can be regarded as "hot spot areas" for nitrate pollution. This concerns many regions in the northern and western parts of North Rhine-Westphalia (Münsterland, Lower Rhine Embayment) as well as in the southwest of Rhineland-Palatinate (Upper Rhine Valley) and in the Neuwieder Becken. In the Münsterland, the hotspot areas are exclusively due to high agricultural N balance surpluses originating from livestock farming (see Figure 2B). In the Lower Rhine Embayment, the Upper Rhine Valley, and the Neuwieder Becken, N emissions from soil are moderate. In these regions, high nitrate concentrations are also caused by low leachate rates. Particular high values of 150 mg NO_3/L and more are found in some areas in the Upper Rhine Valley and in the Vorderpfalz due to the combination of special crops cultivation and low leachate rates. In contrast, extended areas showing nitrate concentrations in the leachate <25 mg NO_3/L occur in both federal states, especially in extensively farmed low mountain regions, where relatively low N-emissions from soil are considerably diluted by high leachate rates.

As Figure 3B shows, the majority of areas where modelled nitrate concentrations in leachate exceed 50 mg NO_3/L are in groundwater bodies that are in poor chemical status due to nitrate. Qualitatively, this is a first indication of the model's accuracy. Less good agreement is seen in the Münsterland core area, where groundwater bodies in good status due to nitrate are juxtaposed with extensive areas of modelled nitrate concentrations in leachate exceeding 50 mg NO_3/L. As, however, aquifers with high nitrate degradation potential are located in the groundwater bodies of this region, larger differences are expected and do not challenge the modelled nitrate concentrations in the leachate.

4. Comparison of Modelled Nitrate Concentrations in the Leachate with Observation Data

Measured nitrate concentration values from soil depth profiles, suction probes, and lysimeters are, in general, the most suitable data to validate nitrate concentrations in the leachate. In practice, however, there is a lack of monitoring networks at the Federal State level that systematically record nitrate concentrations in soils at a sufficient, i.e., regionally representative, number of measuring points.

Modelled nitrate concentrations in the leachate can be further directly compared to nitrate concentrations in groundwater if excess N_2 measurements [35,36] have been carried out. Excess N_2 measurements allow for the determination of degraded nitrate concentration and thus a backward calculation of the initial nitrate concentration of the recharged groundwater without denitrification influence. With this method, in principle, all groundwater monitoring sites can be used for the estimation of nitrate concentrations in recharged groundwater, and thus for validating modelled nitrate concentrations in leachate [37]. A preselection of monitoring sites with oxidizing groundwater is not necessary in this case. However, systematic N_2/Ar sampling at a state scale in Germany is still in its early stages [38] and may be available in some areas, but—again—not at a federal state scale.

Due to these data limitations, a direct plausibility check of modelled nitrate concentrations in the leachate for larger areas, i.e., at the level of Federal States, is therefore often not possible. Measured nitrate concentrations from groundwater monitoring stations are available for most regions in sufficient numbers. However, its usability for checking the plausibility of modelled nitrate concentrations in the leachate has not been assessed so far. We assume that measured nitrate concentrations in groundwater are suitable for testing the reliability of modelled nitrate concentrations in leachate if suitable groundwater monitoring sites have been identified in advance using a number of preselection criteria.

5. Preselection Criteria for Groundwater Monitoring Stations

5.1. Sampling Depth and Type of Measuring Point

The quality of near-surface groundwater reflects the quality of the leachate that has infiltrated into an aquifer. Consequently, only measured nitrate concentrations from upper aquifers or shallow sources should be included in the validity check.

Furthermore, for groundwater monitoring sites, the type of well was considered. Measured values from groundwater extraction wells are, in general, unsuitable for this kind of plausibility check. Due to pumping processes, groundwater abstraction wells display water from different withdrawal depths, so that a reference to the nitrate concentration at the groundwater surface and to the nitrate concentration in the leachate, respectively, is usually missing.

For the remaining groundwater monitoring wells, the depth range of the filter applies as an additional preselection criterion to ensure that only monitoring sites with filter depths of less than 20 m below groundwater level or from the upper third of shallow aquifers are selected for the plausibility check.

5.2. Denitrification Potential in Groundwater

Since denitrification in groundwater is not relevant for the modelled nitrate concentrations in leachate but may be significant for the observed nitrate concentration in near-surface groundwater, observed nitrate concentration in monitoring wells may be significantly lower than nitrate concentration in the leachate [14]. Consequently, in regions with denitrification in the aquifer, the observed nitrate concentrations may already have been reduced or completely eliminated by denitrification processes during the passage of the groundwater on its way to the monitoring site. Conversely, the measured nitrate concentrations in the groundwater of aquifers displaying oxidative groundwater can be regarded as suitable for checking the plausibility of modelled nitrate concentrations in leachate, since significant nitrate degradation processes in aquifers with oxidative conditions can be excluded.

An assessment of the concentration ranges of essential redox parameters at the monitoring sites is, therefore, an indispensable preselection criterion to exclude that discrepancies between modelled nitrate concentrations in the leachate and measured nitrate concentrations in groundwater are erroneously attributed to model-related causes. Rather simple indicators can be used to identify the redox status of groundwater samples from monitoring stations (see Table 2).

Table 2. Concentration ranges of essential redox parameters for identifying redox status and denitrification conditions in aquifers [10–13,26,31,39–48].

Parameter	Reduced Groundwater	Oxidized Groundwater
Nitrate	<1 mg NO$_3$/L	various
Iron (II)	>0.2 mg/ Fe (II)/L	<0.2 mg/ Fe (II)/L
Manganese (II)	>0.05 mg Mn (II)/L	<0.05 mg Mn (II)/L
Oxygen	<2 mg O$_2$/L	>2 mg O$_2$/L
DOC	>1.75 mg DOC/L	<1.75 mg DOC/L

Basic prerequisites for denitrification in groundwater are low oxygen concentrations and the presence of organic carbon and/or iron sulfide compounds (pyrite) in an aquifer. A typical nitrate-degrading groundwater usually shows high contents of divalent iron, manganese, and/or organic carbon (>0.1 mg Fe(II)/L, >0.05 mg Mn(II)/L, and >0.75 mg DOC/L), while in general, only low nitrate and oxygen contents (<2 mg/L) can be observed [32]. Non-nitrate-degrading groundwater is usually characterized by high oxygen concentrations and low Iron (II), Manganese (II), and DOC contents.

In order to include only monitoring sites showing oxidizing groundwater conditions in the plausibility check, all monitoring sites indicating reducing groundwater conditions for at least three of the five parameters listed in Table 2 were excluded from further evaluations.

5.3. Inflow Area of Measuring Point

The entire inflow area of a groundwater monitoring site contributes to the nitrate concentration observed at that site. It can therefore be expected that the nitrate concentration in the leachate modelled for the specific raster cell where a monitoring station is located will deviate from the observed nitrate concentration of that monitoring station.

Therefore, the plausibility check was carried out under consideration of the inflow area of the observation sites using averaged modelled nitrate concentrations. The databases for delineating the inflow area of groundwater measuring points exactly are usually not available in the required resolution at the federal state level. For pragmatic reasons, we used a state-wide available model of the groundwater surface. As long as nitrate degradation potentials in the unsaturated zone below the soil and in the aquifer can be excluded, the median nitrate concentration in the leachate from the inflow area of a monitoring site should correspond to the observed nitrate concentrations in the groundwater of this monitoring site.

In order to determine the inflow areas of measuring points, groundwater flow directions (calculation of the maximum gradient to one of the eight adjacent cells) and the corresponding catchment areas can be derived from digital groundwater surfaces using a flow accumulation algorithm [49] implemented in the tool r.watershed of GIS GRASS (see Figure 4).

Groundwater inflow areas of measuring points derived in this way represent contiguous areas of all grid cells whose water flows into a respective groundwater measuring point. Accordingly, the median of the nitrate concentrations in the leachate from the individual raster cells within the inflow area of an observation point is derived and compared to the observed nitrate concentrations in the groundwater.

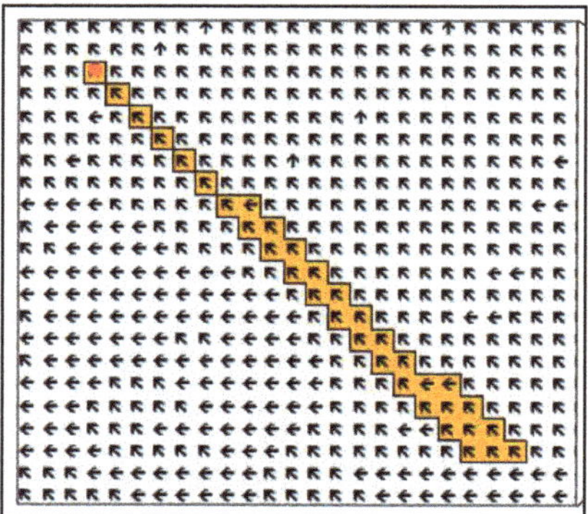

Figure 4. Example of a groundwater measuring point (red) with corresponding catchment area (orange) derived from groundwater flow directions (arrows).

The same procedure was applied for the delineation of catchment areas of shallow springs, however, not based on groundwater surfaces but digital elevation models (DGM10 for NRW and DGM25 for Rhineland-Palatinate, respectively) for deriving flow directions and measuring point-specific catchment areas. Occasionally, inflow areas exceeding a distance of 1 km to a measuring point have been identified by the flow accumulation algorithm. In this case, the inflow areas have been buffered accordingly.

6. Preselected Groundwater Monitoring Sites from the Federal State Groundwater Databases

Data collected from two federal states' groundwater databases were used for checking the magnitude and spatial representability of the modelled nitrate concentration in leachate. Figure 5A shows locations of measuring points in NRW and RLP. For NRW, nitrate values from 4839 monitoring stations from the period 2006 to 2015 were available from the NRW state groundwater database. For RLP, data from a total of 1428 monitoring sites from the time period 2006–2017 have been provided by the RLP State Agency for the Environment. The spatial distribution of these groundwater measuring points is quite heterogeneous, i.e., in regions with high-yielding groundwater resources (Lower Rhine Bight, Upper Rhine Münsterland), it is significantly higher than in regions with low-yielding groundwater resources (Rhenish Slate Mountains). In accordance with the above-mentioned preselection criteria, only groundwater samples were used for further evaluations when any of the following constraints were met:

1. Groundwater samples originate from monitoring stations with filter depths of less than 20 m below groundwater level;
2. Groundwater samples were taken from the upper third of shallow aquifers;
3. Samples stem from shallow springs;
4. Samples show oxidative groundwater conditions;
5. Inflow areas of groundwater measuring points and catchment areas of springs, respectively, could be delineated.

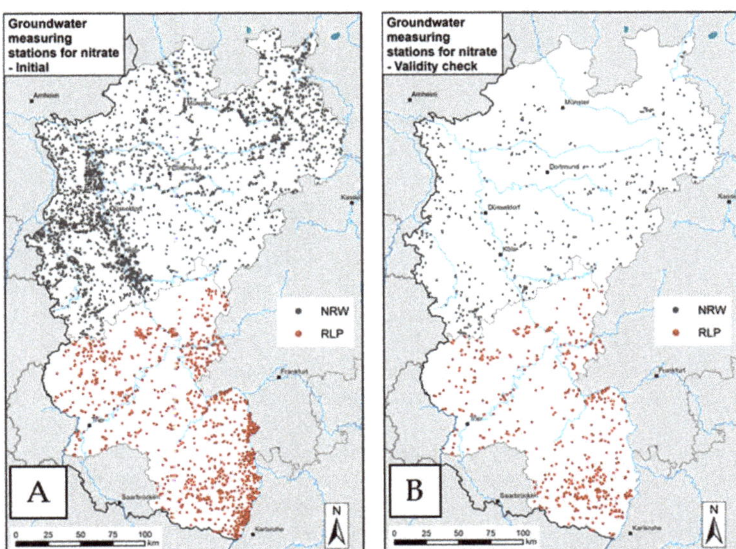

Figure 5. Initially available monitoring sites from state measuring networks in North Rhine-Westphalia (NRW) and Rhineland-Palatinate (RLP) (**A**) and preselected monitoring points included in the plausibility check (**B**).

Applying all these preselection criteria, the initially available number of monitoring stations has been reduced from 6267 to 1119, i.e., ca. 515 for NRW and 614 for RLP. The locations of these sites are shown in Figure 5B. In the vast majority of cases, missing information about the sampling depth or filter depth of measuring points has been the reason for discarding monitoring sites. Still, the spatial distribution of the preselected groundwater monitoring sites in both German federal states is quite homogeneous with the exception of the northern part of NRW. For this area (the Münsterland), a significant number of groundwater monitoring sites are available (see Figure 5A); most of them, however, were excluded from the plausibility check as the groundwater shows reducing conditions.

Modelled nitrate concentration in the leachate represents a hypothetical value that refers to the leachate rates of a—usually 30-year—hydrological reference period. This is performed with the aim to limit the influence of hydrological dry or wet years on the nitrate concentration in the leachate. The nitrogen sources, above all, the N balance surpluses from agriculture and the atmospheric N deposition, are also considered as moving averages of 3–5-year reference periods with the aim to consider average conditions. Measured nitrate concentrations in groundwater, however, are random samples, which may be strongly influenced by hydrological conditions (e.g., rate of infiltration into groundwater) and N emissions (type of cultivated crops grown, date of fertilizer application) in the vicinity of the monitoring site at or near the time of measurement. Accordingly, for the preselected groundwater monitoring stations and shallow springs, the question arose as to whether monitoring sites for which only a few individual samples are available have to also be excluded from the plausibility checks.

7. Plausibility Checks Based on Preselected Monitoring Data from Federal State Groundwater Observation Data

When modelling nitrate concentrations in the leachate at a state scale, the aim cannot be to reproduce the value measured at a specific point in time at a specific location. This is not to be expected due to the limited site-specific accuracy of the input data at the federal state level described in Section 3.

Since a comparison of measured point values and modelled raster values is misleading, it must be defined how a "good" or "bad" match can be defined. As a minimum requirement, it can certainly be stated that a large-scale model should primarily reflect the spatial patterns. This means that regions with high nitrate concentrations ("hot-spot" regions) must be represented just as representatively as regions with low concentrations. In terms of absolute values, however, it is acceptable if there are significant differences.

If, for example, in a region with predominantly low concentrations, the measured value is 1 mg NO_3/L and the modelled value is 3 mg NO_3/L, then the agreement is excellent, although the measured value is overestimated by a factor of 3. In the upper concentration range, even a high absolute deviation can still be considered as a good agreement. If, for example, a measured nitrate concentration is 126 mg NO_3/L and the modelled value is 150 mg NO_3/L, then the deviation is 24 mg/L, but the local pollution situation is still very well represented in the model.

In order to come to a systematic, comprehensible assessment of the model validity, the observed and measured values were first classified. Subsequently, the compliance of the measured and observed values to the respective classes was assessed. In cases where the modelled and observed values fell in the same class, the agreement was considered to be very good. The more the classes of the modelled and measured values deviate from each other, the worse the agreement is.

The difference between the classes was used as a measure for this purpose. If all measuring points are considered, a distribution of the class differences of measured and calculated values is obtained. In the ideal case of a perfect model, all class differences would be zero; in practice, the class differences are distributed around a mean value and with a certain scatter. From these parameters, conclusions can be drawn about both the validity of the model and possible under- or overestimations.

For this type of evaluation, the class widths must be defined. Basically, the ideal class width of a histogram results from the number of measured values [50]. Since a very different number of measured values was available for the evaluations, a uniform class width of 25 mg NO_3/L was defined after a series of tests to ensure comparability. The comparison was, therefore, based on seven classes (0–25, 2550, ... > 150 mg NO_3/L).

Figure 6 provides a spatial overview of the results of the comparison in both German states. Dots show in ca. 500-fold superelevation the concentration ranges of the measured values at the 1119 preselected measuring points. The same class widths and the same color gradation were used to represent the modelled mean long-term nitrate concentration in the leachate of the individual rater cells.

Figure 7 additionally shows the comparison of the class widths of the modelled nitrate concentrations in leachate and the measured nitrate concentrations in groundwater as frequency distributions. In this context, the agreement is denoted as being good, if observed and measured concentrations are in the same category; fair, if there is a deviation of one category; poor if the deviation is of two categories; and bad if there is a deviation of more than two classes.

Figure 6 shows that the compliance between the modelled nitrate concentrations in leachate and the measured nitrate concentrations in groundwater is predominantly good to very good for the selected class widths. The frequency distribution in Figure 7 shows more precisely that the modelled nitrate concentrations in the leachate are for 58% of the 1119 groundwater monitoring sites in the same concentration class range. For another 27% of the monitoring sites, the modelled values fall in the next higher or next lower class range, which is still acceptable. For about 15% of the monitoring stations, however, significant deviations of more than one class width occur.

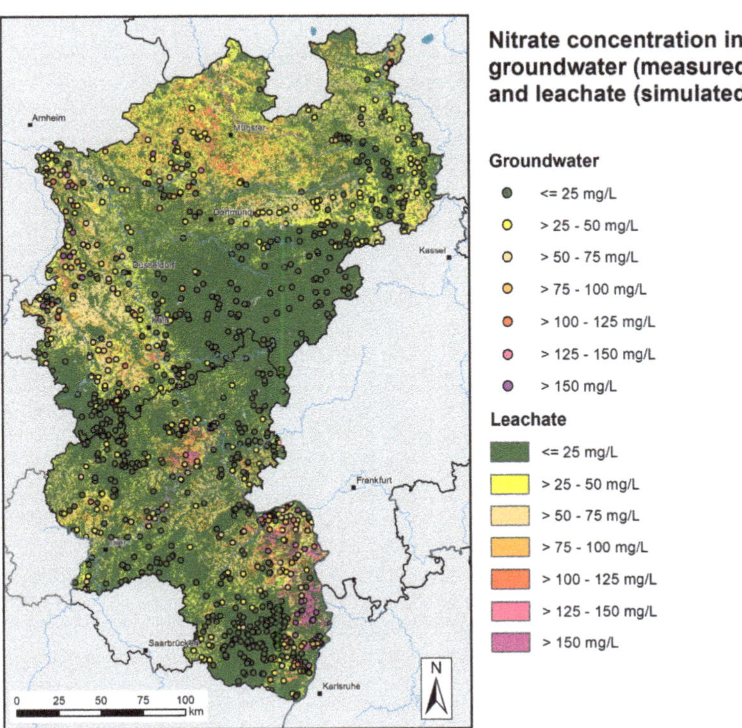

Figure 6. Spatial representation of class widths of modelled nitrate concentrations in leachate with measured nitrate concentrations in groundwater in the Federal States of NRW and RLP.

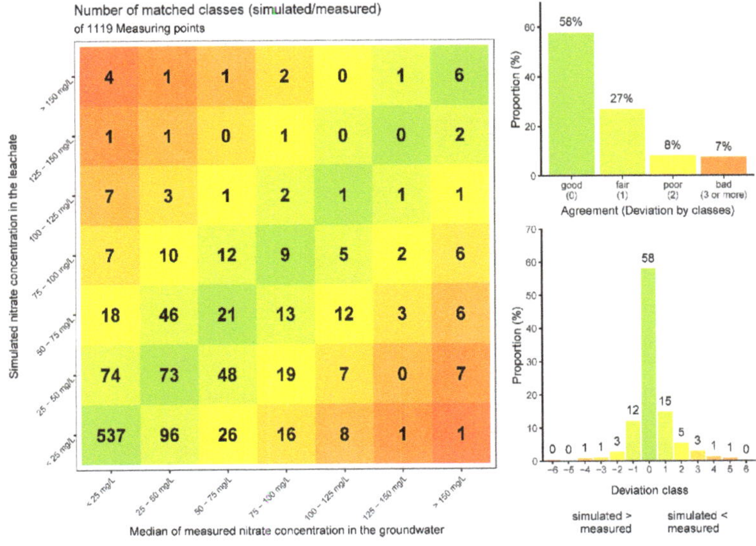

Figure 7. Frequency distribution of simulated nitrate concentrations in the leachate and observed nitrate concentrations in groundwater.

7.1. Influence of Number of Individual Samples

Against this background, we analyzed whether values from groundwater monitoring sites for which long-term time series on nitrate concentration are available show a higher agreement with modelled nitrate concentrations in leachate than groundwater monitoring sites with only a few individual samples. For measuring points with long-term time series an average value has been derived, for which it can be assumed that measuring dates with high nitrate concentrations in groundwater and measuring dates with low nitrate concentrations in groundwater have been approximated. In aquifers without significant denitrification potentials, this mean value may be of the same order of magnitude as the modelled mean long-term nitrate concentration in the leachate. Conversely, as the number of samples decreases, the deviation from the modelled mean long-term nitrate concentrations in leachate may randomly increase.

In order to determine whether the 1119 preselected monitoring sites should also have a minimum number of measured values above which a comparison with modelled nitrate concentrations in leachate is appropriate, frequency distributions with successively decreasing numbers of measured values per monitoring site were established. Table 3 shows mean values and standard deviations of frequency distributions of successively decreasing numbers of measured values, starting with measuring points with more than 22 measured values and ending with measuring points with more than 2 measured values for time periods of 10 and 12 years, respectively.

Table 3 illustrates that, contrary to expectations, neither the mean value of deviation classes nor the standard deviation of deviation classes show significantly more discrepancies as the number of samples per measuring point decreases. Thus, it seems to be appropriate—at least for the evaluated data collected—to include all measuring points with at least one measurement in the plausibility checks of modelled nitrate concentrations in leachate.

Table 3. Statistics of "deviation classes" (difference of classified average simulated nitrate concentration in leachate and classified median of measured nitrate concentration in groundwater) and frequency of successively decreasing numbers of measured values per measuring station.

Number of Measured Values Per Measuring Point		<22	<18	<14	<10	<6	1
Number of measuring points		1075	919	865	672	408	236
Statistics of deviation classes — Mean	$\{-6, -5, \dots 5, 6\}$	0.16	0.11	0.10	0.09	0.13	0.02
SD		1.63	1.59	1.59	1.58	1.63	1.62
Mean (absolute)	$\{0, \dots 5, 6\}$	0.70	0.65	0.64	0.62	0.60	0.53
SD (absolute)		1.38	1.36	1.36	1.35	1.42	1.44

7.2. Influence of Land Use Types

In a next step, modelled and observed concentrations were compared differentiated to four main types of land use, namely arable land, grassland, forest, and urban areas, derived from the digital land cover model LBM-DE2015. Decisive for the assignment of a monitoring station to a specific land use type was the dominant land use in the respective inflow area of a monitoring station and catchment area of a spring, respectively. The comparison of modelled nitrate concentrations in leachate and observed nitrate concentrations in groundwater is presented as frequency distributions in Figure 8 and on the basis of values of 25-percentile, median, and 75-percentile in Table 4.

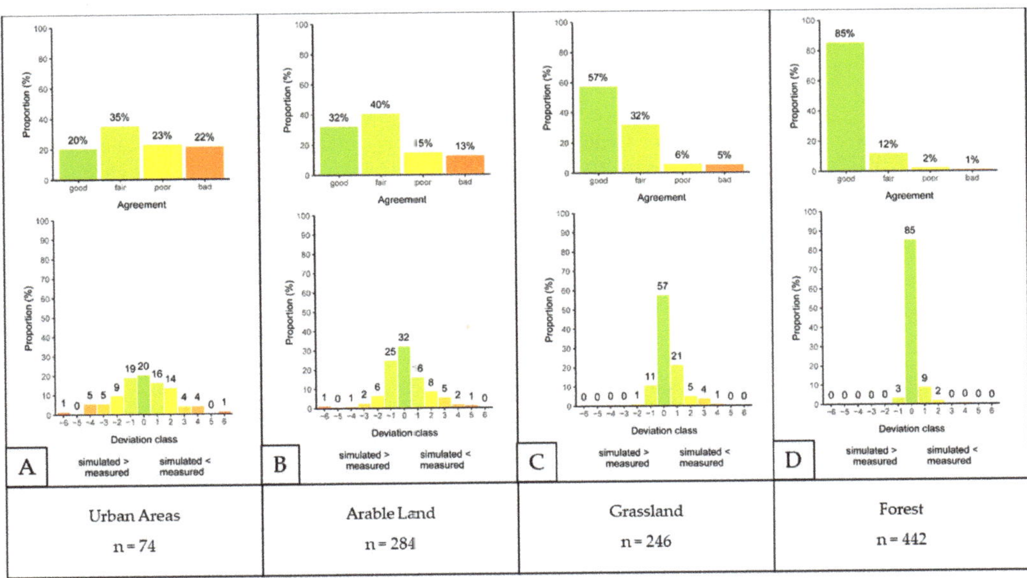

Figure 8. Comparison of overall agreement (top) and agreement for single deviation classes (down) (difference of classified average simulated nitrate concentration in leachate and classified median of measured nitrate concentration in groundwater) differentiated according to the main types of use: urban (**A**), arable land (**B**), grassland (**C**), and forest (**D**).

Table 4. Number of measuring points (n), first quartile (Q1), median, and third quartile (Q3) for simulated and observed nitrate concentrations for the land use types.

	Urban Areas		Arable Land		Grassland		Forest	
	sim	obs	sim	obs	sim	obs	sim	obs
n	74		284		246		442	
Q1	13.0	10.8	32.8	27.9	8.5	9.5	7.9	4.0
Median	29.5	34.5	48.5	46.4	15.8	19.0	10.0	8.4
Q3	54.5	53.8	68.3	69.8	26.2	37.0	13.7	16.3

Figure 8A shows the comparison of the modelled nitrate concentrations in the leachate of monitoring sites for the land use category of urban areas. After preselection, only 74 monitoring sites from urban areas remained for the comparison. For about 20% of these monitoring sites, a good agreement between the modelled nitrate concentrations in the leachate and the measured nitrate concentrations in the groundwater was found. While a deviation of one class was found for another 35% of the measuring points, clear over- and underestimations were found for 43% of the measuring points.

High deviation in both ascending and descending direction is probably due the homogeneous distribution of the "N-emissions from leaking urban systems" over the entire urban areas of the respective communities when calculating the nitrate concentration in the leachate, whereas the highest N emission occurs probably along leaking urban systems and decreases with increasing distance from them. Still, the distribution of simulated and measured values represented by median, Q1 and Q3 in Table 4 shows that there is no sign of a clear systematic over- or underestimation.

Figure 8B shows the agreement of the modelled nitrate concentrations in the leachate of all monitoring sites for the land use category of arable land. A total of 284 monitoring sites from arable land remained for comparison after preselection. While for about 32%

of these monitoring sites, a good compliance between the classes of the modelled nitrate concentrations in the leachate and the measured nitrate concentrations in the groundwater was found, 40% of the monitoring sites show a satisfactory fair compliance, i.e., the simulated nitrate concentration in the leachate is one class above or below the class of the observed nitrate concentrations in groundwater. Such deviations can be expected for the land use category arable land and interpreted as an indication of small-scale deviations in the fertilization level and/or in the cultivation of crops that could not be represented in the model.

In this context, it should be taken into account that for modelling at a federal states level, agricultural N-balance surpluses have been determined at the community level [14]. Model calculations on nitrate concentrations in the leachate for regions where locally collected N-balance surplus data could be used in parallel with data available at the community level have already shown that the discrepancies to measured values in groundwater are indeed smaller when using locally collected fertilization data [11,46]. Against this background, a deviation of one class in the plausibility check for the land use category arable land can still be regarded as spatially representative so that it is neither necessary nor purposeful to recalibrate the model. The statistical parameters Q1 (first quartile) and median in Table 4 show that the modelled values slightly overestimate the observed values. In order to improve the agreement of measured and simulated nitrate concentrations in future model applications, the coefficients used for assessing denitrification in soil may be increased.

Figure 8C compares nitrate concentrations at all monitoring sites where grassland is the main land use category. After applying the above-mentioned preselection criteria, 246 monitoring sites were available for comparison. Of these monitoring sites, 57% show a good agreement in the nitrate concentration classes, while 32% over- or underestimate the concentrations by one class. The frequency distribution shows a higher number of groundwater monitoring sites where the modelled nitrate concentrations in leachate are lower than the measured nitrate concentrations in groundwater. Accordingly, the concentrations at the Q1, median, and Q3 show a slight underestimation of the modelled values. It is likely that the coefficient for considering N immobilization in soil under grassland has been set slightly too low in the model. Nevertheless, the overall agreement of modelled and observed nitrate concentrations for the land use category of grassland can be regarded as very representative.

Figure 8D compares the nitrate concentration classes of simulated and measured monitoring sites where forest is the main land use category. Out of a total 442 monitoring sites in this category, 85% show a good and 12% a fair agreement. Since there is no N source other than atmospheric N deposition, the good agreement between the modelled nitrate concentrations in the leachate and the observed nitrate concentrations in groundwater is an indirect confirmation of the accuracy of the other factors contributing to the nitrate concentration in the leachate, mainly the modelled leachate rates and the coefficients accounting for N immobilization of forest soils.

8. Plausibility Checks Using Monitoring Data from the Area of Responsibility of the Erftverband

Complementary to the monitoring data from the monitoring networks of the federal states, monitoring data from the area of responsibility of an important water board in NRW were used for the plausibility check. More specifically, monitoring data from 1086 groundwater measuring points and wells from the period 2013 to 2018 from the area of responsibility of the Erftverband, i.e., an area of approx. 4200 km^2, were evaluated. This high number of suitable monitoring sites can be explained by the comprehensive documentation of the monitoring stations operated by the Erftverband with regard to hydrogeological conditions, filter positions, land use, and soil conditions. Figure 9 gives an overview of the comparison in the Erftverband region.

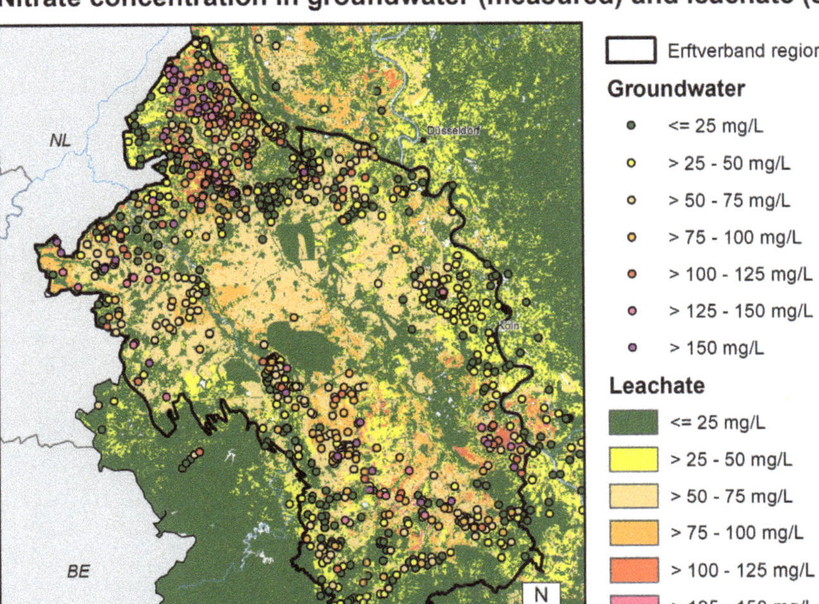

Figure 9. Spatial representation of class widths of modelled nitrate concentrations in leachate with measured nitrate concentrations in groundwater in the area of responsibility of the Erftverband.

Same as in the previous section, the concentration class ranges of the measured nitrate concentrations in groundwater (dots) and the modelled nitrate concentrations in the leachate for the 100 m grids are presented in the same class widths and in the same color gradations. Model results and measured values in groundwater show a very good agreement with regard to both the level of nitrate concentrations and the spatial distribution, although the hydrological and pedological, but also the agricultural, site conditions are different and also change on a small scale.

The very profound knowledge of local site conditions (soil, hydrogeology, land use, etc.) in the area of responsibility of the Erftverband offered a unique possibility to carry out selective verifications of modelled nitrate concentrations in the leachate, specifically in the case of larger deviations between modelled and measured values. The following figures show exemplary some sub-regions with significant discrepancies. Using these regions as examples, on the one hand, causes for deviations are pointed out and on the other hand, the implemented solutions in the model are presented.

Figure 10 shows a comparison of modelled nitrate concentration in the leachate (underlying colors) with measured nitrate concentrations in groundwater (spots) in an area of the former lignite opencast mine Zukunft-West. There, measured values in groundwater (<25 mg NO_3/L) were systematically overestimated by modelled nitrate concentrations in leachate (25–75 mg NO_3/L) in the first model run (Figure 10A). Evaluation of the concentration ranges of the essential redox parameters (Table 2) at the measuring points showed reduced groundwater conditions. Therefore, the discrepancy between the modelled nitrate concentrations in the leachate and the measured nitrate concentrations in the groundwater can be attributed to the denitrification processes of the groundwater taking place in this area.

Nitrate concentration in groundwater (measured) and leachate (simulated)

Zukunft-West (former lignite opencast mine)

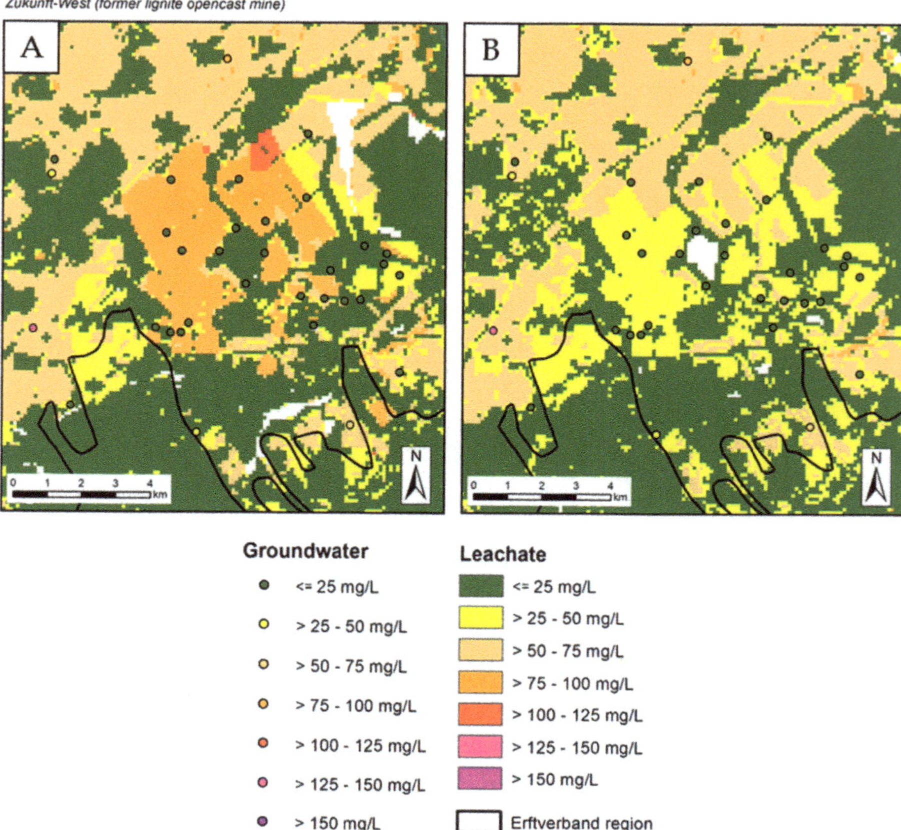

Figure 10. Comparison of the nitrate concentration in the leachate (grids) with measured nitrate concentrations in groundwater (dots) for an area of the former opencast mine Zukunft-West before (**A**) and after (**B**) adapting the denitrification conditions of the soil class "soil made of backfill material".

A comprehensive analysis of the databases used for modelling has shown that the 1:50,000 NRW soil map indicates the category "soils made of backfill material" for the area in the center of Figure 10 displaying nitrate concentrations in the leachate between 75 and 100 mg NO_3/L. Due to an assumed low organic carbon content, a low denitrification potential (nitrate degradation of max. 30 kg N/ha a) had been assigned to this soil category in the first model run.

Based on the Erftverband's profound knowledge of the regional soil conditions, the assumed denitrification potential of this soil was corrected. More specifically, the high organic carbon content that these soils actually have has been taken into account by reclassifying the denitrification conditions of the soil unit "soils made of backfill material" from low to high (nitrate degradation potential of max. 100 kg N/ha a). This adjustment reduced the deviations between the model results and the measured values in a subsequent model run (Figure 10B). This, again, was the reason to adapt the denitrification conditions of all "soils made of backfill material" in the lignite mining areas in the Lower Rhine Embayment and the Rhenish lignite mining area accordingly. As Figure 10B shows, the modelled nitrate concentrations in leachate are in the range between 25 and 50 mg NO_3/L, i.e., still one concentration class higher than the observed values in groundwater. The still existing discrepancy between the modelled nitrate concentrations in the leachate and the measured nitrate concentrations in the groundwater can be attributed to the denitrification processes of the groundwater taking place in this area.

The same principle, only in reverse, has been applied in the lignite mining areas to the soil unit pseudogley. The regional knowledge of the Erftverband allowed, in this case, downgrading of the assumed denitrification potential of the pseudogley from high to low. Due to the influence of mining, the pseudogley indicated in the 1:50,000 soil map has not been affected by waterlogging for a long time. Thus, the assumed temporary or seasonal waterlogging due to poor drainage does not occur anymore. In reality, the pseudogley soils occurring in the lignite mining area are well-aerated, i.e., the occurrence of anaerobic conditions promoting denitrification in the soil is very unlikely. By downgrading the denitrification conditions of the pseudogley, the nitrate concentrations in the leachate of these soils showed a good agreement with the measured values in the second model run.

The two examples illustrate that deviations between model results and measured values may be explainable by a regional blurring of the input data (here, the 1:50,000 soil map) available at the federal state scale. Selective improvements of model input parameters are possible, but require profound regional expert knowledge.

Figure 11 shows a generally good agreement between modelled nitrate concentrations in the leachate and the measured nitrate concentrations in groundwater for an area south of Cologne. However, large deviations occur for the Alfter region (Figure 11, southeastern part of the map section). There, modelled nitrate concentrations in leachate of <10 mg NO_3/L are contrasted by a few measured values from groundwater at a level of >100 mg NO_3/L.

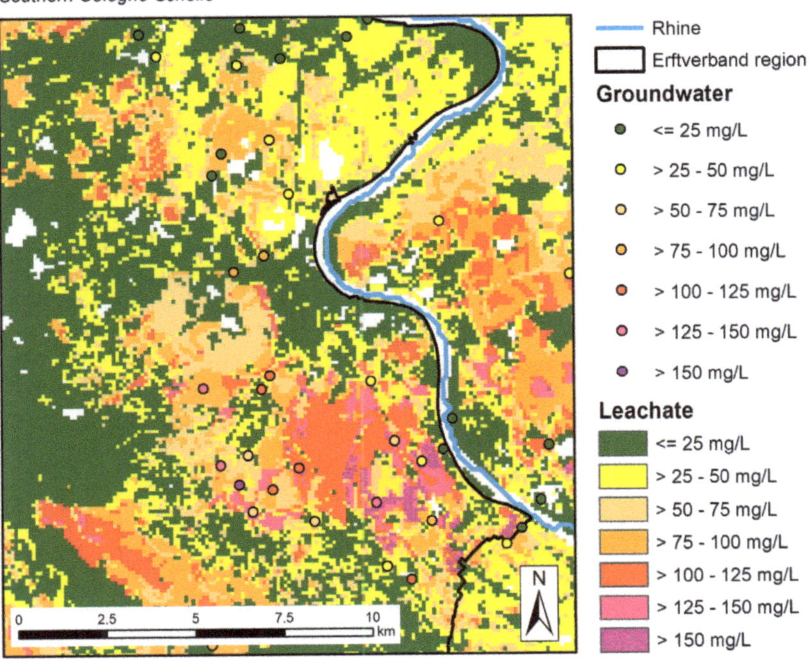

Figure 11. Comparison of the modelled nitrate concentration in the leachate (grids) with measured nitrate concentrations in groundwater (dots) for an area south of Cologne.

The detailed analysis of input parameters revealed for this area a regional blurring of the input data of the agricultural N-balance model RAUMIS. Whereas, in general, an extensive form of agriculture dominates in the Alfter region, the groundwater measuring points with the high nitrate concentrations are located in the vicinity of the few field plots where special crops are cultivated. Consequently, the high nitrate concentrations in groundwater mirror the high N-emissions from these plots. The example illustrates again the dependence of the model results from the input data available for the modelling.

For data protection reasons, agricultural N balances in NRW have been calculated with the RAUMIS model as average values at the community level. Thus, the (high) N-balance surplus of the special crop cultivation of the individual field plots in the Alfter region has been averaged out at the community level. The smaller the area of special crop cultivation compared to the total agriculturally used area in a community, the smaller the mean N balance surplus for this community and the bigger the difference to the N balance surplus of the individual field plot obtained. Exactly this situation applies to the municipality of Alfter. Special crop cultivation comprises <5% of the agriculturally used area of this community, whereas >95% of the agriculturally used area displays an extensive land use. Consequently, as the averaged N balance surplus has been used as input for the subsequent modelling of nitrate concentrations in the leachate, the modelled values are significantly lower than the observed high values in groundwater in the vicinity of the field plots with special crops. The only way to avoid misinterpretations and regional blurring in the model results is to increase the spatial resolution of N-balance surplus calculations, e.g., at the field plot level.

The example of the Alfter region also shows how much the regional agreement between the modelled and measured values may be influenced by the location of measuring points. In the Alfter region, measuring points were installed in the vicinity of field plots, where special crops are cultivated. Cultivation of special crops, however, is actually not representative in this region in comparison to the dominant, rather extensive form of agriculture.

Large deviations between the modelled and the measured values were also found for individual measuring points for the northern Rurscholle (Figure 12A) and for the northwestern Venloer Scholle (Figure 12B). There, modelled nitrate concentrations in the leachate in the range of 75–100 mg NO_3//L contrast with nitrate concentrations in groundwater in the range of >150 mg NO_3/L.

A detailed analysis of the model parameters for this region has shown that the agricultural statistics included in the RAUMIS model do not contain any information on the illegal import of liquid manure from the neighboring Netherlands. The lack of this important additional N source in the calculation of the nitrate concentration in the leachate is an explanation for the deviation from the observed higher nitrate concentrations in groundwater. Another reason may be, as already discussed with the example of the Alfter region, the non-consideration of field plot-specific N emissions from special crops in the N balance surpluses at the community level.

Nitrate concentration in groundwater (measured) and leachate (simulated)

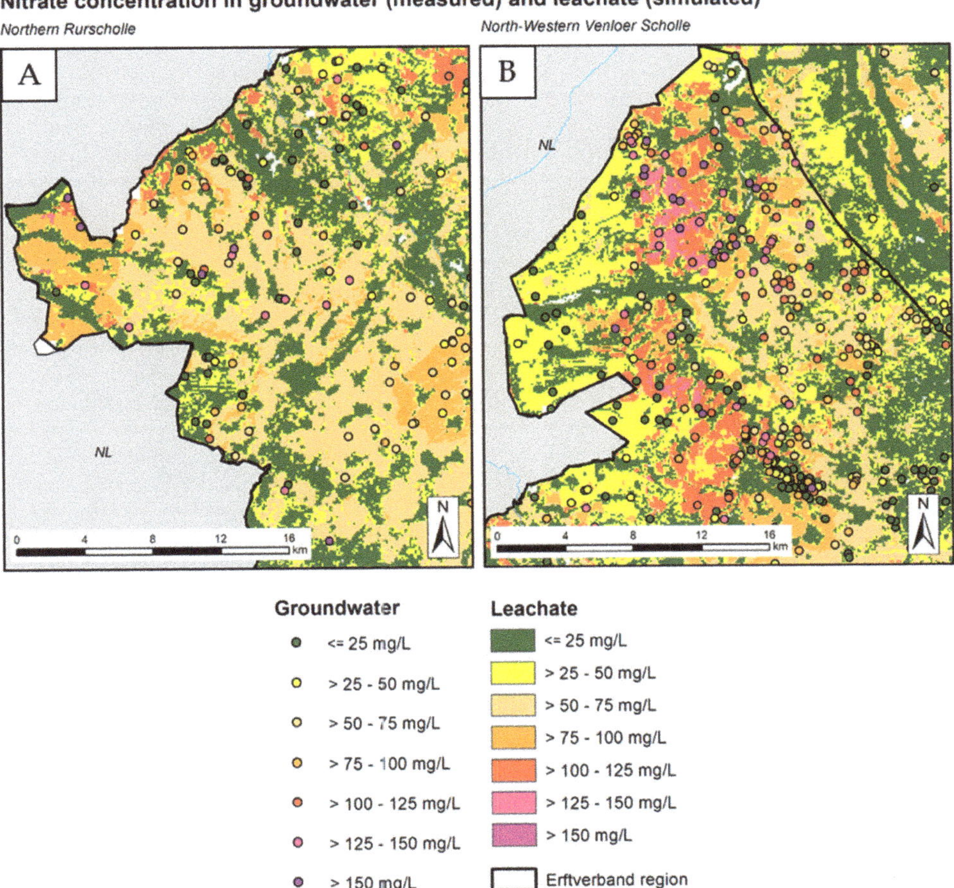

Figure 12. Comparison of the modelled nitrate concentration in the leachate (underlying colors) with measured nitrate concentrations in groundwater (dots) for the northern Rurscholle (**A**) and the north-western Venloer Scholle (**B**).

9. Discussion and Conclusions

In Germany, the modelled nitrate concentration in leachate plays a central role in the dimensioning of nitrogen mitigation measures to achieve the groundwater quality objective according to the specific requirements of the EU Water Framework Directive [8] as well as within the context of the recently adopted general administrative regulation for the designation of nitrate-polluted areas [6]. Against this special practice-related background, the regional representativity and plausibility of modelled nitrate concentrations in the leachate calculated with the model system RAUMIS-mGROWA-DENUZ has been assessed for the federal German states of North Rhine-Westphalia and Rhineland-Palatinate as example regions. Peculiarity and novelty were to use measured nitrate concentrations in groundwater for this purpose.

Results of the plausibility check showed an overall good correspondence in the dimension and the spatial representation of the modelled and the observed nitrate concentrations across the dominant land use categories in both Federal German States. Four general conclusions can be drawn from this finding:

1. There is a clear indication that modelled nitrate concentrations in the leachate are a useful reference framework for assessing large-scale nitrate contamination of groundwater.

2. Apparently, the RAUMIS-mGROWA-DENUZ model system used to calculate the nitrate concentrations in the leachate in this study is capable of reliably representing the relationships and influencing factors that determine the nitrate concentration of the leachate on the scale of Federal States or countries. An application of this model system for the assessment of large-scale nitrate contamination of groundwater beyond the case study region can be recommended and has already been realized, e.g., [46,51].

3. Profound knowledge about local hydrological or agricultural site conditions in the area of responsibility of the Erftverband has shown that local particularities in the area surrounding a monitoring site, e.g., caused by a specific land use or by the occurrence of a sub-scale soil type, are often the reasons for deviations between modelled nitrate concentrations in the leachate and observed nitrate concentrations in groundwater. Based on this finding, we conclude that before recalibrating a nitrate leaching model with the goal of achieving a good fit to observed values, the input data to the model should be carefully reviewed.

4. The measured nitrate concentrations in groundwater from official monitoring networks like the ones used in this study for the plausibility check are available for many regions worldwide [52]. The results of this study indicate that such observed nitrate concentrations in groundwater can provide a solid data source for checking the plausibility of modelled nitrate concentrations in the leachate, provided that certain preselection criteria are met:

- Monitoring data are from the upper aquifer, since only this aquifer experiences direct nitrate inputs with groundwater recharge;
- Monitoring data originate from aquifers with oxidative properties as nitrate inputs are usually reduced by denitrification processes in aquifers with reducing properties;
- Monitoring data are from groundwater monitoring wells or shallow wells to exclude the effect of active pumping, i.e., the mixing of groundwater from different aquifer depths;
- Geographical references for the comparison of modelled and measured values are the inflow areas of the measuring point.

After applying these preselection criteria, nitrate concentrations from groundwater monitoring networks can be recommended for this purpose. This issue is of general practical importance as nitrate concentrations from (official) groundwater monitoring networks are usually not exploited for checking the validity of modelled nitrate concentrations in leachate.

Author Contributions: Conceptualization, F.W.; methodology, all authors; software, T.W.; validation, T.W., N.C.; formal analysis, F.W.; investigation, T.W.; resources, M.E., P.K., N.C.; data curation, T.W.; writing—original draft preparation, F.W.; writing—review and editing, T.W., R.K., N.C., M.E.; visualization, T.W.; supervision, F.W.; project administration, F.W.; funding acquisition, F.W., P.K. All authors have read and agreed to the published version of the manuscript.

Funding: This study received funding by the Ministry of Environment, Agriculture, Nature and Consumer Protection, North Rhine-Westphalia and the Ministry for Environment, Energy, Food and Forestry, Rhineland Palatinate.

Institutional Review Board Statement: Not applicable.

Informed Consent Statement: Not applicable.

Data Availability Statement: Data was obtained from Rhineland-Palatinate State Agency for the Environment in Mainz, North Rhine Westphalia State Office for Nature, Environment and Consumer Protection in Duisburg and from the area of responsibility of the Erftverband in Bergheim. Data in this study was used with the permission of these institutions. Data sharing is not applicable to this article.

Acknowledgments: The authors acknowledge the support by Stephan Sauer from the Rhineland-Palatinate State Agency for the Environment in Mainz.

Conflicts of Interest: The authors declare no conflict of interest. The funding institutions had no role in the design of the study; in the collection, analyses, or interpretation of data; in the writing of the manuscript, or in the decision to publish the results.

References

1. Nitrate Directive. Council Directive of 12 December 1991 Concerning the Protection of Waters against Pollution Caused by Nitrates from Agricultural Sources (91/676/EEC), 1991. Available online: http://data.europa.eu/eli/dir/1991/676/2008-12-11 (accessed on 30 January 2020).
2. Rechtssache C-543/16. Klage, Eingereicht am 27 October 2016—Europäische Kommission/Bundesrepublik Deutschland. Available online: http://curia.europa.eu/juris/document/document.jsf?docid=186628&ode=req&pageIndex=1&dir=&occ=first&part=1&text=&doclang=DE&cid=7291173 (accessed on 30 January 2020).
3. Rechtssache C-543/16. Urteil des Gerichtshofs (Neunte Kammer) vom 21. Juni 2018—Europäische Kommission/Bundesrepublik Deutschland. Available online: http://curia.europa.eu/juris/document/document.jsf;jsessionid=60251F5986268DC961CE7FA3089BF044?text=&docid=204843&pageIndex=0&doclang=DE&mode=req&dir=&occ=first&part=1&cid=7291173 (accessed on 30 January 2020).
4. Nitratbericht. Gemeinsamer Bericht der Bundesministerien für Umwelt, Naturschutz und nukleare Sicherheit (BMU) sowie für Ernährung und Landwirtschaft (BMEL). 2020, 167 S. Available online: https://www.bmu.de/fileadmin/Daten_BMU/Download_PDF/Binnengewaesser/nitratbericht_2020_bf.pdf (accessed on 30 January 2020).
5. DüV (Düngeverordnung). Verordnung über die Anwendung von Düngemitteln, Bodenhilfsstoffen, Kultursubstraten und Pflanzenhilfsmitteln nach den Grundsätzen der Guten Fachlichen Praxis beim Düngen (Düngeverordnung vom 26. Mai 2017 (BGBl. I S. 1305), die durch Artikel 1 der Verordnung vom 28. April 2020 (BGBl. I S. 846) geändert worden ist). Available online: http://www.gesetze-im-internet.de/d_v_2017/index.html (accessed on 30 January 2020).
6. AVV Gebietsausweisung. Allgemeine Verwaltungsvorschrift zur Ausweisung von mit Nitrat Belasteten und Eutrophierten Gebieten. 2020, 41 S. Available online: https://www.bmel.de/SharedDocs/Downloads/DE/Glaeserne-Gesetze/Kabinettfassung/avv-gebietsausweisung.pdf?__blob=publicationFile&v=3 (accessed on 30 January 2020).
7. Bundesrat. Bundesrat—993. Sitzung—18. September 2020, Tagesordnungspunkt 79. Available online: https://.www.bundesrat.de/SharedDocs/downloads/DE/plenarprotokolle/2020/Plenarprotokoll-993.pdf?blob=publicationFile&v=2 (accessed on 30 January 2020).
8. EG-WRRL. Richtlinie 2000/60/EG des Europäischen Parlaments und des Rates vom 23. Oktober 2000 zur Schaffung eines Ordnungsrahmens für Maßnahmen der Gemeinschaft im Bereich der Wasserpolitik (Wasserrahmenrichtlinie). Available online: https://eur-lex.europa.eu/resource.html?uri=cellar:5c835afb-2ec6-4577-bdf8-756d3d694eeb.0003.02/DOC_1&format=PDF (accessed on 30 January 2020).
9. EG-GWR. Richtlinie 2006/118/EG des europäischen Parlaments und des Rates vom 12. Dezember 2006 zum Schutz des Grundwassers vor Verschmutzung und Verschlechterung. Available online: https://eurlex.europa.eu/-LexUriServ/LexUriServ.do?uri=OJ:L:2006:372:0019:0031:DE:PDF (accessed on 30 January 2020).
10. Kunkel, R.; Eisele, M.; Schäfer, W.; Tetzlaff, B.; Wendland, F. Planning and implementation of nitrogen reduction measures in catchment areas based on a determination and ranking of target areas. *Desalination* **2008**, *226*, 1–12. [CrossRef]
11. Kunkel, R.; Herrmann, F.; Kape, H.-E.; Keller, L.; Koch, F.; Tetzlaff, B.; Wendland, F. Simulation of terrestrial nitrogen fluxes in Mecklenburg-Vorpommern and scenario analyses how to reach N-quality targets for groundwater and the coastal waters. *Environ. Earth Sci.* **2017**, *76*, 146. [CrossRef]
12. Wendland, F.; Behrendt, H.; Gömann, H.; Hirt, U.; Kreins, P.; Kuhn, U.; Kunkel, R.; Tetzlaff, B. Determination of nitrogen reduction levels necessary to reach groundwater quality targets in large river basins: The Weser basin case study, Germany. *Nutr. Cycl. Agroecosyst.* **2009**, *85*, 63–78. [CrossRef]
13. Wendland, F.; Heidecke, C.; Keller, L.; Kreins, P.; Kuhr, P.; Tetzlaff, B.; Trepel, M.; Wagner, A. Räumlich differenzierte Quantifizierung der Stickstoffeinträge ins Grundwasser und die Oberflächengewässer in Schleswig-Holsteins. *KW* **2014**, *6*, 327–332.
14. Wendland, F.; Bergmann, S.; Eisele, M.; Gömann, H.; Herrmann, F.; Kreins, P.; Kunkel, R. Model-based analysis of nitrate concentration in the leachate—The North Rhine-Westfalia case study, Germany. *Water* **2020**, *12*, 550. [CrossRef]
15. LAWA. Empfehlungen für eine harmonisierte Vorgehensweise zum Nährstoffmanagement (Defizitanalyse, Nährstoffbilanzen, Wirksamkeit landwirtschaftlicher Maßnahmen) in Flussgebietseinheiten. *LAWA-Produktdatenblätter WRRL* **2017**, *35–37*, 42.
16. Fraters, D.; van Leeuwen, T.; Boumans, L.; Reijs, J. Use of long-term monitoring data to derive a relationship between nitrogen surplus and nitrate leaching for grassland and arable land on well-drained sandy soils in the Netherlands. *Acta Agr. Scand. B S P* **2005**, *65*, 144–154. [CrossRef]
17. Dalgaard, T.; Hansen, B.; Hasler, B.; Hertel, O.; Hutchings, N.J.; Jacobsen, B.H.; Stoumann Jensen, L.; Kronvang, B.; Olesen, J.E.; Schjørring, J.K.; et al. Policies for agricultural nitrogen management—Trends, challenges and prospects for improved efficiency in Denmark. *Environ. Res. Lett.* **2014**, *9*, 16p. [CrossRef]

18. Flussgebiete NRW.: Regional hoch aufgelöste Quantifizierung der diffusen Stickstoff- und Phosphoreinträge ins Grundwasser und die Oberflächengewässer NRWs 2020. Available online: https://www.flussgebiete.nrw.de/regional-hoch-aufgeloeste-quantifizierung-der-diffusen-stickstoff-und-phosphoreintraege-ins-4994#:~{}:text=Mit%20den%20Ergebnissen%20aus%20dem%20Kooperationsprojekt%20GROWA%2B%20NRW,auf%20die%20Aufstellung%20der%20Bewirtschaftungsplanung%20und%20des%20Ma%C3%9Fnahmenprogramms (accessed on 30 January 2020).

19. Groenendijk, P.; Heinen, M.; Klammler, G.; Fank, J.; Kupfersberger, H.; Pisinaras, V.; Gemitzi, A.; Peña-Haro, S.; García-Prats, A.; Pulido-Velazquez, M.; et al. Performance assessment of nitrate leaching models for highly vulnerable soils used in low-input farming based on lysimeter data. *Sci. Total Environ.* **2014**, *499*, 463–480. [CrossRef]

20. Šimunek, J.; Šejna, M.; Saito, H.; Sakai, M.; van Genuchten, M.T. The HYDRUS-1D software package for simulating the movement of water, heat, and multiple solutes in variably saturated media, Version 4.0. In *HYDRUS Software Series 3*; Department of Environmental Sciences, University of California Riverside: Riverside, CA, USA, 2008; Volume 3, p. 315.

21. Manevski, K.; Børgesen, C.D.; Li, X.; Andersen, M.N.; Abrahamsen, P.; Hu, C.; Hansen, S. Integrated modelling of crop production and nitrate leaching with the Daisy model. *MethodsX* **2016**, *3*, 350–363. [CrossRef] [PubMed]

22. Colombani, N.; Mastrocicco, M.; Vincenzi, F.; Castaldelli, G. Modeling Soil Nitrate Accumulation and Leaching in Conventional and Conservation Agriculture Cropping Systems. *Water* **2020**, *12*, 1571. [CrossRef]

23. Arnold, J.G.; Srinivasan, R.; Muttiah, S.; Williams, J.R. Large-area hydrologic modeling and assessment: Part I. *Model Dev. J. Am. Water Resour.* **1998**, *34*, 73–89. [CrossRef]

24. Arheimer, B.; Dahné, J.; Donnelly, C.; Lindström, G.; Strömqvist, J. Water and nutrient simulations using the HYPE model for Sweden vs. the Baltic Sea basin-influence of input-data quality and scale. *Hydrol. Res.* **2012**, *43*, 315–329. [CrossRef]

25. Fuchs, S.; Scherer, U.; Wander, R.; Behrendt, H.; Venohr, M.; Opitz, D. Berechnung von Stoffeinträgen in die Fließgewässer Deutschlands mit dem Modell MONERIS—Nährstoffe, Schwermetalle, und polyzyklische aromatische Kohlenwasserstoffe. *UBA-Texte* **2010**, *45*, 243p.

26. Heidecke, C.; Hirt, U.; Kreins, P.; Kuhr, P.; Kunkel, R.; Mahnkopf, J.; Schott, M.; Tetzlaff, B.; Venohr, M.; Wagner, A.; et al. Endbericht zum Forschungsprojekt "Entwicklung eines Instrumentes für ein flussgebietsweites Nährstoffmanagement in der Flussgebietseinheit Weser" AGRUM+-Weser. Braunschweig: Johann Heinrich von Thünen-Institut. *Thünen. Rep.* **2015**, *21*, 380. [CrossRef]

27. Herrmann, F.; Keller, L.; Kunkel, R.; Vereecken, H.; Wendland, F. Determination of spatially differentiated water balance components including groundwater recharge on the Federal State level—A case study using the mGROWA model in North Rhine-Westphalia (Germany). *J. Hydrol. Reg. Stud.* **2015**, *4*, 294–312. [CrossRef]

28. Schaap, M.; Hendriks, C.; Kranenburg, R.; Kuenen, J.; Segers, A.; Schlutow, A.; Nagel, H.D.; Ritter, A.; Banzhaf, S. PINETI-3: Modellierung atmosphärischer Stoffeinträge von 2000 bis 2015 zur Bewertung der ökosystem-spezifischen Gefährdung von Biodiversität durch Luftschadstoffe in Deutschland. *UBA-Texte* **2018**, *79*, 149.

29. Wendland, F.; Behrendt, H.; Hirt, U.; Kreins, P.; Kuhn, U.; Kuhr, P.; Kunkel, R.; Tetzlaff, B. Analyse von Agrar- und Umweltmaßnahmen zur Reduktion der Stickstoffbelastung von Grundwasser und Oberflächengewässer in der Flussgebietseinheit Weser. *Hydrologie und Wasserbewirtschaftung* **2010**, *54*, 231–244.

30. Schäfer, W.; Höper, H.; Müller, U. Diffuse Nitrat- und Phosphatbelastung—Ergebnisse der Bestandsaufnahme der EUWRR in Niedersachsen. *Geoberichte* **2007**, *2*, 3–32.

31. Wendland, F. Die Nitratbelastung in den Grundwasserlandschaften der "alten" Bundesländer (BRD). *Forschungszentrum Jülich, Berichte aus der Ökologischen Forschung* **1992**, *2*, 150p.

32. MKULNV. *Entwicklung und Stand der Abwasserbeseitigung in Nordrhein-Westfalen. Auflage*; Ministerium für Klimaschutz, Umwelt, Landwirtschaft, Natur- und Verbraucherschutz Nordrhein-Westfalen: Düsseldorf, Germany, 2014; Volume 17, 74p.

33. LAWA-AH. Recommendations for the Establishment of Flood Hazard Maps and Flood Risk Maps. In Proceedings of the Adopted at the 139th LAWA General Meeting, Dresden, Germany, 25–26 March 2010.

34. DEA-Datendrehscheibe. 2016. Available online: https://www.elwasweb.nrw.de/elwas-web/index.jsf (accessed on 30 January 2020).

35. Weymann, D.; Well, R.; Flessa, H.; von der Heide, C.; Deurer, M.; Meyer, K.; Konrad, C.; Walther, W. Groundwater N$_2$O emission factors of nitrate-contaminated aquifers as derived from denitrification progress and N$_2$O accumulation. *Biogeosciences* **2008**, *5*, 1215–1226. [CrossRef]

36. Vogel, J.C.; Talma, A.S.; Heaton, T.H.E. Gaseous nitrogen as evidence for denitrification in groundwater. *J. Hydrol.* **1981**, *50*, 191–200. [CrossRef]

37. Rüppel, C.; Eschenbach, W.; Meyer, K. Messung des Exzess-N$_2$ im Grundwasser mit der N$_2$/Ar-Methode als neue Möglichkeit zur Prioritätensetzung und Erfolgskontrolle im Grundwasserschutz. Niedersächsischer Landesbetrieb für Wasserwirtschaft, Küsten- und Naturschutz. *Grundwasser* **2012**, *15*, 31.

38. Eschenbach, W.; Budziak, D.; Elbracht, J.; Höper, H.; Krienen, L.; Kunkel, R.; Meyer, K.; Well, R.; Wendland, F. Möglichkeiten und Grenzen der Validierung flächenhaft modellierter Nitrateinträge ins Grundwassermit der N$_2$/Ar-Methode. *Grundwasser* **2018**, *23*, 125–139. [CrossRef]

39. Obermann, P. *Hydrochemische/Hydromechanische Untersuchungen zum Stoffgehalt von Grundwasser bei landwirtschaftlicher Nutzung*; Besondere Mitteilungen zum Deutschen Gewässerkundlichen Jahrbuch: Dusseldorf, Germany, 1981; pp. 1–217.

40. Leuchs, W. Geochemische und mineralogische Auswirkungen beim mikrobiellen Abbau organischer Substanz in einem anoxischen Porengrundwasserleiter. *Z. Dtsch. Geol. Ges.* **1988**, *139*, 415–423.

1. Merz, C.; Steidl, J.; Dannowski, R. Parameterization and regionalization of redox based denitrification for GIS-embedded nitrate transport modeling in Pleistocene aquifer systems. *Environ. Geol.* **2009**, *58*, 1587–1599. [CrossRef]
2. *DVWK-Regeln zur Wasserwirtschaft. Entnahme- und Untersuchungsumfang von Grundwasserproben*; Paul Parey: Hamburg/Berlin, Germany, 1992; Volume 128, p. 36.
3. Hannappel, S. Die Beschaffenheit des Grundwassers in den hydrogeologischen Strukturen der neuen Bundesländer. Berliner Geowiss. *Abhandlungen* **1996**, *A182*, 151.
4. Hölting, B.; Coldewey, W.G. *Hydrogeologie: Einführung in die Allgemeine und Angewandte Hydrogeologie*; Spektrum: Stuttgart, Germany, 2013; Volume 8, p. 384.
5. Ackermann, A.; Heidecke, C.; Hirt, U.; Kreins, P.; Kuhr, P.; Kunkel, R.; Mahnkopf, J.; Schott, M.; Tetzlaff, B.; Venohr, M. Der Modellverbund AGRUM als Instrument zum landesweiten Nährstoffmanagement in Niedersachsen. *Thünen Rep.* **2015**, *37*, 283.
6. Wendland, F.; Keller, L.; Kuhr, P.; Kunkel, R.; Tetzlaff, B. *Regional differenzierte Quantifizierung der Nährstoffeinträge in das Grundwasser und in die Oberflächengewässer Mecklenburg-Vorpommerns unter Anwendung der Modellkombination GROWA-DENUZ-WEKU-MEPhos*; Endbericht zum Forschungsprojekt im Auftrag des Landesamts für Umwelt, Naturschutz und Geologie Mecklenburg-Vorpommern: Frankfurt, Germany, 2015; p. 300.
7. Kuhr, P.; Haider, J.; Kreins, P.; Kunkel, R.; Tetzlaff, B.; Vereecken, H.; Wendland, F. Model Based Assessment of Nitrate Pollution of Water Resources on a Federal State Level for the Dimensioning of Agro-environmental Reduction Strategies: The North Rhine-Westphalia (Germany) Case Study. *Water Resour. Manag.* **2013**, *27*, 885–909. [CrossRef]
8. Kunkel, R.; Bach, M.; Behrendt, H.; Wendland, F. Groundwater-borne nitrate intakes into surface waters in Germany. *Water Sci. Technol.* **2004**, *49*, 11–19. [CrossRef]
9. Ehlschläger, C. Using the AT Search Algorithm to Develop Hydrologic Models from Digital Elevation Data. In Proceedings of the International Geographic Information Systems (IGIS) Symposium '89, Baltimore, MD, USA, 18–19 March 1989; pp. 275–281.
50. Kunkel, R.; Voigt, H.-J.; Wendland, F.; Hannappel, S. Die natürliche, ubiquitär überprägte Grundwasserbeschaffenheit in Deutschland. *Schriften des Forschungszentrums Jülich, Reihe Umwelt/Environment* **2004**, *47*, 204.
51. Andelov, M.; Kunkel, R.; Uhan, J.; Wendland, F. Determination of nitrogen reduction levels necessary to reach groundwater quality targets in Slovenia. *J. Environ. Sci.* **2014**, *26*, 1806–1818. [CrossRef] [PubMed]
52. IGRAC. *Groundwater Monitoring Programmes: A Global Overview of Quantitative Groundwater Monitoring Networks*; IGRAC: Delft, The Netherlands, 2020.

 *water*

Article

Investigating the Effects of Agricultural Water Management in a Mediterranean Coastal Aquifer under Current and Projected Climate Conditions

Vassilios Pisinaras *, Charalampos Paraskevas and Andreas Panagopoulos

Soil & Water Resources Institute, Hellenic Agricultural Organization, Gorgopotamou, Sindos, 57400 Thessaloniki, Greece; paraskevasb@gmail.com (C.P.); a.panagopoulos@swri.gr (A.P.)
* Correspondence: v.pisinaras@swri.gr; Tel.: +30-2310-798-790

Abstract: Coastal delta plains are areas with high agricultural potential for the Mediterranean region because of their high soil fertility, but they also constitute fragile systems in terms of water resources management because of the interaction of underlying aquifers with the sea. Such a case is the Pinios River delta plain located in central Greece, which also constitutes a significant ecosystem. Soil and Water Assessment Tool (SWAT) and SEAWAT models were combined in order to simulate the impact of current water resources management practices in main groundwater budget components and groundwater salinization of the shallow aquifer developed in the area. Moreover, potential climate change impact was investigated using climate data from Regional Climate Model for two projected periods (2021–2050 and 2071–2100) and two sea level rise scenarios (increase by 0.5 and 1 m). Modeling results are providing significant insight: although the contribution of the river to groundwater inflows is significant, direct groundwater recharge from precipitation was found to be higher, while capillary rise constitutes a major part of groundwater outflows from the aquifer. Moreover, during the simulation period, groundwater flow from the aquifer to the sea were found to be higher than the inflows of seawater to the aquifer. Regarding climate change impact assessment, the results indicate that the variability in groundwater recharge posed by the high variability of precipitation during the projected periods is increasing the aquifer's deterioration potential of both its quantity and quality status, the latter expressed by the increased groundwater Cl^- concentration. This evidence becomes more significant because of the limited groundwater storage capacity of the aquifer. Concerning sea level rise, it was found to be less significant in terms of groundwater salinization impact compared to the decrease in groundwater recharge and increase in crop water needs.

Keywords: seawater intrusion; Soil and Water Assessment Tool; SEAWAT model; irrigation management; groundwater; climate change; sea level rise

Citation: Pisinaras, V.; Paraskevas, C.; Panagopoulos, A. Investigating the Effects of Agricultural Water Management in a Mediterranean Coastal Aquifer under Current and Projected Climate Conditions. *Water* **2021**, *13*, 108. https://doi.org/10.3390/w13010108

Received: 11 December 2020
Accepted: 30 December 2020
Published: 5 January 2021

1. Introduction

Globally, the agricultural sector constitutes the dominant water consumer, as about 80% of the total water consumption is accounted to agriculture [1], while according to Rost et al. [2], irrigation water use, abstracted from rivers, lakes and aquifers has been estimated to be about 70% of total human blue water consumption. Irrigation demand is estimated to be higher in Mediterranean region and especially in the south and the east part, in which irrigation accounts for 74% and 81% of the total water withdrawals, respectively [3]. Therefore, the relation between agricultural production and water resources is direct, especially in arid and semi-arid areas, such as located in the south and east Mediterranean, where agricultural production is largely dependent on irrigation. Taking into account: (a) the expected population growth which will increase the agricultural production needs, (b) the anticipated reduction in water resources availability in Mediterranean region due to climate change [4–6] and (c) the overexploitation and overall poor water resources management [7], the necessity for effective water management in agricultural areas becomes very important. More specifically, according to Molden [8], crop water requirements and therefore agricultural water demand is expected to be almost doubled by 2050, assuming that the current status of water productivity remains stable. Nevertheless, according to Olesen et al. [9] increasing irrigation amounts in Mediterranean region will possibly not be a viable option due to water resources availability reduction, as a consequence of total runoff and groundwater recharge reduction. Moreover, Garrote et al. [10] indicate that a reduction on future maximum potential water withdrawal for irrigation in the south Mediterranean-European countries is expected, which was estimated to be higher for basins in Iberian Peninsula and Greece.

Especially for groundwater management in Mediterranean region related to agriculture, two more critical aspects are identified: (a) Considering the fact that groundwater constitutes the primary irrigation water source for a significant part of the Mediterranean region, the importance of groundwater for irrigated land becomes vital. As indicated by Garrido and Iglesia [11] and Fornes et al. [12], in most cases, almost all groundwater extracted in Mediterranean countries is used for irrigation. (b) Due to the long shoreline of Mediterranean region, a significant part of fertile agricultural land is developed in coastal deltaic systems in which coastal aquifers constitutes a significant source of irrigation water. Because of the hydraulic connection between coastal aquifers and the sea, overexploitation can potentially lead to seawater intrusion and therefore to groundwater quality deterioration. According to Mazi et al. [13], several aquifer systems located along the Mediterranean coastline are significantly impacted by seawater intrusion, while according to Nixon et al. [14], seawater intrusion has affected large areas of the Mediterranean coastline in Italy, Spain and Turkey. The impact of seawater intrusion in groundwater quality is expected to increase in Mediterranean due to increased groundwater abstractions driven by increased irrigation water requirements. Stigter et al. [15] indicated that groundwater quality deterioration because of seawater intrusion is expected to increase due to climate change effects in three coastal aquifers located in Morocco, Portugal and Spain. Haj-Amor et al. [16] simulated an increase in average aquifer salinity located in a Tunisian coastal oasis from 4.2 dS m^{-1} in 2018 to about 5.3 dS m^{-1} in 2050.

Considering the above, effective water resources management in coastal agricultural areas of the Mediterranean region is more crucial than ever in order to cope and potentially adapt to climate change effects and thus maintain sustainability. Water systems modeling is inevitably one of the most effective tools in water resources management due to the fact that based on assumptions, the complex processes and mechanisms taking place in a water system can be simulated and represented in a realistic manner. Especially for climate change impact studies, water models give the opportunity to incorporate climate data into the corresponding processes and therefore assess and quantify the impacts of climate change. One of the most effective approaches for simulating groundwater salinization processes is numerical modeling. Such an approach has also been applied in Mediterranean coastal aquifers. Despite the fact that this approach requires significant computational resources, numerical modeling and especially variable density models, provide a more realistic representation of groundwater salinization processes [17]. Alcolea et al. [18] combined a surface water balance model with the open-source finite elements code SUTRA [19] in order to simulate the dynamics of an unconfined aquifer discharging in a lagoon located in Spanish Mediterranean coast. The model gave significant insight towards the understanding of the link between the aquifer and the lagoon. SUTRA was also applied by Haj-Amor et al. [16] in order to simulate climate change effects in an aquifer located in a Tunisian coastal oasis. Hugman et al. [20] applied a density-coupled flow and transport model in a coastal aquifer located in southern Portugal and concluded that the adverse effects of climate change in saltwater intrusion are attenuated by the slow rate of movement of the freshwater-saltwater interfaces. Stigter et al. [15] applied the FEN code, which consists of well-known groundwater flow codes in three coastal aquifers located in Morocco, Portugal and Spain in order to simulate climate change impacts. Siarkos and Latinopoulos [21] applied the finite difference code SEAWAT to an overexploited coastal aquifer located in north Greece. SEAWAT model has been also applied in Nile Delta Aquifer [22], in a coastal aquifer in Lebanon [23] and in the aquifer systems located in Apulia region, Italy [24,25].

The present study aims to investigate the impacts of current water resources management practices in the quantity and salinity status (expressed as Cl^-) of an environmentally sensitive, coastal aquifer located in central Greece, focusing on potential climate change effects. This is achieved by applying sequentially two well established modeling codes, namely Soil and Water Assessment Tool (SWAT) [26,27] and SEAWAT [28] models with climate data from a carefully selected high resolution Regional Climate Model (RCM) for two projected periods indicating high variability in terms of precipitation and temperature variation. Moreover, two sea level rise scenarios are tested in order to identify and assess potential impacts on quantity and salinity status of the study area aquifer.

2. Materials and Methods

2.1. Study Area Description

The location and boundary of the Pinios River Deltaic Plain (PRDP), as well as the geological regime of the surrounding area is presented in Figure 1. PRDP covers an area of about 75 km^2 and is situated at the downstream-most part of Pinios River basin, which constitutes the largest fully developing basin in Greece (11,000 km^2). The economic, social and environmental significance of PRDP is high, as the agricultural and touristic activities developed in PRDP are significantly supporting the local society, while the basin has been included in NATURA2000 network (GR1420002) and is a designated international Important Bird Area.

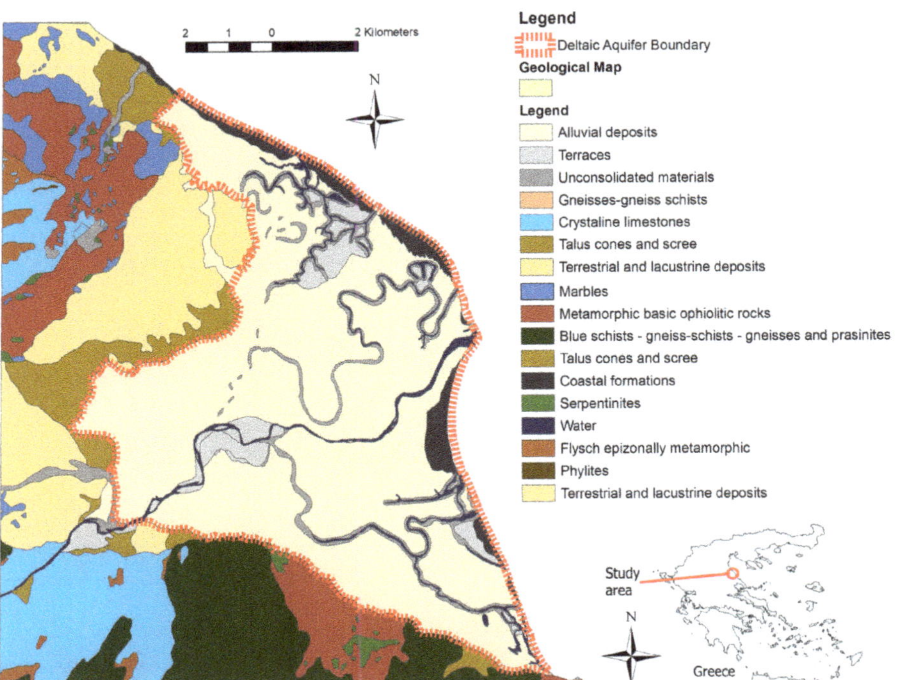

Figure 1. Location and geological map [29] of the study area.

Two major geological formations are found in the wider study area, including sequences of folded alpine formations and plio-quaternary deposits in which PRDP is situated.

More specifically, Neogene terrestrial and lacustrine deposits spread at the northwestern boundary of PRDP, which are considered to be the bedrock of the overlying Pleistocene and Holocene sediments. Concerning Pleistocene deposits, they are located at the western boundary of PRDP and they include talus cones and screes. Holocene is represented by alluvial deposits that are dominant in PRDP and considered hydrologically significant coastal formations situated along the coast, and unconsolidated material along the old and recent Pinios river course. Metamorphic basic ophiolithic rocks and blue gneiss-schists and prasinites outcrops are dominant at the gern margins of the basin.

With regard to the hydrogeological conditions of PRDP, three major hydrogeological units are identified according to the study of Alexopoulos et al. [30], in which geophysical investigation methods were applied: (a) An upper hydrogeological unit in which an unconfined aquifer occurs, (b) a middle hydrogeological unit which indicates very low permeability and therefore serves as an aquitard and (c) a lower hydrogeological unit in which a confined aquifer is present. The unconfined aquifer exists within the alluvial deposits and its thickness is up to 10–15 m. The general trend observed for alluvial deposits is to become finer shifting from the margins of the plain (at the west) towards the coast to the east, with subsequent impact on the hydraulic properties of the unconfined aquifer. Coarser fractions, at least towards the upstream part of the deltaic plain, are indicated along the river course as a result of river sediment deposition process. A typical alluvial stratigraphic structure occurs, hence, continuous alterations of fine (silt and clay) and coarse sediments are found. The aquitard consists mainly of fine material (clay-marl composition) and its thickness varies between 30 and 35 m, while it clearly discretizes the overlying and underlying aquifers. With regard to the lower hydrogeological unit (the confined aquifer), it consists mainly of sandstones and compacted conglomerates possibly of the Neogene sequence, which crops out and is observed at the western part of the wider study area. Since there is inadequate hydrogeological information for the confined aquifer, and groundwater abstractions from it are very limited, the present study focuses on the unconfined aquifer only.

The unconfined aquifer of PRDP was, almost exclusively, used to serve irrigation needs of the deltaic plain. Therefore, according to the local farmers' information, more than 600 small diameter groundwater wells of depth ranging between 6 and 15 m were used in order to cover irrigation demands. This water resources management motif has progressively changed the last 15–20 years. Pinios River surface waters nowadays serve a significant part of the irrigation needs, especially at the western and southern part of the deltaic plain, while groundwater is used complementarily and in conjunction with Pinios River surface water. Land use distribution in the deltaic plain resulted from CORINE2000 land cover and crop spatial distribution data as provided by the Hellenic Payment and Control Agency for Guidance and Guarantee Community Aid, is presented in Figure 2. More than 75% of the deltaic plain area is covered by agricultural land. The dominant crop in PRDP is corn (20.11%), followed by wheat (17.27%) and sunflower (13.73%). Other crops such as kiwi fruit, cotton, alfalfa and olives are also cultivated in the study area and are covering a considerable portion of agricultural land.

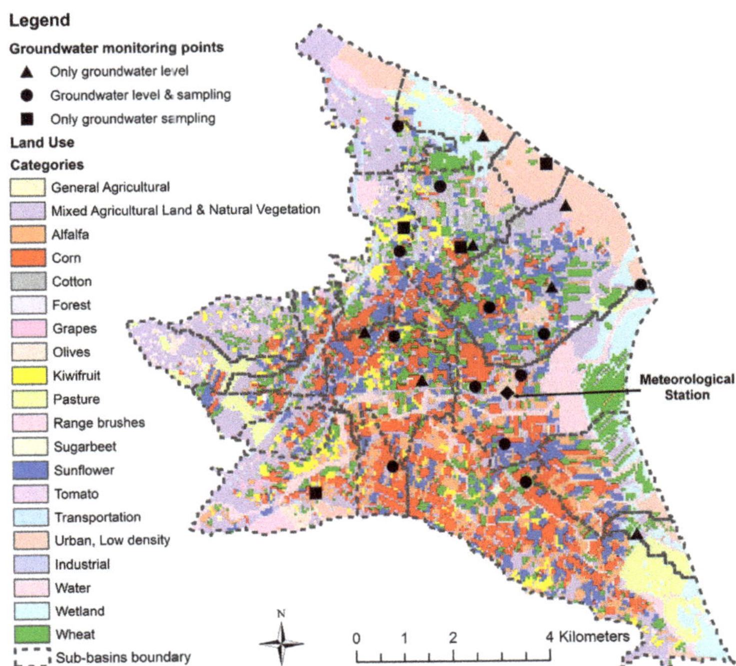

Figure 2. Land use map of PRDP. Groundwater monitoring network and SWAT model sub-basins are also illustrated.

2.2. Modeling

Two models were combined and applied in order to reliably simulate the several components of the hydrologic budget of the aquifer system. The core of the modeling framework is the SEAWAT code [28] which was used in order to simulate groundwater flow, groundwater budget and seawater intrusion in the coastal aquifer. SEAWAT constitutes a combination of MODFLOW [31] and MT3DMS [32] codes developed in order to simulate three-dimensional, variable-density, transient groundwater flow and pollution transport in porous media and has been globally applied in numerous aquifers. The groundwater flow equation is solved using an implicit finite-difference approximation while several implicit or explicit finite-difference methods can be applied in order to solve the solute transport equation.

The second model included in the modeling framework is the Soil & Water Assessment Tool (SWAT) model [26,27]. SWAT is one of the most widely applied watershed management models that it is physically-based and semi-distributed. It was initially developed to be applied in large, ungauged watersheds under complex soil, land use and management conditions, while it simulates a wide range of processes including surface and subsurface hydrology, crop and vegetative growth, pesticide transport and fate and nutrient transport and cycling in streams, soils and crop uptake [33]. SWAT was chosen to be included in the present modeling framework for three reasons. The first reason is that SWAT gives the opportunity to simulate actual crop growth and subsequently crop water needs by incorporating the local-specific cultivation practices. Crop growth in SWAT is simulated by a simplified version of EPIC model [34] according to which the concept of accumulated heat units is used to simulate the phenological plant development. The calculation of potential biomass production is made according to Monteith and Moss approach [35], while water, temperature and nutrient stress can potentially affect crop growth. Therefore, the actual water amount consumed by crops in an agricultural area, which constitutes the major water sink, can be estimated by a physically based approach and in essence, accounting for the specific land and water management conditions.

The second reason is that SWAT simulates the land phase of the hydrologic cycle incorporating all the related components and therefore groundwater recharge from the vadose zone can be computed. The third reason is that SWAT incorporates the estimation of capillary rise from the shallow aquifers to the soil profile for which there is strong evidence that it constitutes a significant component of the PRDP phreatic aquifer budget. Capillary rise in SWAT is empirically calculated as the product of revap coefficient and potential evapotranspiration and it is controlled by the amount of water existing in the shallow aquifer. In order to estimate the potential of capillary rise to contribute to crop water needs, HYDRUS 1-D [36] was applied for several combinations of crops, soil profiles and groundwater table depth. HYDRUS 1-D internal pedotransfer functions were used in order to predict soil hydraulic parameters. HYDRUS 1-D has been applied for the same purposes in several studies [37–39]. SEAWAT and SWAT have been effectively applied together in the study of Chang et al. [40] in order to simulate the effects of climate change and urbanization on groundwater resources in a small barrier island located in the USA.

2.3. Data Collection

Except from the geological, hydrogeological and land use data presented above, a comprehensive wells' census was carried out in the study area, that resulted in the establishment of a monitoring network in which monthly groundwater level measurements and seasonal groundwater sampling was performed for the period October 2013–September 2015 (Figure 2). This network consists of: (a) seventeen wells in which groundwater level measurements and sampling was performed, (b) seven wells in which only groundwater level measurements were conducted and (c) four wells in which only groundwater sampling was performed. Due to the very mild slopes observed in PRDP, the location of all monitoring wells was recorded using a high accuracy Leica Viva GS08 GPS system in order to achieve high precision altitude values hence reliable absolute groundwater level elevation. Chloride concentrations were determined at the collected groundwater samples. Moreover, topographical, weather and meteorological data were collected in order to support SWAT application. Therefore, a Digital Elevation Model (DEM) produced from the interpolation of elevation contours of 1:5000 scale topographic maps was used. Also, detailed soil data for the PRDP were incorporated that include more than 40 soil profiles [41]. Weather data including precipitation, temperature, wind speed, relative humidity and global solar radiation were gathered from a local meteorological station (Figure 2), which is located at the center of the deltaic plain and is considered representative of the climate conditions at PRDP. Since the current study focuses on agriculture, information on cultivation practices was collected from local agricultural cooperatives and farmers including sowing and harvest day of crops, tillage practices, crop yields, as well as irrigation practices and applied irrigation water quantities.

2.4. Climate Change Impacts

In order to investigate the potential climate change impacts on PRDP, data from RACMO2 Regional Climate Model (RCM) [42] driven by ECHAM5-r3 Global Circulation Model (GCM) were used. This RCM was incorporated in the framework of ENSEMBLES project and it was implemented in the current study due to its better performance in simulating climate conditions compared to other models. More specifically, as stated by Karali et al. [43] and based on ENSEMBLES [44], RACMO2 has been found to present the highest accuracy in simulating climate and extremes in Mediterranean region, when compared to the other climate models included in ENSEMBLES project dataset. Deidda et al. [45] investigated the performance of 14 RCMs in representing precipitation and temperature variation over six Mediterranean catchments and their results demonstrated that RACMO2 is included in the best options for the four catchments while it was indicated as good option for the other two. Moreover, Kostopoulou et al. [46] compared datasets produced by RCMs of ENSEMBLES project against E-OBS gridded dataset and concluded that RACMO2 is satisfactorily reproducing extreme temperature and precipitation patterns in the Balkan Peninsula. Furthermore, this RCM has been efficiently used in several studies for climate change impact assessment on water resources around Greece [47–49]. Two time periods were considered in order to investigate the potential climate change impacts on PRDP: the period 2021–2050 representing near future and the period 2071–2100 representing far future.

Sea level rise is expected to be one of the most severe impacts of climate change. As indicated by Intergovernmental Panel on Climate Change (IPCC) [50] and depending on the emissions scenario, global mean sea level is likely to increase on the average by 0.24–0.30 m for the period 2046–2065 and by 0.40–0.63 for the period 2081–2100. The likely range of global mean sea level rise for the highest emissions scenario (RCP8.5) is 0.45–0.82 m for the period 2081–2100. Other studies are indicating higher sea level rise values, such as this of Rahmstorf [51], according to which sea level rise values of 0.5 to 1.4 m are projected by year 2100, while Hansen [52] mentions that sea level rise is probable to be significantly larger than the range presented in most studies. Ketabchi et al. [53] reviewed studies investigating the impacts of sea level rise in coastal aquifers. The majority of the studies included are applying sea level rise up to 1 m, while sea level rise by 1 m constitutes the most frequently applied scenario. Based on the above, two scenarios were considered for the study area, assuming 0.5 m and 1.0 m sea level rise. These scenarios were incorporated in SEAWAT in a simplified way and more specifically by increasing the hydraulic head in Constant Head boundary along the coast. Both sea level rise scenarios were applied for both future periods.

3. Results

3.1. SWAT Model Application

Based on topographic, soil and land use data, PRDP was divided into 20 sub-basins which were further divided into 384 Hydrologic Response Units (HRUs). Moreover, cultivation and irrigation practices, as well as capillary rise as estimated by HYDRUS-1D application were introduced in SWAT. Since SWAT was originally developed to be applied in ungauged basins, it has been found to demonstrate satisfactory performance when applied in basins for which calibration data was not available or the calibration dataset was of limited quantity or quality [33,54–56]. Nevertheless, model calibration has to be applied when it is possible in order to decrease the degree of uncertainty in model results. The most common approach in hydrological models' calibration is the adjustment of several parameters in ranges restricted by the physical boundaries of each, in order to achieve satisfactory match between observed and simulated river discharge. This approach is not applicable for the purposes of the present study since PRDP contributes only with less than 0.7% to the total area of Pinios River basin, and combined to the very mild slopes, the overall contribution of PRDP in surface runoff is very hard to be identified in Pinios main river course. Moreover, some other streams located in PRDP are indicating surface runoff only after severe precipitation events.

Considering the above, SWAT application in PRDP was based on actual evapotranspiration (ETa) calibration. This parameter was chosen due to the fact that it constitutes the major water sink in PRDP and moreover it can be estimated with a satisfactory degree of accuracy. Reference evapotranspiration (ETo) was calculated using the Penman-Monteith formula [57]. Then, potential evapotranspiration for each crop (ETc) cultivated in PRDP, which corresponds to the evapotranspiration that would have occurred under full crop development, was calculated as the product of crop coefficient (Kc) and ETo. This method which was introduced by Jensen [58] and further developed by Doorenbos and Pruitt [59] and Allen et al. [60] gives the ability to adjust ET based on specific crops and local conditions and therefore to have a satisfactory approach of ETc. Kc values after Papazafeiriou [61] and Galanopoulou-Sendouka [62], that are representative of the Greek cultivation environment, were used. Considering the fact that actual crop yields in PRDP during the calibration period (2013–2015) were very close to full development crop yields, as stated by local agricultural cooperatives and farmers, it was assumed that ETa was also very close to ETc. Consequently, ETc for each crop was used as the basis for ETa calibration in SWAT.

Two groups of parameters were adjusted in order to perform SWAT model calibration in PRDP. The first group includes parameters that relate evapotranspiration process and soil. Two parameters were identified to significantly affect actual evapotranspiration, namely the soil evaporation-compensation factor (ESCO) and the plant uptake compensation factor (EPCO). ESCO controls the depth distribution of soil evaporative demands and values of 0.7–0.8 were found to fit better in PRDP. EPCO controls the depth within the soil profile from which plant water uptake can occur and values of 0.95–1.0 were found to improve matching between ETa and ETc. The second group includes parameters that control plant growth and subsequently crop water uptake and actual evapotranspiration.

ET calibration results are presented in Table 1. With regard to corn, the average simulated ETa was found to be very close to the corresponding ETc, as their difference was found to be 3.9 mm or 0.7%. Similarly, the corresponding results for cotton are satisfactory (difference of 3.4 mm or 0.6%). On the average, higher differences between simulated ETa and ETc were noted for sunflower (14.4 mm or 2.7%) and winter wheat (32.1 mm or 10.9%). The highest differences were found for alfalfa (55.0 mm or 7.3%) and kiwi fruit (48.3 mm or 6.6%). As expected, the above results indicate that for all crops the average simulated ETa was found to be lower than ETc. This fact can be attributed to the following: a) intrinsically, the simulated ETa has incorporated the real cultivation and water management practices which may deviate from the nominal conditions assumed in ETc calculation and b) SWAT incorporates pressures resulting from lack of water and nutrients, as well as from the high or low temperature, thus reducing the actual evapotranspiration.

Table 1. Comparison of ETa for the dominant crops of the study area, as simulated by SWAT, and ETc as calculated using the Kc approach for two cultivation periods (2014–2015).

| | | **Actual Evapotranspiration (ETa) (mm per Cultivation Period)** | | | | | |
		Corn	Cotton	Sunflower	Wheat	Alfalfa	Kiwi Fruit
	No. of HRUs	57	12	64	56	32	19
	Average	611.2	582.4	539.5	262.6	702.9	684.4
SWAT model	Median	612.2	591.4	540.1	262.4	703.5	681.5
	Min	540.1	543.1	520.0	226.0	667.0	669.1
	Max	636.2	598.2	551.3	274.0	716.9	701.0
		Crop Reference Evapotranspiration (ETc) (mm per Cultivation Period)					
Kc approach		615.1	585.8	553.9	294.7	757.9	732.7

3.2. SEAWAT Model Application

The model grid as well as boundary conditions assigned in PRDP phreatic aquifer are presented in Figure 3. The study area was discretized in 25,080 cells of 50 × 50 m size and the whole model grid area was 62.7 km². SEAWAT was applied on the phreatic aquifer, the thickness of which varies between 5 and 15 m, while it was simulated as a single layer due to its small thickness and in order to reduce computational effort. The hydraulic interaction of the aquifer with the sea (eastern boundary) was simulated with Dirichlet (or first-type) boundary condition by implementing the Constant Head package of MODFLOW code. Head value was assumed to be 0 m above mean sea level (amsl), while chloride concentration was also kept constant at 21 g/L. The hydraulic communication of the aquifer with talus cones and screes at the west was simulated with Cauchy (or third-type) boundary condition by implementing the General Head package of MODFLOW code. Head values from neighboring wells were used to assign the relevant parameter, while the initially estimated conductance values, on the basis of the boundary geometry and the hydrogeological properties of the aquifer matrix and the bounding formations, were fine trimmed during calibration. The RIVER package of MODFLOW code (Cauchy boundary condition), was employed to account for the hydraulic interaction of the aquifer with the main hydrographic network. River stage and geometry were assigned based on measurements at several points, while river conductance was calibrated following the initially introduced estimates. Finally, capillary rise, as estimated by SWAT for each sub-basin, was introduced in the simulation using the evapotranspiration (ET) package of MODFLOW. Due to the fact that the pumping program and potential of each production well in the aquifer is not known, groundwater abstractions were also incorporated in the ET package at the sub-basin scale, calculated on the basis of irrigation demands resulted from SWAT. To this end, it was assumed that the totally used irrigation water is abstracted from groundwater at the areas where no surface water irrigation network occurs.

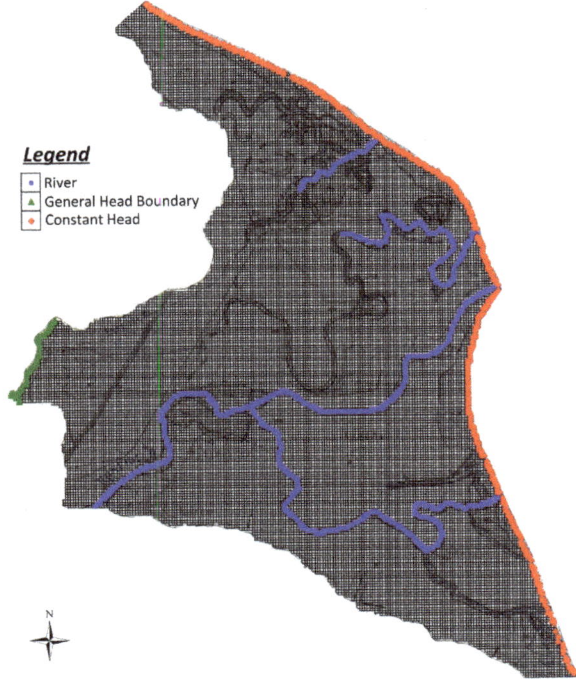

Figure 3. Simulation grid of PRDP phreatic aquifer along with key boundary conditions. Where no specific outer boundary condition is indicated, no flow conditions are implied.

SEAWAT model calibration was based on the methodological framework of Siarkos and Latinopoulos [21], according to which steady-state (only groundwater flow), false transient and transient simulations are involved in the calibration procedure. With regard to transient simulation, monthly stress periods were used for a two-year model application period (October 2013–September 2015). The transient model calibration period was from October 2013 to September 2014, while the validation period was from October 2014 to September 2015. Eleven monthly groundwater level datasets and 3 seasonal groundwater chloride datasets were used for each simulation period (calibration and validation). SEAWAT model application performance was made using typical indices such as Mean Error (ME), Mean Absolute Error (MAE), Root Mean Squared Error (RMSE) and normalised RMSE (NRME). Moreover, scatter diagrams were used in which simulated quantities are plotted against observed ones and a linear regression line is fitted through, resulting in line slope (SLP) and correlation coefficient (R^2) calculation.

Regarding aquifer's hydraulic parameters, they were initially assigned based on pumping test data available and were further adjusted during the calibration process. Hydraulic conductivity was found to range between 1.2 and 5 m/d with a trend to decrease on the west-east direction, while specific yield values varied between 0.024 and 0.089. With regard to pollutant transport model, longitudinal dispersivity was calibrated at 5 m, while transverse dispersivity, molecular diffusion and effective porosity were calibrated at 0.1, 1×10^{-10} m^2/d and 0.08, respectively.

Transient model calibration and validation results are summarized in Table 2 and Figure 4. Model performance was evaluated for each dataset and the results demonstrated satisfactory matching between simulated and observed groundwater levels and chloride concentrations.

Table 2. Statistical indices of model performance evaluation during calibration and validation periods.

Groundwater Level			Groundwater Cl$^-$ Concentration		
Index	Calibration	Validation	Index	Calibration	Validation
No of Observations	198	136	No of Observations	53	53
SLP	0.985	1.019	SLP	0.877	0.959
R^2	0.992	0.985	R^2	0.966	0.963
ME (m)	−0.068	−0.061	ME (mg/L)	−7.238	−6.255
MAE (m)	0.280	0.450	MAE (mg/L)	14.745	20.781
RMSE (m)	0.417	0.608	RMSE (mg/L)	32.077	41.129
NRMSE (%)	1.855	2.517	NRMSE (%)	6.179	7.058

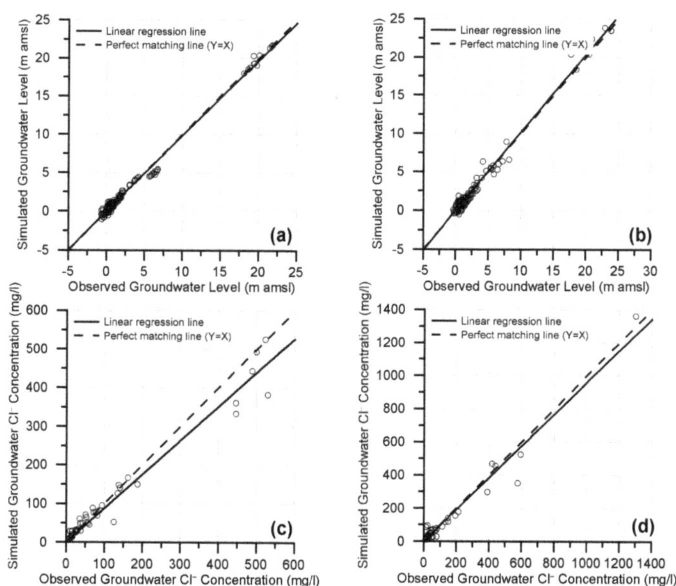

Figure 4. Scatter plots of observed versus simulated groundwater levels for calibration (**a**) and validation (**b**) periods. The corresponding plots for groundwater chloride concentrations are also presented for calibration (**c**) and validation (**d**).

More specifically, with regard to groundwater flow simulation, groundwater level R^2 values were found to be very close to 1 (0.992 for calibration and 0.985 for validation periods, respectively) while ME was found to be less than 0.1 m both for calibration and validation periods. MAE was 0.28 m for calibration and 0.45 for validation. Both RMSE and NRMSE values were satisfactory, since RMSE value for calibration was 0.417 m and 0.608 m for validation, while the corresponding values for NRMSE were 1.855% and 2.517%, respectively. The values of SLP were very close to 1 and therefore considered as satisfactory. The fact that SLP value for calibration period was lower than 1 (0.985), indicates that the model slightly underestimates groundwater levels, while for the validation the model slightly overestimates groundwater levels (SLP = 1.019). The corresponding statistical metrics for groundwater Cl^- concentrations are also indicating satisfactory model performance. R^2 was 0.966 for calibration and 0.963 for validation periods, respectively, and therefore very close to 1, which corresponds to the perfect matching between observed and simulated values. ME values were -7.238 mg/L and -6.255 mg/L for calibration and validation periods, respectively, while the corresponding MAE values were 14.745 mg/L and 20.781 mg/L. RMSE and NRMSE values were also found in satisfactory levels: RMSE was 32.077 mg/L for calibration and 41.129 mg/L for validation, while the corresponding values for NRMSE were 6.179 and 7.058%. SLP values were lower but close to 1, thus indicating underestimation of groundwater Cl^- concentrations both for calibration and validation periods.

The water budget of PRDP phreatic aquifer for the hydrological years 2014 and 2015 is presented in Figure 5. Groundwater recharge from surface water percolation constitutes the major inflow for PRDP aquifer during both years, since it accounts for 5.165 Mm^3 for year 2014 and 8.796 Mm^3 for year 2015, corresponding to 88% and 95% of total inflows, respectively. Inflows from Pinios River main course and streams located in PRDP are contributing to groundwater budget with 0.52 Mm^3 for year 2014 and 0.368 Mm^3 for year 2015, corresponding to 9% and 4% of total inflows, respectively. Seawater intrusion accounts for 0.156 Mm^3 during 2014 and 0.098 Mm^3 during 2015 or 3% and 1% of total inflows, respectively. Inflows from the scree cones can be considered as negligible.

With regard to outflows from PRDP phreatic aquifer, the major outflow includes capillary rise and groundwater pumping, accounting for 5.435 Mm^3 or 86% of the total outflows for year 2014, while the corresponding values for year 2015 are 5.965 Mm^3 or 77%. The next most significant outflow was found to be groundwater discharge to the hydrographic network, which accounts for 0.663 Mm^3 or 10% of total outflows for year 2014 and for 1.256 Mm^3 or 16% of total outflows for year 2015. With regard to groundwater discharge to the sea, it was estimated for year 2014 at 0.191 Mm^3 or 3.5% of total outflows, while the corresponding values for year 2015 were 0.37 Mm^3 and 16%, respectively. Outflows to the scree cones were found to be negligible for year 2014 and less than 2% of total outflows for year 2015.

3.3. Water Budget of PRDP Aquifer under Projected Climate Conditions

Before the presentation and analysis of projected water balance and variation of groundwater chloride concentration, the temporal variation of total annual precipitation and average annual temperature for the two projected periods are presented in Figure 6, as deduced from the results of RACMO2 RCM, driven by ECHAM5-r3 Global Circulation Model (GCM). The results demonstrate a significant decrease trend of 5.2 mm/year for total annual precipitation during the period 2021–2050, while for the period 2071–2100 a very small increase trend of 0.37 mm/year is indicated. Average total annual precipitation for the period 2021–2050 was 615 mm and very close to the corresponding value for the period 1961–1990 (625 mm). The corresponding value for the period 2071–2100 was found about 25% lower (469 mm), thus indicating significantly decreased precipitation during the latter period. Moreover, when counting years with total precipitation lower than 400 mm, which could be considered as threshold for severely dry conditions in the study area, 3 years are counted for the period 1971–1990, 4 years for the period 2021–2050 and 12 years for the period 2071–2100. Even more interestingly, for the reference and the near-future projection (2021–2050) periods, only single year occurrence of annual precipitation of lees than 400 mm is noted. On the contrary, for the projection period 2071–2100, five cases may be observed where 2 or more consecutive dry years (annual precipitation less than

400 mm) occur. Therefore, much drier conditions are indicated for the period 2071–2100 compared to the period 2021–2050. With regard to average annual temperature variation, increase trends are clearly indicated for both projected periods with almost equal rates (0.05 °C/year). Despite the almost equal increase trends, the average annual temperature for the period 2071–2100 (18.0 °C) is much higher than the corresponding value of the period 2021–2050 (15.6 °C). The average annual temperature for the period 1961–1990 was 13.9 °C, thus indicating temperature increase by 1.7 °C for the period 2021–2050 and by 4.1 °C for the period 2071–2100.

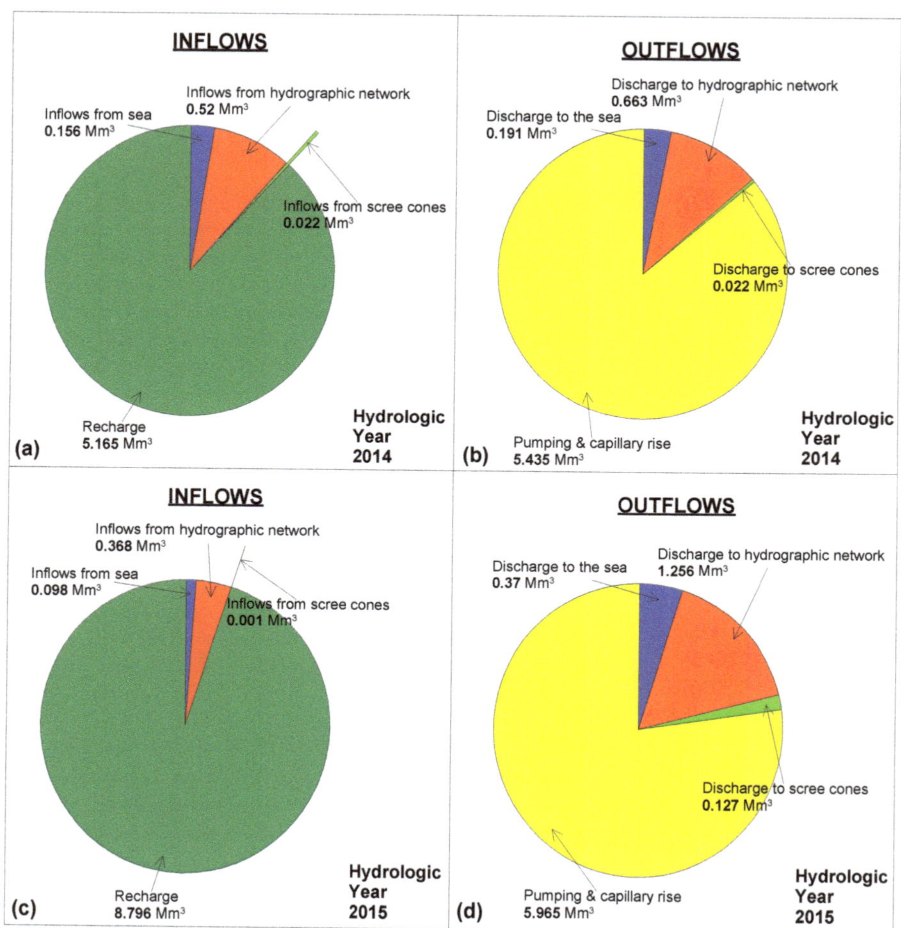

Figure 5. Graphical representation of groundwater balance of PRDP phreatic aquifer for hydrological years 2014 (**a**,**b**) and 2015 (**c**,**d**). Change in groundwater storage is not included.

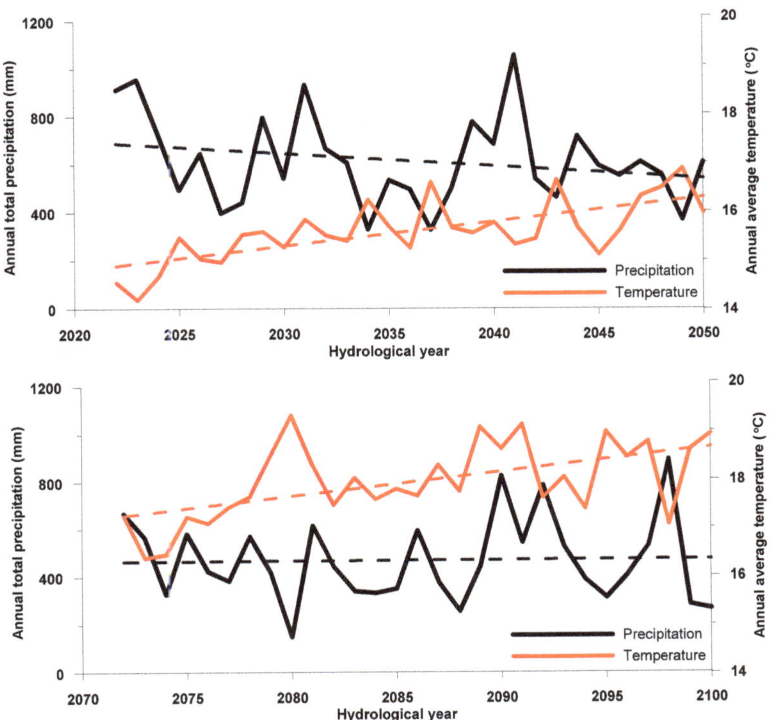

Figure 6. Time series of total annual precipitation (mm) and average annual temperature for the two projected periods, under the adopted RCM.

The annual variation of aquifer water budget elements for the two projected periods is presented in the box-plots of Figure 7. With regard to inflows, median direct groundwater recharge for the period 2021–2050 is estimated at 7.93 Mm^3/year, while the corresponding estimated value for the period 2071–2100 is almost half (4.2 Mm^3/year). The minimum and maximum annual groundwater recharge values for the two projected periods are almost equal, likewise the extent of the interquartile range. Nevertheless, it is interesting to mention that the upper bound of the interquartile range for the period 2071–2100 is close to the median of the period 2021–2050; This in turn, indicates considerable variability of annual groundwater recharge, which in absolute values is significantly lower during the period 2071–2100, compared to 2021–2050. Annual median seawater inflows for the period 2021–2050 is estimated at 0.14 Mm^3/year while for the period 2071–2100 it is more than triple (0.385 Mm^3/year). Moreover, interquartile ranges of the two projected period almost do not coincide, thus indicating significantly different temporal variation pattern of seawater intrusion for the two periods. A similar variation pattern is also presented for the inflows from the hydrographic network, according to which median annual values for the period 2071–2100 is almost triple compared to those of the period 2021–2050 (Figure 7a), while interquartile ranges of the two projected periods almost do not coincide.

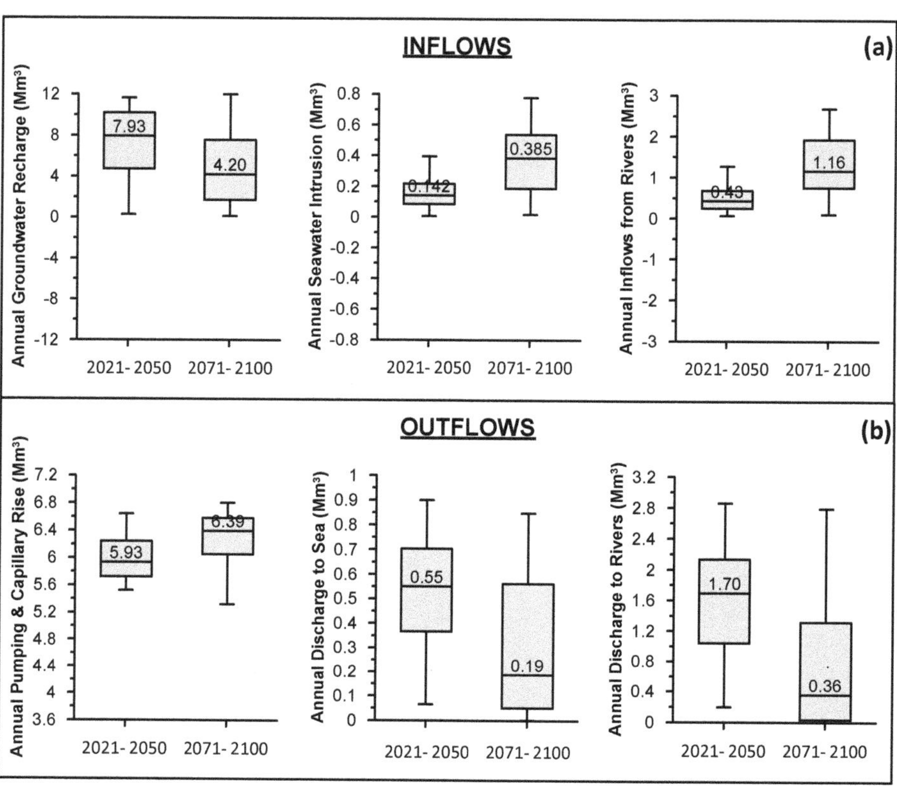

Figure 7. Boxplots presenting the annual variation of major: (**a**) inflows and (**b**) outflows for the two projected periods.

The variation of aquifer budget outflows during the two projected periods are presented in Figure 7b. Annual outflows in the form of groundwater abstraction and capillary rise for the satisfaction of irrigation demands were found to be on the median 5.9 Mm³/year for the period 2021–2050, while for the period 2071–2100 8.5% higher (6.4 Mm³/year). With regard to annual groundwater discharge to the sea, its simulated median value was 0.55 Mm³/year for the period 2021–2050, while for the period 2071–2100 it was estimated about two times lower (0.19 Mm³/year). The difference in annual groundwater discharge to rivers between the two projected periods was even higher, since the corresponding values were 1.7 and 0.36 Mm³/year for the periods 2021–2050 and 2071–2100, respectively.

3.4. Groundwater Chloride Concentration in PRDP Aquifer under Projected Climate Conditions

The variation of annual groundwater Cl^- concentration in PRDP aquifer for the two projected periods is presented in Figure 8 (no sea level rise, ΔH0). Median annual groundwater Cl^- concentration was found to be 544 mg/L for the period 2021–2050, while for the period 2071–2100 the corresponding value was 1,716 mg/L. The minimum and maximum annual groundwater Cl^- concentration values for the period 2021–2050 were 369 mg/L and 780 mg/L, respectively, while the corresponding values for the period 2071–2100 were 404 mg/L and 2539 mg/L. Moreover, the interquartile range for the period 2021–2050 was found to be 405 mg/L to 602 mg/L, while for the period 2071–2100 the interquartile range was 862 mg/L to 2145 mg/L.

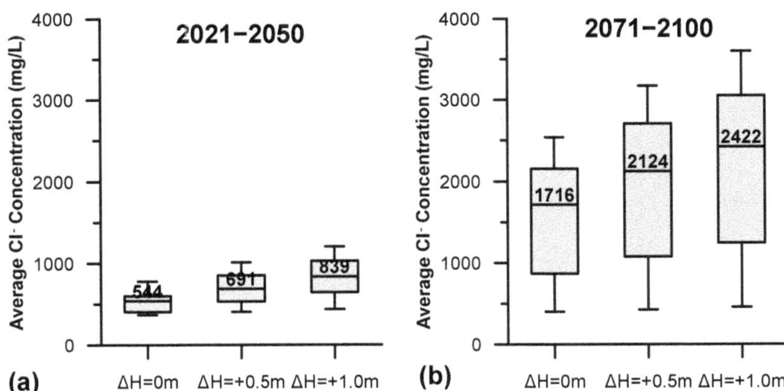

Figure 8. Boxplots presenting the annual variation of groundwater Cl^- concentration for the periods (**a**) 2021–2050 and (**b**) 2071–2100 under no sea level rise (ΔH0), sea level rise of 0.5 m (ΔH0.5) and sea level rise of 1 m (ΔH1).

The significant differences in the water budget of PRDP aquifer between the two projected periods are further reflected in annual groundwater Cl^- concentration. The median annual groundwater Cl^- concentration for the period 2071–2100 was found to be more than triple compared to the period 2021–2050 (Figure 7b as opposed to Figure 7a, respectively), while groundwater Cl^- variation pattern during the two periods is completely different, as demonstrated by the fact that the interquartile ranges do not coincide at all. This is the impact of seawater intrusion on groundwater salinization because of the decrement of precipitation by 24% and the subsequent decrement of groundwater recharge by about 50%, accompanied by the increment of groundwater outflows for irrigation and capillary rise by 8% compared to period 2021–2050.

The spatial distribution of groundwater Cl^- concentration at the end of the two projected periods is presented in Figure 9. At the end of hydrologic year 2050, the major part of PRDP deltaic aquifer was found to be almost completely unaffected from seawater intrusion, while groundwater Cl^- concentration values above 2000 mg/L are mainly observed along the coast in a zone of no more than 200 m width. This is very close to the current spatial distribution of groundwater Cl^- concentration, thus demonstrating that the current water management status does not significantly impact groundwater salinization under the climate conditions predicted by the adopted RCM scenario. The corresponding spatial distribution at the end of the hydrologic year 2100 demonstrate that, although the western part of the aquifer seems to be unaffected from seawater intrusion, the eastern half presents considerable deterioration since the zone with groundwater Cl^- concentration > 2000 mg/L has been extended up to about 500 m, especially at the southern coastal part, while groundwater Cl^- concentrations up to 1000 mg/L were simulated for the central part of the aquifer.

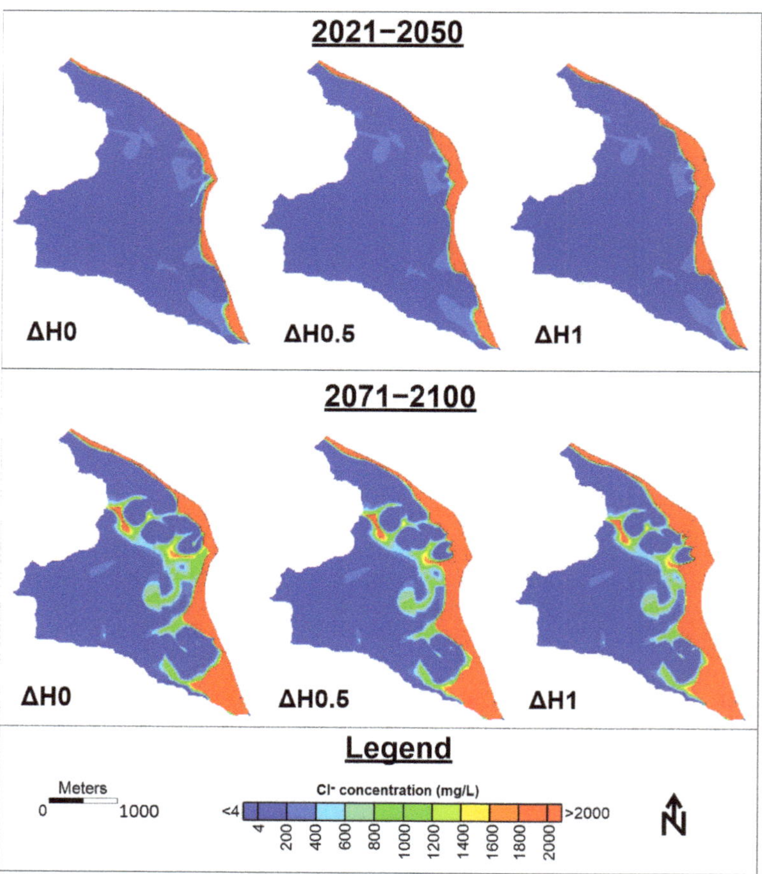

Figure 9. Spatial distribution of groundwater Cl^- concentration at the end of periods 2021–2050 and 2071–2100 for no sea level rise (ΔH0), sea level rise of 0.5 m (ΔH0.5) and sea level rise of 1 m (ΔH1).

3.5. Effects of Sea Level Rise Scenarios on Water Budget and Groundwater Chloride Concentration

The simulated effects of sea level rise by 0.5 m (ΔH0.5) and 1 m (ΔH1) on critical groundwater budget elements for the two projected periods are presented in Figure 10. The effects of sea level rise in seawater intrusion was found to be significant, since seawater intrusion volume was found to be on the median increased by 0.1 Mm^3/year (or 71%) under ΔH0.5 scenario and by 0.24 Mm^3/year (or 157%) under the ΔH1 scenario for the period 2021–2050. Significant increase in seawater intrusion volume was also simulated for the period 2071–2100, since seawater intrusion volume was found to be on the median increased by 0.18 Mm^3/year (or 57%) under ΔH0.5 scenario and by 0.39 Mm^3/year (or 100%) under the "ΔH1" scenario. Inflows from rivers indicated decrease trend for both projected periods. More specifically, they were found to be on the median decreased by 0.05 Mm^3/year (or 11.6%) under ΔH0.5 scenario and by 0.08 Mm^3/year (or 18.6%) under the ΔH1 scenario for the period 2021–2050, compared to no sea level rise (ΔH0). The corresponding decreases for the period 2071–2100 were 0.07 Mm^3/year (or 6.0%) under ΔH0.5 scenario and by 0.12 Mm^3/year (or 10.4%) under the ΔH1 scenarios.

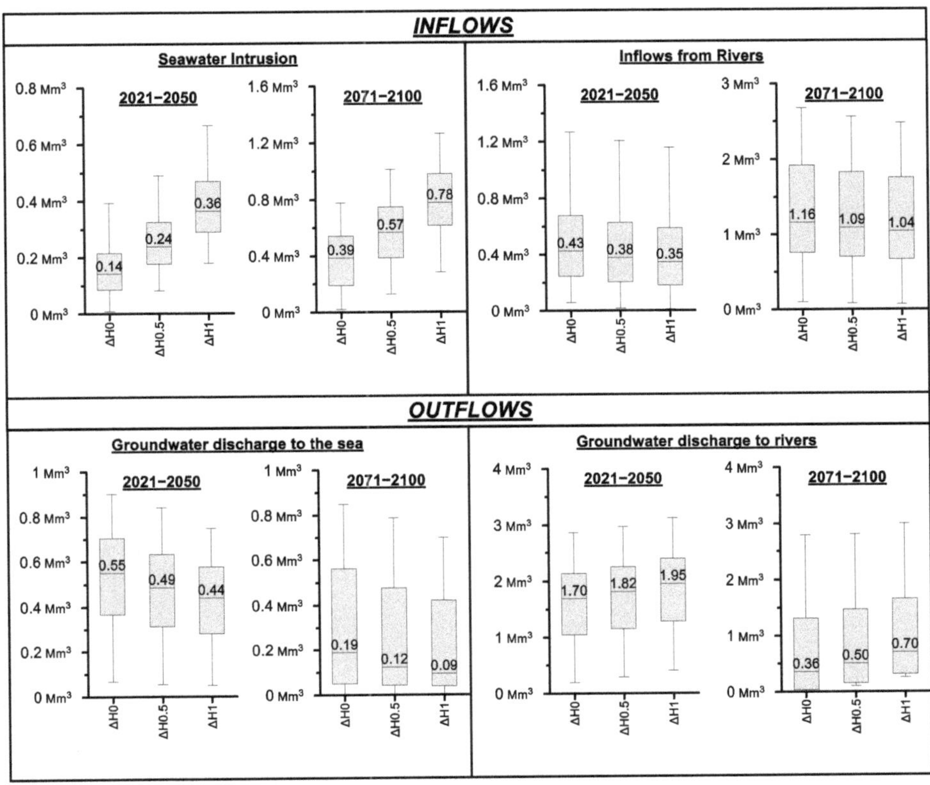

Figure 10. Boxplots presenting the annual simulated variation of inflows and outflows for the two projected periods under no sea level rise (ΔH0), sea level rise of 0.5 m (ΔH0.5) and sea level rise of 1 m (ΔH1).

Regarding sea level rise impact on outflows from PRDP, groundwater discharge to the sea demonstrates decrease trend for both scenarios and both projected periods, while groundwater discharge to rivers demonstrate increase trend. In detail, for the period 2021–2050, groundwater discharge to the sea was found to be on the median decreased by 0.06 Mm³/year (or 10.9%) under ΔH0.5 scenario and by 0.11 Mm³/year (or 20%) under the ΔH1 scenario compared to no sea level rise. The corresponding decrease for the period 2071–2100 was 0.07 Mm³/year (or 36.9%) under ΔH0.5 scenario and 0.1 Mm³/year (or 52.6%) under the ΔH1 scenario. With regard to groundwater discharge to rivers during the period 2021–2050, it was simulated to be increased by 0.12 Mm³/year (or 7.1%) under ΔH0.5 scenario and 0.25 Mm³/year (or 14.7%) under ΔH1 scenario, compared to no sea level rise scenario. Accordingly, for the period 2071–2100, groundwater discharge to rivers was simulated to be increased by 0.14 Mm³/year (or 38.9%) under ΔH0.5 scenario and 0.34 Mm³/year (or 94.4%) under ΔH1 scenario, compared to no sea level rise scenario.

The spatial distribution of groundwater Cl^- concentration at the end of the two projected periods for all three sea level rise scenarios is presented in Figure 9. An expansion of the high Cl^- concentration zones is observed which is found to be higher for the central coastal part of the aquifer and, as expected, higher for the ΔH1 scenario compared to the other 2 scenarios.

4. Discussion and Conclusions

4.1. Models Application

The statistic indices used for the assessment of SEAWAT model calibration and validation in PRDP phreatic aquifer indicated satisfactory matching between simulated and observed parameters (groundwater level and chloride concentration). It has to be mentioned that uncertainty in model performance is considerable in the present application originating not only from uncertainty on model conceptualization and the aquifer's hydraulic parameters' distribution and values, but also from the short duration of groundwater level and chloride concentration time series (2 years). Although the temporal density of measurements within the 2-year period is relatively high (monthly measurements for groundwater level and seasonal for Cl^- concentrations), seawater intrusion constitutes a slow process and therefore long times series are increasing model calibration efficiency and therefore reducing uncertainty. According to Werner et al. [17], data scarcity in relation to seawater intrusion is widely observed and it is one of the reasons for lacking seawater intrusion studies on the global scale. Nevertheless, a wide range of studies are offered in the literature which are based on datasets similar to the one used in the present study [63–69].

With regard to SWAT model application in PRDP, it was calibrated using ETa and assuming that ETa was also equal to ETc. Although this assumption is simplistic, it was based on the fact that actual crop yields were very close to the full development crop yields during the model calibration period. Therefore, despite the uncertainty that such an assumption incorporates in the modeling procedure, it can provide a good quantitative estimate of the land phase of the hydrologic cycle in PRDP watershed. Moreover, ETa has been proved to be an efficient calibration parameter in areas (mainly agricultural) in which ETa constitutes the major hydrologic budget component, when river discharge data is absent [70–72].

4.2. Water Budget under Current Climate Conditions

The aquifer budget presented above reveals new evidence and is quantifying hydrological processes in PRDP, for which only qualitative information and approaches were existing until now. Previous studies suggested that there is hydraulic interaction between PRDP aquifer and the Pinios River, but the interaction pattern (recharging or discharging) is variable, depending on the exact location and the season [73,74]. Our results are justifying that there is hydraulic interaction between the Pinios River and the other river courses across PRDP, since both inflows and outflows are simulated. Inflows from surface waters are mainly observed during the dry period for which groundwater levels are lower than surface water levels due to capillary rise and pumping. Outflows from the aquifer to river courses are observed mainly during the wet period since direct groundwater recharge from precipitation is rising groundwater above surface water levels. Moreover, the variation of inflows and outflows related to surface waters was found to be high during the simulation period, thus indicating that the interaction of surface and ground waters in PRDP is highly dynamic and in line with the corresponding variation in climate conditions affecting direct groundwater recharge. Although not clearly mentioned in previous studies, Pinios River was thought to be the factor that dominates in aquifer budget and controls it. Our results indicate that although the hydraulic interaction is clear, the contribution of the river to the water budget is significant but not dominant.

According to the results presented above, the dominant inflow to PRDP phreatic aquifer is direct groundwater recharge from precipitation. Although the dominant inflow (about 90% of total inflows), groundwater recharge for PRDP phreatic aquifer was found to be 11.1% and 19.9% of total precipitation for years 2014 (620 mm) and 2015 (589 mm), respectively. According to Lambrakis et al. [75], direct groundwater recharge for Glafkos alluvial plain was estimated at 20% of total annual precipitation. Pisinaras et al. [76] estimated direct groundwater recharge rates varying between 10.3% and 13.3% of total annual precipitation in Xanthi's alluvial plain, while groundwater recharge rates for other alluvial aquifers in Greece are also reported in this study, ranging between 9.5% and 15% of total annual precipitation.

With regard to hydraulic interaction of the study area aquifer with the scree cones located at the western part of PRDP, it could be expected to constitute a significant inflow due to the fact that these formations are typically indicating high groundwater potential. Nevertheless, our findings indicate that the contribution of scree cones in PRDP groundwater budget is practically negligible. This finding comes in agreement with hydrolithological information about the study area, which indicate that these formations are expected to be of poor groundwater potential with hydraulic conductivity values ranging between 10^{-7} and 10^{-5} m/sec. Except from the scree cones, the hydraulic interaction of study area aquifer with the sea was also found to be limited, while outflow of phreatic aquifer to the sea was found to be higher than the corresponding inflow. The restricted hydraulic interaction is attributed to: (a) the low permeability sand dunes developed in the central part of the coastal zone as described by Alexopoulos et al. [30] and (b) to the climate and hydrological conditions prevailing during the simulation period.

The major outflow from the phreatic aquifer corresponds to groundwater pumping and capillary rise from the saturated to unsaturated zone. Similarly to other agricultural areas [38,77,78] capillary rise was found to significantly contribute to the satisfaction of crop water requirements. The experimental runs conducted with HYDRUS-1D model, indicated capillary rise contribution ranging up to 280 mm per cultivation period for the crops cultivated in PRDP. These results come in agreement with those of Babajimopoulos et al. [79], according to which groundwater contribution to crop water requirements were 283.8 mm for irrigated maize in the plain of Thessaloniki, located adjacent to PRDP.

4.3. Projected Climate Change Impacts

Projected climate change signal for the study area, as retrieved by RCM data, indicated: (a) a decrease in precipitation by 1.6% and an average increase in temperature by 1.7 °C for the period 2021–2050 and (b) an average decrease in annual precipitation by about 25% and an average increase in annual temperature by 4.1 °C for the period 2071–2100. Comparing to previous studies for the whole Pinios River basin, a decrease in precipitation by 5% for the period 2021–2040 and by 25% for the period 2081–2100 were reported by Arampatzis et al. [47], in the study of which data from three RCMs was incorporated. With regard to temperature, the results of the aforementioned study indicated increase by 1.7 °C for the period 2021–2040 and increase by 4 °C for the period 2081–2100. Zanis et al. [80] used the PRUDENCE dataset consisting of nine RCMs, and their results for central-eastern Greece indicated an annual precipitation decrease by 15.5% and an average annual temperature increase by 4.0 °C for the period 2071–2100. In conclusion, two highly different scenarios are created according to which precipitation is practically stable and temperature is moderately increased for the period 2021–2050 and precipitation is significantly decreased and temperature is significantly decreased for the period 2071–2100.

The significant differences in the major aquifer budget elements between the two periods indicate the impact of the corresponding differences in temperature, but mainly in precipitation variation. Despite the fact that, on the median, precipitation decrease for the period 2071–2100 was 25%, the corresponding decrease in groundwater recharge was almost double (47%). This can be attributed not only to the decreased precipitation volume but also to changes in precipitation intensities within the hydrological year, as well as changes in wet and dry spells distribution. According to Pisinaras [49], different groundwater recharge volumes should be expected even when wet spells are of equal duration and precipitation intensity but the duration of intervening dry spells is different. Indeed, during the second climate projection period studied (2071–2100) there are consecutive periods of dry spells consisting of at least two consecutive years of low annual precipitation (below 400 mm). These differences are impacting the complex soil water balance and therefore soil water percolation and groundwater recharge. Moreover, increased ET due to temperature increase constitute another reason for the significant decrease of groundwater recharge.

When expressing groundwater recharge as percentage of precipitation, its value for the period 2021–2050 was 17.9% while for the period 2071–2100 it was 13.3%. This fact comes to underline that the simplistic approach of assigning a constant percentage of precipitation for groundwater recharge may significantly affect groundwater budget results and their interpretation, especially in climate change impact assessment studies.

With regard to the hydraulic interaction of PRDP phreatic aquifer with the sea, the results indicate two contrasting situations for the two projected periods: (a) significantly higher groundwater discharge to the sea compared to seawater intrusion during the period 2021–2050 and (b) significantly higher seawater intrusion compared to groundwater discharge to the sea during the period 2071–2100. The above demonstrate the notably higher water quality deterioration potential during the period 2071–2100 due to seawater intrusion caused by decreased groundwater recharge and increased crop water demand. This is also indicated by the fact that during the period 2021–2050, seawater intrusion constitutes about 1.7% of the total inflows, while during the period 2071–2100, it raises to 6.7%. Moreover, the significant impact of change in precipitation variation in seawater intrusion is also underlined, since the decrease of precipitation by 25% during the period 2071–2100 and the subsequent changes in hydrologic cycle resulted, on the median, in 4 times higher seawater intrusion compared to the period 2021–2050 and expansion of the high groundwater salinity area by about 250 m. Despite the fact that the current high groundwater salinity area is covered mainly by touristic houses and facilities, the aforementioned expansion will pose groundwater of high salinity in the cultivated area, thus increasing the possibility of agricultural land deterioration or abandonment.

Similarly contrasting results are demonstrated for the hydraulic interaction of PRDP phreatic aquifer with Pinios River and the other courses. During the period 2021–2050, groundwater discharge to rivers was about four times higher than the corresponding inflows, while during the period 2071–2100, groundwater discharge to rivers was about 3 times lower. This is attributed to the fact that groundwater level is for the most time lower than river water level, as a result of reduced groundwater recharge caused by reduced precipitation, increased irrigation demands and in conjunction to occurrence of prolonged consecutive dry spells.

Since water budget of PRDP phreatic aquifer is controlled by groundwater recharge and subsequently from precipitation, the significantly decreased precipitation during the period 2071–2100 resulted in significantly decreased groundwater recharge and therefore in significant and contrasting changes in water budget, compared to period 2021–2050. Taking into account the results of several climate change impact assessment studies compiled for Pinios River basin which indicate decreasing trend in precipitation and high inter-annual variability [47,80,81], the anticipated effects in PRDP phreatic aquifer budget have to be strongly considered for effective water management, mainly due to the fact that: (a) precipitation affects groundwater recharge, which constitutes the major inflow and subsequently the whole water budget, (b) PRDP phreatic aquifer potential and buffer capacity, hence resilience to climate change phenomena, is relatively restricted because of its hydraulic characteristics and limited thickness. Considering the increased by about 9% groundwater abstractions, water stress potential for PRDP phreatic aquifer is further increased. Since groundwater abstractions from PRDP phreatic aquifer are fully allocated to irrigation, decreasing irrigation water demand would act beneficially towards decreasing the water stress of the aquifer. Sprinkler irrigation systems are dominating in PRDP, which can be substituted by the much more efficient drip irrigation systems. Moreover, for some crops such as cotton and corn which are cultivated in PRDP, deficit irrigation has been proved to be very effective for decreasing irrigation water consumption with a negligible loss on crop production [82,83].

The fact that PRDP phreatic aquifer presents limited groundwater storage capacity indicate the necessity for short-term effective water resources management in order to avoid significant deterioration of both its quality and quantity status, especially under the high variability in groundwater recharge posed by the high variability of precipitation.

Shallow aquifers such as this of PRDP could be of high importance both from the environmental and socio-economic point of view due to the fact that these aquifers: (a) support the ecosystems developed especially in deltaic areas through a wide range of hydro-environmental processes such as baseflow and (b) constitute a technically and economically feasible solution for water abstraction in order to support the agricultural production and therefore local food security and economy. Another aspect that has not been investigated in the present study but has to be mentioned as critical for shallow aquifers and the interacting ecosystems is the influence of air temperature increase in groundwater temperature. Taylor and Stefan [84] indicate that in the extreme case of atmospheric carbon dioxide doubling, groundwater temperature could be increased by 4 °C in the Minneapolis/St. Paul metropolitan area. Kurylyk et al. [85] simulated temperature increase up to 3.6 °C of groundwater discharge to the adjacent river for the period 2045–2065, while they emphasize the influence of aquifer dimensions in thermal response of shallow, unconfined aquifers to climate change. Moreover, the projected dramatic reduction in outflow to Pinios River is suggested to have a serious adverse effect on the riparian zone of the river, thus impacting not only on the abiotic but also on the biotic factors of the regional ecosystem. Last but not least, soil salinization, which constitutes a common problem for the Mediterranean coastal areas [86], will potentially increase because of the increasing salt content of irrigation water and the increasing irrigation water demand. Soil salinization constitutes one of the major threats for Mediterranean agriculture and desertification. For instance, Zekri et al. [87] estimated that the currently applied irrigation water management practices in Batinah region of Oman will result in 46% loss of the cropland.

With regard to sea level rise, both scenarios indicated further increment of groundwater salinity for both projected periods. When comparing seawater intrusion volumes for the two projected periods under the no sea level rise scenario to the impact of sea level rise it can be concluded that the impact of groundwater recharge decrement is higher than sea level rise by 0.5 or 1 m. This finding comes in agreement with the study of Stigter et al. [15], according to which sea level rise was found to be less significant than the decrease in groundwater recharge and increase in crop water needs under climate change conditions for three Mediterranean aquifers. Nevertheless, when assessing the combined effects of high sea level rise and high groundwater recharge decrease ($\Delta H1$ scenario for the period 2071–2100) it is remarkable, since it indicated more than 5 times higher seawater intrusion volumes, compared to the $\Delta H0$ scenario for the period 2021–2050. Considering that land-surface inundation has not been incorporated in the simulation, the aquifer deterioration potential from sea level rise could be higher.

5. Conclusions

SWAT and SEAWAT models were implemented in the PRDP in order to simulate and quantify the effects of current agricultural water management practices to the groundwater budget and groundwater salinization status of the underlying shallow, unconfined aquifer under current and projected climate conditions. Despite the uncertainty incorporated in the modeling process, the corresponding results revealed significant insight on the hydrogeologic/hydraulic behavior of the aquifer: (a) direct groundwater recharge from precipitation was found to be the dominant inflow to PRDP aquifer, (b) aquifer-river hydraulic interaction is variable depending both on the exact location and season, (c) capillary rise from the saturated to the unsaturated zone was found to significantly contribute to the satisfaction of crop water requirements.

Despite the fact that during the calibration/validation period, the aquifer was not found to be affected by seawater intrusion, as indicated from the simulation of Cl^- concentrations with the SEAWAT model, the corresponding results for the projected period 2071–2100 indicate that groundwater salinization potential could be significantly increased mainly because of the significant decrease of precipitation which leads to direct groundwater recharge decrease. The aquifer is further stressed due to the increased groundwater abstractions needed for the satisfaction of the increased crop water requirements. Sea level rise was found to further contribute to increasing groundwater salinization but nevertheless, the significant decrease in precipitation was found to have a higher impact on groundwater budget and subsequently to groundwater salinization.

Considering the above and taking into account the fact that shallow aquifers such as this of PRDP could be of high importance both from the environmental and socio-economic point of view, both short and long-term effective water resources management strategies have to be developed in order to avoid significant deterioration of both its quality and quantity status, especially under the high variability in groundwater recharge posed by the high variability of precipitation. Maintaining sufficient crop production and simultaneously ecosystem's functions and integrity should be the priority axis followed towards the development of sustainable, integrated water resources management strategies.

Author Contributions: Writing—original draft preparation, V.P.; writing—review and editing, C.P. and A.P.; modeling, V.P. and C.P.; conceptualization, V.P., C.P. and A.P.; methodology, V.P., C.P. and A.P.; data curation, V.P. and C.P.; supervision, A.P.; funding aquisition, A.P. All authors have read and agreed to the published version of the manuscript.

Funding: This research work was elaborated in the framework of the project "AGROCLIMA", under the national action "SYNERGASIA 2011: Partnerships of Production and Research Institutions in Focused Research and Technology Sectors" that is co-funded by the European Regional Development Fund of the EU and National Resources of Hellenic Government.

Institutional Review Board Statement: Not applicable.

Informed Consent Statement: Not applicable.

Data Availability Statement: The data presented in this study are available on request from the corresponding author.

Acknowledgments: The author would like to acknowledge the fruitful comments of the four anonymous reviewers who provided beneficial suggestions that improved the quality of this paper.

Conflicts of Interest: The authors declare no conflict of interest.

References

1. Shiklomanov, I.A.; Rodda, J.C. *World Water Resources at the Beginning of the Twenty-First Century*; Cambridge University Press: Cambridge, UK, 2003; ISBN 9780521617222.
2. Rost, S.; Gerten, D.; Bondeau, A.; Lucht, W.; Rohwer, J.; Schaphoff, S. Agricultural Green and Blue Water Consumption and Its Influence on the Global Water System. *Water Resour. Res.* **2008**, *44*, 44. [CrossRef]
3. Thivet, G.; Fernandez, S. *Water Demand Management: The Mediterranean Experience*; Global Water Partnership (GWP): Stockholm, Sweden, 2012; ISBN 9185321885.
4. Bates, B.C.; Kundzewicz, Z.; Wu, S.; Palutikof, J. *Climate Change and Water, IPCC Technical Paper VI*; Intergovernmental Panel on Climate Change (IPCC): Geneva, Switzerland, 2008; ISBN 9789291691234.
5. Alpert, P.; Hemming, D.; Jin, F.; Kay, G.; Kitoh, A.; Mariotti, A. *The Hydrological Cycle of the Mediterranean*; Springer Science and Business Media: Berlin/Heidelberg, Germany, 2013; Volume 50, pp. 201–239.
6. Mariotti, A.; Zeng, N.; Yoon, J.-H.; Artale, V.; Navarra, A.; Alpert, P.; Li, L.Z.X. Mediterranean Water Cycle Changes: Transition to Drier 21st Century Conditions in Observations and CMIP3 Simulations. *Environ. Res. Lett.* **2008**, *3*, 044001. [CrossRef]
7. De Fraiture, C.; Wichelns, D. Satisfying Future Water Demands for Agriculture. *Agric. Water Manag.* **2010**, *97*, 502–511. [CrossRef]
8. Molden, D. *Water for Food, Water for Life: A Comprehensive Assessment of Water Management in Agriculture*; Earthscan: London, UK, 2007; ISBN 9781844073979.
9. Olesen, J.; Trnka, M.; Kersebaum, K.; Skjelvåg, A.; Seguin, B.; Peltonensainio, P.; Rossi, F.; Kozyra, J.; Micale, F. Impacts and Adaptation of European Crop Production Systems to Climate Change. *Eur. J. Agron.* **2011**, *34*, 96–112. [CrossRef]
10. Garrote, L.; Iglesias, A.; Granados, A.; Mediero, L.; Martín-Carrasco, F. Quantitative Assessment of Climate Change Vulnerability of Irrigation Demands in Mediterranean Europe. *Water Resour. Manag.* **2014**, *29*, 325–338. [CrossRef]
11. Garrido, A.; Iglesias, A. Groundwater's Role in Managing Water Scarcity in the Mediterranean Region. In Proceedings of the International Symposium on Groundwater Sustainability (ISGWAS), Alicante, Spain, 24–27 January 2006; pp. 113–138.
12. Fornés, J.M.; La Hera, Á.; Llamas, M.R. The Silent Revolution in Groundwater Intensive Use and Its Influence in Spain. *Hydrol. Res.* **2005**, *7*, 253–268. [CrossRef]
13. Mazi, A.; Koussis, A.D.; Destouni, G. Intensively Exploited Mediterranean Aquifers: Resilience to Seawater Intrusion and Proximity to Critical Thresholds. *Hydrol. Earth Syst. Sci.* **2014**, *18*, 1663–1677. [CrossRef]
14. Nixon, S.; Trent, Z.; Marcuello, C.; Lallana, C. *Europe's Water: An Indicator-Based Assessment*; European Environment Agency: Copenhagen, Denmark, 2003.

5. Stigter, T.Y.; Nunes, J.P.; Pisani, B.J.; Fakir, Y.; Hugman, R.; Li, Y.; Tome, S.M.; Ribeiro, L.F.; Samper, J.; Oliveira, R.; et al. Comparative Assessment of Climate Change and Its Impacts on Three Coastal Aquifers in the Mediterranean. *Reg. Environ. Chang.* **2014**, *14*, 41–56. [CrossRef]

6. Haj-Amor, Z.; Acharjee, T.K.; Dhaouadi, L.; Bouri, S. Impacts of Climate Change on Irrigation Water Requirement of Date Palms under Future Salinity Trend in Coastal Aquifer of Tunisian Oasis. *Agric. Water Manag.* **2020**, *228*, 105843. [CrossRef]

7. Werner, A.D.; Bakker, M.; Post, V.E.A.; Vandenbohede, A.; Lu, C.; Ataie-Ashtiani, B.; Simmons, C.T.; Barry, D. Seawater Intrusion Processes, Investigation and Management: Recent Advances and Future Challenges. *Adv. Water Resour.* **2013**, *51*, 3–26. [CrossRef]

8. Alcolea, A.; Contreras, S.; Hunink, J.E.; García-Aróstegui, J.L.; Jimenez-Martinez, J. Hydrogeological Modelling for the Watershed Management of the Mar Menor Coastal Lagoon (Spain). *Sci. Total. Environ.* **2019**, *663*, 901–914. [CrossRef] [PubMed]

9. Voss, C.I.; Provost, A.M. SUTRA: A model for 2D or 3D Saturated-Unsaturated, Variable-Density Ground-Water Flow with Solute or Energy Transport. *Water-Resour. Investig. Rep.* **2002**, 4231. [CrossRef]

10. Hugman, R.; Stigter, T.Y.; Costa, L.; Monteiro, J.-P. Numerical Modelling Assessment of Climate-Change Impacts and Mitigation Measures on the Querença-Silves Coastal Aquifer (Algarve, Portugal). *Hydrogeol. J.* **2017**, *25*, 2105–2121. [CrossRef]

11. Siarkos, I.; Latinopoulos, P. Modeling Seawater Intrusion in Overexploited Aquifers in the Absence of Sufficient Data: Application to the Aquifer of Nea Moudania, Northern Greece. *Hydrogeol. J.* **2016**, *24*, 2123–2141. [CrossRef]

12. Abd-Elhamid, H.F.; Abdelaty, I.; Sherif, M. Evaluation of Potential Impact of Grand Ethiopian Renaissance Dam on Seawater Intrusion in the Nile Delta Aquifer. *Int. J. Environ. Sci. Technol.* **2019**, *16*, 2321–2332. [CrossRef]

13. Khadra, W.M.; Stuyfzand, P.J. Simulation of Saltwater Intrusion in a Poorly Karstified Coastal Aquifer in Lebanon (Eastern Mediterranean). *Hydrogeol. J.* **2018**, *26*, 1839–1856. [CrossRef]

14. Romanazzi, A.; Gentile, F.; Polemio, M. Modelling and Management of a Mediterranean Karstic Coastal Aquifer under the Effects of Seawater Intrusion and Climate Change. *Environ. Earth Sci.* **2015**, *74*, 115–128. [CrossRef]

15. Romanazzi, A.; Polemio, M. Modellazione Degli Acquiferi Carsici Costieri a Supporto Della Gestione: IL Caso Del Salento (PUGLIA). *Ital. J. Eng. Geol. Environ.* **2013**, *2013*, 65–83. [CrossRef]

16. Arnold, J.G.; Srinivasan, R.; Muttiah, R.S.; Williams, J.R. Large Area Hydrologic Modeling and Assessment Part I: Model Development. *JAWRA J. Am. Water Resour. Assoc.* **1998**, *34*, 73–89. [CrossRef]

17. Arnold, J.G.; Kiniry, J.R.; Srinivasan, R.; Williams, J.R.; Haney, E.B.; Neitsch, S.L. *Soil and Water Assessment Tool, Input/Output Documentation—Version 2012*; Texas Water Resources Institute: College Station, TX, USA, 2012; Available online: https://swat.tamu.edu/media/69296/swat-io-documentation-2012.pdf. (accessed on 31 December 2020).

18. Langevin, C.D.; Thorne Jr., D.T.; Dausman, A.M.; Sukop, M.C.; Guo, W. SEAWAT Version 4: A Computer Program. for Simulation of Multi-Species Solute and Heat Transport. *Tech. Methods* **2008**, 6. [CrossRef]

19. Katsikatsios, G.; Migiros, G. *Rapsani Sheet—Geological Map in Scale 1:50,000*; Institute of Geology and Mineral Exploration: Athens, Greece, 1982.

20. Alexopoulos, J.; Matiatos, J.; Dilalos, S.; Vassilakis, E.; Panagopoulos, A.; Ghionis, G.; Poulos, S. Investigation of the Phreatic Aquifer Development at the Pinios Delta Basin (Thessaly), through a Combination of Geophysical and Hydrogeological Data. In Proceedings of the 10th Congress of the Hellenic Geographical Society, Mytilene, Greece, 22–24 October 2014; pp. 1130–1139.

21. McDonald, M.G.; Harbaugh, A.W. *A Modular Three-Dimensional Finite-Difference Ground-Water Flow Model*; U.S. Department of Energy Office of Scientific and Technical Information: Oak Ridge, TN, USA, 1988.

22. Zheng, C.; Wang, P.P. *MT3DMS: A Modular Three-Dimensional Multispecies Transport Model for Simulation of Advection, Dispersion and Chemical Reactions of Contaminants in Groundwater Systems, Documentation and User's Guide*; Engineer Research and Development Center: Washington, DC, USA, 1999.

23. Srinivasan, R.; Zhang, X.; Arnold, J.G. SWAT Ungauged: Hydrological Budget and Crop Yield Predictions in the Upper Mississippi River Basin. *Trans. ASABE* **2010**, *53*, 1533–1546. [CrossRef]

24. Neitsch, S.L.; Arnold, J.G.; Kiniry, J.R.; Williams, J.R. *Soil and Water Assessment Tool, Theoritical Documentation*; Version 2009; Texas A&M University: College Station, TX, USA, 2011.

25. Monteith, J.L. Climate and the Efficiency of Crop Production in Britain. *Philos. Trans. R. Soc. Lond. B* **1977**, *281*, 277–294. [CrossRef]

26. Šimůnek, J.; Šejna, M.; Saito, H.; Sakai, M.; van Genuchten, M.T. *The HYDRUS-1D Software Package for Simulating the One-Dimensional Movement of Water, Heat, and Multiple Solutes in Variably-Saturated Media*; Department of Environmental Sciences, University of California: Riverside, CA, USA, 2009.

27. Akhtar, F.; Tischbein, B.; Awan, U.K. Optimizing Deficit Irrigation Scheduling Under Shallow Groundwater Conditions in Lower Reaches of Amu Darya River Basin. *Water Resour. Manag.* **2013**, *27*, 3165–3178. [CrossRef]

28. Awan, U.K.; Tischbein, B.; Martius, C. A GIS—Based Approach for Up-Scaling Capillary Rise from Field to System Level under Soil-Crop-Groundwater MIX. *Irrig. Sci.* **2014**, *32*, 449–458. [CrossRef]

29. Khakbaz, B.; Imam, B.; Hsu, K.; Sorooshian, S. From Lumped to Distributed via Semi-distributed: Calibration Strategies for Semi-distributed Hydrologic Models. *J. Hydrol.* **2012**, 61–77. [CrossRef]

30. Chang, S.W.; Nemec, K.; Kalin, L.; Clement, T.P. Impacts of Climate Change and Urbanization on Groundwater Resources in a Barrier Island. *J. Environ. Eng.* **2016**, *142*, 4016001. [CrossRef]

31. Toulios, L.; Lelentzis, T.; Lipimenou, E. *Soil Survey of Pinios River Delta*; Hellenic National Research Foundation: Larisa, Greece, 1997.

42. Van Meijgaard, E.; Van Ulft, L.H.; Van De Berg, W.J.; Bosveld, F.C.; Van Den Hurk, B.J.J.M.; Lenderink, G.; Siebesma, A. *The KNMI Regional Atmospheric Climate Model RACMO Version 2.1*; Koninklijk Nederlands Meteorologisch Instituut: De Bil, The Netherlands, 2008.
43. Karali, A.; Hatzaki, M.; Giannakopoulos, C.; Roussos, A.; Xanthopoulos, G.; Tenentes, V. Sensitivity and Evaluation of Current Fire Risk and Future Projections Due to Climate Change: The Case Study of Greece. *Nat. Hazards Earth Syst. Sci.* **2014**, *14*, 143–153. [CrossRef]
44. *ENSEMBLES RCM—Specific Weights Based on Their Ability to Simulate the Present Climate Calibrated for the ERA40-Based Simulations*; Deliverable 3.2.2 of ENSEMBEL Project; ENSEMBLES RCM: Cincinnati, OH, USA, 2009.
45. Deidda, R.; Marrocu, M.; Caroletti, G.N.; Pusceddu, G.; Langousis, A.; Lucarini, V.; Puliga, M.; Speranza, A. Regional Climate Models' Performance in Representing Precipitation and Temperature over Selected Mediterranean Areas. *Hydrol. Earth Syst. Sci.* **2013**, *17*, 5041–5059. [CrossRef]
46. Kostopoulou, E.; Giannakopoulos, C.; Hatzaki, M.; Tziotziou, K. Climate Extremes in the NE Mediterranean: Assessing the E-OBS Dataset and Regional Climate Simulations. *Clim. Res.* **2012**, *54*, 249–270. [CrossRef]
47. Arampatzis, G.; Panagopoulos, A.; Pisinaras, V.; Tziritis, E.; Wendland, F. Identifying Potential Effects of Climate Change on the Development of Water Resources in Pinios River Basin, Central Greece. *Appl. Water Sci.* **2018**, *8*, 51. [CrossRef]
48. Pisinaras, V.; Wei, Y.; Bärring, L.; Gemitzi, A. Conceptualizing and Assessing the Effects of Installation and Operation of Photovoltaic Power Plants on Major Hydrologic Budget Constituents. *Sci. Total. Environ.* **2014**, *493*, 239–250. [CrossRef] [PubMed]
49. Pisinaras, V. Assessment of Future Climate Change Impacts in a Mediterranean Aquifer. *Glob. NEST J.* **2016**, *18*, 119–130. [CrossRef]
50. IPCC. Climate Change 2014: Synthesis Report. In *Contribution of Working Groups I, II and III to the Fifth Assessment Report of the Intergovernmental Panel on Climate Change*; IPCC: Geneva, Switzerland, 2014.
51. Rahmstorf, S. A Semi-Empirical Approach to Projecting Future Sea-Level Rise. *Science* **2007**, *315*, 368–370. [CrossRef] [PubMed]
52. Hansen, E.J. Scientific Reticence and Sea Level Rise. *Environ. Res. Lett.* **2007**, *2*, 024002. [CrossRef]
53. Ketabchi, H.; Mahmoodzadeh, D.; Ataie-Ashtiani, B.; Simmons, C.T. Sea-Level Rise Impacts on Seawater Intrusion in Coastal Aquifers: Review and Integration. *J. Hydrol.* **2016**, *535*, 235–255. [CrossRef]
54. Gitau, M.W.; Chaubey, I. Regionalization of SWAT Model Parameters for Use in Ungauged Watersheds. *Water* **2010**, *2*, 849–871. [CrossRef]
55. Omani, N.; Srinivasan, R.; Lee, T. *Estimating Sediment. and Nutrient Loads of Texas Coastal Watersheds with SWAT A Case Study of Galveston Bay and Matagorda Bay*; Final Report for the Texas Water Development Board; Texas A&M University: College Station, TX, USA, 2012. Available online: https://www.twdb.texas.gov/publications/reports/contracted_reports/doc/1004831012_SWAT.pdf (accessed on 31 December 2020).
56. Prabhanjan, A.; Rao, E.P.; Eldho, T. Application of SWAT Model and Geospatial Techniques for Sediment-Yield Modeling in Ungauged Watersheds. *J. Hydrol. Eng.* **2015**, *20*, 6014005. [CrossRef]
57. Monteith, J.L. Evaporation and Environment. *Symp. Soc. Exp. Biol.* **1965**, *19*, 205–234.
58. Jensen, M.E. Water Consumption by Agricultural Plants (Chapter 1). In *Water Deficits and Plant Growth*; Kozlowski, T.T., Ed.; Academic Press: New York, NY, USA, 1968; pp. 1–22.
59. Doorenbos, J.; Pruitt, W.O. *Guidelines for Predicting Crop Water Requirements, FAO Irrigation and Drainage Paper No. 24*; Food and Agriculture Organization (FAO): Rome, Italy, 1977; ISBN 9251002797.
60. Allen, R.G.; Pereira, L.S.; Raes, D.; Smith, M. Crop. *Evapotranspiration: Guidelines for Computing Crop Water Requirements. FAO Irrigation and Drainage Paper 56*; Food and Agriculture Organization of the United Nations: Rome, Italy, 1998; ISBN 9251042195.
61. Papazafeiriou, Z. *Crop. Irrigation Needs*; Zitis Publishing: Thessaloniki, Greece, 1999; ISBN 960-431-580-3. (In Greek)
62. Galanopoulou-Sendouka, S. *Industrial Crops*; Stamoulis Publishing: Athens, Greece, 2002.
63. Cobaner, M.; Yurtal, R.; Dogan, A.; Motz, L.H. Three Dimensional Simulation of Seawater Intrusion in Coastal Aquifers: A Case Study in the Goksu Deltaic Plain. *J. Hydrol.* **2012**, *2012*, 262–280. [CrossRef]
64. Datta, B.; Vennalakanti, H.; Dhar, A. Modeling and Control of Saltwater Intrusion in a Coastal Aquifer of Andhra Pradesh, India. *Hydro. Res.* **2009**, *3*, 148–159. [CrossRef]
65. Nofal, E.; Amer, M.; El-Didy, S.; Fekry, A.M. Delineation and Modeling of Seawater Intrusion into the Nile Delta Aquifer: A New Perspective. *Water Sci.* **2015**, *29*, 156–166. [CrossRef]
66. Chang, Y.; Hu, B.X.; Xu, Z.; Li, X.; Tong, J.; Chen, L.; Zhang, H.; Miao, J.; Liu, H.; Ma, Z. Numerical Simulation of Seawater Intrusion to Coastal Aquifers and Brine Water/Freshwater Interaction in South Coast of Laizhou Bay, China. *J. Contam. Hydrol.* **2018**, *215*, 1–10. [CrossRef] [PubMed]
67. Mansour, A.Y.S.; Baba, A.; Gunduz, O.; Şimşek, C.; Elçi, A.; Murathan, A.; Sözbilir, H. Modeling of Seawater Intrusion in a Coastal Aquifer of Karaburun Peninsula, Western Turkey. *Environ. Earth Sci.* **2017**, *76*, 775. [CrossRef]
68. Zhao, J.; Lin, J.; Wu, J.; Yang, Y.; Wu, J. Numerical Modeling of Seawater Intrusion in Zhoushuizi District of Dalian City in Northern China. *Environ. Earth Sci.* **2016**, *75*, 1–18. [CrossRef]
69. Zeng, X.; Wu, J.; Wang, D.; Zhu, X. Assessing the Pollution Risk of a Groundwater Source Field at Western Laizhou Bay under Seawater Intrusion. *Environ. Res.* **2016**, *148*, 586–594. [CrossRef]
70. Emam, A.R.; Kappas, M.; Linh, N.H.K.; Renchin, T. Hydrological Modeling and Runoff Mitigation in an Ungauged Basin of Central Vietnam Using SWAT Model. *Hydrology* **2017**, *4*, 16. [CrossRef]

71. Immerzeel, W.W.; Gaur, A.; Zwart, S.J. Integrating Remote Sensing and a Process-Based Hydrological Model to Evaluate Water Use and Productivity in a South Indian Catchment. *Agric. Water Manag.* **2008**, *95*, 11–24. [CrossRef]
72. Cheema, M.J.M.; Immerzeel, W.; Bastiaanssen, W. Spatial Quantification of Groundwater Abstraction in the Irrigated Indus Basin. *Ground Water* **2014**, *52*, 25–36. [CrossRef]
73. Matiatos, I.; Paraskevopoulou, V.; Lazogiannis, K.; Botsou, F.; Dassenakis, M.; Ghionis, G.; Alexopoulos, J.D.; Poulos, S.E. Surface–Ground Water Interactions and Hydrogeochemical Evolution in a Fluvio-Deltaic Setting: The Case Study of the Pinios River Delta. *J. Hydrol.* **2018**, *561*, 236–249. [CrossRef]
74. Panagopoulos, A.; Kotsopoulos, S.; Kalfountzos, D.; Alexiou, I.; Evangelopoulos, A.; Belesis, A. *Supplementary Environmental Acts of Reg. 2078/92/EU-Study of Natural Resources and Factors Influencing the Yield and the Quality Characteristics of Agricultural Areas in Thessaly*; Hellenic National Research Foundation: Thessaloniki, Greece, 2001.
75. Lambrakis, N.J.; Voudouris, K.; Tiniakos, L.N.; Kallergis, G.A. Impacts of Simultaneous Action of Drought and Overpumping on Quaternary Aquifers of Glafkos Basin (Patras region, western Greece). *Environ. Earth Sci.* **1997**, *29*, 209–215. [CrossRef]
76. Pisinaras, V.; Petalas, C.; Tsihrintzis, V.A.; Karatzas, G.P. Integrated Modeling as a Decision-Aiding Tool for Groundwater Management in a Mediterranean Agricultural Watershed. *Hydrol. Process.* **2013**, *27*, 1973–1987. [CrossRef]
77. Soppe, R.; Ayars, J. Characterizing Ground Water Use by Safflower Using Weighing Lysimeters. *Agric. Water Manag.* **2003**, *60*, 59–71. [CrossRef]
78. Kahlown, M.; Ashraf, M.; Haq, Z.-U. Effect of Shallow Groundwater Table on Crop Water Requirements and Crop Yields. *Agric. Water Manag.* **2005**, *76*, 24–35. [CrossRef]
79. Babajimopoulos, C.; Panoras, A.; Georgoussis, H.; Arampatzis, G.; Hatzigiannakis, E.; Papamichail, D. Contribution to Irrigation from Shallow Water Table under Field Conditions. *Agric. Water Manag.* **2007**, *92*, 205–210. [CrossRef]
80. Zanis, P.; Kapsomenakis, I.; Philandras, C.; Douvis, K.; Nikolakis, D.; Kanellopoulou, E.; Zerefos, C.; Repapis, C. Analysis of an Ensemble of Present Day and Future Regional Climate Simulations for Greece. *Int. J. Clim.* **2009**, *29*, 1614–1633. [CrossRef]
81. Vasiliades, L.; Loukas, A.; Patsonas, G. Evaluation of a Statistical Downscaling Procedure for the Estimation of Climate Change Impacts on Droughts. *Nat. Hazards Earth Syst. Sci* **2009**, *9*, 879–894. [CrossRef]
82. Tsakmakis, I.; Kokkos, N.; Pisinaras, V.; Papaevangelou, V.; Hatzigiannakis, E.; Arampatzis, G.; Gikas, G.; Linker, R.; Zoras, S.; Evagelopoulos, V.; et al. Operational Precise Irrigation for Cotton Cultivation through the Coupling of Meteorological and Crop Growth Models. *Water Resour. Manag.* **2016**, *31*, 563–580. [CrossRef]
83. Tsakmakis, I.; Kokkos, N.; Gikas, G.D.; Pisinaras, V.; Hatzigiannakis, E.; Arampatzis, G.; Sylaios, G. Evaluation of Aquacrop Model Simulations of Cotton Growth under Deficit Irrigation with an Emphasis on Root Growth and Water Extraction Patterns. *Agric. Water Manag.* **2019**, *213*, 419–432. [CrossRef]
84. Taylor, C.A.; Stefan, H.G. Shallow Groundwater Temperature Response to Climate Change and Urbanization. *J. Hydrol.* **2009**, *375*, 601–612. [CrossRef]
85. Kurylyk, B.L.; MacQuarrie, K.T.B.; Voss, C.I. Climate Change Impacts on the Temperature and Magnitude of Groundwater Discharge from Shallow, Unconfined Aquifers. *Water Resour. Res.* **2014**, *50*, 3253–3274. [CrossRef]
86. Libutti, A.; Monteleone, M. Soil vs. Groundwater: The Quality Dilemma. Managing Nitrogen Leaching and Salinity Control under Irrigated Agriculture in Mediter-Ranean Conditions. *Agric. Water Manag.* **2017**, *186*, 40–50. [CrossRef]
87. Zekri, S.; Madani, K.; Bazargan-Lari, M.R.; Kotagama, H.; Kalbus, E. Feasibility of Adopting Smart Water Meters in Aquifer Management: An Integrated Hydro-Economic Analysis. *Agric. Water Manag.* **2017**, *181*, 85–93. [CrossRef]

Article

Drought Index as Indicator of Salinization of the Salento Aquifer (Southern Italy)

Maria Rosaria Alfio, Gabriella Balacco *, **Alessandro Parisi, Vincenzo Totaro** and **Maria Dolores Fidelibus**

Dipartimento di Ingegneria Civile, Ambientale, del Territorio, Edile e di Chimica (DICATECh), Politecnico di Bari, 70125 Bari, Italy; mariarosaria.alfio@poliba.it (M.R.A.); alessandro.parisi@poliba.it (A.P.); vincenzo.totaro@poliba.it (V.T.); mariadolores.fidelibus@poliba.it (M.D.F.)
* Correspondence: gabriella.balacco@poliba.it; Tel.: +39-08-0596-3791

Received: 17 May 2020; Accepted: 3 July 2020; Published: 6 July 2020

Abstract: Salento peninsula (Southern Italy) hosts a coastal carbonate and karst aquifer. The semi-arid climate is favourable to human settlement and the development of tourism and agricultural activities, which involve high water demand and groundwater exploitation rates, in turn causing groundwater depletion and salinization. In the last decades these issues worsened because of the increased frequency of droughts, which emerges from the analysis of Standardized Precipitation Index (SPI), calculated during 1949–2011 on the base of monthly precipitation. Groundwater level series and chloride concentrations, collected over the extreme drought period 1989–1990, allow a qualitative assessment of groundwater behaviour, highlighting the concurrent groundwater drought and salinization.

Keywords: drought; precipitation; SPI; groundwater salinization; karst

1. Introduction

In 2017 the European Environment Agency (EEA) [1] indicated that droughts are projected to increase in frequency, duration and severity in most of Europe, while the greatest increase is expected for Southern Europe. Under the A2 emissions scenario of IPCC [2] it is expected that all of Italy will go through significant drying and that precipitation will decrease by about 10% to over 40% in the summer [3]. Records show that the average temperature of Europe [4] has risen by 0.95 °C over the last century (1901–2001) and that climate change has caused a steepening of precipitation and temperature gradients resulting in wetter conditions in northern regions and drier conditions in southern areas. Thus, the climate change (i.e., changes in precipitation, total runoff, temperature, potential evapotranspiration) and recurrent drought periods will significantly affect freshwater resources stored in rivers, lakes and aquifers.

The decrease in potential groundwater recharge in Southern Europe [5] should have a severe impact on the availability of freshwater resources for drinking and irrigation uses. Thus, concerning water resources, climate change may have a series of cascading consequences and originate feedback loops, as well as changes in land use [6] that lead to a variation of evapotranspiration. It should be expected that the projected reduction of soil moisture storage will first produce a decrease of groundwater recharge and, later, a decline of groundwater levels and discharge.

In this framework droughts play a key role in understanding environmental complex dynamics [7]. Consistently to the different phenomenologies, they can be classified into four categories [8–10]: (i) meteorological, which refers to a lack of precipitation in a large area and over a long period of time; (ii) agricultural (also called soil moisture), which depends on a deficiency of soil moisture, usually in the root zone; (iii) hydrological, associated to negative anomalies in surface and sub-surface water; and (iv) socio-economic, due to a failure of water resources systems concerning water demands and ecological or health-related impacts. When a combination of the above-mentioned types of droughts occurs at the same time in a certain area it can generate a so-called groundwater drought [11].

Generally, by using frequency analysis of historical data, a groundwater drought is defined as the lack of groundwater, expressed in terms of recharge, storage or hydraulic heads in a certain area and over a particular period of time [12] compared to "normal" conditions (average amount or level). However, an increase of groundwater abstraction occurs during prolonged dry periods or droughts: in the Mediterranean regions, the quite diffuse imbalance between water-demand and water availability (especially due to agricultural sector) may then enhance naturally occurring droughts. Over-exploitation reduces groundwater quantity, leaving aquifers without an efficient storage to cover dry periods [13].

Notwithstanding the potential and serious drawbacks of "superficial" droughts and the knowledge that the effects of primary meteorological drought events are destined to propagate on the entire water cycle [12], groundwater is considered a resilient resource during periods of lower than average rainfall. During the initial phases of a drought, indeed, groundwater can provide relatively resilient water supplies and will sustain surface flows through groundwater baseflow [14]. On the other hand, groundwater may be highly vulnerable to protracted droughts, since groundwater storage may need a longer time to be restocked and recover in comparison to surface water resources as a drought starts to break.

If we consider the complex character of the dependence between groundwater and groundwater-dependent (natural and urban) ecosystems, a worsening in quality and quantity of the former can generate cascading consequences and crises on the latter. In light of water management, such complex and strong interconnections and the possible high delays in the onset of a groundwater drought compared to the superficial drought, when unidentified, raise serious issues about the potential secondary emergencies and cascading vulnerabilities on groundwater-dependent systems.

In the Mediterranean area most of the population resides in the coastal zone, relying on groundwater in coastal aquifers. They are characterized by fresh groundwater floating on salt water due to a different fluid density and may also contain great amounts of water of very good quality, which are, today, subject to alarming salinization processes [15] due to exploitation and/or reduced recharge. Under a decrease of water levels, the transition zone expands, thus reducing the thickness of fresh water with a concurrent increase in its salt content. Depending on the aquifer scale, groundwater takes different times to recover to the previous water level. However, even if after a drought period a normal wet period leads to recovered levels, groundwater quality often remains compromised because exploitation normally does not stop.

An exponential rise of groundwater drawings favours and accelerates this phenomenon due to the attempt to bridge the gap between the increasing water demand and the water availability.

Groundwater in the coastal aquifer of Salento (Puglia, Southern Italy) is highly vulnerable to salinization because of the structure of the aquifer, where discontinuities and karst forms are ways of fast and deep intrusion for seawater and saltwater [16]. In Salento, droughts may easily propagate their effects to the coastal aquifer, worsening qualitative and quantitative status of groundwater and causing cascade crisis [17]. This situation is currently threatened by climate change, which is leading to an increase in groundwater exploitation as in other areas typified by a high level of urbanization and low natural availability of water resources [18]. In the Adriatic and Ionian coastal zones of Salento there is, in fact, an increasing aridity, which impacts on water demand for irrigated agriculture [19].

This study aims as first at representing the drought scenario of the karst coastal aquifer of Salento (Puglia region, Southern Italy) between 1949 and 2011. Then, in correspondence with a period characterized by extreme droughts, the study will show and discuss data about the response of groundwater at four monitoring wells, concerning both its quantitative and qualitative status, with the aim of understanding what the relationships between meteorological and groundwater droughts are. Section 2 describes the meaning and main characteristics of the current drought indicators, while Section 3 illustrates the case study of Salento, with details about dataset problems. Section 4 deals with the methodology adopted for computing SPI and resulting trends; Section 5 shows results and Section 6 discusses related hydrogeological implications, highlighting the main findings. Finally, paragraph 7 outlines the main conclusions of this study.

2. Drought Indicators

Literature proposes numerous methods for calculating meteorological drought indices aimed at quantifying and comparing drought severity, as well as its duration and extent across a certain area and during years. Nowadays, there is not a common method that is suitable for all circumstances and users [20].

The Standardised Precipitation Index (SPI) is among the most used meteorological indices [21]: the World Meteorological Organization considers SPI the best suitable indicator of wetness or dryness conditions [22]. The calculation of SPI requires monthly precipitation, ideally for a continuous period of at least 30 years; it consists of a normalized index with zero mean and standard deviation of one, obtained by fitting a gamma distribution to long-term records of monthly precipitation to represent the relationship of probability to precipitation. Usually, accumulation periods of precipitation are used to estimate the index for different timescales, typically 1, 3, 6, 12 and 24 months; at longer timescales, drought frequencies decrease and consequently drought durations increase. Thus, accumulation precipitations of: (i) 1–3–6 months are used to account agricultural drought; (ii) 12 months to evaluate hydrological drought; and (iii) 24-months to define socio-economic impacts [23]. Successively, to obtain SPI, the corresponding cumulative probability distribution is computed and transformed to the standard normal distribution. Positive values of SPI indicate wet conditions, while negative values refer to dry conditions; an extreme drought occurs when SPI is less than −2. The SPI approach includes several strengths: the use of a unique input data (precipitation), although it requires a long and continuous precipitation time series, the possibility to estimate the index for a variety of timescales, and the feasibility of comparison with other indicators.

The more recently proposed Standardised Precipitation-Evapotranspiration Index (SPEI) [24,25] is similar to SPI in the mathematical structure. Differently from SPI, that uses the precipitation as input, SPEI is calculated from normalized accumulated climatic water balance anomalies, defined as the difference between precipitation and potential evapotranspiration (PET). Then, accumulated water balance is transformed to probabilities and finally converted to the standard normal distribution for computing drought index values. It is important to remark that potential evapotranspiration is the amount of evaporation and transpiration that would occur if a sufficient water source is available. It can be calculated with three approaches: (i) the Thornthwaite method [26], which is the simplest method because it needs only of monthly mean temperature registrations and latitude of the site; (ii) the Hargreaves method [27], that computes the monthly reference evapotranspiration (ET0) of a grass crop and requires minimum and maximum temperature registrations and latitude of the site; and (iii) Penman–Monteith method, that according to Allen et al. [28] calculates ET0 of a hypothetical reference crop, known minimum and maximum temperature registrations and time series of monthly mean daily external radiation, monthly mean daily wind speeds at 2 m height, monthly mean daily bright sunshine hours and monthly mean cloud cover in percentage. The original formulation of SPEI [24] suggests using Thornthwaite method to calculate potential evapotranspiration because of its simplicity, but previous studies demonstrate that this approach underestimates PET in arid and semiarid region. Thus, the Food and Agriculture Organization of the United Nations (FAO) and the American Society of Civil Engineers (ASCE) advised the use of the Penman–Monteith formulation [29,30]. Furthermore, with regard to the Hargreaves formulation, as for the Thornthwaite method, it needs limited data, and it is demonstrated that at monthly and annual timescales PET estimates do not differ significantly from the Penman–Monteith equations, with differences less than 2 mm per day [30]. SPI and SPEI indices are statistically interpretable, representing the number of standard deviations from typical accumulated precipitation, or climatic water balance, for a given location and time of year [31].

The Rainfall Anomaly Index (RAI) was developed by Van Rooy [32]. The RAI calculation can be developed for weekly, monthly or annual timescales, according to dry period frequency: in areas with short dry periods, a smaller timescale is used than in areas with long, dry periods. RAI is calculated arranging precipitations in descending order and selecting the ten highest values and the ten lowest ones from which the average values are computed to represent the thresholds for respectively positive and negative anomalies.

The Palmer Drought Severity Index (PDSI), developed by Palmer [33] for providing an index based on drought severity, allows the comparison of droughts with different time and spatial scales. It belongs to agricultural drought indicators, although many authors classify it as a meteorological one [34,35]. This index evaluates droughts according to the quantity of water stored in the unsaturated zone; therefore, it takes into account precipitation, evapotranspiration and soil moisture. Palmer method starts with a monthly or weekly water balance, using precipitation and temperature time series and contains many assumptions, which make it quite involved [35]. To account for soil moisture, the soil model considers two layers and two simplifications: the former is that the top layer can store 25 mm of water and the underlying layer has an available capacity according to local soil characteristics, whereas the latter is that all water in the first layer is used before the second layer starts leaving water.

All these indicators are climate-linked: they only highlight the influence of precipitation and air temperature on groundwater, but do not represent what may occur to this system. For the calculation of groundwater droughts, the Groundwater Resource Index (GRI) [36] was tested in Calabria, Southern Italy. It represents the normal distribution of the simulated groundwater storage of the Calabria region for 40-years of simulated data. Simulated data were generated by a hydrological model, based on precipitation, air temperature and air pressure as data input. Mendicino et al. [36], by comparing GRI with the SPI of 6, 12 and 24 months, find that the GRI was a better indicator than SPI for droughts in the Mediterranean area. They concluded that this approach appears a stable operative support for decision making when severe droughts and water scarcity problems occur because it is able to account for different information regarding meteorological, hydrological and agricultural aspects.

Regarding groundwater droughts, groundwater level time series can be converted into the Standardised Ground-water level Index (SGI) [37] using a non-parametric normal score transformation of groundwater level data for each calendar month. These monthly estimations are then merged to form a continuous index that is built on the SPI approach. This methodology shows how qualitative information on groundwater use and annual long-term averages help getting a better understanding of an asymmetric impact of groundwater use on groundwater droughts [38].

The requested variables for computing the above-described indices are summarized in Table 1.

Table 1. Indices and linked variables.

Index	Hydrological Variable	Timescale
Standard Precipitation Index (SPI)	Rainfall	Monthly
Standardised Precipitation-Evapotranspiration Index (SPEI)	Rainfall, Temperature	Monthly
Rainfall Anomaly Index (RAI)	Rainfall	Weekly, monthly or annual
Palmer Drought Severity Index (sc-PDSI)	Rainfall, PET	Monthly
Palmer Drought Severity Index (PDSI)	Rainfall, Evaporation	Monthly

3. Study Area and Dataset

3.1. Geological and Hydrogeological Framework

The Salento Peninsula is located in the southeastern part of Puglia region (Southern Italy) and extends from Ionian to Adriatic Sea (Figure 1). It covers 2760 km^2, and its limits roughly coincide with those of Lecce Province. The Salento peninsula belongs to the Apulian carbonate platform that is composed by well-bedded succession of Jurassic-Cretaceous carbonate rocks, with a thickness varying from about 3–5 km. The geological basement of the Salento Peninsula is composed by limestone and dolomitic limestone of Cretaceous age, which outcrop in large areas (Figure 1). The covers are characterized by clay, sand and calcarenite of Miocene to Pleistocene age [39]. The basement is characterized by structural highs and lows separated by sub-vertical normal, strike-slip and oblique-slip faults of Plio-Pleistocene age, striking NNW–SSE and subordinately NW–SE [40].

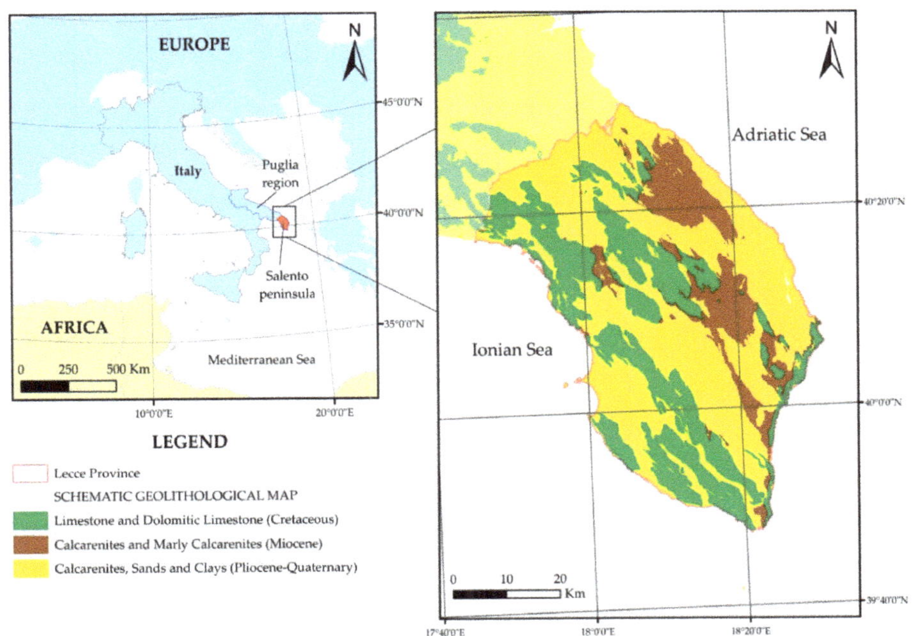

Figure 1. Location of Salento Peninsula and schematic geological map.

The Mesozoic carbonate basement represents the main deep aquifer, which is bordered by the sea. As reported by Portoghese et al. [41], precipitation in Salento is about 638 mm/year (with reference to the period 1951–2002), where 60% is lost due to evapotranspiration, 118 mm/year represent the runoff, 34 mm/year the irrigation and only 132 mm/year recharge the aquifer. Recharge is mainly of focused type because of the existence of hundreds of endorheic basins, as also occurs in the adjacent Murgia karst coastal aquifer [42]. The endorheic basins convert the internal runoff to effective infiltration, unless cascading effects between basins: considering that the surface of endorheic basins occupies more than the 40% of the total area of Salento, it is clear that they play the main role in driving both recharge and pollution transport processes. The covers may often contribute to the recharge of deep aquifer through lateral stratigraphic contacts and tectonic discontinuities. In the geological time, the Salento aquifer has been affected by vadose, water table, and transition zone karst processes, favoured by lithology and fractures under the combined effect of tectonics and glacio-eustatic oscillations [43]: thus, Salento currently shows, in addition to endorheic basins, karst plains, fracture zones, dolines, sinkholes, and karst sub-horizontal levels. These elements constitute an interconnected discontinuity system, which determines a high anisotropy of the hydraulic conductivity, with a mean high permeability at regional scale. Freshwater, with a salt content varying between 0.2 and 0.5 g/L [16], floats on saltwater of marine origin as a lens because of different density; groundwater discharges through coastal springs (sub-aerial and submarine, concentrated and diffuse), with Total Dissolved Solids (TDS) varying between 3.5 and 20 g/L. Due to the presence of low permeability carbonate units and/or tectonics dislocation of the basement, saturated zone is often found under mean sea level. However, the hydraulic continuity of the rock framework allows the development of an extensive groundwater flow system, with a consistent areal distribution of matter and heat, and hydraulic interdependence of different aquifer areas. Water levels reach maximum values of 4.5 m a.s.l. in the

NW and SE sectors of the peninsula, where the maximum fresh groundwater thickness is around 120 m. The hydraulic gradient is about 0.02‰.

Taken as a whole, the aquifer represents a complex system, which combines the complexity of a tectonic karst with that of a coastal aquifer. Its detailed features are outlined in Fidelibus and Pulido-Bosch [16], who explain how the interplay of surface and subsurface features and hydrogeological coastal conditions determine from place to place different groundwater vulnerability, posing consequent different monitoring and management questions.

Starting from the 1960s, population began growing with a concurrent economic development mainly based on tourism, agriculture and small family-size manufacturing activities. Due to the high permeability of karst surface, the region cannot rely on superficial waters: thus, these changes were totally supported by the availability of groundwater, which still represents the unique water resource to satisfy drinking and, especially, irrigation demand. Starting from the 1960s, after the onset of deep drilling techniques, which allowed exploiting the carbonate karst coastal aquifer, thousands of authorized and unauthorized irrigation wells exploit groundwater. Exploitation continuously increased in the time, causing a relentless increase of salinization. Figure 2 shows the earliest distribution of irrigation, not-irrigation and urban zones for the Salento Peninsula referred to the summer season 1997 (data on land use are from SIGRIA—Information System for Water Management for Irrigation [44].

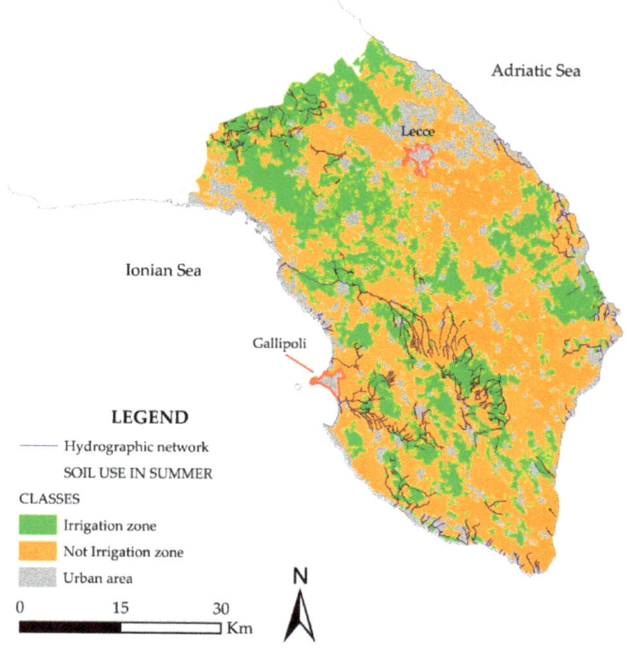

Figure 2. Land use map of Salento peninsula derived by SIGRIA data on land use for the summer 1997.

3.2. Dataset

Monthly average temperatures and rainfall data for Salento peninsula were provided by the Civil Protection of the Puglia Government [45]. Figure 3 shows the location of considered gauges, while in Table 3 their main features are summarized.

Time series of the hydrological input variables required to evaluate drought indices (Table 1) must satisfy a fundamental criterion, i.e., a record continuity not less than 30 years. From the analysis of rainfall records it emerged how the longest time window useful for the study refers to monthly rainfall in the time interval 1949–2011 only for four stations (Gallipoli, Minervino di Lecce, Nardò, and Otranto) out of the 17 located in Salento

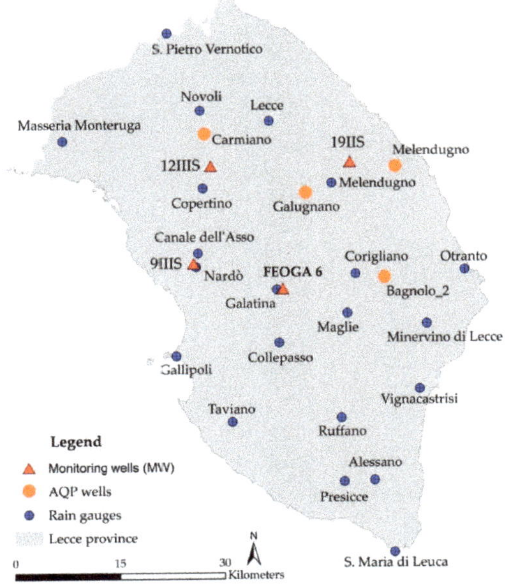

Figure 3. Rain gauge stations (Table 3), Monitoring Wells (Table 2), and wells of the Acquedotto Pugliese (AqP) potable net considered in the study.

Table 2. MW well features.

Well Name	Latitude[1]	Longitude[1]	Distance from the Closest Sea (km)	Well-head Elevation (m a.m.s.l.)	Well Depth (m a.m.s.l.)	Water Strike (m a.m.s.l.)	Static Level at Drilling (m a.m.s.l.)	Saturated Thickness b_a (m) crossed by Screens b_a (m)
Feoga-6	40.135	18.178	11.55	91.7	−134	−61.3	4.5	72.7
19IIS	40.293	18.299	7.5	35.7	−191.2	−179.2	3.7	12
12IIIS	40.295	18.065	14.15	42.6	−19.8	2.6	2.6	22.4
9IIIS	40.171	18.029	4.9	39.9	−29.1	1.3	1.3	30.4

[1] Coordinates in Decimal Degrees (WGS84).

For temperature, instead, there are only fragmented records that prevent the analysis of drought indices that require the temperature-derived data. This is a relevant issue when dealing with all datasets, since it requires the reconstruction of missing data. However, the importance of the topic would require a proper and deep assessment of each method that is out of the scope of this paper. On this basis, we decided to use only reliable data and focus the attention on the SPI index.

Data about water level series refer to four monitoring wells (MW, location in Figure 3). As to their location, the 19IIS and 9IIIs MWs are closer to the Adriatic and Ionian coast, respectively, while the Feoga-6 and 12IIIs MWS are located between the two coasts. All belong to the northern part of the Salento aquifer. Water level series refer to historical records, which are uniquely available in the territory, adequately covering part of the period of precipitation series. Information on their location and technical features are reported in Table 2. The table shows that they are of different depth; two of them have a water strike elevation below mean sea level, corresponding to the elevation of top of the carbonate basement. The saturated thickness refers to the length of well equipped with screens. During 1973–1995, water levels were measured manually with monthly frequency, with different time gaps.

Table 3. Rain gauges stations.

ID	Station Name	Latitude [1]	Longitude [1]	Height (m a.s.l.)
1	Copertino	40.26667	18.05000	34
2	Galatina	40.13417	18.16778	73
3	Gallipoli	40.05444	17.99444	31
4	Lecce	40.35028	18.16667	78
5	Maglie	40.10083	18.28389	77
6	Masseria Monteruga	40.33389	17.81722	72
7	Minervino di Lecce	40.08361	18.41667	98
8	Nardò	40.16694	18.03333	43
9	Novoli	40.36694	18.05056	37
10	Otranto	40.15028	18.48417	52
11	Presicce	39.88389	18.26667	114
12	Ruffano	39.96750	18.26667	125
13	S.Maria di Leuca	39.78417	18.35000	65
14	S. Pancrazio Salentino	40.41667	17.83361	62
15	S. Pietro Vernotico	40.46750	18.00083	36
16	Taviano	39.96750	18.08361	61
17	Vignacastrisi	40.00056	18.40000	94

[1] Coordinates in decimal degrees (WGS84).

Wells belong today to different public organizations: however, in the period 1973–1995 measures were carried out by operators from a same institution with similar technical means and with reference to a constant well-head having an accurate value of the elevation. Archive data do not give indications about instrumentation and type of cables. However, the accuracy in water level measures should be around 5–10 cm, which is good considering the length of cables to be used in most Salento wells. However, rather than absolute values we have considered the "variation" of water levels, which is effective even with a fair degree of accuracy. Technical sheet of drillings also report data on simple discharge tests, but little is known about the conditions of tests. Since it is not possible to check the field historical data, we generally used provided values.

Figure 3 also shows the location of four wells tapping the Salento aquifer for drinking purposes: they belong to the regional potable net (Acquedotto Pugliese, AqP). The study considers the series of chloride concentrations measured in groundwater samples drawn from these wells in pumping condition for the period 1980–2012 by AqP laboratories; data were collected with low (yearly or three-monthly) frequency in the first year (data from paper registrations) and with monthly frequency up to the end of the series.

4. Methodologies

4.1. Data Analysis

As stated in the previous paragraph, only monthly rainfall data in 1949–2011 were available for this study. Moving from some of the most recent climate studies on the Salento peninsula [46,47], we decided to investigate these time series not only in the entire interval, but also in two sub-periods, namely 1949–1979 and 1980–2011. This choice was also motivated by a visual analysis of time series and the opportunity of analysing two time series of comparable length.

Furthermore, the analysis of the rainfall dataset was also aimed at detecting the possible presence of statistically significant trends, following a practice widely diffuse in the scientific literature.

To provide a quantitative assessment of any trend in annual rainfall for the period 1949–2011, we implemented Mann–Kendall test [48,49].

Based on the null hypothesis of absence of trends, given a time series of N independent data $x = [x_1, x_2, \ldots, x_N]$, this trend test relies on the computation of the following statistics:

$$S = \sum_{i=1}^{N-1} \sum_{j=i+1}^{N} \mathrm{sgn}\left(x_j - x_i\right) \tag{1}$$

For $N \geq 8$, statistic S can be retained asymptotically normal distributed with zero mean and variance V that, if there are t_r ties of length r, can be expressed as [50]:

$$V = \frac{N(N-1)(2N+5) - \sum_{r=1}^{N} t_r r(r-1)(2r+5)}{18} \tag{2}$$

According to these statements, the Mann-Kendall test is performed using the variable Z:

$$Z = \begin{cases} \dfrac{S-1}{\sqrt{V(S)}} & S > 0 \\ 0 & S = 0 \\ \dfrac{S+1}{\sqrt{V(S)}} & S < 0 \end{cases} \tag{3}$$

which allows an easy calculation of the p-value, to be compared with the fixed level of significance. In order to take into account the effect that serial correlation can exerts on output of the test, the corrected version of the test proposed by Hamed and Rao [51] was applied.

4.2. SPI

To evaluate hydrological impacts and to explore the drought variation at inter-annual timescales, SPI was calculated aggregating monthly precipitation data over an accumulation period of 12 months [23,52]. According to McKee et al. [21], drought intensity is arbitrarily defined for values of SPI in relation to the categories shown in Table 4.

Table 4. SPI drought categories [1].

SPI Values	Drought Category
0 to −1.00	Mild drought
−1.00 to −1 50	Moderate drought
−1.50 to −2 00	Severe drought
≤−2.00	Extreme drought

[1] McKee et al. [21].

Afterwards, the long-term records of precipitations (X) were converted into log-normal values after removing zero values, to calculate the statistic U according to the following equation:

$$U = \ln(\overline{X}) - \frac{\sum \ln X}{N} \tag{4}$$

where N is the number of observations.

The statistic U was then used for the calculation of two shape parameters (α and β) of the gamma distribution with:

$$\alpha = \frac{\overline{X}}{\beta} \tag{5}$$

$$\beta = \frac{1 + \sqrt{1 + \frac{4U}{3}}}{4U} \tag{6}$$

These shape parameters were then used to compute the basic equation of gamma distribution:

$$0G(X) = \frac{\int_0^X X^{\alpha-1}e^{-\frac{X}{\beta}}}{\beta^{\alpha}\Gamma(\alpha)}dx \tag{7}$$

Regarding zero observations, a new cumulative probability function was introduced

$$H(X) = q + (1 - q)G(X) \tag{8}$$

where q represents the percentage of zero-values in the long-terms records of precipitations. This new probability function was transformed into a standard normal random variable with mean zero and variance of one, in which the created random variable is the value of the SPI.

The R package SPEI [24] has been used to evaluate the SPI index.

4.3. Specific Capacity and Specific Capacity Index

Well drilling technical sheets report results of discharge tests, which allow calculation of specific capacity and specific capacity index.

Specific capacity (S_c) [53] is partly a function of the aquifer transmissivity showing its same dimension ($L^2 \cdot T^{-1}$). It is generally reported as yield per unit of drawdown:

$$S_c = \frac{Q}{s_w}$$

where Q is the pumping rate ($L^3 \cdot T^{-1}$) and s_w is the drawdown (change in hydraulic head) in the well (L). Specific capacity can be normalized to aquifer thickness by using the specific-capacity index [54]:

$$S_i = \frac{S_c}{b_a}$$

where b_a is the aquifer thickness. Specific-capacity index has the same dimension as hydraulic conductivity ($L \cdot T^{-1}$). S_i is useful especially when wells do not cross all the aquifer thickness, as in the case of concerned wells. In the study S_i is calculated normalizing the specific capacity to the length of the open (uncased) borehole that intercept the saturated zone in each well as suggested by Siddiqui and Parizek [55].

S_c and S_i are used in study as "analogous" of transmissivity and hydraulic conductivity respectively, considering, with the due caution, the order of magnitude of their values for the comparison of the properties of the saturated zones intercepted by the selected wells.

5. Results

5.1. Rainfall and SPI Index Analysis

The mean values of annual rainfall for the whole time series and for the two partial sub-series, 1949–1979 and 1980–2011, are reported in Table 5. For the sites of Gallipoli and Nardò there are no significant differences between the selected time windows; on the contrary, for Minervino di Lecce and Otranto the averages values seem showing discordances.

Table 5. Mean values of annual rainfall (mm).

Time Windows	Gallipoli	Minervino di Lecce	Nardò	Otranto
1949–2011	573.8	842.3	611.4	814.7
1949–1979	579.1	886.1	617.4	854.4
1980–2011	568.6	799.8	605.7	776.3

On the contrary, compared to the Mann–Kendall test on a yearly basis there does not seem to be a statistically significant trend at 5%. This is in agreement with the results found by D'Oria et al. [46]. It should be remarked that these usual applications of trend tests are mainly carried out with the evaluation of only type I error of the test (rejecting the null hypothesis when it is true), neglecting the type II error (non-rejecting the null hypothesis when it is false). In particular, a complete interpretation of test application needs to include the evaluation of the power of the test, whose importance was shown by several studies (e.g., [56–58]). However, the simple evaluation of type I error provides useful information for users.

Figure 4 shows the estimated SPI for each of the four study sites in the period 1949–2011.

As for the rainfall, the percentage of drought periods is calculated for each station with reference to both the whole period 1949–2011, and two sub-periods 1949–1979 and 1980–2011. In the first case, the percentage of droughts refers to the number of years of mild, moderate, severe and extreme droughts on the total number of values of the reference period (Table 6). This percentage is lightly greater than 50% for all stations, reaching the highest value in the station of Otranto (54.8%); the percentage for the Nardò station does not show any significant discordancy (50.2%).

Some anomalies appear considering the two sub-periods. In 1949–1979 the percentage of droughts for the Gallipoli, Minervino di Lecce and Otranto stations is lower than 50% (47.7, 42.4 and 45.7 respectively), while the percentages change abruptly in the following period, reaching values of 54.4, 61.7 and 63.3. These percentages denote an increase of drought periods between 1980 and 2011 compared to the previous 30 years. For the Nardò station, the distribution of droughts in the two sub-periods is substantially the same.

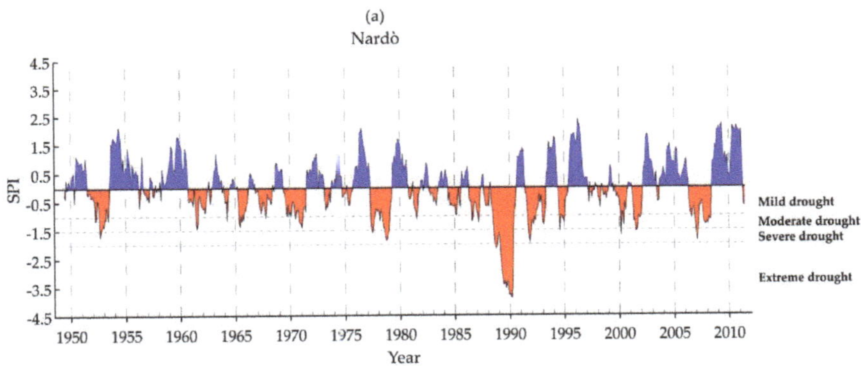

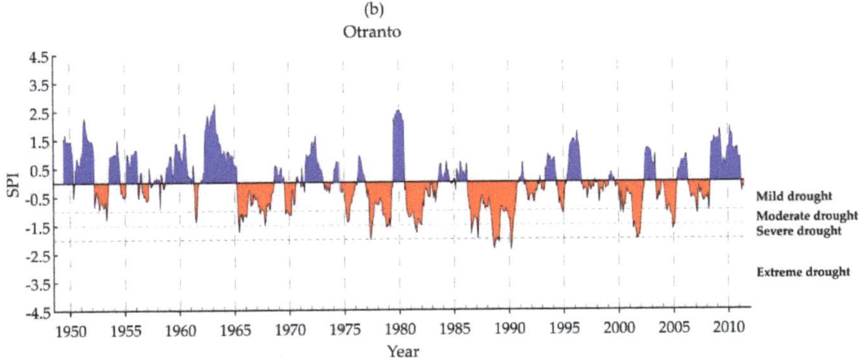

Figure 4. *Cont.*

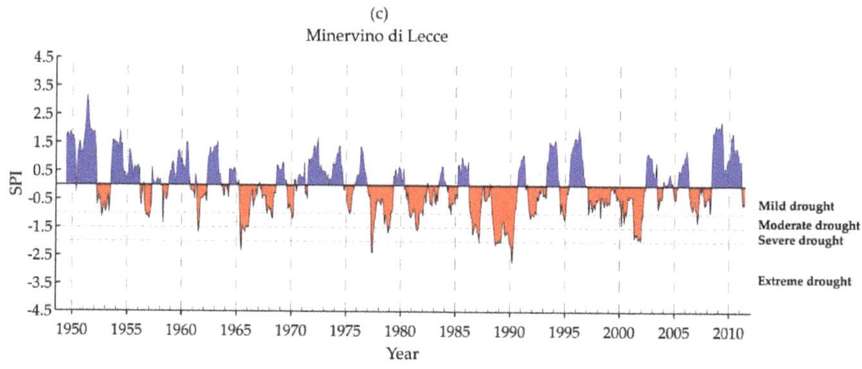

(c)
Minervino di Lecce

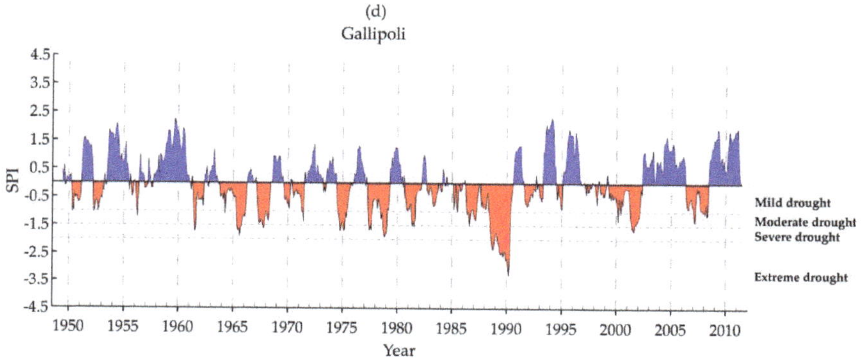

(d)
Gallipoli

Figure 4. SPI in the period 1949–2011, evaluated, for (**a**) Nardò, (**b**) Otranto, (**c**) Minervino di Lecce and (**d**) Gallipoli. The year marks are centred on 1 July of each year.

Table 6. Distribution (%) of different types of droughts.

		Raingauge Stations			
		Gallipoli	Minervino	Nardò	Otranto
	1949–2011	51.14	52.35	50.20	54.77
Time Periods	1949–1979	47.65	42.38	49.58	45.71
	1980–2011	54.43	61.72	50.78	63.28

		Gallipoli		Minervino		Nardò		Otranto	
Time Periods		1949–1979	1980–2011	1949–1979	1980–2011	1949–1979	1980–2011	1949–1979	1980–2011
	Mild	68.60	75.12	75.16	70.89	78.21	64.62	74.55	69.55
SPI Drought	Moderate	18.60	11.48	16.99	13.92	17.32	19.49	20.61	19.34
categories	Severe	12.79	4.78	5.88	10.97	4.47	7.18	4.24	7.41
	Extreme	0.00	8.61	1.96	4.22	0.00	8.72	0.61	3.70

Table 6 also shows the percentage of mild, moderate, severe, and extreme droughts within the whole drought periods. From these last data an important increase of the percentage of extreme droughts clearly emerges, moving from a nearly complete lack (as for Gallipoli and Nardò) to a significant presence for all sites. The percentage of severe droughts increases as well in the second sub-period compared to the first, with the exception of Gallipoli. This station shows, in fact, a decrease in severe droughts: however, the sum of severe and extremes droughts before and after 1980 is substantially unchanged. Furthermore, for this station, a decline can also be observed for moderate droughts, in contrast to mild ones. Regarding the latter, relevant modifications affect other sites. The percentages of the extreme droughts in the other sites suggest the presence of a general shift from mild to more concentrated and heavy events (with different intensities), which is clearly shown by the drought occurred in 1989–1990.

All four sites in this two-year period showed extreme droughts during March 1989 and September–October 1990. Severe and some extreme droughts, more or less contemporaneous, occurred starting from 1978 to 1988. Other moderate to severe droughts and a few extreme droughts appear over September 2001–August 2002. Considering the SPI patterns, we selected for further analyses the period 1989–1990, being the most affected by extreme droughts within the considered time window of 1949–2011.

5.2. Drought Effects on the Salento Aquifer

The severe and extreme meteorological droughts that occurred during 1989–1990 reflect on groundwater levels measured at the Feoga-6, 12IIIS, 9IIIS and 19IIS monitoring wells (Figure 5, well locations are shown in Figure 3). Figure 5 also shows the SPI for the four rain-gauge stations for the same time window.

The largest water level measurement period (1979–1994) relates to the Feoga-6 MW. Unluckily, the available measurements of the water level were performed manually, and measures are not regularly distributed in that time. This prevents any statistical correlation between water levels and SPI values. Figure 5 allows, however, a qualitative reading of the relationships between droughts and water levels. All the water level patterns highlight a decrease of water level over the period 1989–1990, which is characterised by meteorologically severe and extreme droughts (Figure 5a). The decrease is quite abrupt for 19IIS (80 cm, between March and April 1989, Figure 5c) and 9IIIS (70 cm, between January and the end of March 1989, Figure 5d). Unfortunately, for the well 9IIIS there is a lack of measurements just in February 1989: this gap prevents a precise attribution of the onset of the water level decrease: however, both wells show a prompt response just after the extreme drought period, which starts on February 1989 and lasts till May 1989.

Well 12IIIS (Figure 5c) mimics the behaviour of 19IIS, however showing a decrease of only 15 cm between February and March 1989. The water level of the Feoga-6 MW shows a different behaviour. It shows a continuous decrease in the time, which starts from the beginning of the period of moderate and severe droughts, occurred between February and September 1987, and ending in correspondence with the termination of the period of extreme drought 1989–1990 (Figure 5b).

The increase of SPI during 1988 causes a groundwater level increase in all wells; however, the increase for the Feoga-6 is short and modest. The next increase of SPI after the end of 1990 is reflected instead in all water levels: however, even if SPI increases in 1991, showing only mild droughts and rare moderate droughts between 1992 and 1993, this does not allow the water levels to recover the values shown before 1989 for Feoga-6, 19IIS and 9IIIs. On the contrary, the water level of 12IIIS increases well beyond the previous value.

Understanding the reasons of the different behaviour of water level under the effects of droughts in the examined wells is not immediate, due to the interplay among the natural complexity of the aquifer and the human pressures on groundwater. Data on specific capacity and specific capacity index can help outlining the different hydrogeological conditions of Salento aquifer in the zones where the wells are drilled.

Table 7 shows the calculated values of S_c (m^2/s) and S_i (m/s) for the four considered wells. S_c ranges from 0.8×10^{-3} to 6.5×10^{-2} m^2/s, while S_i varies between 1×10^{-5} and 2.1×10^{-3} m/s. Even if such values have to be considered only as "analogous" of transmissivity and hydraulic conductivity, respectively, having the same dimensions, they agree with literature data. As an example, mean hydraulic conductivity measured in boreholes in karst masses may range between 10^{-8} and 10^{-3} m/s because of the variable influence of macro-fractures and the karst network [59], while typical hydraulic conductivity of karst conduits ranges between 1 and 10 m/s.

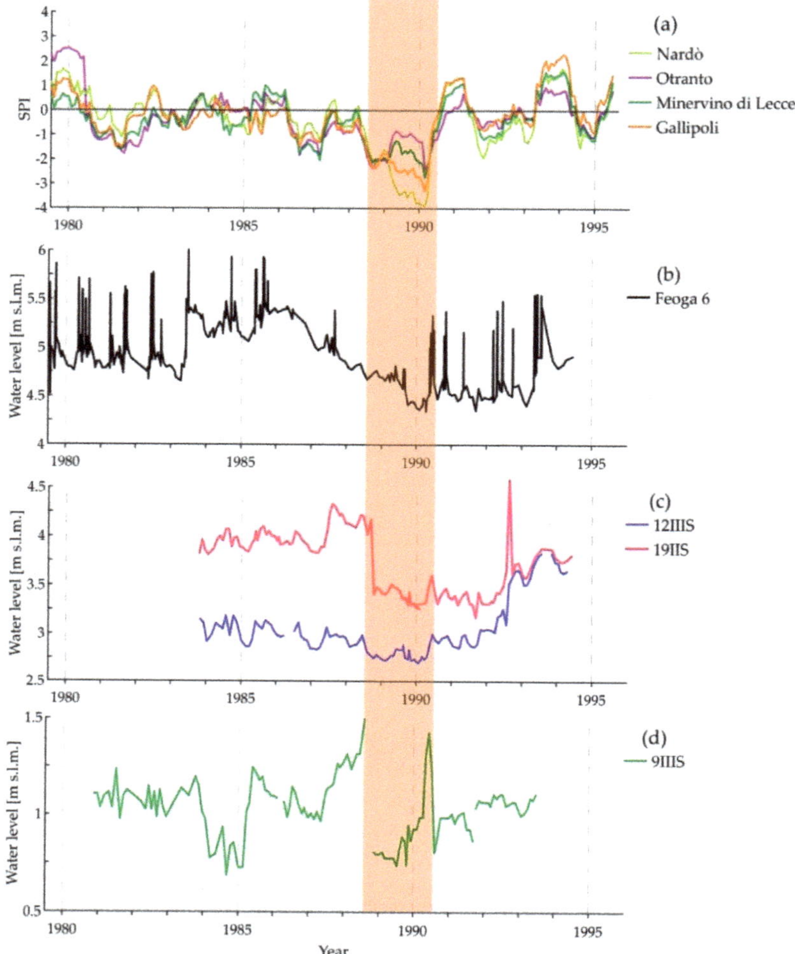

Figure 5. SPI calculated for precipitation measured between 1979–1995 at the Nardò, Otranto, Minervino di Lecce and Gallipoli rain gauge stations (**a**) compared to the water level time series over 1979–1994 of the Feoga-6 (**b**), 12IIIS and 19IIS (**c**), and 9IIS (**d**) MWs respectively. The coloured bar covers the period 1 January 1989–31 December 1990; the date ticks are centred on 1 July of the correspondent year. The location of gauge stations and MWs is in Figure 3.

Table 7. Specific capacity and specific capacity index values.

Well Name	Specific Capacity S_c	Specific Capacity Index S_i
	m^2/s	m/s
Feoga-6	0.8×10^{-3}	1×10^{-5}
19IIS	2.1×10^{-3}	1.8×10^{-4}
12IIIS	3.7×10^{-2}	1.6×10^{-4}
9IIIS	6.5×10^{-2}	2.1×10^{-3}

Feoga-6 shows the lowest value of S_c, with a S_i of one or two orders of magnitude lower than the S_i of 9IIIS and 12IIIS wells. 9IIIS show the highest S_c and S_i. On the basis of these values, the reaction of MW Feoga-6 to the succession of droughts and the long time after the end of drought period before the start of water level recovering, and the length of groundwater drought (eight years) can be explained by a low transmissivity of the aquifer in the zone of the well, where groundwater circulates in a "confined" condition. The other wells promptly react to the extreme droughts, while showing a more resilient behaviour when moderate and severe droughts occur: for these wells the period of groundwater drought is lower (less than two years) than for Feoga-6 MW and the water level more rapidly recovers from the extreme drought.

In addition to the drought and the differences in the hydrological properties of the aquifer, to explain the water level variations we should also consider the role of exploitation and effective infiltration. The current and past total amounts of groundwater exploitation are not easy to assess because the official data do not include the exploitation from the thousands of abusive wells used in the irrigated areas. Since 60% of the total amount of groundwater exploitation is for irrigation [41], we will use, for the aims of the discussion, an evaluation of the total yearly amount of exploitation for irrigation referred to the entire Salento aquifer (Figure 6c) carried out by the G-Mat hydrological model [60] for the period 1970–2002 [41]. Figure 6b shows, for the same time window, the evolution of the yearly precipitation and evapotranspiration, and yearly recharge (Figure 6c) modelled with the same hydrological model [41]; groundwater stress (GWS) (Figure 6a) relates to the ratio between yearly irrigation and yearly effective infiltration [41]. Figure 6a shows the SPI values of the considered gauge stations. Figure 6d also shows groundwater chloride concentrations measured in samples from a few drinking wells located close to the considered MWs (location in Figure 3).

In correspondence with the extreme drought period of 1989–1990, recharge decreases from 54.4 to 38 mm, while irrigation increases from 44 to 61 mm (Figure 6c), causing in 1990 a GWS around 1.6 (Figure 6a). This high level of stress shows the role of exploitation for irrigation, which increases only during the low recharge periods, in worsening the effects of meteorological droughts on groundwater. The high value of GWS corresponds to a minimum for all groundwater level series of Figure 5.

Due to the coastal nature of Salento aquifer, groundwater, alongside with the decrease of water levels, shows salinization. In fact, if the reaction of inland aquifers to precipitation shortage (with a consequent recharge drought) normally consists in droughts in the different parts of the hydrogeological system, with "head drought" and "discharge and groundwater droughts" [10], in coastal aquifers a "recharge drought", apart from varying the quantitative status of groundwater, disturbs the freshwater-seawater equilibrium as well, with a consequent worsening of its qualitative status (groundwater salinization). Figure 6d shows an increase of chloride concentrations in correspondence with the extreme drought period 1989–1990. In the following year the recharge increases, while concurrently irrigation decreases, bringing GWS back to acceptable values. However, chloride concentrations do not come back to previous levels like the water levels shown in Figure 5. Moderate-severe droughts appear between 1991 and 1992 with an increase of GWS over 0.5. The next moderate and severe drought period occurs over November 2001–August 2002, during which the GWS again increases up to 0.5 because of a new serious imbalance between recharge and irrigation.

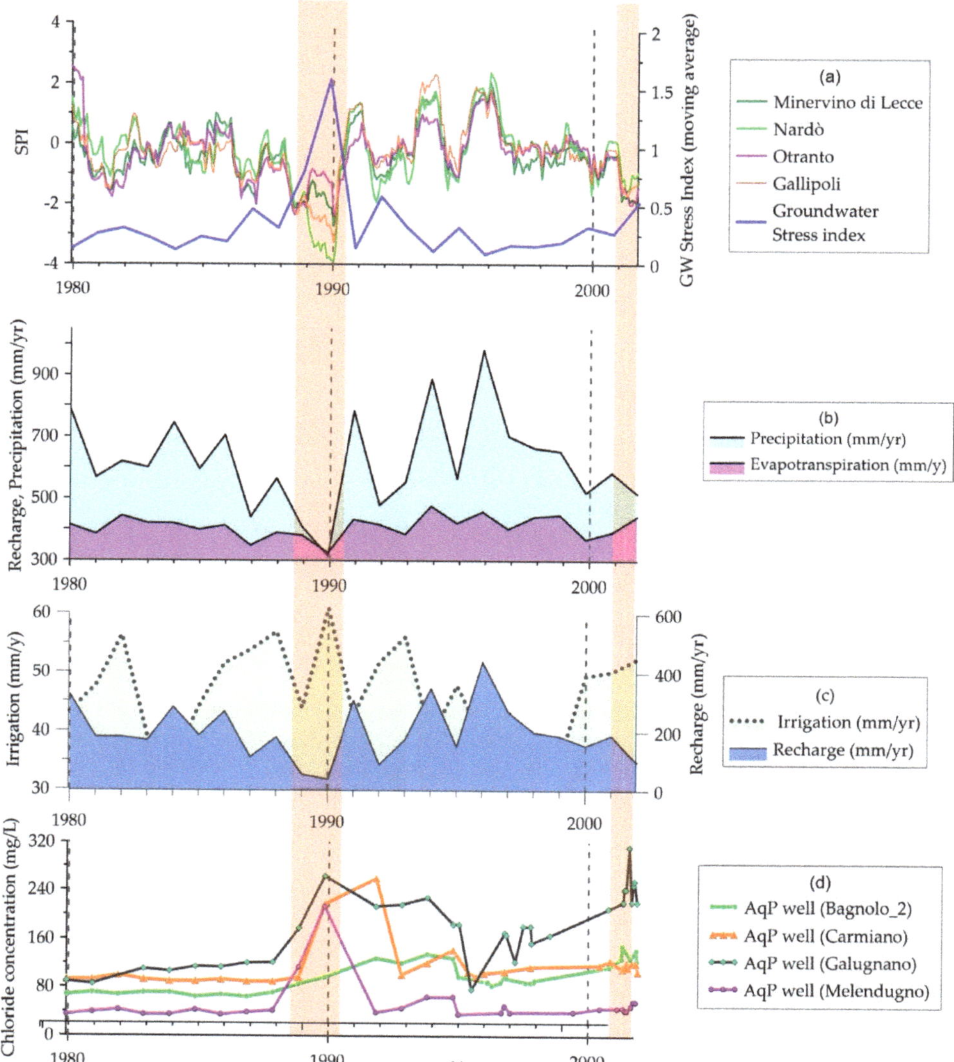

Figure 6. SPI (**a**) calculated for precipitation measured between 1980–2002 at the Nardò, Otranto, Minervino di Lecce and Gallipoli rain gauge stations compared to (**a**) GWS, (**b**) yearly precipitation, evapotranspiration and (**c**) irrigation and recharge. GWS and data shown in (**b**) and (**c**) refer to the entire Salento aquifer. (**d**) Chloride concentrations measured in samples from AqP wells tapping groundwater for potable use. The coloured bars cover the periods 1 January 1989–31 December 1990 and November 2001–August 2002; the date ticks are centred on 1 July of the correspondent year. Location of gauge stations and AqP wells is shown in Figure 3.

6. Discussion

The SPI data highlight that, over 1980–2011, the percentage of drought periods increased compared to the previous 30 years; moreover, all four SPI patterns clearly denote a period of extreme droughts during March 1989 and September–October 1990. The lack of continuity in temperature series prevented the calculation of more complex indices different from the SPI. This is a precipitation-based drought index that neglects the role of temperature and evapotranspiration on drought conditions, while considering that the other variables have no temporal trend. The use of SPI data might underestimate droughts in the Apulia region in light of the trends toward warmer and marginally drier conditions during 1951–2005 [47] especially observed in the spring season series between 1961 and 2006 [61]. However, SPI data clearly allowed identifying that the worse scenery of drought in the period 1970–2010 is the period 1989–1990. Thus, the relationships between the meteorological droughts and groundwater behaviour were examined, focusing on this drought period.

The available historical data on groundwater monitoring required to this aim are of different origin and not systematic enough to allow the verification of statistical correlations. They, however, provided evidence of some significant agreements among the hydrogeological and SPI series patterns. As to groundwater levels, they do not recover pristine levels after droughts following a new phase of recharge. Actually, during the recharge periods subsequent to a drought, hydrogeological reserves should be naturally recovered. The reason of a delayed recovering of water levels may be that exploitation continues during and after the drought period at a rate that is not compatible with groundwater recovery. However, SPI indicates that other drought periods later occur, which could cause the superposition of new periods of "recharge drought" (and GW stress) on the first delayed condition of "groundwater drought". This way, the negative effects of a succession of drought periods could not be recovered by the contribution of new volumes of effective infiltration due to the invariance of anthropic pressure. In explaining the length of "groundwater droughts", the complexity of the karst aquifer adds to the complexity of the coupling of natural droughts and human drivers: thereby, some parts of the Salento aquifer demonstrate to be more sensitive to the succession of droughts because of their low permeability and/or transmissivity.

Concerning chloride concentrations, their evolution after the extreme drought period 1989–1990, while showing the positive effects of recharge variations, also indicates that the following moderate and severe droughts do not allow recovering the low concentrations typical of the period 1970–1989. This, with respect to water level patterns, indirectly highlights a groundwater drought period that lasts well beyond the extreme drought period 1989–1990. Data on chloride concentrations after 2002 (not shown) indicate that concentrations continue to be disturbed by the occurrence of other moderate and severe droughts (Figure 6a). Unfortunately, data from the hydrological model are based on real measurements only up to 2002, while after 2002 data are projected on the basis of downscaling the projection of climate models [41]: this prevents any further reliable comparison.

The chloride concentration patterns seem to signal that the drought period of 1989–1990 has taken groundwater in an alternative overexploited state, probably not recoverable in light of increased drought frequency and the permanent pressure of exploitation. The situation evokes the occurrence during 1989–1990 of a so-called critical transition [62], a sort of point of no return and settlement on a new equilibrium situation, however worse than the previous one.

7. Conclusions

Some conclusions can be drawn from the results.

First, the groundwater levels of all the examined wells show a decrease under the drought impact: however, the decrease is not attributable to the occurrence of droughts, but rather to the coupling of "recharge droughts" and concurrent increases of exploitation. Exploitation, indeed, increases just with water shortage, causing high GWS: this finding coincides with the conclusions of a few studies who already outlined that the primary enemy of groundwater resources is not climate change, but groundwater exploitation [63,64].

A second conclusion concerns the length of "groundwater drought" periods. The evolution of groundwater levels and chloride concentrations related to wells exploited for drinking purposes generally indicate a delay of groundwater droughts beyond the main extreme drought period of 1989–1990. In addition to the persistent pressure of the irrigation, this delay can be also linked to the superposition of different periods of "recharge drought", which reduce water storage and prevent the water level from recovering. This means that the response of groundwater to each single drought depends on the prior state, as normally occurs in complex systems.

As a third conclusion, the study demonstrates that groundwater droughts have cascade effects on groundwater quality. Really, due to the nature of complex system and non-linear behaviour, we believe that, in the Salento coastal aquifer, the gradual change in the system drivers, such as the extreme drought period of 1989–1990, has brought groundwater to a "tipping point" [62], which is a sort of "catastrophic bifurcation point" due to a decrease in groundwater resilience. Human drivers and climate change do not promise any future improvement: on the base of global forecasts of climate change, Portoghese et al. [41] estimate, for the next 50 years, even worse periods of GW stress in Salento, especially linked to the change of precipitation patterns, which could trigger the occurrence of other tipping points.

With all the caution due to the lack of efficient (in time and space) monitoring data, what is shown indicates that climate change, while making important changes in the relationships between all the terms of the hydrogeological balance, is not the main cause of the deterioration of the aquifer, but that the lack of control of the exploitation plays the fundamental role. Moreover, what seems clear is that groundwater responds to events in a delayed manner, moving forward the negative effects of actions (past and present).

All of the above draws attention to a significant issue concerning the methods and parameters used in environmental monitoring, which in most cases is expensive but not tuned to complex goals. The recognition of relationships among significant environmental elements requires long-term monitoring records. Unfortunately, there is a general lack of systematic studies and records that prevent catching the complexity of such relationships because monitoring mainly focuses on water quality for compliance with regulatory issues. Really, we should monitor simultaneously and in the same sites, and for consistent periods, different elements of the water cycle (drivers) and specific parameters able to describe the evolution of climate and groundwater salinization. This type of approach would provide important information about the potential non-linearity in groundwater behaviour, lag times between causes and effects, complex feedbacks and non-linear interaction between components.

Author Contributions: Conceptualization, G.B., V.T. and M.D.F.; Methodology, G.B., V.T. and M.D.F.; Data Curation, A.P. and M.R.A.; Writing-Review & Editing G.B., M.R.A., A.P., V.T. and M.D.F.; Resources, G.B. and M.D.F.; Supervision, M.D.F. All authors have read and agreed to the published version of the manuscript.

Funding: The present investigation was partially carried out with support from the Puglia Region (POR Puglia FESRFSE 2014–2020) through the "T.E.S.A."—Tecnologie innovative per l'affinamento Economico e Sostenibile delle Acque reflue depurate rivenienti dagli impianti di depurazione di Taranto Bellavista e Gennarini—project.

Acknowledgments: The authors thank AQP for providing technical support and data. In addition, the authors thank the anonymous reviewers for their valuable comments.

Conflicts of Interest: The authors declare no conflict of interest.

References

1. European Environment Agency. *Climate Change Adaptation and Disaster Risk Reduction in Europe. Enhancing Coherence of the Knowledge Base, Policies and Practices*; European Environment Agency: Copenhagen, Denmark, 2017. [CrossRef]

2. Nakicenovic, N.; Swart, R. *IPCC Special Report on Emissions Scenarios (SRES)*; Working Group III, Intergovernmental Panel on Climate Change (IPCC); Cambridge University Press: Cambridge, UK, 2000.

3. Giorgi, F.; Im, E.S.; Coppola, E.; Diffenbaugh, N.S.; Gao, X.J.; Mariotti, L.; Shi, Y. Higher Hydroclimatic Intensity with Global Warming. *J. Clim.* **2011**, *24*, 5309–5324. [CrossRef]

4. IPCC. *Climate Change 2007: Impacts, Adaptation and Vulnerability*; Contribution of Working Group II to the fourth Assessment Report of the Intergovernmental Panel; IPCC: Geneva, Switzerland, 2007.

5. Hiscock, K.; Sparkes, R.; Hodgson, A.; Martin, J.L.; Taniguchi, M. Evaluation of future climate change impacts in Europe on potential groundwater recharge. *Geophs. Res. Abstr.* **2008**, *10*, EGU2008-A-10211.

6. Figorito, B.; Tarantino, E.; Balacco, G.; Fratino, U. An object-based method for mapping ephemeral river areas from worldview-2 satellite data. In *Remote Sensing for Agriculture, Ecosystems, and Hydrology XIV*; International Society for Optics and Photonics: Edinburgh, UK, 2012; Volume 8531, p. 85310B.

7. Loukas, A.; Vasiliades, L.; Dalezios, N.R. Intercomparison of Meteorological Drought Indices for Drought Assessment and Monitoring in Greece. In Proceedings of the International Conference on Environmental Science and Technology, Lemnos Island, Greece, 8–10 September 2003; Volume B, pp. 484–491.

8. Van Loon, A.F.; Van Huijgevoort, M.H.J.; Van Lanen, H.A.J. Evaluation of drought propagation in an ensemble mean of large-scale hydrological models. *Hydrol. Earth Syst. Sci.* **2012**, *16*, 4057–4078. [CrossRef]

9. Chang, K.Y.; Xu, L.; Starr, G.; Paw U, K.T. A drought indicator reflecting ecosystem responses to water availability: The Normalized Ecosystem Drought Index. *Agric. For. Meteorol.* **2018**, *250–251*, 102–117. [CrossRef]

10. Wanders, N.; Van Lanen, H.A.J.; Van Loon, A.F. *Indicators for Drought Characterization on a Global Scale*; Watch Tech. Rep. 24; Wageningen Universiteit: Wageningen, The Netherlands, 2010.

11. Van Lanen, H.A.J.; Peters, E. Definition, Effects and Assessment of Groundwater Droughts. In *Drought and Drought Mitigation in Europe Europe*; Jorgen, V., Vogt, F.S., Eds.; Springer-Science+Business Media, BV: Berlin, Germany, 2000; pp. 49–61.

12. Van Loon, A.F. Hydrological drought explained. Wiley Interdiscip. *Rev. Water* **2015**, *2*, 359–392.

13. Van Loon, A.F.; Stahl, K.; Di Baldassarre, G.; Clark, J.; Rangecroft, S.; Wanders, N.; Gleeson, T.; Van Dijk, A.I.J.M.; Tallaksen, L.M.; Hannaford, J.; et al. Drought in a human-modified world: Reframing drought definitions, understanding, and analysis approaches. *Hydrol. Earth Syst. Sci.* **2016**, *20*, 3631–3650. [CrossRef]

14. Hughes, J.D.; Petrone, K.C.; Silberstein, R.P. Drought, Groundwater Storage and Stream Flow Decline in Southwestern Australia. *Geophys. Res. Lett.* **2012**. [CrossRef]

15. Leduc, C.; Pulido-Bosch, A.; Remini, B.; Massuel, S. Changes in Mediterranean groundwater resources. In *The Mediterranean Region under Climate Change*; IRD, Ed.; IRD: Marseille, France, 2016; pp. 328–333. ISBN 978-2-7099-2219-7.

16. Fidelibus, M.D.; Pulido-Bosch, A. Groundwater temperature as an indicator of the vulnerability of Karst coastal aquifers. *Geosciences* **2019**, *9*, 23. [CrossRef]

17. Parisi, A.; Monno, V.; Fidelibus, M.D. Cascading vulnerability scenarios in the management of groundwater depletion and salinization in semi-arid areas. *Int. J. Disaster Risk Reduct.* **2018**, *30*, 292–305. [CrossRef]

18. Alsumaiei, A.A. Monitoring Hydrometeorological Droughts Using a Simplified Precipitation Index. *Climate* **2020**, *8*, 19. [CrossRef]

19. Passarella, G.; Bruno, D.; Lay-Ekuakille, A.; Maggi, S.; Masciale, R.; Zaccaria, D. Spatial and temporal classification of coastal regions using bioclimatic indices in a Mediterranean environment. *Sci. Total Environ.* **2019**, *700*, 134415. [CrossRef] [PubMed]

20. Lloyd-Hughes, B. The impracticality of a universal drought definition. *Theor. Appl. Climatol.* **2014**, *117*, 607–611. [CrossRef]

21. McKee, T.B.; Doesken, N.J.; Kleist, J. The relationship of drought frequency and duration to time scales. In Proceedings of the Eight Conference on Applied Climatoligy, Anaheim, CA, USA, 17–22 January 1993.

22. World Meteorological Organization. *Experts Agree on a Universal Drought Index to Cope with Climate Risks*; Press release No. 872; World Meteorological Organization: Geneva, Switzerland, 2009.

23. Potop, V.; Boroneanţ, C.; Možný, M.; Štěpárek, P.; Skalák, P. Observed spatiotemporal characteristics of drought on various time scales over the Czech Republic. *Theor. Appl. Climatol.* **2014**, *115*, 563–581. [CrossRef]

24. Vicente-Serrano, S.M.; Beguería, S.; López-Moreno, J.I. A multiscalar drought index sensitive to global warming: The standardized precipitation evapotranspiration index. *J. Clim.* **2010**, *23*, 1696–1718. [CrossRef]

25. Beguería, S.; Vicente-Serrano, S.M.; Reig, F.; Latorre, B. Standardized precipitation evapotranspiration index (SPEI) revisited: Parameter fitting, evapotranspiration models, tools, datasets and drought monitoring. *Int. J. Climatol.* **2013**, *34*, 3001–3023. [CrossRef]

26. Thornthwaite, C.W. An Approach toward a Rational Classification of Climate. *Geogr. Rev.* **1948**, *38*, 54–94. [CrossRef]

27. Hargreaves, G.H. Defining and using reference evapotranspiration. *J. Irrig. Drain. Eng.* **1994**, *120*, 1132–1139. [CrossRef]

28. Allen, R.G.; Pereira, L.S.; Raes, D.; Smith, M. *Crop Evapotranspiration: Guidelines for Computing Crop Requirements*; Irrig. Drain. Pap.; United Nations FAO: Rome, Italy, 1998; Volume 56.

29. Walter, I.A.; Allen, R.G.; Elliot, R.; Jensen, M.E.; Itenfisu, D.; Mecham, B.; Howell, T.A.; Snyder, R.; Brown, P.; Echings, S.; et al. ASCE's standardized reference evapotranspiration equation. In *Proceedings National Irrigation Symposium*; Evans, R.G., Benham, B.L., Trooien, T.P., Eds.; ASAE: Phoenix, AZ, USA, 2000; pp. 209–215.

30. Droogers, P.; Allen, R.G. Estimating reference evapotranspiration under inaccurate data conditions. *Irrig. Drain. Syst.* **2002**, *16*, 33–45. [CrossRef]

31. Stagge, J.H.; Tallaksen, L.M.; Gudmundsson, L.; Van Loon, A.F.; Stahl, K. Candidate Distributions for Climatological Drought Indices (SPI and SPEI). *Int. J. Climatol.* **2015**, *35*, 4027–4040. [CrossRef]

32. Van Rooy, M.P. A Rainfall anomaly index (RAI) independent of time and space. *Notos* **1965**, *14*, 43–48.

33. Palmer, W.C. *Meteorological Drought*; U.S. Department of Commerce: Weather Bureau, Washington, DC, USA, 1965; Volume 45.

34. Wells, N.; Goddard, S.; Hayes, M.J. A Self-Calibrating Palmer Drought Severity Index. *J. Clim.* **2004**, *17*, 2335–2351. [CrossRef]

35. Alley, W.M. The Palmer Drought Severity Index: Limitations and assumptions. *J. Clim. Appl. Meteorol.* **1984**, *23*, 1100–1109. [CrossRef]

36. Mendicino, G.; Senatore, A.; Versace, P. A Groundwater Resource Index (GRI) for drought monitoring and forecasting in a mediterranean climate. *J. Hydrol.* **2008**, *357*, 282–302. [CrossRef]

37. Bloomfield, J.P.; Marchant, B.P. Analysis of groundwater drought building on the standardised precipitation index approach. *Hydrol. Earth Syst. Sci.* **2013**, *17*, 4769–4787. [CrossRef]

38. Wendt, D.E.; Van Loon, A.F.; Bloomfield, J.P.; Hannah, D.M. Asymmetric impact of groundwater use on groundwater droughts. *Hydrol. Earth Syst. Sci.* **2020**. in review.

39. Ciaranfi, N.; Pieri, P.; Ricchetti, G. Note alla carta geologica delle Murge e del Salento (Puglia centro—meridionale). *Mem. Della Soc. Geol. Ital.* **1988**, *41*, 449–460.

40. Gambini, R.; Tozzi, M. Tertiary Geodynamic Evolution of the Southern Adria Microplate. *Terra Nov.* **1996**, *8*, 593–602. [CrossRef]

41. Portoghese, I.; Bruno, E.; Dumas, P.; Guyennon, N.; Hallegatte, S.; Hourcade, J.; Nassopoulos, H.; Pisacane, G.; Struglia, M.V.; Vurro, M. Impacts of Climate Change on Freshwater Bodies: Quantitative Aspects. In *Advances in Global Change Research 50*; Springer: Dordrecht, The Netherlands, 2013; pp. 241–304. ISBN 9789400757813.

42. Fidelibus, M.D.; Balacco, G.; Gioia, A.; Iacobellis, V.; Spilotro, G. Mass transport triggered by heavy rainfall: The role of endorheic basins and epikarst in a regional karst aquifer. *Hydrol. Process.* **2016**, *31*, 394–408. [CrossRef]

43. Canora, F.; Fidelibus, D.; Spilotro, G. Coastal and Inland Karst Morphologies Driven by Sea Level Stands: A GIS Based Method for Their Evaluation. *Earth Surf. Process. Landf.* **2012**, *37*, 1376–1386. [CrossRef]

44. INEA, Stato Dell'irrigazione in Puglia. 2000. Available online: https://sigrian.crea.gov.it/wp-content/uploads/2019/03/Irrigazione_Puglia.pdf (accessed on 11 May 2020).

45. Protezione Civile Puglia—Centro Funzionale Decentrato. Available online: https://protezionecivile.puglia.it/centro-funzionale-decentrato/ (accessed on 11 May 2020).

46. D'Oria, M.; Tanda, M.G.; Todaro, V. Assessment of local climate change: Historical trends and RCM multi-model projections over the Salento Area (Italy). *Water* **2018**, *10*, 978. [CrossRef]

47. Lionello, P.; Congedi, L.; Reale, M.; Scarascia, L.; Tanzarella, A. Sensitivity of typical Mediterranean crops to past and future evolution of seasonal temperature and precipitation in Apulia. *Reg. Environ. Change* **2013**, *14*, 2025–2038. [CrossRef]

48. Mann, H.B. Nonparametric Tests against Trend. *Econometrica* **1945**, *13*, 245–259. [CrossRef]

49. Kendall, M.G. *Rank Correlation Methods*; Griffin: London, UK, 1975; ISBN 0852641990.

50. Yue, S.; Pilon, P.; Phinney, B.; Cavadias, G. The influence of autocorrelation on the ability to detect trend in hydrological series. *Hydrol. Process.* **2002**, *16*, 1807–1829. [CrossRef]

51. Hamed, K.H.; Rao, A.R. A modified Mann-Kendall trend test for autocorrelated data. *J. Hydrol.* **1998**, *204*, 182–196. [CrossRef]

52. Tan, C.; Yang, J.; Li, M. Temporal-spatial variation of drought indicated by SPI and SPEI in Ningxia Hui Autonomous Region, China. *Atmosphere* **2015**, *6*, 1399–1421. [CrossRef]

53. Mace, R.E. *Estimating Transmissivity Using Specific-Capacity Data*; Bureau of Economic Geology, University of Texas at Austin: Austin, TX, USA, 2001.

54. Davis, S.N.; DeWiest, R.J.M. *Hydrogeology*; John Wiley and Sons: New York, NY, USA, 1966.

55. Siddiqui, S.H.; Parizek, R.R. Hydrogeologic Factors Influencing Well Yields in Folded and Faulted Carbonate Rocks in Central Pennsylvania. *Water Resour. Res.* **1971**, *7*, 1295–1312. [CrossRef]

56. Yue, S.; Pilon, P.; Cavadias, G. Power of the Mann-Kendall and Spearman's rho tests for detecting monotonic trends in hydrological series. *J. Hydrol.* **2002**, *259*, 254–271. [CrossRef]

57. Vogel, R.M.; Rosner, A.; Kirshen, P.H. Brief communication: Likelihood of societal preparedness for global change: Trend detection. *Nat. Hazards Earth Syst. Sci.* **2013**, *13*, 1773–1778. [CrossRef]

58. Totaro, V.; Gioia, A.; Iacobellis, V. Numerical investigation on the power of parametric and nonparametric tests for trend detection in annual maximum series. *Hydrol. Earth Syst. Sci.* **2020**, *24*, 473–488. [CrossRef]

59. Kovács, A. Geometry and Hydraulic Parameters of Karst Aquifers: A Hydrodynamic Modeling Approach. Ph.D. Thesis, Université de Nechatel, Neuchâtel, Switzerland, 2003.

60. Portoghese, I.; Uricchio, V.; Vurro, M. A GIS tool for hydrogeological water balance evaluation on a regional scale in semi-arid environments. *Comput. Geosci.* **2005**, *31*, 15–27. [CrossRef]

61. Toreti, A.; Desiato, F.; Fioravanti, G.; Perconti, W. Seasonal Temperatures over Italy and Their Relationship with Low-Frequency Atmospheric Circulation Patterns. *Clim. Chang.* **2010**, *99*, 211–227. [CrossRef]

62. Scheffer, M. Critical Transitions in Nature and Society. *Choice Rev. Online* **2009**. [CrossRef]

63. Ferguson, G.; Gleeson, T. Vulnerability of Coastal Aquifers to Groundwater Use and Climate Change. *Nat. Clim. Chang.* **2012**, *2*, 342–345. [CrossRef]

64. Vörösmarty, C.J.; Green, P.; Salisbury, J.; Lammers, R.B. Global Water Resources: Vulnerability from Climate Change and Population Growth Contemporary Population Relative to Demand per Discharge. *Science* **2000**, *289*, 284–288. [CrossRef] [PubMed]

MDPI
St. Alban-Anlage 66
4052 Basel
Switzerland
Tel. +41 61 683 77 34
Fax +41 61 302 89 18
www.mdpi.com

Water Editorial Office
E-mail: water@mdpi.com
www.mdpi.com/journal/water

Geo-Environmental Approaches for the Analysis and Assessment of Groundwater Resources at Catchment-Scale